ELEMENTS OF
PHYSICAL GEOGRAPHY

TO the Student: *A Study Guide for the textbook is available through your college bookstore under the title* Study Guide to Accompany **Elements of Physical Geography,** *4th edition by Arthur N. Strahler and Alan H. Strahler. The Study Guide can help you with course material by acting as a tutorial, review and study aid. If the Study Guide is not in stock, ask the bookstore manager to order a copy for you.*

ELEMENTS OF
PHYSICAL GEOGRAPHY

Fourth Edition

Arthur N. Strahler

Alan H. Strahler

John Wiley & Sons

New York   Chichester   Brisbane   Toronto   Singapore

Library of Congress Cataloging in Publication Data:

Strahler, Arthur Newell, 1918-
    Elements of physical geography.

    Bibliography: p.
    Includes index.
    1. Physical geography.   I. Strahler, Alan H.
I. Title.

GB54.5.S79   1989   910'.02   88-20863
ISBN 0-471-61647-8

Printed in the United States of America

10 9 8 7 6 5 4 3 2 1

Throughout this book, illustrations with credit reading *Based on Goode Base Map*
use as the base map Goode Map No. 201HC World Homolosine, copyrighted by
the University of Chicago. Used by permission of the Geography Department,
University of Chicago.

PREFACE

Elements of Physical Geography, Fourth Edition, is written for use in one-semester and one-quarter survey courses. Both content and structure are designed for students who are following general education programs in nonscience fields and who need only an overview of the subject. In keeping with this objective, we stress important concepts and basic facts, treating them descriptively and minimizing the use of technical terms. Emphasis is on ways in which the physical environment influences human activity, both directly and indirectly. These influences act through climate, landforms, soils, and vegetation.

If we are to make physical geography more than a mere collection of science topics, and if we are to give students an insight into the nature and goals of professional geography, we must empha-size spatial distribution of physical environmental variables and their interactions. Thus, we stress the global patterns of climate, landforms, soils, and vegetation; we explain these patterns as simply as possible in terms of natural processes.

We call attention to contemporary trends in physical geography that serve to bring greater unity and purpose to physical geography. One of these is the identification and analysis of natural flow systems of energy and matter, particularly in the atmosphere and hydrosphere. Thus we give special attention to the global budgets and balances of radiation and water. Another such trend is toward a climatology geared more closely to the soil-water balance and the availability of soil water to plants. Following this trend allows us to evaluate freshwater resources of the lands explicitly in terms of the annual and seasonal water surplus or water deficit. Exposure to the concept of the soil-water balance also helps the student to understand the modern Comprehensive Soil Classification System, which uses soil-water regimes as a major factor in differentiating soil orders and suborders.

Physical geography is closely involved with analysis of human impact on the physical environment. This deep involvement is understandable because physical geography integrates most of the diverse factors contributing to environ-mental changes. Wherever relevant, we include material on the impact of human activities on the environment and on resources that are renewable through inorganic and organic cycling of energy and matter.

This Fourth Edition is significantly changed in content and structure in ways designed to simplify the text and to make it more appealing to the students. Carefully planned deletions serve to maintain a uniform technical level, eliminating points of undue difficulty and complexity. Insertions of new topics both enrich the text and lend interest to standard topics. How these improvements have been achieved is detailed in the following paragraphs.

Perhaps the most conspicuous change affecting the entire range of physical geography topics is the transfer of the complex schematic diagrams and text on flow systems of energy and matter to an appendix. These abstract diagrams require an understanding of the various kinds of energy transformation and changes of state of matter—principles of physics that present formidable problems to many students. In an appendix, this material remains available but does not break up the continuity of the main text.

Our Introduction emphasizes the nature and goals of physical geography, using two brief case studies to illustrate relationships between humans and their environment. Chapter 1, introducing the earth as a sphere turning under the sun's rays, remains essentially unchanged. The next nine chapters cover processes and forms within the atmosphere and hydrosphere. One significant change we have made is to eliminate former Chapter 2 (Earth's Oceans and Atmosphere) and place its essential topics at appropriate positions in following chapters. Certain of the peripheral topics in physical oceanography have been deleted. Otherwise, Chapters 2 through 6 remain largely unchanged in the presentation of basic principles of atmospheric science. A brief description of the worldwide weather phenomenon known as *El Niño* has been added, while environmental topics such as carbon-dioxide and climate change, the ozone layer, and acid precipitation have been updated.

Two new ancillary topics have been added to these early chapters. Appended to Chapter 1 is a brief section entitled "Maps for Physical Geography," presenting some elementary principles of cartography. Appended to Chapter 2 is a new section on remote sensing in geography, explaining the techniques used and illustrating the kinds of imagery generated.

A major improvement will be seen in the treatment of climate, now presented in three chapters. Chapter 7, entitled "The Global Scope of Climate," now presents the principles of climate description and classification, combining the conventional descriptive parameters—temperature and precipitation—with those of the soil-water balance. The venerable Köppen climate system, still in demand by some instructors, remains in this chapter. Description of the global climates is now expanded to fill two new chapters (8 and 9), in which the presentation is greatly enriched and illuminated by inclusion of descriptions of the general landscape associated with each climate. These include paragraphs on the associated soils, natural vegetation, agricultural systems, and plant resources. The environmental features are described in simple language, free of classification systems. Some of the added text sections and landscape photographs are transferred from the final two chapters of the previous edition (Environmental Regions).

Chapter 10, Runoff and Water Resources, remains in its transitional position, leading into eight geology/geomorphology chapters (11 through 18), which are largely unchanged. A few topics formerly in the final environmental-regions chapters have been inserted into these chapters at appropriate locations. Some important improvements have been made in the plate-tectonics part of Chapter 12 by introducing arc-continent collisions and accreted terranes, along with improved maps and diagrams. Revision of Chapter 13 (Volcanic and Tectonic Landforms) includes new material on foreland fold belts and the East African rift-valley system and an updating of the text on earthquake hazards and prediction.

Important changes have involved the four former chapters on soils and vegetation. What were two brief chapters on soils have now been joined into one (Chapter 19), but with content unchanged. Natural vegetation is now covered in a single chapter (20). Former Chapter 21 (The Flow of Energy and Materials in the Biosphere) has been eliminated, along with much of the material on ecosystems, the food chain, ecosystem energetics and productivity, and agricultural ecosystems. The remaining topics (photosynthesis, primary production, materials cycling) have been included in the single chapter, dealing largely with plant structure and environment and the biomes and formation classes. The world map of natural vegetation has been simplified to conform with a somewhat simplified treatment of the formation classes.

The subject of environmental regions, formerly in two final chapters, has been consolidated into a single chapter. This move was made because few instructors used the environmental-regions concept as a synthesis of physical geography. However, for those few who make this a required topic covered in their lectures, we retain the complete system and its accompanying world maps of environmental regions, but without the many descriptive sections on landforms, soils, vegetation, and agriculture that now have found places in earlier chapters.

In sum, this revised edition will prove more inviting to students and easier to cope with because of a more uniform flow of essential content. What we have retained upholds the same high standard of thorough coverage of basic processes and concepts, expressed in language appropriate to a college-level science course.

Not the least of the features of this new edition is the improvement in color photos throughout the book. Many have been reprocessed for better clarity and color rendition. Some have been replaced with new subjects. As always, photos have been chosen primarily to display accurately the scientific attributes of the subjects under discussion, aesthetics being a secondary consideration.

As before, a revised *Study Guide* provides chapter abstracts including explanations of essential concepts, facts, and terms, presented in a question/answer format for intensive review, along with sample test questions. The revised *Instructor's Manual* contains teaching suggestions and background information, lists of supplementary readings, available visual aids and exercise materials, and a full set of achievement tests that may be used or adapted as needed.

Acknowledgments

We acknowledge with thanks the contribution of a large group of geography instructors—over 70 in all—who completed a publisher's questionnaire prepared by the authors, setting forth in detail the entire revision plan. Numerous comments and suggestions attached to the questionnaire by many of the respondents focused our attention on a wide range of possibilities for improving the book. These inputs were carefully studied by the authors and the answers to specific questions tallied. Each

important change we have implemented stands supported by a substantial majority of the respondents.

We would also like to thank the following persons who reviewed the entire manuscript of the revised edition, making numerous corrections and proposing many desirable improvements:

Guy Q. King,
Assistant Prof., Dept. of Geography
The Univ. of New Mexico

Richard M. MacKinnon
Allan Hancock College

Gary Peters,
Prof. & Chair, Dept. of Geography
California State Univ., Long Beach

Donald E. Petzold,
Faculty Research Assoc.,
The University of Maryland

Albert M. Tosches,
Assist. Prof., Dept. of Geography
Salem State College

Douglas Wheeler
Dept. of Geography,
Utah State University

Thomas B. Williams,
Assist. Prof., Dept. of Geography
Indiana University

Arthur N. Strahler
Alan H. Strahler

ABOUT THE AUTHORS

ARTHUR N. STRAHLER (b. 1918) received his B.A. degree in 1938 from the College of Wooster, Ohio, and his Ph.D. degree in geology from Columbia University in 1944. He is a fellow of the Geological Society of America and the Association of American Geographers. He was appointed to the Columbia University faculty in 1941, serving as Professor of Geomorphology from 1958 to 1967 and as Chair of the Department of Geology from 1959 to 1962. He is the author of several widely used textbooks of physical geography, environmental science, and the earth sciences.

ALAN H. STRAHLER (b. 1943) received his B.A. degree in 1964 and his Ph.D. degree in 1969 from The Johns Hopkins University, Department of Geography and Environmental Engineering. His published research is in the fields of plant geography, forest ecology, quantitative methods, and remote sensing. He has held academic appointments at the University of Virginia, the University of California at Santa Barbara, and Hunter College of the City University of New York, and is now Professor of Geography at Boston University. He is a coauthor of several textbooks on physical geography and environmental science.

ABBREVIATED CONTENTS

EXPANDED CONTENTS

INTRODUCTION
THE HUMAN ENVIRONMENT

THE LIFE LAYER   The focus of physical geography is on the life layer, a shallow zone at the surface of the lands and oceans.

INORGANIC EARTH REALMS   The life layer draws its qualities from three inorganic earth realms—atmosphere, hydrosphere, and lithosphere.

THE BIOSPHERE   The life layer includes most of the world's organic life, representing the biosphere.

ENVIRONMENTAL REGIONS   Within the life layer of the lands we recognize distinct environmental regions, each with its special life-supporting qualities.

SHUFFLING quickly through the chapters of this book, pausing now and then to look at illustrations and read a few figure captions or some lines of text, you may be tempted to ask "Isn't physical geography simply a collection of excerpts from several of the earth sciences and life sciences?" You may have noticed that the chapter on storms (Chapter 6) looks as if it belonged in a textbook of weather science (meteorology). Our section on volcanoes in Chapter 13 could have been taken from a textbook of geology. Our description of photosynthesis by plants (Chapter 20) may seem to have been lifted right out of a biology textbook.

If these are your first impressions of the makeup of physical geography they are at least partly correct, but they are also incomplete. Some important basic concepts of physical geography set it apart from a mere collection of natural science topics. Let us first find out what is really unique about physical geography and how it fits into a geographer's view of the meaning of modern geography and the objectives that geographers are trying to reach.

Modern geography is deeply involved

with the study of ways in which human beings and their institutions are spread over the surface of the earth. Because many human activities are influenced by the physical environment in which they take place, the study of physical geography is a key to understanding the different cultural patterns that have evolved over many centuries in habitable areas of the earth.

The Human Habitat

The lands of the earth comprise the habitat of the human species and all other terrestrial forms of life. The *habitat* of any plant or animal is the physical environment in which it is most likely to be found. Physical geography brings together and interrelates the important elements of our physical environment that make up the human habitat. While emphasizing features of the environment that are most important to human survival, physical geography also deals with the environments of other life forms—both plants and animals—for we are dependent on them for food.

The Life Layer

The focus of physical geography is on the *life layer,* a shallow zone of the lands and oceans containing most of the world of organic life, or *biosphere.* Quality of that life layer is the major concern of physical geography. By quality we mean the sum of the physical factors that make the life layer habitable for all forms of plants and animals, but most particularly for the human species.

The quality of the physical environment of the land is established by factors, forces, and inputs coming from both the atmosphere above and the solid earth below. The *atmosphere,* a gaseous shell surrounding the solid earth, dictates climate, which governs the exchange of heat and water between atmosphere and ground. The atmosphere also supplies vital elements—carbon, hydrogen, oxygen, and nitrogen—needed to sustain all life of the lands.

The solid earth, or *lithosphere,* forms the stable platform for the life layer and is also shaped into landforms. These relief features—mountains, hills, and plains— bring another dimension to the physical environment and provide varied habitats for plants. The solid earth is also the basic source of many nutrient elements, without which plants and animals cannot live. These elements pass from rock into the shallow soil layer, where they are held in forms available to organisms.

Water, another of the essential materials of life, permeates the life layer, the overlying atmosphere, and the underlying solid earth. In all its forms, water on the earth constitutes the *hydrosphere.* Our study of physical geography can be described in the broadest of terms as a study of the atmosphere, hydrosphere, and lithosphere in relation to the biosphere.

Environmental Regions

Acting together, the inputs of energy and materials into the life layer from atmosphere and solid earth determine the quality of the environment and the richness or poverty of organic life it can support. Thus we can recognize *environmental regions,* each with its particular qualities for life support.

A given environmental region usually has certain definite locations on the globe in terms of latitude and continental position. It has a characteristic combination of soil type and native plant cover and offers a certain set of opportunities for the development of vital supplies of fresh water and food.

Some environmental regions are richly endowed with water and food; others are very poorly endowed. The poorly endowed environments are too cold, too dry, or too rocky to support much life. A major goal of physical geography is to evaluate each environmental region in terms of its life-support capacity.

Our Impact on the Environment

An understanding of physical geography is vital to planning for survival of the earth's rapidly expanding human population. Survival will depend not only on how much fresh water and food is available, but also on protecting the environment from forms of pollution and destruction that will reduce the capacity of the land to furnish those necessities. Here we encounter another of the important goals of physical geography: to evaluate the impact of human activities on the natural environment.

Physical geography has always been at the heart of environmental studies, because physical geography is strongly oriented toward understanding how the natural environment shapes and is shaped by the world's expanding population.

A Plan for Study

Our plan of study of physical geography is first to take a broad view of our planet as a sphere bathed in the sun's rays. We then examine the atmosphere and oceans and learn how they gain and lose energy from the sun's rays. Following a study of the interactions between these gaseous and liquid layers we can evaluate global climate and water resources so vital to life on earth.

Next, the solid earth occupies our attention as we review geological principles essential to an understanding of major features of the earth's crust. We then investigate the configuration of the earth's land surface—its landforms and the processes that shape them. Now the stage is set for the role of organic activities in the biosphere; these strongly influence the soil layer, which supports the natural vegetation of the lands.

In our final chapter, all the ingredients of physical geography are brought together in a review of environmental regions of the globe. As we assess the life-support capabilities of each region, physical geography takes on new meaning and becomes a realistic base on which to build new plans for human survival.

Before beginning the systematic treatment of physical geography, we present two case studies to illustrate the nature of geography. The first case takes us back two centuries into American history, for it deals with the impact of farming on the land in colonial and postrevolutionary times on the eastern seaboard. The second case study deals with a foreign land in modern times; it tells of drought, famine, and death in a harsh environment in North Africa. Both cases illustrate how the physical environment of a region modifies human activities and how, in turn, those activities modify the environment.

Trouble at Mount Vernon—
A Farm in Distress

As the wave of American Bicentennial celebrations recedes further into the past, many romantic notions about the successes of our founding fathers remain in our minds. We visualize George Washington, retired happily on his Virginia estate Mount Vernon, reaping a richly deserved bounty from expansive farmlands under his personal supervision. We picture Thomas Jefferson at Charlottesville, managing his fields from his eyrie atop Monticello, while at the same time keeping a telescopic sight on his brainchild under construction, the new campus of the University of Virginia. These romantic visions fade under reality when we learn that fertility of farmlands in the colonies was declining rapidly, even as the new nation was struggling to stay alive. Ignorance of a simple point of soil science was responsible—acid soils need lime.

Without lime to correct soil acidity, essential nutrients for healthy crop growth cannot be retained in the soil, even though natural fertilizers are added. We read that George Washington conserved animal manure to spread on his fields, and that he had his field hands bring rich mud from creeks and marshes to spread on the soil to bring in a new supply of nutrients. A modern historian, Avery Craven, an authority on the agricultural history of that region, tells us that in 1834, 35 years after Washington's death, a visitor to Mount Vernon declared that "a more widespread and perfect agricultural ruin could not be imagined."

Another American, Edmund Ruffin (1794–1865), is credited with solving the mystery of failing agricultural fertility of the eastern seaboard. Ruffin owned lands at Coggins' Point on the coastal plain of Virginia. As with others, his land was rapidly declining in crop yields in the early 1800s. He tried many experiments to stop the decline, but application of manure had little effect and clover would not grow to enrich the soil. Quite by chance, Ruffin obtained a copy of Davy's *Agricultural Chemistry*, published in 1813. Despite Ruffin's lack of formal education in science, he was quick to grasp the significance of one statement: "any acid matter . . . may be ameliorated by application of quicklime."

So it came about that on a February morning in 1818, Ruffin directed his field hands to haul marl from pits in low areas of his lands. (Marl is a soft lime mud that occurs widely as sedimentary strata on the eastern coastal plain.) The workers spread two hundred bushels of marl over several acres of newly cleared ridge land of poor quality. In the spring, Ruffin planted this area in corn to test the effect of the marl. In the words of historian Avery Craven, this is what happened: "Eagerly he waited. As the season advanced, he found reason for joy. From the very start the plants on marled ground showed marked superiority, and at harvest time they yielded an advantage of fully forty per cent. The carts went back to the pits. Fields took on fresh life. A new era in agricultural history of the region had dawned."* In 1832, Ruffin published his findings in a work titled *An Essay on Calcareous Manures.* His advice was outspokenly opposed, but time showed him to be right.

A geographer, reading this brief anecdote in history, will not be satisfied with the simple statement that the use of lime saved the agricultural resources of the eastern seaboard. The colonists who settled this region were largely from England, where their ancestors had farmed continuously and successfully for centuries, using the land for growing grains for their food and forage for their dairy animals. Like all farmers of Western Europe these colonists knew the necessity of applying animal manure as a soil fertilizer. What, then, was different about the physical environment of the American seaboard that resulted in rapid deterioration of the soils?

A geographer thinks first about the character of the American climate and soil and the geologic history of the landscape. Physical geography can demonstrate that a

* Avery Craven (1932), *Edmund Ruffin, Southerner,* Appleton, New York, p. 55.

Harrowing the soil, sowing the seed. (New York Public Library, Picture Collection.)

unique combination of these basic factors was responsible for the near disaster that beset the American agriculturists. Soils over much of northwestern Europe were formed on freshly ground mineral matter left by the great ice sheets; nutrients needed by crops are abundant in those soils. In contrast, upland soils of the eastern seaboard from Virginia to Georgia have been continually exposed to a leaching process in a mild, moist climate for tens of thousands of years. These soils have lost the ability to generate adequate quantities of mineral nutrients. Throughout this book we will explain the factors of climate, soils, and geologic history that bear on this case. In Chapter 19 we will summarize our conclusions from a geographer's point of view.

Data Source: Emil Truog (1938), Putting soil science to work, *Journal of the American Society of Agronomy,* vol. 30, pp. 973–985.

Drought in the Sahel— The Sahara Desert Creeps Southward

Several West African nations lie in a perilous climatic belt called the Sahel. Because this belt lies along the southern border of the great Sahara Desert of North Africa, it is also referred to as the sub-Sahara region. Seven countries occupy much of the Sahel in western Africa: Senegal, Mauritania, Mali, Burkina Faso, Niger, Nigeria, and Chad. All these countries were struck a severe blow by drought, which began in 1968, became particularly severe in 1971 and 1972, was a major human catastrophe by 1974, and then was temporarily alleviated in 1975. Farther eastward, the drought zone extends across Sudan and into the highlands of Ethiopia.

The drought zone of the West African Sahel is, for the most part, a tropical grassland. It has a feast-or-famine climate. There is a short rainy season when the sun rides high in the sky (June, July, August), but a long dry season when the sun is low in the sky (November through April). To the north lies year-round drought of the Sahara Desert; to the south is a savanna region having a much longer wet season with much heavier rainfall. Drought in the Sahel means a dearth in the annual rains on which the growth of grasses depends. Two groups of humans live on the natural resources of the Sahel: nomadic herders and grain farmers. Both groups depend on the annual rains to turn the landscape

At the height of the Sahelian drought vast numbers of cattle had perished and even the goats were hard pressed to survive. Trampling of the dry ground prepared the region for devastating soil erosion in rains that eventually ended the drought. (Alain Nogues/Sygma.)

dug wells can also supply water needs throughout the long dry season. But when the rains have failed for three, four, or five successive years, the subsurface water reserve is depleted. Water holes and shallow wells run dry.

In the worst stages of the Sahel drought nomads were forced to sell the remaining cattle that were their sole means of subsistence. Reduced to the status of refugees, these miserable humans crowded into camps close to the major cities. Here they might be fortunate enough to receive handouts of grain flown in from other nations on desperate relief missions. Despite relief supplies, tens of thousands of people died of starvation and disease in the Sahel. It is estimated that some 5 million cattle perished in the Sahel drought. In the hardest-hit areas 90 percent of the cattle died. The population of Ethiopia, far to the east in the same climatic belt, also suffered heavily from this drought, with a death toll estimated at more than 50,000.

As a geographer reads this brief news account of a human disaster, some searching questions come to mind. The news media used headlines reading "Sahara Desert creeps southward," implying that the great tropical desert is expanding and that the climate in the Sahel is changing. Yet, when the rainfall records are examined for several decades prior to the recent drought, there appears a pattern of cycles consisting of periods of drought interspersed with periods of ample rainfall. There were, in fact, severe droughts in 1910–1913 and in the 1940s. Geographers were reluctant to interpret the Sahel drought of 1971–1974 as a permanent climate change. The arrival of ample rains in 1975 supported the theory that rainfall in this region follows a cyclic pattern. But more drought years followed, and the deficiency steadily decreased from 1978 through 1984. Geographers then began to suspect that a long-term climate change is in progress, superimposed on the cyclic pattern.

There is more to the picture than a cyclic pattern of rainfall. The geographer is aware that over many decades, European managers of the African colonies that are now the Sahelian nations were responsible for some far-reaching changes. Grain was imported to prevent starvation. Improved

green, rejuvenating the grasses on which their cattle graze and supplying the soil moisture needed for the annual crop of grain. When the rains fail to appear, there are no grain crops to harvest and the cattle starve for lack of forage.

Water in another form is vital to people of the Sahel—fresh water to drink. Neither humans nor cattle survive long without drinking water. Streams of the Sahel flow copiously during the brief rainy season, but rapidly shrink as the dry season sets in. Water remains trapped in sand and gravel below the surface. This water fills holes scraped in low points in the stream beds. Water holes of this sort are essential to survival of humans and their cattle. Shallow

nutrition and medical aid may also have reduced the death rate. Thus the human population swelled from one decade to the next. Boring of deep water wells to tap underground sources provided additional water for crop irrigation and for the support of cattle, so that herds increased greatly in size. Did the land become overburdened by the increased load of human and animal populations? Several factors must be weighed carefully. Your study of physical geography will include investigations of the processes that bring rainy and dry seasons to the Sahel, the availability of soil water to plants in this remarkable tropical climate, and the kinds of plants that can cope with the harsh environment. You will also learn how overgrazing and the cutting of trees can lead to erosion of precious soil. In Chapter 20, we make a reassessment of the factors that conspired to bring what may be an irreversible change for the worse in the quality of the life layer across the Sahel.

Data Sources: John H. Douglas (1974), The omens of famine, *Science News,* vol. 105, pp. 306–308; Nicholas Wade (1974), Sahelian drought: No victory for Western aid, *Science,* vol. 185, pp. 234–237; John G. Lockwood (1986), The causes of drought with particular reference to the Sahel, *Progress in Physical Geography,* vol. 10, no. 1, pp. 111–119; Michael H. Glantz (1987), Drought in Africa, *Scientific American,* vol. 256, no. 6, pp. 34–40.

CHAPTER

OUR SPHERICAL HABITAT

1

OUR EARTH IN SPACE   Our physical environment is conditioned by several astronomical phenomena; these serve as environmental controls.

A SPHERICAL EARTH   The earth is approximately spherical in form, but to prove this fact is not as easy as it might seem.

EARTH ROTATION   The earth spins on its axis and, at the same time, travels in a nearly circular path around the sun.

THE GEOGRAPHER'S GRID   Humans have invented the geographic grid, a network of imaginary lines to pinpoint the locations of all features of the earth's surface.

GLOBAL TIME   Another culture system is global time, determined by the unceasing westward procession of imaginary hour meridians.

CORIOLIS   Earth rotation influences our environment through the mysterious Coriolis effect.

THE SEASONS   While its axis remains tilted, the earth revolves about the sun to give us the astronomical seasons.

I N this age of orbiting satellites, the spherical form of the earth is such an obvious fact that we may have difficulty imagining ourselves living in ancient times when the extent and configuration of the earth were entirely unknown. In that day, to sailors of the Mediterranean Sea, on their ships and out of sight of land, the sea surface looked perfectly flat and seemed to be terminated by a circular horizon. Based on this perception, the sailors might well have inferred that the earth has the form of a flat disk and that, if they traveled to its edge, their ships might fall off. Even so,

these sailors could have sensed optical phenomena suggesting that the sea surface is not flat, but curved so as to be upwardly convex like a small part of the surface of a sphere. Given sufficiently acute eyesight and very clear air, these sailors could observe that as another vessel moved away from theirs, it seemed to pass below the horizon, so that the upper sails and rigging might remain visible even though the hull was out of sight. This phenomenon is easily observed today with the use of a telescope (Figure 1.1) and is a crude proof of the earth's curvature.

FIGURE 1.1 Seen through a telescope, a distant ship seems to be partly submerged.

A second phenomenon that might have been put to use in reasoning about the earth's form is one we have all observed. After the sun has set below our horizon, its rays continue to illuminate clouds above us, or the peaks of high mountains, should any be near us. This effect could be explained by a curved earth surface and is perhaps the reason that scholars among the ancient Greeks believed the earth to be spherical. Pythagoras (540 B.C.) and associates of Aristotle (384–322 B.C.) held this view. They had also speculated on the length of the earth's circumference, but with highly erroneous guesses.

The constancy of gravity over the earth's surface might be used in an experiment to prove the earth's sphericity. If we first assume that Newton's law of gravitation is valid, it follows that a given object should register the same weight at all places over the earth's surface. Using a spring balance as the scales, we might travel widely over the earth, repeatedly weighing a small mass of iron and recording the values. If they proved to be unvarying, we could conclude that we have taken our measurements at points all equidistant from the earth's center of mass, and thus we are on a spherical surface. Actually, this same experiment, carried out with great precision and using highly refined instruments, has shown that the earth's true figure departs slightly from that of a true sphere.

Eratosthenes Measures the Earth

It remained until about 200 B.C. for Eratosthenes, head of the library at Alexandria, Egypt, to perform a direct measurement of earth circumference based on a sound principle of geometry. He observed that on a particular date of the year (close to summer solstice, June 21), at Syene, a place located on the upper Nile River far to the south, the sun's rays at noon shone directly on the floor of a deep vertical well. In other words, the sun at noon was in the zenith point in the sky, and its rays were perpendicular to the earth's surface at that point on the globe (Figure 1.2). At Alexandria, on the same date, the rays of the noon sun made an angle with respect to the vertical. The magnitude of this angle was one-fiftieth of a complete circle, or seven and one-fifth degrees ($7\frac{1}{5}° \times 50 = 360°$).

Eratosthenes needed only to know the north–south distance between Syene and Alexandria to calculate the earth's circumference; he would simply multiply the ground distance by fifty to obtain the circumference. In those days distances between cities were only crude estimates, based on travelers' reports. Eratosthenes took the distance to be 5000 stadia, and so evaluated the earth's circumference as 250,000 stadia. It is not easy to rate his results in terms of accuracy, since we are

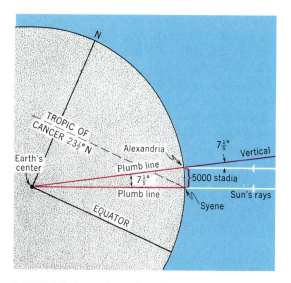

FIGURE 1.2 Eratosthenes' method of measuring the earth's circumference.

not sure of the length equivalent of the distance unit (the stadium) that he used. If it were the Attic stadium, equivalent to 185 m (607 ft) in modern English units, the circumference would come to about 43,000 km (26,700 mi). Considering that the true circumference is close to 40,000 km (25,000 mi), Eratosthenes' results seem offhand to be remarkably good.

From Eratosthenes' classic experiment, it is an easy step to design an astronomical method of measuring the earth's figure using star positions instead of the sun. We need only to select a north–south line, whose length can be measured directly on level ground by surveying means. This line should be several tens of kilometers long. At the ends of the line, the angular position of any selected star can be measured at its highest point above the horizon or with respect to the vertical, using a level bubble or a plumb bob as a means of establishing a true horizontal or vertical reference. The difference in angular positions of the star will be equal to the arc of the earth's circumference lying between the ends of the measured line. This very procedure is believed to have been followed by ninth-century Arabs. Their measurements were probably much more accurate than those of Eratosthenes but, because the units of measure are not known in modern equivalents, their work cannot be checked.

In the many centuries following Eratosthenes' work Western science lay in stagnation. Then, about 1615, Willebrord Snell, a professor of mathematics at the University of Leiden, developed methods of precise distance and angular measurement that he applied to the problem of the earth's circumference. His work presaged the era of scientific geodesy ("geodesy" comes from the Greek word meaning "to divide the earth") and led to remarkably accurate measurements of the earth's figure about a century later.

Earth Rotation and the Geographic Grid

The earth spins like a top on an axis; this is a phenomenon called *rotation*. One earth-turn with respect to the sun defines the *solar day*. We have arbitrarily assigned a value of

24 hours to the mean duration of the solar day, and so all cultural aspects of human life can be regulated and coordinated.

The direction of earth rotation can be determined by using one of the following rules: (1) Imagine yourself looking down on the north pole of the earth; the direction of turning is counterclockwise. (2) Place your finger on a point on a globe near the equator and push eastward. You will cause the globe to rotate in the correct direction (Figure 1.3). Earth rotation affects our environment in two quite different ways: one is cultural; the other is physical.

Taking first the cultural influences of earth rotation, the spin of the earth on an axis allows us to set up a system designating the location of points and the directions of lines on the sphere. Without such a system human society would, indeed, be lost in the literal sense of the word. First, the axis of spin determines the *north pole* and *south pole*, which are fixed points of reference. Second, as any given point on the earth surface (except the pole points) moves with the earth's spin, the point generates a curved line. With the completion of one earth-turn with respect to fixed stars the line becomes a full circle, known as a *parallel of latitude* (Figure 1.4).

The largest parallel of latitude lies midway between the two poles and is designated the *equator*. As a unique circle on the sphere, the equator is a fundamental reference line for measuring position of points on the globe. Parallels of latitude, as true east–west lines, define the cardinal directions east and west on the globe.

If the earth is imagined to be sliced in two by a plane passing through the poles, a circle oriented at right angles to the plane

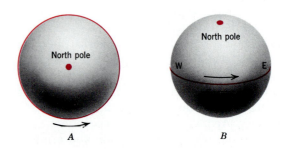

FIGURE 1.3 The direction of rotation of the earth can be thought of as (A) counterclockwise at the north pole, or (B) eastward at the equator.

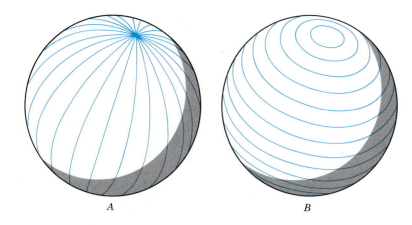

FIGURE 1.4 (A) Meridians of longitude. (B) Parallels of latitude.

A   B

of the equator is generated. The half-part of the circle, connecting one pole to the other, is known as a *meridian of longitude* (Figure 1.4). Meridians can be produced in any desired number, and a meridian can be found to pass through any point on the earth's surface. Meridians as true north–south lines define the cardinal directions north and south.

The numbers of parallels and meridians are infinite, just as the number of points on the sphere is infinite. Every point on the globe has its unique combination of one parallel and one meridian; their intersection defines the position of the point. The total system of parallels and meridians forms a network of intersecting circles, and this is called the *geographic grid*.

south latitude. Figure 1.5 shows how latitude is measured. The point *P* lies at latitude 50 degrees north, which we can abbreviate to lat. 50° N. Notice that latitude is actually measured along the meridian that passes through the point; it is an arc of that meridian.

Longitude is a measure of the position of a point eastward or westward with respect to a chosen reference meridian, called the *prime meridian*. As Figure 1.5 shows, longitude is the arc, measured in degrees, between a given point and the prime meridian. This arc can be measured east-to-west along the parallel that passes through the point. The point *P* lies at longitude 60 degrees west (long. 60° W). The prime meridian is almost universally

Latitude and Longitude

To communicate effectively about any phase of geography—whether it be physical or cultural—a numerical system must be fitted to the geographic grid. Master it well, because it will appear a thousand times in this book.

Latitude is a measure of the position of a given point in terms of the angular distance between the equator and the poles. Latitude is an indicator of how far north or south of the equator a given point is situated. Latitude is measured in degrees of arc from the equator (0°) toward either pole, where the value reaches 90°. All points north of the equator—in the northern hemisphere, that is—are designated as north latitude; all points south of the equator—in the southern hemisphere—are designated as

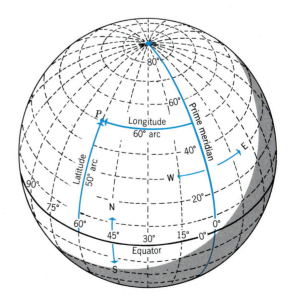

FIGURE 1.5 The point *P* has a latitude of 50° N, a longitude of 60° W.

12

accepted as that which passes through the old location of the Royal Observatory at Greenwich, near London, England, and is referred to as the Greenwich meridian. This meridian has the value long. 0°. The longitude of any given point on the globe is measured eastward or westward from this meridian, whichever is the shorter arc. Longitude may thus range from 0° to 180°, east or west. When both the latitude and longitude of a place are given, it is accurately and precisely located with respect to the geographic grid.

Statements of latitude and longitude tell nothing about distances in kilometers or miles. However, you can easily make rough conversions from degrees of latitude into kilometers. One degree of latitude is approximately equivalent to 111 km (69 mi) of surface distance in the north–south direction. This value can be rounded off to 110 km (70 mi) for multiplying in your head. For example, if you live at lat. 40° N (on the 40th parallel north), you are located about 40 × 110 = 4400 km north of the equator. East–west distances cannot be converted so easily from degrees of longitude into kilometers because of the convergence of meridians in the poleward direction. Only at the equator is a degree of longitude equivalent to 111 km (69 mi). At lat. 60° N or S, one degree of longitude is reduced to half its equatorial value, or about 56 km (35 mi).

Global Time

Besides generating the geographic grid, the rotation of a spherical earth has a cultural impact through the phenomenon of global time. Our clocks run by the sun's schedule, but the sun sends us only a single set of parallel rays. This means that what is noon for persons on one side of the globe is midnight for persons on the opposite side.

Because the earth turns through 360° of longitude every 24 hours, it must turn 15° of longitude every hour, or one degree of longitude every 4 minutes. We therefore find it convenient to state that 1 hour of time is the equivalent of 15° of longitude. This equality forms the basis for all calculations concerning time belts of the globe.

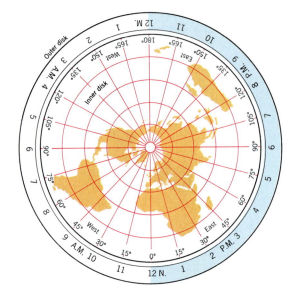

FIGURE 1.6 This working model of global time zones uses a movable inner disk that can be set to any given hour at any selected standard meridian.

A working model of global time relations is illustrated in Figure 1.6. Two disks of different radius are attached at their centers in such a way that one disk can be turned while the other remains still. On the inner disk radii are drawn to represent 15° meridians of a globe seen from a point above the north pole. On the outer disk hour-numbers are positioned to represent the time net. As a further refinement, the inner disk may be a map on which one can estimate time differences between various countries around the world.

As set in Figure 1.6, it is noon on the Greenwich meridian (0° longitude), while it is midnight on the 180° meridian in mid-Pacific. Figure 1.6 also illustrates global *standard time*, based on *standard meridians* spaced 15° apart around the globe. With reference to the Greenwich meridian, which is the world standard for reckoning time, 12 *time zones* lie in the eastern hemisphere and 12 in the western hemisphere. However, the 12th zone is shared by the two hemispheres. That half of the zone lying on the Asiatic side of the 180th meridian is exactly one full calendar day later than the half on the American side. For this reason, the 180th meridian defines the position of the International Date Line.

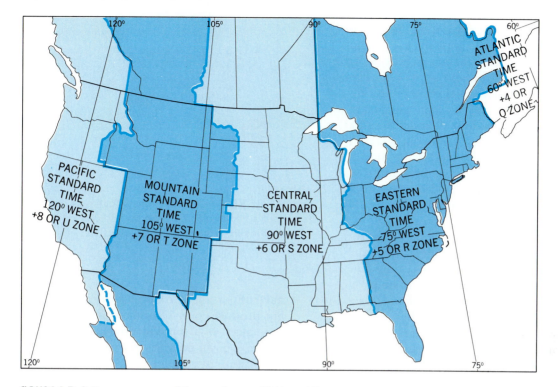

FIGURE 1.7 A time-zone map of the contiguous 48 United States and southern Canada. Figures with plus signs tell the time difference between the zone and Greenwich time. Zones are also designated on a global basis by letters of the alphabet.

The United States encompasses six time zones. Their names and standard meridians of west longitude are:

Eastern	75°
Central	90°
Mountain	105°
Pacific	120°
Alaska–Hawaii	150°
Bering	165°

Had it been carried out precisely, the system would have resulted in belts extending exactly 7½° east and west of each standard meridian, but a glance at the map (Figure 1.7) shows that great liberties have been taken in locating the boundaries. Wherever the time-zone boundary could conveniently be located along some already existing and widely recognized line, this was done. Natural physiographic boundaries have been used. For example, the Eastern time–Central time boundary line follows Lake Michigan down its center, and the Mountain time–Pacific time boundary follows a ridge-crest line also used by the Idaho–Montana state boundary. Most frequently, the time-zone boundary follows state and county boundaries.

Because many human activities, especially in urban areas, start well after sunrise but continue long after sunset, it is desirable to set forward the hours of daylight so as to utilize them to best advantage. A considerable saving in electric power can be made in summer when the early morning daylight period, wasted while schools, offices, and factories are closed, is transferred to the early evening when the large majority of persons are awake and busy. The adjusted time system is known as *daylight saving time* and is obtained by setting ahead all timepieces by 1 hour.

Physical Effects of Earth Rotation

The physical effects of earth rotation are truly profound in terms of the environmental processes. First, and perhaps most obvious, is that rotation imposes a

14

daily, or diurnal, rhythm on many phenomena to which plants and animals respond. These phenomena include light, heat, air humidity, and air motion. Plants respond to the daily rhythm by storing energy during the day and releasing it at night. Animals adjust their activities to the daily rhythm, some preferring the day, others the night for food-gathering activities. The daily cycle of input of solar energy and a corresponding cycle of air temperature will be important topics for analysis in Chapters 2 and 3.

Second, as we will find in the study of the earth's systems of winds and ocean currents, earth rotation causes the flow paths of both air and water to be consistently turned in direction; toward the right hand in the northern hemisphere and toward the left in the southern hemisphere. This phenomenon goes by the name of the Coriolis effect; we will investigate it further in Chapter 4.

A third physical effect of the earth's rotation is also important. Because the moon exerts its gravitational attraction on the earth, while at the same time the earth is turning with respect to the moon, a set of forces known as the tide results. Tidal forces induce a rhythmic rise and fall of the ocean surface, and these motions in turn cause water currents of alternating direction to flow in shallow waters of the coastal zone. To a grain farmer in Kansas, the ocean tide has no significance whatsoever, but for the clam digger and charter boat captain on Cape Cod, the tidal cycle is a clock regulating their daily activities. For many kinds of plants and animals of saltwater estuaries tidal currents are essential to maintain a suitable life environment. The tide and its currents are discussed in Chapter 17.

The Earth in Orbit around the Sun

Simultaneously with its rotation on an axis, the earth is in motion in an orbit around the sun. This motion is called *revolution*. As a first approximation, the orbit can be considered a circle with the sun lying at the center. To be exact, the orbit is an ellipse in which the sun occupies one focus. The elliptical form of the orbit does, in fact,

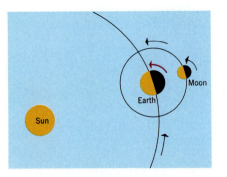

FIGURE 1.8 Viewed as if from a point over the earth's north pole, the earth both rotates and revolves in a counterclockwise direction.

have important astronomical consequences but produces only secondary environmental influences. By using the circular orbit as a model, we will not miss out on any important environmental influences on life processes. The earth completes its circuit about the sun in about 365¼ days, a period known as the *tropical year*. The year is defined by astronomers in other ways as well, but these are not important here.

Every four years the extra one-fourth day difference between the tropical year and the calendar year of 365 days totals nearly one whole day. By inserting a 29th day in February, every leap year, we are able to correct the calendar with respect to the tropical year. Further minor corrections are necessary to perfect this system.

In its orbit the earth revolves in such a direction that if we imagine ourselves in space, looking down on the earth and sun so as to see the north pole of the earth, the earth is traveling counterclockwise around the sun (Figure 1.8). It is further worth noting that nearly all planets and their major satellites in our solar system have the same direction of rotation and revolution, suggesting that their motions were imparted to them at a time when they and the sun were originally formed.

The important point about the period of the earth's revolution is that it sets the timing for climatic seasons that profoundly influence life on earth. To understand how seasons are generated, we must take another astronomical step in which we examine the relationship between the earth's axis and the plane of the orbit.

Tilt of the Earth's Axis

First, imagine the earth's axis to be exactly perpendicular with the plane in which the earth revolves about the sun. Astronomers call the plane containing the earth's orbit the *plane of the ecliptic*. Under these imagined conditions the earth's equator would lie exactly in the plane of the ecliptic. The sun's rays, which furnish all energy for life processes on earth, would always strike the earth most directly at a point on the equator. In fact, at the equator, the rays would be exactly perpendicular to the earth's surface at noon. The rays would always just graze the north and south poles. Conditions on any given day would be exactly the same as conditions on every other day of the year (assuming the orbit to be circular). In other words, there would be no seasons.

In reality, the earth's axis is not perpendicular to the plane of the ecliptic; it is inclined by a substantial angle of tilt, measuring almost exactly 23½° from the perpendicular. Figure 1.9 shows this axial tilt in a three-dimensional perspective drawing. The angle between axis and ecliptic plane is then 66½° (90° − 23½° = 66½°).

To proceed, we must couple the fact of axial tilt with a second fact: the earth's axis, while always holding the angle 66½° with the plane of the ecliptic, maintains a fixed orientation with respect to the stars. The north end of the earth's axis points constantly toward Polaris, the North Star. To help in visualizing this movement, hold a globe so as to keep the axis tilted at 66½° with respect to the horizontal. Move the globe in a small horizontal circle, representing the orbit, at the same time keeping the axis pointed at the same point on the ceiling.

Solstice and Equinox

What is the consequence of these two facts? (1) The earth's axis keeps a fixed angle with the plane of the ecliptic; and (2) the axis always points to the same place among the stars. You will find that at one point in its orbit the north polar end of the earth's axis leans toward the sun; at an opposite point in the orbit the axis leans away from the sun. At the two intermediate points, the axis leans neither toward nor away from the sun (Figure 1.10). Next, consider the four critical positions in detail.

On June 21 or 22 the earth is so located in its orbit that the north polar end of its axis leans at the maximum angle 23½° toward the sun. The northern hemisphere is tipped toward the sun. This event is named the *summer solstice.* Six months later, on December 21 or 22, the earth is in an equivalent position on the opposite point in its orbit. At this time, known as the *winter solstice,* the axis again is at a maximum inclination with respect to a line drawn to the sun, but now it is the southern hemisphere that is tipped toward the sun.

Midway between the dates of the solstices occur the *equinoxes,* at which time the earth's axis makes a 90° angle with a line drawn to the sun, and neither the north nor south pole has any inclination toward the sun. The *vernal equinox* occurs on March 20 or 21; the *autumnal equinox* occurs on September 22 or 23. Conditions are identical on the two equinoxes as far as earth–sun relationships are concerned, whereas on the two solstices the conditions of one are the exact reverse of the other.

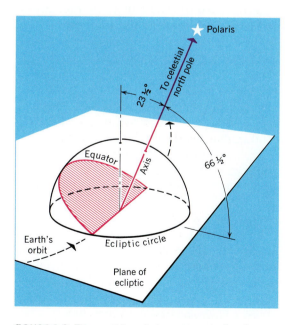

FIGURE 1.9 The earth's axis keeps an angle of 66½° with the plane in which its orbit lies.

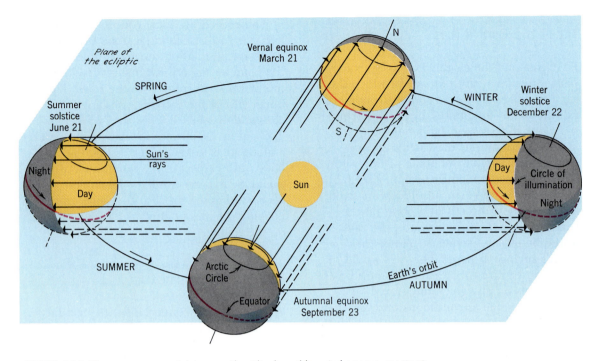

FIGURE 1.10 The seasons result because the tilted earth's axis keeps a constant orientation in space as the earth revolves about the sun.

Consider first the conditions at the equinoxes, since this is the simplest case. Figure 1.11 shows that the earth is at all times divided into two hemispheres with respect to the sun's rays. One hemisphere is lighted by the sun; the other lies in darkness. Separating the hemisphere is a circle, the *circle of illumination;* it divides day from night. You will notice that at either equinox the circle of illumination cuts precisely through the north and south poles. Looking at the earth from the side, as in Figure 1.11, conditions at equinox are essentially as follows. The *subsolar point*—the point at which the sun's noon rays are perpendicular to the earth—is located exactly at the equator. Here the angle between the sun's rays and the earth surface is 90°. At both poles, the sun's rays graze the surface. As the earth turns, the equator receives the maximum intensity of solar energy; the poles receive none. At an intermediate latitude, such as 40° N, the rays of the sun at noon make an angle with the surface equal to 90° minus the latitude, or 50°.

Next, examine winter solstice conditions, as shown in three dimensions in Figure

1.12. Because the maximum inclination of the axis is away from the sun, the entire area lying inside the *arctic circle,* lat. 66½° N, is on the dark side of the circle of illumination. Even though the earth rotates through a full circle during one day, this area poleward of the arctic circle remains in darkness. Conditions in the southern

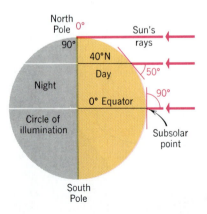

FIGURE 1.11 Equinox conditions. From this viewpoint the earth's axis appears to have no inclination.

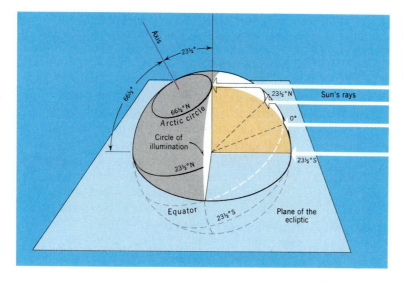

FIGURE 1.12 At winter solstice the entire area between the arctic circle and the north pole is in darkness throughout the 24-hour period.

hemisphere during winter solstice are shown in Figure 1.13. All the area lying south of the *antarctic circle,* lat. 66½° S, is under the sun's rays and enjoys 24 hours of day. The subsolar point has shifted to a point on the *tropic of capricorn,* lat. 23½° S.

At summer solstice, conditions are exactly reversed from those of winter solstice. As Figure 1.13 shows, the subsolar point is now on the *tropic of cancer,* lat. 23½° N. Now the region poleward of the arctic circle experiences a 24-hour day; that poleward of the antarctic circle experiences a 24-hour night. From one solstice to the next, the subsolar point has shifted over a latitude range of 47°.

The progression of changes from equinox to solstice to equinox and back to solstice initiates the *astronomical seasons:* spring, summer, autumn, and winter. These are labeled on Figure 1.10. The inflow of energy from the sun is thus varied in an annual cycle, and climatic seasons are generated. In Chapter 2 we will study the seasonal changes in solar energy reaching various latitude zones of the earth.

Our Spherical Habitat

Looking back over what we have covered in this chapter, certain concepts emerge that will be drawn upon in coming chapters.

Earth rotation generates the diurnal rhythms of the environment, because solar

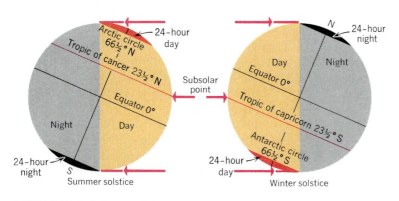

FIGURE 1.13 Solstice conditions.

energy is applied from a single source and its parallel rays can be intercepted by only a hemisphere. Life processes show a strong response to the diurnal cycle of energy input. Earth rotation has the effect of turning the direction of motion of air and water. This Coriolis effect profoundly influences the surface environment. Earth rotation with respect to the moon generates the ocean tide and its ceaseless rhythmic water motions essential to the life of the shallow coastal waters.

The accident of a rotational axis inclined from the ecliptic plane generates a seasonal rhythm of changes in solar energy of the life layer and is a major factor in causing the environment to differ from one latitude zone to another.

In this study of the astronomical controls of environment we visualize the planet as a perfectly smooth, solid sphere of uniform surface quality, exposed to the near vacuum of outer space. In the next chapter we must take a step toward realism and recognize that our planet has an envelope of air—the atmosphere—an extensive water layer—the oceans—and large patches of exposed solid rock—the continents.

THE EARTH'S SURFACE ON MAPS

Maps play an essential role in the study of physical geography, because much of the information content of geography is stored and displayed on maps. Map literacy—the ability to read and understand what a map shows—is a basic requirement for the day-to-day functioning of all educated persons. Maps appear in almost every issue of a newspaper and in nearly every TV newscast. Most of you routinely use highway maps and street maps. You now need to supplement the map-reading skills you have already acquired with some additional information on the subject of *cartography,* the science of maps and their construction.

Map Projections

Our first problem is to find satisfactory ways to transfer the spherical network of parallels and meridians shown on a globe to the flat surface of a map. Cartographers have struggled with this problem for centuries, driven by the need of the early navigators to have usable charts in their discovery voyages. Despite all the efforts of the finest mathematicians and cartographers, the quest for a perfect chart has ended in frustration. It's simply impossible to transform a spherical surface to a flat (planar) surface without some violation of the true surface as a result of cutting, stretching, or otherwise distorting the information that lies on the sphere.

A *map projection* is an orderly system of parallels and meridians used as a base on which to draw a map on a flat surface. Perhaps the simplest of all map projections is a grid of perfect squares: horizontal lines are parallels; vertical lines are meridians. This grid is often used in modern computer-generated world maps displaying data that consist of a single number for each square, representing one degree or 10 degrees of latitude and longitude. A grid of this kind can show the true spacing (approximately) of the parallels, but it fails to show how the meridians converge toward the two poles. This grid fails dismally in high latitudes and the map usually has to be terminated at about 70° to 80° north and south.

Early attempts to find satisfactory map projections made use of a familiar concept. Imagine the spherical earth grid as a cage of wires (a kind of birdcage). A brilliant light source can be placed at the center of the sphere and the image of the wire grid cast upon a surface outside the sphere. Think, too, of a reading lamp on which a shade is mounted. Basically, three kinds of "lampshades" can be used, as shown in Figure 1.14.

First is a flat paper disk balanced on the north pole. The shadow of the wire grid on this plane surface will appear as a combination of concentric circles (parallels) and radial straight lines (meridians). Here we have a polar-centered projection. Second

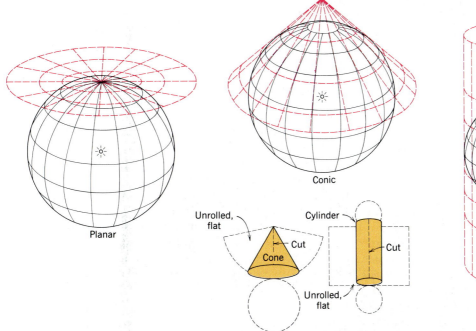

Planar

Conic

Unrolled, flat

Cut

Cone

Cylinder

Cut

Unrolled, flat

Cylindric

FIGURE 1.14 Simple ways to generate map projections. Rays from a central light source cast shadows of the spherical geographic grid on target screens. The conical and cylindrical screens can be unrolled to become flat maps. (A. N. Strahler.)

is a cone of paper resting point-up on the wire grid. The cone can be slit down the side, unrolled, and laid flat to produce a map that is some part of a full circle. This is called a *conic projection*. Parallels are arcs of circles; meridians are radiating straight lines. Third, a cylinder of paper can be wrapped around the wire sphere so as to be touching all around the equator. When slit down the side along a meridian, the cylinder can be unrolled to produce a *cylindrical projection*, which is a true rectangular grid.

Take note that none of these three projection methods can show the entire earth grid, no matter how large a sheet of paper is used to receive the image. Obviously, if the entire earth grid is to be shown, some quite different system must be devised. Many such alternative solutions have been proposed.

Here, we will concentrate on three useful map projections: a polar projection, a cylindrical projection, and one that uses a special mathematical principle. All three appear in many places throughout later chapters. It is important that you know both the advantages and the shortcomings of each of the three.

Figures 1.15, 1.16, and 1.17 show the three projections we selected. First is the *polar stereographic projection*, an ancient and honored grid, essential today for many scientific uses—for example, in weather

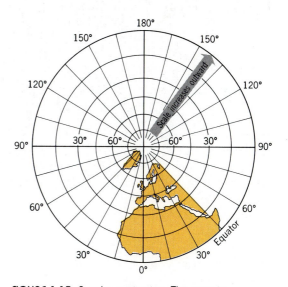

FIGURE 1.15 A polar projection. The map is centered on the north pole. All meridians are straight lines radiating from the center point, and all parallels are concentric circles. This particular network is of the stereographic type, which is conformal.

maps of the polar regions. Second is the *Mercator projection,* a classic navigator's map invented in 1569 by Gerhardus Mercator. It has never gone out of style and is essential for many uses of modern science. Third is the *Goode projection,* named for its designer, Dr. J. Paul Goode. It has special qualities not found in the other two projections and we use it for many important maps throughout this book. Not the least of these special qualities is that it can show the entire earth's surface.

The polar stereographic projection (Figure 1.15) can be centered on either the north or south pole. Spacing of the parallels increases quite rapidly outward from the center, so that the map is usually cut off to show only one hemisphere, with the equator forming the outer edge. Because the intersections of the parallels with the meridians always form true right angles, this projection shows the true shapes of all small land areas. If you were to draw a

perfect small circle anywhere on the globe, it would reproduce as a perfect circle on this map. Projections with this unique property are called *conformal projections.*

The Mercator projection (Figure 1.16) is a unique form of the rectangular grid. Meridians are equidistantly spaced, but the spacing of the parallels with higher latitude is increased according to a mathematical formula such that at 60° the spacing is double that at the equator. Farther poleward, the spacing increases rapidly and the map must be cut off at some arbitrary parallel, such as 80° N.

The Mercator projection is of the conformal type, but with very special properties. A straight line drawn anywhere on the map and in any direction oblique to the grid lines is a line of constant compass direction. This kind of line is called a *rhumb line,* or *loxodrome* (Figure 1.16). A navigator can simply draw a line between any two points on the map and measure the

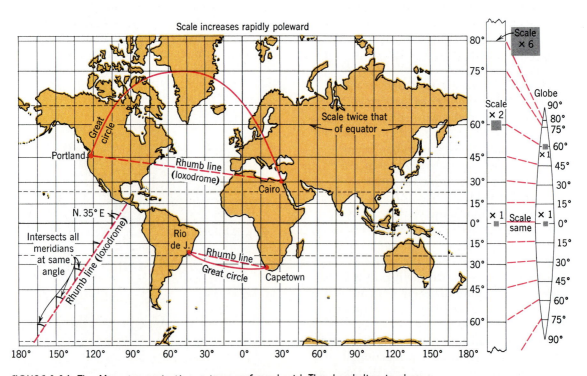

FIGURE 1.16 The Mercator projection, a true conformal grid. The rhumb line is always a straight line and holds a constant compass direction between any two points. The great circle is the shortest distance between any two points on the globe. The diagram at the right shows that map scale increases rapidly into higher latitudes. At lat. 60° the scale is double the equatorial scale; at lat. 80° the scale is six times greater than at the equator.

direction angle (bearing) with a protractor. Once set on that compass direction, the ship or plane is held to the same compass bearing to reach the final point. Of course, the rhumb line is not necessarily the shortest distance between two points. Instead, a *great circle* is actually the shortest path on the globe, but this may appear as a curving line on the map and may (falsely) seem to be a much longer distance. (A great circle is the surface trace of a plane passed through the center of a sphere. The equator is an example. Every meridian is one-half of a great circle.)

Because the Mercator projection shows the true cardinal directions of all lines on the map, it is chosen to depict all sorts of line features, among them flow lines of winds and ocean currents, directions of orientation of crustal lines (such as chains of volcanoes), and any lines of equal values, such as lines of equal air temperature or equal air pressure. This explains why the Mercator projection is chosen for our maps depicting atmospheric phenomena. These maps are supplemented by polar stereographic maps for the high latitudes.

The Goode projection uses two sets of mathematical curves (sine curves and ellipses) to form its meridians (Figure 1.17).

These converge to meet at the two poles and the entire globe can be shown. The straight, horizontal parallels make it easy to scan across the map at any given level to compare regions most likely to be similar in climate. (The same can be said for the Mercator projection.)

The Goode projection has a most important property with regard to the way in which it presents areas of the earth's surface. This network is an *equal-area projection,* on which the size (areal extent) of any small area is correctly scaled, no matter where it falls on the map. If we draw a very small circle on a sheet of clear plastic and move it over all parts of the Goode world map, the circle will always enclose an area with a constant value in square kilometers or square miles. For that reason, we select the Goode map to show geographical features that occupy surface areas. Examples are maps of regions of varieties of soil types or climate types.

The Goode map suffers from a serious defect: it distorts the shapes of areas, particularly in high latitudes and at the far right and left of the network. To minimize this defect, Dr. Goode split his map apart on selected meridians. Each separate sector of the map is centered on a vertical

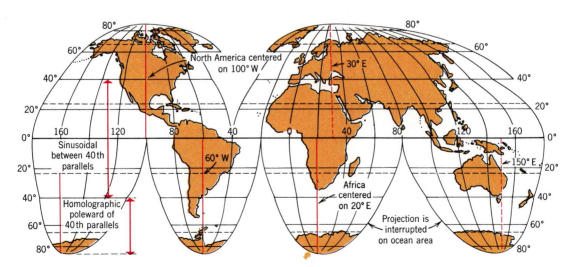

FIGURE 1.17 Goode's interrupted homolosine projection. This equal-area projection makes use of two well-known older projections. The sinusoidal projection is used between lat. 40° N and S and the homolographic projection is used poleward of lat. 40° N. (Copyright © by the University of Chicago. Used by permission of the University of Chicago Press.)

meridian. This type of split map is called an *interrupted projection.* Although the interrupted projection reduces shape distortion a great deal, it has the drawback of separating parts of the earth's surface that are actually close together, particularly in the high latitudes.

Keep in mind that every map projection has both useful qualities and serious defects. Always select a projection so that the useful qualities match the intended use as closely as possible, but remain aware of the defects and of the possible misconceptions those defects can cause.

Scales of Globes and Maps

All globes and maps depict the earth's features in much smaller size than the true features they represent. Globes are intended in principle to be perfect scale models of the earth itself, differing from the earth only in size. The *scale* of a globe is therefore the ratio between the size of the globe and the size of the earth, where "size" is some measure of length or distance (but not of area or volume).

Take, for example, a globe 20 cm (about 8 in.) in diameter, representing the earth, with a diameter of about 13,000 km. The scale of the globe is the ratio between 20 cm and 13,000 km. Dividing 13,000 by 20, this ratio reduces to a scale stated as follows: One centimeter on the globe represents 650 kilometers on the earth. This relationship holds true for distances between any two points on a globe.

Scale is more usefully stated as a simple fraction, termed the *fractional scale,* or *representative fraction.* It can be obtained by reducing both earth and globe distances to the same unit of measure, which in this case is centimeters. (There are 100,000 centimeters in one kilometer.) We go through the following calculations:

$$\frac{1\text{ cm on globe}}{650\text{ km on earth}} = \frac{1\text{ cm}}{650 \times 100,000}$$

$$= \frac{1}{65,000,000}$$

For convenience in printing, this fraction may also be written as 1:65,000,000. The advantage of the representative fraction is that it is entirely free of any specified units

of measure, such as the foot, mile, meter, or kilometer. Persons of any nationality can understand the fraction, regardless of their language or units of measure.

Being a true-scale model of the earth, a globe has a constant scale everywhere on its surface, but this is not true of a map projection drawn on a flat surface. In flattening the curved surface of the sphere to conform to a plane surface, all map projections stretch the earth's surface in a nonuniform manner, so that the map scale changes from place to place. So we can't say about any world map: "Everywhere on this map the scale is 1:65,000,000." It is, however, possible to select a meridian or parallel—the equator, for example—for which a fractional scale can be given, relating the map to the globe it represents. For example, in Figure 1.16, the scale of the Mercator projection along its equator is about 1:325,000,000.

Small-Scale and Large-Scale Maps

From maps on the global scale, showing the geographic grid of the entire earth or a full hemisphere, we turn to maps that show only small sections of the earth's surface. These maps of large scale are capable of carrying the enormous amount of geographic information that is available and must be shown in a convenient and effective manner. For practical reasons, maps are printed on sheets of paper usually less than a meter (3 ft) wide, as in the case of the ordinary highway map or navigation chart. Bound books of maps—atlases, that is—consist of pages usually no larger than 30 by 40 cm (12 by 16 in.), whereas maps found in textbooks and scientific journals are even smaller.

Actually, it is not simply the size of map sheet or page that determines how much information a map can carry; instead, it is the scale on which the surface is depicted. Keep in mind that the relative magnitude of two different map scales is determined according to which representative fraction is the larger and which the smaller. For example, a scale of 1:10,000 is twice as large as a scale of 1:20,000. Don't make the mistake of supposing that the fraction with the larger denominator represents the

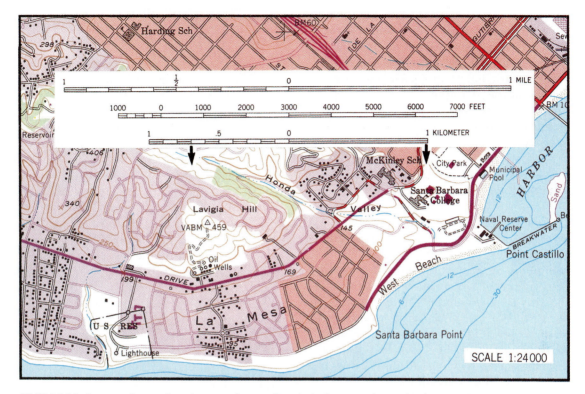

FIGURE 1.18 Portion of a modern, large-scale map for which three graphic scales have been provided. (U.S. Geological Survey.)

larger scale. If in doubt, ask yourself "Which fraction is the larger: 1/4 or 1/2?"

In the case of globes and maps of the entire earth, the fractional scale may range from 1:100,000,000 to 1:10,000,000; these are *small-scale maps*. *Large-scale maps* have fractional scales generally greater than 1:100,000. Stated in words, this fraction means: "One centimeter on the map represents one kilometer on the ground."

Most large-scale maps carry a *graphic scale*, which is a line marked off into units representing kilometers or miles. Figure 1.18 shows a portion of a large-scale map on which sample graphic scales in miles, feet, and kilometers are superimposed. Graphic scales make it easy to measure ground distances.

Suppose we now compare large-scale maps having the same outer dimensions (rectangles of the same size) but having different scales. Let us compare the ground areas covered by each rectangle. Figure 1.19

shows three such maps, all having the same outer dimensions but representing scales of 1:20,000, 1:10,000, and 1:5,000, respectively. Although map B is double the scale of map A, it shows a ground area only one-fourth as great. Map C is on a scale four times larger than that of map A, but it encloses a ground area only one-sixteenth as great. The rule we derive is that ground area within a map of given outer

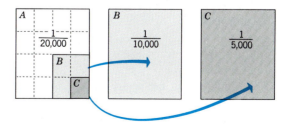

FIGURE 1.19 As scale is made larger, the area shown within the map boundaries rapidly decreases.

dimensions varies inversely with the square of the change in scale. For example, if the scale is reduced to one-third the original value, the area shown increases to nine times the original value. Obviously, greatly increasing the map scale allows a cartographer to show a vastly greater quantity of information on a map sheet of a fixed size.

Informational Content of Maps

Up to this point, our investigation of cartography has yielded a geographic grid with only the barest outlines of the continents. Information conveyed by a map projection grid system is limited to one category only: absolute location of points on the earth's surface. What other categories of information can be shown on the flat map sheet? The number of categories is almost infinite, since the map is capable of carrying information about anything that occupies a particular location at a specified time. Thousands of book pages of fine print would be needed to list the categories of information—"things"—that can be shown on maps. For the needs of physical geography, which is focused primarily on the physical environment of life in the surface layer, the list is greatly reduced, but still enormously long.

One way to classify information shown on maps used in physical geography is in terms of time. This may seem strange, since maps freeze time. Maps can't show time itself, but they can show the effects of time. First, we can recognize maps showing information that does not change with time. More realistically, we are showing things that do not change appreciably or measurably in time spans of, say, decades or even centuries. In this category are fixed physical objects, both natural and cultural (human-made). Examples are exposed rock formations, hills, valleys, soil types, political boundaries, highways, quarries, churches, and courthouses.

Because such time-bound features are fixed in location, many kinds of features can be shown on one map, i.e., it can be a multipurpose map. Map sheets published by national governments are usually multipurpose maps. A good example is

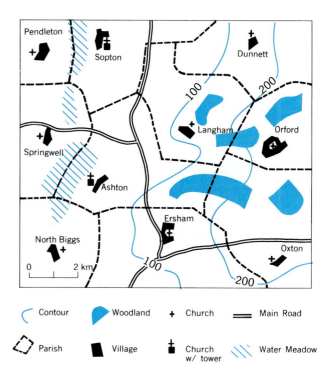

FIGURE 1.20 Planimetric map of an imaginary area with 10 villages. (After J. P. Cole and C. A. M. King, *Quantitative Geography*, p. 37, Figure 2.3. Copyright © 1968 by John Wiley & Sons, London. Used by permission.)

shown in Figure 1.20. Using a great variety of symbols, patterns, and colors, these maps carry a high information content available on demand to the general user, who may select and use only one information category, or theme, at any one time. That such maps are produced and distributed in vast numbers bears witness to the great longevity of most of the information they carry. Nevertheless, these multipurpose maps represent reality at a fixed point in time. As such, they are historical documents.

In contrast to the multipurpose map is the *thematic map*, which shows only one data category. Examples are legion: distribution of the human population, boundaries of election districts, geologic age of exposed rock, elevation of the land surface above sea level, number of days of frost in the year, and speed of winds in the upper atmosphere.

Among the thematic maps we can recognize classes distinctly related to the factor of time. There are classes of data that change continuously with time. Examples:

population of humans and wind speeds at a high level in the atmosphere. Maps of such changing entities show them frozen in time at a specified time or within a short time interval during which observations are made. Population maps show the data of a census, taken over several weeks or a year. Maps of wind speed freeze the atmospheric motion at a given point in time using observations taken simultaneously within a span of minutes.

Entities that change rapidly with time, such as wind speed, air temperature, precipitation, or river flow, can also be averaged over a substantial period of time, yielding an average, or mean value that is comparatively constant through time. In the case of air temperature, there are constant fluctuations in its value and these occur in daily and seasonal rhythms. We can average out these time fluctuations, yielding in succession a daily mean (average), a monthly mean, a yearly mean, and finally the mean over a 20- or 50-year period of record. The distribution patterns of these means can be displayed on a map. The thematic data of physical geography presented on maps throughout this book fall in all the time-span related categories mentioned in the foregoing.

Map Symbols

Points and lines on maps must have a certain minimum width in order to be visible, thus they become symbolic of the ideal mathematical concepts behind them. To express points as symbols, they can be renamed "dots." A dot may be of any diameter that seems appropriate, and an open circle can also serve to indicate a point. Broadly defined, a dot can be any small device to show point location; it might be a letter, a numeral, or a little picture of the object it represents (see "church with tower" in Figure 1.20). The name for a line symbol remains unchanged, i.e., a line; it can be varied in width and can be single or double. The line can also consist of a string of dots or dashes. A specific area of the planimetric surface can be renamed a "patch." (The word "area" is more properly left to the mathematical concept of a surface of infinite extent or the measure of

area.) The patch can be shown simply by a line marking its edge, or it can be depicted by a distinctive pattern or a solid color. Patterns are highly varied. Some consist of tiny dots and others of parallel, intersecting, or wavy lines.

A map consisting of dots, lines, and patches can carry a great deal of information, as Figure 1.20 shows. In this case, the map uses two kinds of dot symbols (both symbolic of churches), three kinds of line symbols, and four kinds of patch symbols (if parishes are considered a set of contiguous patches). Altogether, nine categories of information are offered. Line symbols freely cross patches, and dots can appear within patches. Two different kinds of patches can overlap.

The relation of map symbols to map scale is of prime importance in cartography. Maps of very large scale, along with architectural and engineering plans, can show objects to their true outline form. As map scale is decreased, representation becomes more and more generalized, becoming an analog (for example, church shown by box with cross). Ultimately the analog gives way to a pure symbol using the dot, line, or patch. In physical geography an excellent example is the depiction of a river, such as the lower Mississippi. Figure 1.21 shows the river channel at three scales, starting with a detailed plan, progressing to a double-line analog that generalizes the channel form, and ending with a single-line symbol. As generalization develops, the details of the river banks and channel bends are simplified as well. The level of depiction of fine details is described by the term *resolution,* which has much the same meaning as in photography and telescopic observation. Maps of large scale have much greater resolving power than maps of small scale.

An interesting kind of map that manipulates the width of the line symbol is the *flow map,* on which the function of the line is to indicate direction and volume of channeled movement or flow of some specific form of matter or energy. The line becomes a solid band, its width scaled in proportion to flow volume. (For an example, see Figure 10.11.) In effect, the flow map is a kind of pictorial bar graph, designed for quick visual appraisal.

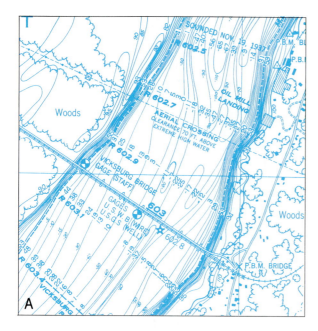

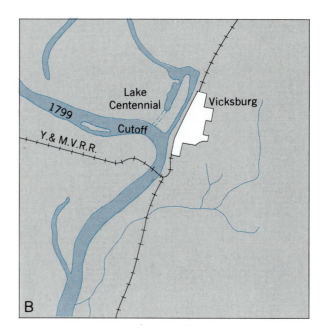

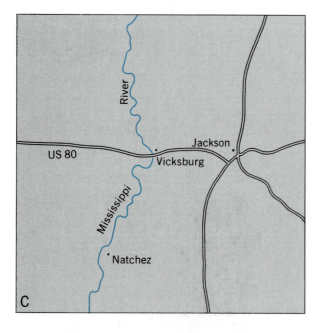

FIGURE 1.21 Maps of the Mississippi River on three scales. (A) 1:20,000. Channel contours give depth below mean water level. (B) 1:250,000. Waterline only shown to depict channel. (C) 1:3,000,000. Channel shown as solid line symbol. (U.S. Army Corps of Engineers.)

Presenting Numerical Data on Thematic Maps

We turn now to methods of showing numerical (quantitative) data on thematic maps. In physical geography, much of the information collected about particular areas is in the form of numbers. The numbers might represent readings taken from a scientific instrument at various places throughout the study area. A simple example is the collection of weather data, such as air temperature, air pressure, wind speed, and amount of rainfall. Another example is the set of measurements of the elevation of a land surface, given in meters above sea level. Another category of information consists merely of the presence or absence of a quantity or attribute. In terms of numbers, this boils down to either one or zero (1 for "present" and 0 for "absent"). In such cases, rather than entering ones and zeroes on the map, we can simply place a dot to mean "present," so that when entries are completed, we have before us a field of scattered dots (Figure 1.22).

In some scientific programs, measurements can be taken uniformly, for example, at the centers of grid squares laid over the map. For many classes of data, however, the locations of the observation points are predetermined by a fixed and nonuniform set of observing stations. For example, data of weather and climate are collected at stations typically located at airports, so you take them where you find them. Whatever the sampling method used, we end up with an array of numbers or dots on the field of the base map.

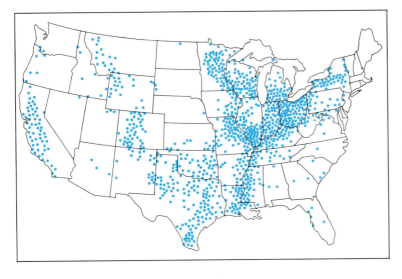

FIGURE 1.22 A dot map showing the distribution of soils of the order Alfisols in the United States. A moving circle was used to generate the dots. (From Philip J. Gersmehl, *Annals of the Assoc. of Amer. Geographers*, vol. 67, p. 426, figure 7. Copyright © 1977 by the Association of American Geographers. Used by permission.)

Let us suppose that the data we have available consist of more or less randomly scattered dots, much as in Figure 1.22. Given a field of large numbers of dots, the usual procedure is to place a square grid overlay on the map. The number of dots falling within a given square is entered as the single representative value for that square; it is assumed to be located at the center point of the square. The result is an orderly set of numbers giving the density of dots in each square. Figure 1.23*A* shows density numbers for each of 100 grid squares. (On the other hand, each number might simply represent the average value of some quantity, such as air temperature in degrees Celsius.)

We can now assign a particular area pattern or color to each square, using a scale of grouped numbers, or value classes. An example is shown in Figure 1.23*B*. Five classes cover the range of observations. In this case, rather than revealing a haphazard or random assemblage of squares of the several classes, the squares group themselves into distinct zones trending diagonally across the field. This grouping shows clearly that the data exhibit a trend to diminish in value from upper left to lower right. The important point is that the boundaries represent abrupt discontinuities, like the vertical risers of a flight of steps. In concept, passing from one zone to the other is like stepping up or down on the flat, horizontal treads of a staircase.

A more sophisticated treatment of the raw data is shown in Figure 1.23*C*. Taking the observations to occupy the center points of the grid squares, lines have been drawn to show the position (locus) of all points on the field having the same integer value. These lines of equal value are called *isopleths* (from the Greek *isos,* "equal," and *plethos,* "fullness" or "quantity"). The term *isarithm* (from the Greek *arithmos,* "number") is also used in scientific writing in essentially the same sense as isopleth.

The concept behind isopleths is that they lie on an imagined continuous surface that slopes down or up from place to place without abrupt steps. Isopleths can depict a sloping plane or a warped surface with hills and valleys. A sloping plane fitted to the data would define a trend from higher to lower values; such a map is sometimes referred to as a *trend surface.* Dashed lines with arrows on Figure 1.23*C* show the direction of the gradient of the sloping surface. Notice that the gradient lines always cross the isopleths at right angles.

Isopleths maps are important in various branches of physical geography. Table 1.1 gives a partial list of isopleths of various kinds used in the earth sciences, together with their special names and the kinds of information they display. Examples are cited from our textbook. A special kind of isopleth, the topographic contour (or isohypse), is shown on the map in Figure 1.18. Topographic contours show the configuration of land surface features, such as hills, valleys, and basins.

Modern cartography has made enormous strides forward in collecting, presenting,

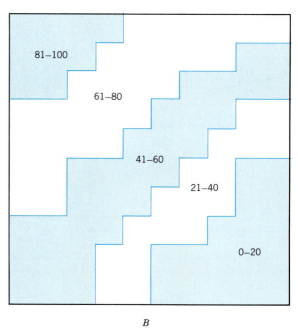

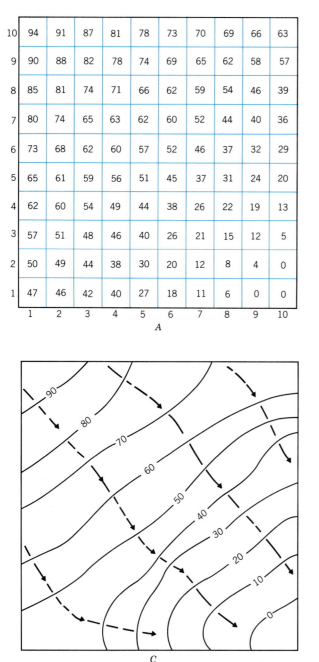

FIGURE 1.23 Presentation of numerical data on maps. (A) Grid of squares. Figures show number of observations falling within a square. (B) Map, showing classes with abrupt discontinuities between. (C) Isopleths drawn on a continuously variable surface. Dashed lines with arrows show direction of gradient.

and analyzing geographic information through the use of data sensed by instruments mounted on high-flying aircraft or orbiting space vehicles and processed by high-speed computers. The information is collected in the digital format universally used in electronic computers. In Chapter 2, we refer to this general subject under the title of "remote sensing."

TABLE 1.1 Examples of isopleths

Name of Isopleth	Greek Root	Property Described	Examples in Figures
Isobar	*baros*, weight	Barometric pressure	4.10
Isotherm	*therme*, heat	Temperature of air, water, or soil	3.14
Isotach	*tachos*, swift	Fluid velocity	4.19
Isohyet	*hyetos*, rain	Precipitation	5.16
Isohypse	*hypso*, height	Elevation	18.11, 18.12

THE METRIC SYSTEM
Metric to English (U.S.)

Rough equivalent: **Precise equivalent:**

Length, Distance

Rough equivalent	Precise equivalent
1 centimeter = 4/10 inch	1 cm = 0.394 in.
1 meter = 40 inches or 3¼ feet	1 m = 39.37 in. or 3.28 ft
1 kilometer = 6/10 mile	1 km = 0.621 mi

Area

Rough equivalent	Precise equivalent
1 square centimeter = 1/7 square inch	$1\ cm^2 = 0.155\ in.^2$
1 hectare = 2½ acres	1 ha = 2.47 a
1 square kilometer = 4/10 square mile	$1\ km^2 = 0.386\ mi^2$

Volume

Rough equivalent	Precise equivalent
1 cubic centimeter = 1/16 cubic inch	$1\ cc,\ 1\ cm^3 = 0.061\ in.^3$
1 cubic meter = 35 cubic feet	$1\ m^3 = 35.3\ ft^3$
1 liter = 1 quart liquid (U.S.)	1 l = 1.057 qt

Weight

Rough equivalent	Precise equivalent
1 gram = 1/30 ounce (avoirdupois)	1 gm = 0.0353 oz
1 kilogram = 2 1/5 pounds (av.)	1 kg = 2.205 lb
1 tonne (metric ton) or 1000 kg = 1 1/10 short tons (U.S.)	1 t = 1.102 t

Temperature

Rough equivalent	Precise equivalent
1 Celsius degree = 9/5 Fahrenheit degrees	1 C° = 1.80 F°

English (U.S.) to Metric

Rough equivalent: **Precise equivalent:**

Length, Distance

Rough equivalent	Precise equivalent
1 inch = 2½ centimeters	1 in. = 2.540 cm
1 foot = 1/3 meter	1 ft = 0.305 m
1 mile = 1½ kilometers	1 mi = 1.609 km

Area

Rough equivalent	Precise equivalent
1 square inch = 6½ square centimeters	$1\ in.^2 = 6.45\ cm^2$
1 acre = 4/10 hectare	1 a = 0.405 ha
1 square mile = 2½ square kilometers	$1\ mi^2 = 2.59\ km^2$

Volume

Rough equivalent	Precise equivalent
1 cubic inch = 16 cubic centimeters	$1\ in.^3 = 16.39\ cc,\ cm^3$
1 cubic foot = 1/35 cubic meter	$1\ ft^3 = 0.0284\ m^3$
1 quart liquid (U.S.) = 1 liter	1 qt = 0.964 l

Weight

Rough equivalent	Precise equivalent
1 ounce (avoirdupois) = 28 grams	1 oz = 28.35 gm
1 pound (av.) = 450 grams, or ½ kilogram	1 lb = 453.6 gm, or 0.454 kg
1 short ton (U.S., 2000 pounds) = 9/10 tonnes (metric tons)	1 t = 0.907 t, or 907 kg

Temperature

Rough equivalent	Precise equivalent
1 Fahrenheit degree = 5/9 Celsius degree	1 F° = 0.556 C°

CHAPTER

2

THE EARTH'S ATMOSPHERE AND RADIATION BALANCE

EARTH REALMS   Three inorganic earth realms—atmosphere, hydrosphere, lithosphere—provide the nonliving substances out of which the living matter of the biosphere is created.

THE ATMOSPHERE   It is a mixture of gases, dominantly nitrogen and oxygen. Carbon dioxide, although a minor constituent, is of great importance to life on earth and to the global climate.

AN ENERGY BALANCE   For our planet as a whole there exists a balance between incoming solar energy and outgoing radiant energy.

ENERGY LOSSES   Part of the incoming solar energy is scattered and reflected back to space, but some is absorbed by the atmosphere; only about half is absorbed by the earth's land and ocean surfaces.

LONGWAVE RADIATION   The warm surface of our earth radiates energy in long wavelengths of a type known as infrared radiation; these warm the atmosphere. Longwave radiation to outer space balances the earth's energy budget.

THE air, sea, and land constitute the major portions of four great material realms, or spheres, that make up the total global environment (Figure 2.1). Three of these realms are inorganic: (1) *atmosphere*, (2) *hydrosphere*, and (3) *lithosphere*. Most of the substances of which they consist are classified by chemists as inorganic matter. The fourth realm, the *biosphere*, encompasses all living organisms of the earth. Because living organisms cannot exist except in a physical environment with which they interact, the biosphere includes parts of the atmosphere, hydrosphere, and lithosphere.

Of the three inorganic spheres, the atmosphere is the gaseous realm. The hydrosphere is the water realm consisting of free water in gaseous, liquid, and solid states; it includes fresh water of the atmosphere and the lands, as well as salt water of the oceans. The lithosphere is the solid realm composed of mineral matter. The three spheres of inorganic matter occupy layerlike shells over the globe because of differences in the density of the three types of substances. Each of the three spheres has a different and distinctive chemical makeup, inherited from its origin

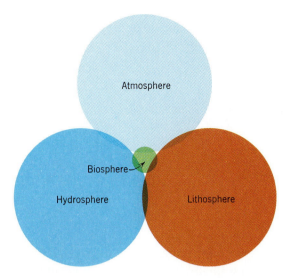

FIGURE 2.1 The earth realms shown as intersecting circles. The large outer circles represent the three great inorganic realms; each one overlaps the other two in a small area, suggesting that some of the substance of each realm is held within the other two. The biosphere, or organic realm, draws its substance from its inorganic environment, represented by the circle overlap into the three surrounding realms. The small diameter of the biospheric circle signifies that the total mass of matter held within the biosphere is but a small fraction of that in the other three realms.

in the geologic past. The biosphere requires materials from all three of the inorganic spheres; these are elements used to form the molecules that make up organic matter.

Atmosphere and Oceans

We live at the bottom of an ocean of air. Humans are air-breathers dependent on favorable conditions of pressure, temperature, and chemical composition of the atmosphere that surrounds them. Humans also live on the solid outer surface of the earth, which they depend on for food, clothing, shelter, and means of movement from place to place. But the air and the land are not two entirely separate realms; they constitute an interface across which there is a continual flux of matter and energy. Our surface environment, the *life layer,* is a shallow but highly complex zone in which atmospheric conditions exert control on the land surface, while at the same time the surface of the land exerts an

influence on the properties of the adjacent atmosphere.

Essentially the same statements apply to the surface of the oceans and the atmospheric layer above it. We utilize the surface of the sea as a source of food and a means of transportation. There is a continual flow of energy and matter between the sea surface and the lower layer of the atmosphere. Here, again, we find an interface of vital concern. The sea influences the atmosphere above it, and the atmosphere influences the sea beneath it.

Our objective in these early chapters is to examine the atmosphere and oceans with particular reference to the air–land and air–sea interfaces that are so vital to humans. To geographers, the distributions of physical properties of the ocean and atmosphere are matters of special interest, concerned as they are with spatial relationships on a global scale. Physical geographers describe and explain the ways in which the environmental ingredients of weather and climate change with latitude and season and with geographical position in relation to oceans and continents. They seek out the broad patterns of similar regions and attempt to define their boundaries and organize them into systems of classes. More important, geographers try to evaluate the environmental qualities of each region, emphasizing the opportunities as well as the limitations of each for future development of natural resources, such as food, water, energy, and minerals.

The principles we will review in this chapter belong largely to an area of natural science known as *meteorology,* the science of the atmosphere. We also touch on *physical oceanography,* the physical-science aspect of the oceans.

Composition of the Atmosphere

The earth's atmosphere consists of a mixture of various gases surrounding the earth to a height of many kilometers. Held to the earth by gravitational attraction, this envelope of air is densest at sea level and thins rapidly upward. Although almost all the atmosphere (97 percent) lies within 30 km (18 mi) of the earth's surface, the upper limit of the atmosphere can be drawn

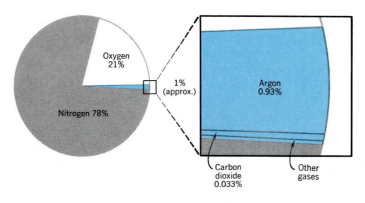

FIGURE 2.2 Components of the lower atmosphere. Figures tell percentage by volume.

approximately at a height of 10,000 km (6000 mi), a distance approaching the diameter of the earth itself. From the earth's surface upward to an altitude of about 80 km (50 mi) the chemical composition of the atmosphere is highly uniform throughout in terms of the proportions of its gases.

Pure, dry air consists largely of nitrogen, about 78 percent by volume, and oxygen, about 21 percent (Figure 2.2). Nitrogen of the atmosphere exists as a molecule consisting of two nitrogen atoms (N_2). Nitrogen gas does not enter easily into chemical union with other substances and can be thought of as primarily a neutral substance. Very small amounts of nitrogen are extracted by soil bacteria and made available for use by plants.

In contrast to nitrogen, oxygen gas (O_2) is highly active chemically and combines readily with other elements in the process of oxidation. Combustion of fuels represents a rapid form of oxidation, whereas certain forms of rock decay (weathering) represent very slow forms of oxidation. Animals require oxygen to convert foods into energy.

The remaining 1 percent of the air is mostly argon, an inactive gas of little importance in natural processes. In addition, there is a very small amount of carbon dioxide (CO_2), amounting to about 0.033 percent. This gas is of great importance in atmospheric processes because of its ability to absorb radiant heat. Carbon dioxide thus augments the warming of the lower atmosphere by radiant energy coming from the sun and from the earth's surface. Green plants, in the process of photosynthesis, utilize carbon dioxide from the atmosphere, converting it with water into solid carbohydrate (Chapter 20).

All the component gases of the lower atmosphere are perfectly diffused among one another so as to give the pure, dry air a definite set of physical properties, just as if it were a single gas.

Our Dynamic Planet

Imagine yourself as a new arrival on the moon at Lunar Base One, a permanently manned research station. Sitting inside your life-support compartment, you gaze through the viewport at a majestic earthrise. The planet is a patchwork of royal blues, verdant greens, and rich browns, superimposed with vast white cloud swirls, framed by the velvet blackness of space (Figure 2.3). From your lunar viewpoint, the earth's environmental realms are easily identified by their colors. You note that the blue of the world oceans—the hydrosphere—dominates the area of the disk. Rich browns characterize the lithosphere—the solid portion of the planet, where it rises above the seas. Vast white cloud expanses dominate the gaseous portion of the planet—the atmosphere. The lush greens of plants are the visible portion of the biosphere, the realm of life on earth, which is confined to the life layer—the interface between land, sea, and air.

The dramatic contrast between the day and night portions of the globe strikes you immediately. The land and water masses of the illuminated side bask in the sunlight, absorbing its warmth and reflecting its light.

FIGURE 2.3 Planet Earth photographed in 1972 by astronauts of the Apollo 17 mission on their way to the moon. On this date the earth was near the December solstice, with its south polar region tilted toward the sun to receive maximum insolation. Africa is clearly outlined in the central and upper part of the disk, its reddish-brown northern and southern desert belts conspicuously free of clouds. The intervening equatorial zone of the African continent reveals green vegetation between patches of clouds. Cyclonic storms form a white chain over the blue of the Southern Ocean. Antarctica, with its blanket of snow and clouds, is solid white. (NASA 72–HC–928.)

Solar rays travel at the speed of light in straight lines and freely penetrate the atmosphere. You visualize the energy transformations (conversions) occurring at the planetary surface: solar energy strikes the earth and is absorbed and converted to heat, increasing the temperature of the life layer. On the shadowed side of the globe, however, the reverse process dominates: the heat energy stored by the oceans and continents is continuously being lost— radiated into outer space. Without sunlight, the earth's surface cools.

As you watch the earth during the next few hours, you notice its slow rotation. The continent of South America glides across the center of the disk to its edge, then disappears. Australia emerges from behind the curtain of night and makes its way out onto the illuminated portion of the disk. Under your eye, each land and water mass takes its turn in the sun, absorbing and reflecting its quota of solar energy. Clearly, Africa and South America receive the lion's share, for they lie near the midline of the earth, where they meet the perpendicular rays of the sun. North America and Europe, on the other hand, are near the edges of the disk and are shortchanged because they intercept the sun's rays at a low angle. Not only are you witnessing great changes or transformations of energy, but you are also seeing that although the sun's energy input to the whole earth is constant, there is a great variation in the quantity received by any given small area of surface, both through time and with position on the globe.

As the days go by (for you are still keeping earth time), you gaze often at your planet, noting the changes. Swirls of clouds form, dissolve, and reform in its atmosphere, making their way eastward across the disk, first obscuring one portion of a continent, then another. This atmospheric turmoil is related to uneven solar heating of the globe and is part of a transport mechanism serving to carry excess heat of the equatorial belt to polar regions, where heat is in short supply. Somehow these great swirls seem tied in with the earth's rotation as it carries the turbulent atmosphere with it. Once again, you are witnessing a set of energy transformations: a portion of the radiant energy of the sun, after being transformed into heat, is now being converted to energy of motion, because moving masses of air and water also represent a form of energy. This energy of matter in motion, kinetic energy, meets with resistance—friction of various sorts— generating heat in the atmosphere and at the earth's land and ocean surfaces. This heat adds to the intensity of radiation of energy back into space.

In fact, from your point of view on Lunar Base One, you might visualize the earth as a single great energy system that receives solar energy as an input while it reflects light energy and radiates heat energy as an output. Within this system, many transformations of energy occur, and each transformation is associated with matter in one form or another.

Solar Radiation

Our sun, a star of about average size and temperature as compared with the overall range of stars, has a surface temperature of about 6000°C (11,000°F). The highly heated, incandescent gas that comprises the sun's surface emits energy in the form of electromagnetic radiation. This form of energy transfer can be thought of as a collection, or spectrum, of waves of a wide range of lengths traveling at the uniform velocity of 300,000 km (186,000 mi) per second. The energy travels in straight lines radially outward from the sun and requires about 8⅓ minutes to travel the 150 million km (93 million mi) from sun to earth.

Although solar radiation travels through space without energy loss, the rays are diverging as they move away from the sun. Consequently, the intensity of radiation within a beam of given cross section (such as 1 sq cm) decreases inversely as the square of the distance from the sun. The earth intercepts only about one two-billionth of the sun's total energy output.

The sun's electromagnetic spectrum can be divided into three major portions, based on wavelengths. *Wavelength* describes the distance separating one wave crest from the next (Figure 2.4). Long waves have a

comparatively large separating distance; short waves are closely spaced. Figure 2.5 shows a scale of wavelengths from short to long lengths. The unit of wavelength is the *micron*. (A micron is one ten-thousandth of a centimeter: 0.0001 cm.) At the left lies the *ultraviolet radiation* portion of the spectrum, with wavelengths ranging from 0.2 to 0.4 micron. In the center is the *visible light* portion in the range of 0.4 to 0.7 micron. At the right lies the *infrared radiation* region, ranging upward of 0.7 micron. The scales used on this graph increase by powers of ten, meaning that the scale becomes tightly compressed, toward both the right and the top of the graph.

We are primarily concerned with the energy of the solar spectrum and how it is distributed. The high curve at the left on the graph of Figure 2.5 shows the relative energy intensity from one end of the sun's spectrum to the other, if it were to be measured at the top of the atmosphere. Energy is shown on the graph on the vertical scale. The units used here do not have to be explained, because we are only interested in making rough comparisons of how intensities are related to the bands of the spectrum. Energy intensity rises rapidly through the ultraviolet portion of the spectrum. This portion accounts for about 9 percent of the total incoming solar energy. Also included in the ultraviolet division are X rays and gamma rays. These are forms of radiation with shorter wavelengths than the ultraviolet and are not shown on the graph. The energy curve peaks in the visible light range, which accounts for about 41 percent of the total energy. The infrared portion of the spectrum accounts for 50 percent of the total energy. Very little energy arrives in wavelengths longer than 2 microns.

Scientists refer to the entire solar spectrum as *shortwave radiation*, because the peak of intensity lies in the shorter wavelengths. In contrast, radiation from earth to outer space lies entirely within the infrared region and is called *longwave radiation*.

The source of solar energy is in the sun's interior. Here, under enormous confining pressure and high temperature, hydrogen is converted to helium. In this nuclear fusion process, a vast quantity of heat is generated and finds its way to the sun's surface.

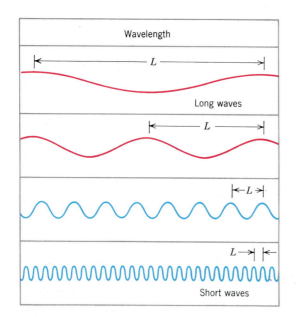

FIGURE 2.4 Wavelength, *L*, is the crest-to-crest distance between successive wave crests.

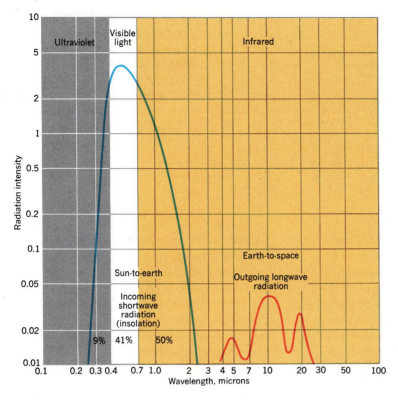

FIGURE 2.5 Intensity of incoming shortwave radiation at the top of the atmosphere (left); longwave radiation from earth to outer space (right).

Because the rate of production of nuclear energy is constant, the output of solar radiation is nearly unvarying. So, at the average distance of earth from the sun, the amount of solar energy received on a unit area of surface held at right angles to the sun's rays is almost unvarying. We are assuming, of course, that the radiation intensity is measured beyond the limits of the earth's atmosphere so that none has been lost. Known as the *solar constant,* this rate of incoming shortwave energy has a value of 2 gram calories per square centimeter per minute. The gram calorie is that quantity of heat required to raise by 1 C° the temperature of 1 gram of pure water. One gram calorie per square centimeter constitutes a unit measure of radiation intensity known as the *langley.* Therefore, we can say that the solar constant is equal to 2 langleys per minute (2 ly/min).

The Global Energy Balance

The flow of energy from sun to earth and then out into space is a complex system. It involves not only radiation, but also energy storage and transport. Both storage and transport of heat occur in the gaseous, liquid, and solid matter of the atmosphere, hydrosphere, and lithosphere. However, we can simplify the study of this total system by first examining each of its parts. We will start with the radiation process itself and develop the concept of a *radiation balance,* in which energy absorbed by our planet is matched by the planetary output of energy into outer space.

Solar energy is intercepted by our spherical planet, and the level of heat tends to be raised. At the same time, our planet radiates energy into outer space, a process that tends to diminish the level of heat energy. Incoming and outgoing radiation processes are simultaneously in action (Figure 2.6). In one place and time more energy is being gained than lost; in another place and time more energy is being lost than gained.

The equatorial region receives much more energy through solar radiation than is lost directly to space. In contrast, polar regions lose much more energy by radiation into space than is received directly from the sun. So mechanisms of energy transfer must be included in the energy system. These

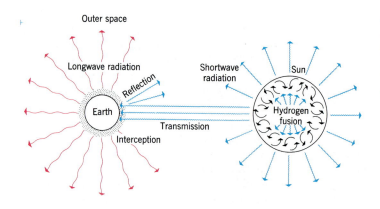

FIGURE 2.6 A pictorial diagram of the global energy balance. (From A. N. Strahler, The life layer, *Jour. of Geography*, vol. 69, p. 72. Copyright © 1970 by *The Journal of Geography*. Used by permission.)

mechanisms must be adequate to export energy from the region of surplus and to carry that energy into the regions of deficiency. On our planet, motions of the atmosphere and oceans act as heat-transfer mechanisms. A study of the earth's energy balance will not be complete until the global patterns of air and water circulation are described and explained in Chapter 4.

Thinking further, we realize that the movement of water through atmosphere and oceans and on the lands comprises a global system of transport of matter. This system is equal in environmental importance to the flow of energy. We also realize that the activities of these two systems are closely intermeshed. The concept of a water balance will be developed in Chapter 10 and will take its place beside the energy balance. Together, these two great flow systems of energy and matter form a single, grand planetary system and permit us to relate and explain many of the environmental phenomena of our earth within a single unified framework.

A systematic approach to the earth's energy balance begins with an examination of the input, or source, of energy from solar radiation. We will trace this radiation as it penetrates the earth's atmosphere and is absorbed or transformed. We then turn to the mechanism of output of energy by the earth as a secondary radiator.

Modeling the Global Energy Balance

Knowing the distinction between incoming shortwave radiation and outgoing longwave radiation, we can model the global energy

balance as an open energy system. Figure 2.7 uses a rectangular box to show the arbitrary boundary of the global system; it lies at the outer limit of the atmosphere.

Input of solar energy is in the form of shortwave radiation, part of which is reflected back directly into space without being transformed or stored. As the shortwave energy penetrates the atmosphere, it is transformed into sensible heat. Thus the temperature of the atmosphere, oceans, and lands tends to rise. (Other forms of energy have been omitted for the sake of keeping the diagram as simple as possible.) Energy stored as sensible heat is then emitted in the form of longwave radiation, involving another energy transformation. The ultimate system output is by longwave radiation into outer space.

Over long periods of time, this system is approximately in a steady state of operation, in which incoming and outgoing forms of radiation are equal and the quantity of stored energy remains constant.

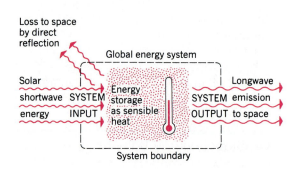

FIGURE 2.7 A schematic diagram of the global radiation balance as an open energy system. (A. N. Strahler.)

Insolation over the Globe

Referring back to Chapter 1 and Figure 1.11, showing equinox conditions, recall that at only one point, the subsolar point, does the earth's spherical surface present itself at right angles to the sun's rays. In all directions away from the subsolar point, the earth's curved surface becomes turned at a decreasing angle with respect to the rays until the circle of illumination is reached. Along that circle the rays are parallel with the surface.

Let us now assume that the earth is a perfectly uniform sphere with no atmosphere. Only at the subsolar point will solar energy be intercepted at the full value of the solar constant, 2 ly/min. We will now use the term *insolation* to mean the interception of solar shortwave energy by an exposed surface. At any particular place on the earth, insolation received in one day will depend on two factors: (1) the angle at which the sun's rays strike the earth; and (2) the length of time of exposure to the rays. These factors are varied by latitude and by the seasonal changes in the path of the sun in the sky.

Figure 2.8 shows that intensity of insolation is greatest where the sun's rays strike vertically, as they do at noon at some

parallel within the belt lying between the tropic of cancer and the tropic of capricorn. With diminishing angle, the same amount of solar energy spreads over a greater area of ground surface. So, on the average, the polar regions receive the least insolation over a year's time.

In Chapter 1 we considered a hypothetical situation in which the earth's axis is perpendicular to the plane of the ecliptic as the earth revolves around the sun. Under such conditions, the poles would not receive any insolation, regardless of time of year, whereas the equator would receive an unvarying maximum. In other words, equinox conditions would prevail year-round. But we know that the earth's axis is not perpendicular to the plane of the orbit. The axis is tilted with respect to the orbital plane by an angle that measures 23½° from the perpendicular (Figure 1.9). Moreover, the axis at all times holds its orientation in space. Consequently, as the earth travels in its orbit about the sun, the tilted globe assumes different positions with respect to the sun's rays (Figure 1.10). At this point, a review of the conditions at solstices and equinoxes may be helpful (Chapter 1).

The subsolar point, representing the sun's noon rays, shifts through a total

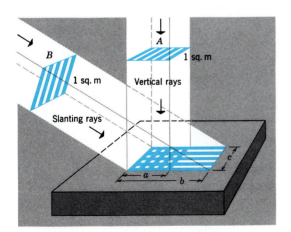

FIGURE 2.8 The angle of the sun's rays determines the intensity of insolation on the ground. The energy of vertical rays A is concentrated in square a, but the same energy in the slanting rays B is spread over a rectangle, b.

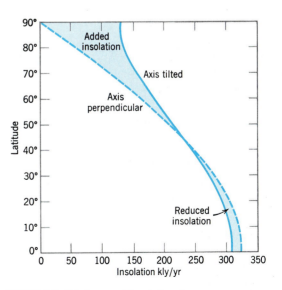

FIGURE 2.9 Total annual insolation from equator to pole (solid line) compared with insolation on a globe with axis perpendicular to the ecliptic plane (dashed line).

38

latitude range of 47° from one solstice to the next. This cycle does not make the yearly total of insolation for the entire globe different from an imaginary situation in which the earth's axis is not inclined, but it does cause a great difference in the quantities received at various latitudes.

The total annual insolation from equator to poles in thousands of langleys (kilo-langleys) per year is shown in Figure 2.9 by a solid line. A dashed line shows the insolation that would result if the earth's axis had no tilt. Notice how much insolation the polar regions actually receive—over 40 percent of the equatorial value. The added insolation at high latitudes is of major environmental importance because it brings heat in summer to allow forests to flourish and crops to be grown over a much larger portion of the earth than would otherwise be the case.

Insolation and the Seasons

Another effect of the axial tilt is to produce seasonal differences in insolation at any given latitude. These differences increase toward the poles, where the ultimate in opposites (six months of day, six of night) is reached. Along with the variation in angle of the sun's rays, another factor operates—

the duration of daylight. At the season when the sun's path is highest in the sky, the length of time it is above the horizon is correspondingly greater. The two factors thus work hand in hand to intensify the contrast between amounts of insolation at opposite solstices. We reinforce this concept with an example.

Figure 2.10 is a three-dimensional diagram showing how the sun's path in the sky changes from season to season. The diagram is drawn for lat. 40° N and is typical of conditions found in midlatitudes in the northern hemisphere. The diagram shows a small area of the earth's surface bounded by a circular horizon. This is the way things look to human beings standing on a wide plain. The earth's surface appears to be flat, and the celestial dome on which the sun, moon, and stars move seems to be a hemispherical dome. Actually, this situation resembles a planetarium with its domed ceiling.

At equinox, the sun rises in the east and sets in the west; at noon, the sun is at a point in the southern sky at an angle of 50° above the horizon. At equinox the sun is above the horizon for just 12 hours, as shown by the hour numbers on its path. At summer solstice the sun is above the horizon for about 15 hours, and its path rides much higher in the sky than at

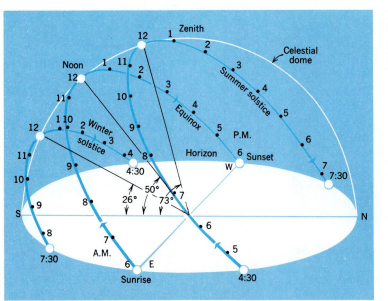

FIGURE 2.10 The sun's path in the sky changes greatly in position and height above horizon from summer to winter. This diagram is for a place located at lat. 40° N.

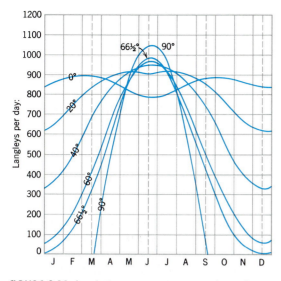

FIGURE 2.11 Insolation curves at various latitudes in the northern hemisphere. (From A. N. Strahler, *The Earth Sciences*, 2nd ed., Harper & Row, New York. Copyright © 1971 by Arthur N. Strahler.)

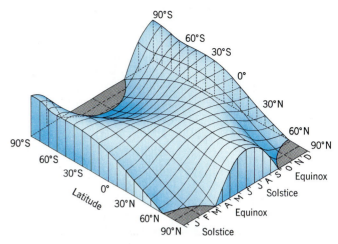

FIGURE 2.12 The effects of both latitude and season of year on the intensity of insolation show on this diagram for the whole globe. At any given latitude and date the relative amount of insolation is proportional to the height of the surface point above the flat base of the block. (After W. M. Davis.)

equinox. At winter solstice the path is low, and the sun is above the horizon for only about 9 hours. Obviously the insolation reaching a square centimeter of horizontal surface is much greater during a day at summer solstice than during a day at winter solstice. When the daily values of insolation are plotted for a full year, they form a wavelike curve on a graph. Figure 2.11 is just such a graph, and the curve for lat. 40° N is the one to which we are referring.

A three-dimensional diagram (Figure 2.12) shows how insolation varies with latitude and with season of year. These diagrams show insolation at the outer limits of the atmosphere and would apply at the ground surface for an earth imagined to have no atmosphere to absorb or reflect radiation. Notice that the equator receives two maximum periods (corresponding with the equinoxes, when the sun is overhead at the equator). There are also two minimum periods (corresponding to the solstices, when the subsolar point shifts farthest north and south from the equator).

At the arctic circle (66½° N), insolation is reduced to nothing on the day of the winter solstice and, with increasing latitude poleward, this period of no insolation becomes longer. All latitudes between the

tropic of cancer (23½° N) and the tropic of capricorn (23½° S) have two maxima and two minima, but one maximum becomes dominant as the tropic is approached. Poleward of the two tropics there is a single continuous insolation cycle with a maximum at one solstice and a minimum at the other.

World Latitude Zones

The angle of attack of the sun's rays determines the flow of solar energy reaching a given unit of the earth's surface and so governs the thermal environment of life at the earth's surface. This concept provides a basis for dividing the globe into latitude zones (Figure 2.13). We don't intend that the specified zone limits be taken as absolute and binding but, instead, that the system be considered as a convenient terminology for identifying world geographical belts throughout this book.

The *equatorial zone* lies astride the equator and covers the latitude belt roughly 10° north to 10° south. Within this zone, the sun throughout the year provides intense insolation, and day and night are of roughly

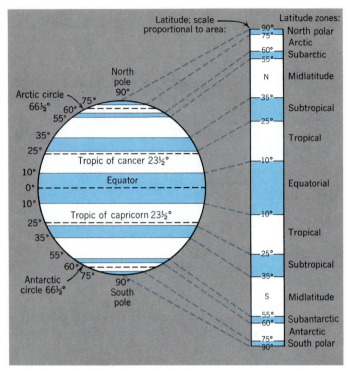

FIGURE 2.13 A geographer's system of latitude zones.

equal duration. Astride the tropics of cancer and capricorn are the *tropical zones*, spanning the latitude belts 10° to 25° north and south. In these zones, the sun takes a path close to the zenith at one solstice and is appreciably lower at the opposite solstice. Thus a marked seasonal cycle exists, but it is combined with a large total annual insolation.

Literary usage and that of many scientific works differ from what is described here; the terms "tropical zone" and "tropics" have been widely used to denote the entire belt of 47 degrees of latitude between the tropics of cancer and capricorn. That is the definition of "tropics" you will find in most dictionaries. Even though correct, that definition is not well suited to a study of our physical environment, because it combines belts of very unlike climatic properties; it lumps the greatest of the world's deserts with the wettest of climates.

Immediately poleward of the tropical zones are transitional regions that have become widely accepted in geographers' usage as the *subtropical zones*. For

convenience, we have assigned these zones the latitude belts 25° to 35° north and south, but it is understood that "subtropical" may be extended a few degrees farther poleward or equatorward of these parallels.

The *midlatitude zones*, lying between 35° and 55° north and south latitude, are belts in which the sun's angle of attack shifts through a relatively large range, so that seasonal contrasts in insolation are strong. Strong seasonal differences in lengths of day and night exist as compared with the tropical zones.

Bordering the midlatitude zones on the poleward side are the *subarctic zone* and *subantarctic zone*, 55° to 60° north and south latitudes. Astride the arctic and antarctic circles, 66½° north and south latitudes, lie the *arctic zone* and *antarctic zone*. These zones have an extremely large yearly variation in lengths of day and night, yielding enormous contrasts in insolation from solstice to solstice. Notice that dictionaries define "arctic" as the entire area from the arctic circle to the north pole and "antarctic" in a corresponding sense for the southern hemisphere.

The *polar zones*, north and south, are circular areas between about 75° latitude and the poles. Here the polar regime of a six-month day and six-month night is predominant. These zones experience the ultimate in seasonal contrasts of insolation.

Insolation Losses in the Atmosphere

As the sun's radiation penetrates the earth's atmosphere, its energy is absorbed or diverted in various ways. At an altitude of 150 km (95 mi), the radiation spectrum possesses almost 100 percent of its original energy but, by the time rays have penetrated to an altitude of 88 km (55 mi), absorption of X rays is almost complete, and some of the ultraviolet radiation has been absorbed as well.

As radiation penetrates into deeper and denser atmospheric layers, gas molecules cause the visible light rays to be turned aside in all possible directions, a process known as *scattering*. Dust and cloud particles in the troposphere cause further scattering, described as *diffuse reflection*. Scattering and

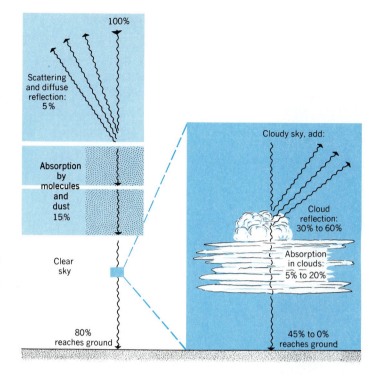

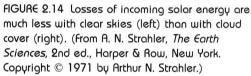

FIGURE 2.14 Losses of incoming solar energy are much less with clear skies (left) than with cloud cover (right). (From A. N. Strahler, *The Earth Sciences*, 2nd ed., Harper & Row, New York. Copyright © 1971 by Arthur N. Strahler.)

diffuse reflection send some energy into outer space and some down to the earth's surface.

As a result of all forms of shortwave scattering, about 5 percent of the total insolation is returned to space and forever lost, as shown in Figure 2.14. Scattered shortwave energy directed earthward is referred to as *down-scatter.*

Another form of energy loss, *absorption,* takes place as the sun's rays penetrate the atmosphere. Both carbon dioxide and water vapor are capable of directly absorbing infrared radiation. Absorption results in a rise of temperature of the air. In this way some direct heating of the lower atmosphere takes place during incoming solar radiation. Although carbon dioxide is a constant quantity in the air, the water-vapor content varies greatly from place to place. Absorption correspondingly varies from one global environment to another.

All forms of direct energy absorption—by air molecules, including molecules of

water vapor and carbon dioxide, and by dust—are estimated to total as little as 10 percent for conditions of clear, dry air to as high as 30 percent when a cloud cover exists. A global average of 15 percent for absorption generally is shown in Figure 2.14. When skies are clear, reflection and absorption combined may total about 20 percent, leaving as much as 80 percent to reach the ground.

Yet another form of energy loss must be brought into the picture. The upper surfaces of clouds are extremely good reflectors of shortwave radiation. Air travelers are well aware of how painfully brilliant the sunlit upper surface of a cloud deck can be when seen from above. *Cloud reflection* can account for a direct turning back into space of from 30 to 60 percent of total incoming radiation (Figure 2.14). Clouds also absorb radiation. So we see that, under conditions of a heavy cloud layer, the combined reflection and absorption from clouds alone can account for a loss of 35 to 80 percent of the incoming radiation and allow from 45 to 0 percent to reach the ground. A world average value for reflection from clouds to space is about 21 percent of the total insolation. Average absorption by clouds is much less—about 3 percent. Figure 2.15 shows these average values.

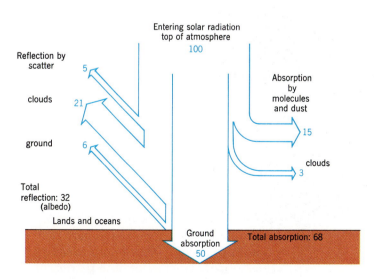

FIGURE 2.15 This schematic diagram shows the global averages of percentages of insolation absorbed and reflected.

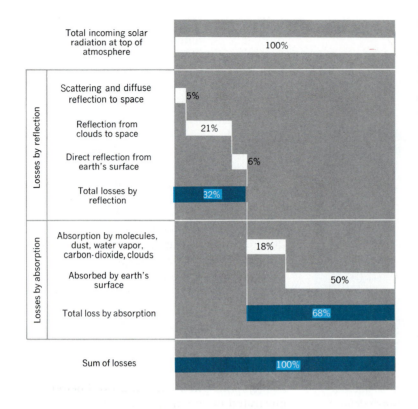

FIGURE 2.16 A tabular summary of the energy losses by reflection and absorption as solar energy penetrates the atmosphere to reach the earth's surface. Data are the same as in Figure 2.15.

The surfaces of the land and ocean reflect some shortwave radiation directly back into the atmosphere. This small quantity, about 6 percent as a world average, may be combined with cloud reflection in evaluating total reflection losses. Figure 2.16 lists the percentages given so far for the energy losses by reflection and absorption. Altogether the losses to space by reflection total 32 percent of the total insolation.

The percentage of radiant energy reflected back by a surface is the *albedo*. For example, a surface that reflects 40 percent of the insolation has an albedo of 40 percent. Albedo is an important property of the earth's surface, because it determines how fast a surface heats up when exposed to insolation. Albedo of a water surface is very low (2 percent) for nearly vertical rays but high for low-angle rays. It is also extremely high for snow or ice (45 to 85 percent). For fields, forests, and bare ground the albedos are of intermediate value, ranging from as low as 3 percent to as high as 25 percent.

Orbiting satellites, suitably equipped with instruments to measure the energy levels of shortwave and infrared radiation, both incoming from the sun and outgoing from the atmosphere and earth's surfaces below, have provided data for estimating the earth's average albedo. Values between 29 and 34 percent have been obtained. The value of 32 percent given in Figure 2.15 lies between these limits.

Continuing down the bar chart in Figure 2.16, we come to the summary of energy losses by absorption. A global figure of 18 percent is given for the combined losses through absorption by molecules, dust, water vapor, carbon dioxide, and clouds. There remains 50 percent of the original solar energy, and this amount is absorbed by the earth's land and water surfaces. We have now accounted for all the energy of insolation. Our next step is to deal with outgoing energy so as to balance the global energy budget.

Longwave Radiation

Any warm substance sends out energy of electromagnetic radiation from its surface. The amount of energy so radiated increases rapidly as temperature goes up. Another important principle is that at low temperatures the radiating substance emits mostly long wavelengths in the infrared region of the spectrum. As temperature rises, the emitted radiation is shifted toward the shorter wavelengths.

The land and ocean surfaces possess heat derived originally from absorption of the sun's rays. These surfaces continually radiate longwave energy back into the atmosphere. The process is known as *ground radiation*. ("Ground" refers here to both land and water surfaces.) As Figure 2.5 shows, longwave radiation consists of infrared wavelengths longer than 3 microns. The atmosphere also radiates longwave energy both downward to the ground and outward into space, where it is lost. Be sure to understand that longwave radiation is quite different from reflection, in which the rays are turned back directly without being absorbed. Longwave radiation from both ground and atmosphere continues during the night, when no solar radiation is being received.

Longwave radiation from the earth's land and ocean surfaces spans a wavelength range of 4 to 30 microns, as shown in Figure 2.5. The emission of longwave radiation into outer space takes place largely in a wavelength band between 8 and 13 microns, often referred to as a *window*. There are also lesser windows in the range of 4 to 6 microns and 17 to 21 microns. These windows show as peaks on the longwave radiation curve in Figure 2.5. Between the windows most of the longwave radiation leaving the ground is absorbed by water vapor and carbon dioxide, and this absorbed energy is converted into sensible heat.

The Global Radiation Balance

As Figure 2.5 shows, the intensity of longwave energy leaving our planet is only a small fraction of the shortwave solar energy. Keep in mind that longwave radiation is constantly emitted from the entire spherical surface of the earth, whereas insolation falls on only one hemisphere. A single hemisphere presents the equivalent of only a cross section of the

sphere to full insolation. Also, do not forget that about 32 percent of the incoming solar radiation is reflected directly back into space. Only the remaining 68 percent that is absorbed must be disposed of by longwave radiation. On the average, year in and year out, for the planet as a whole, the quantity of absorbed solar energy is balanced by an equal longwave emission to space.

Part of the ground radiation absorbed by the atmosphere is radiated back toward the earth surface, a process called *counterradiation*. For this reason, the lower atmosphere with its water vapor and carbon dioxide acts as a blanket that returns heat to the earth. This mechanism helps to keep surface temperatures from dropping excessively during the night or in winter at middle and high latitudes.

Somewhat the same principle is used in greenhouses and in homes using large windows to obtain solar heat. Here, the glass windows permit entry of shortwave energy. This phenomenon is even better illustrated by the intense heating of air in a closed automobile left parked in the sun. Accumulated heat cannot escape by mixing with cooler air outside. Meteorologists use the expression *greenhouse effect* to describe this atmospheric heating principle. Cloud layers are even more important in producing a blanketing effect to retain heat in the lower atmosphere, since they are excellent absorbers and emitters of longwave radiation.

Figure 2.17 puts together the entire planetary radiation balance in a single pictorial diagram. At the left is a summary of incoming shortwave (SW) energy, based on the data of Figures 2.15 and 2.16. At center and right is the diagram of outgoing energy. Atmosphere and ground are depicted as two horizontal bands, each of which can be treated as an isolated *energy subsystem*.

Reading inward from the extreme right-hand part of the diagram, we find that 8 percentage units escape from the ground directly to space as longwave radiation (LW). Upward LW radiation from ground to atmosphere is 90 units, partly compensated for by downward LW counterradiation (77 units), giving a net loss of 13 units for the ground and a corresponding net gain for the atmosphere (90 − 77 =

44

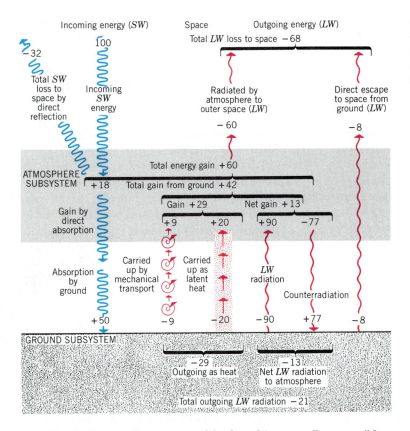

FIGURE 2.17 Diagram of the global radiation balance showing the flow of energy in and out of the two major subsystems.

13). Total LW loss from ground is thus 21 units (13 + 8 = 21).

Now a new concept enters the radiation balance picture. Much of the energy passing from the ground to the atmosphere is carried upward by two mechanisms other than longwave radiation. First, sensible heat is carried upward by turbulent motions of the air. As Figure 2.17 shows, this heat transport amounts to about 9 percentage units. Second, latent heat is carried upward in water vapor that has evaporated from land and ocean surfaces. Latent heat transport accounts for about 20 percentage units. (Latent heat is explained in Chapter 5.)

To summarize, energy leaving the ground subsystem is divided up as follows.

	Percentage units
Longwave radiation (net value)	21
Mechanical transport as sensible heat	9
Transport as latent heat	20
Total	50

Because 50 percentage units represent the amount of energy absorbed by the ground, we have now balanced the annual energy budget at the earth's surface.

Energy gained by the atmosphere subsystem from the ground totals 42 percentage units (13 + 9 + 20 = 42). To this quantity are added 18 units gained by direct absorption of incoming solar energy. Thus, the total atmospheric gain is 60 units (42 + 18 = 60).

The atmosphere subsystem in turn disposes of its 60 units by LW radiation into outer space. Together with direct LW emission from the ground (8 units), the total LW emission to space is 68 percent. Add this figure to 32 percent losses by reflection (the earth's albedo), and the total is 100 percent. The global radiation budget stands balanced.

Latitude and the Radiation Balance

Earlier in this chapter, in relating latitude to insolation, we showed that the tilt of the earth's axis causes a poleward redistribution of insolation, as compared with conditions that would apply if the axis were

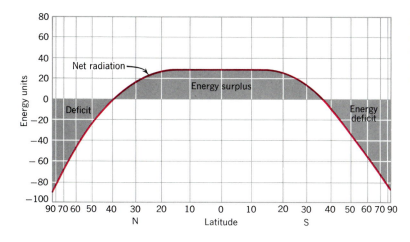

FIGURE 2.18 Net radiation from pole to pole shows two polar regions of energy deficit, matching a large low-latitude region of energy surplus.

perpendicular to the orbital plane (Figure 2.9). Let us now look deeper into the wide range in rates of incoming and outgoing energy in terms of a profile spanning the entire latitude range 90° N to 90° S. In this analysis yearly averages are used so that the effect of seasons is concealed.

We are going to examine the *net radiation,* which is the difference between all incoming energy and all outgoing energy carried by both shortwave and longwave radiation. Our analysis of the radiation balance has already shown that for the entire globe as a unit, the net radiation is zero on an annual basis. However, in some places, energy is coming in faster than it is going out, and the energy balance is a positive quantity, or *energy surplus.* In other places energy is going out faster than it is coming in, and the energy balance is a negative quantity, or *energy deficit.*

Figure 2.18 is a global profile of the net radiation from pole to pole. Between about lat. 40° N and S there is an energy surplus, as you would expect, because at low latitudes insolation is intense throughout the year. Poleward of lat. 40° N and S a deficit sets in and becomes more and more severe as the poles are reached. The two areas labeled "deficit" have a combined area on the graph equal to the area labeled "surplus." In other words, the net radiation for the entire globe is zero.

It is obvious from Figure 2.18 that the earth's radiation balance can be maintained only if heat is transported from the low-latitude belt of surplus to the two high-latitude regions of deficit. This poleward movement of heat is described as *meridional*

transport. "Meridional" means moving north or south along the meridians of longitude. The rate of meridional heat transport is greatest in midlatitudes between the 30th and 50th parallels.

The meridional flow of heat is carried out by circulation of the atmosphere and oceans. In the atmosphere, heat is transported both as sensible heat and as latent heat. Meridional energy transport is explained in Chapters 5 and 6.

Solar Power

Planet earth intercepts solar energy at the rate of 1½ quadrillion megawatt-hours per year. This quantity of energy amounts to about 28,000 times as much as all the energy presently being consumed each year by humans. Thus, we realize that an enormous source of energy is near at hand waiting to be used. That use would simply mean passing some of the flow of solar energy through artificial subsystems within the natural global energy system. Another remarkable virtue of solar energy is that harnessing its use cannot cause an increase in the temperature of the atmosphere and hydrosphere. A major worry we express in Chapter 3 is that the combustion of hydrocarbon fuels will raise the global average temperature, both by emitting large amounts of heat and by raising the level of the carbon dioxide content of the atmosphere. Neither of these concerns applies to solar energy. Adding to these advantages the lack of environmental pollution—no emissions of sulfur dioxide or carbon particles—solar energy becomes a particularly attractive energy source.

Solar radiation provides useful energy in a variety of forms, both direct and indirect. Our concern here is with direct interception and conversion of solar energy in one of two ways: (1) direct absorption by a receiving surface, converting shortwave energy to sensible heat and raising the temperature of the receiving medium; and (2) direct conversion of shortwave energy to electrical energy by the use of solar cells.

Oldest and simplest of the forms of solar energy conversion is the direct interception of the warming rays of the sun by some kind of receiving surface or medium. Applications of this principle range from simple heating of buildings and domestic hot-water supplies to the intense heating of boilers and furnaces by focusing the solar rays on a small target.

A great saving of expenditure of fuel oil and natural gas can be achieved through solar heating of buildings. Each home, school, or office building can have its own solar collecting system so that expensive energy transport, whether by pipeline, truck, or power line can be eliminated. In most cases, the goal of this application is to supplement, rather than to replace, the use of fuels.

The simplest form of solar heating is the use of large glass panes to admit sunlight into a room—the greenhouse principle. Solar rays are admitted during the winter, when the sun's path is low in the sky, but excluded in summer by a suitable roof overhang. Of course, the same large glass panes will result in high heat losses at night and on cold, cloudy days by outgoing longwave radiation, unless thermal drapes or shutters are also used.

Practical solar heating of interior building space and hot-water systems makes use of *solar collectors*. The flat-plate collector consists of a network of metal tubes carrying circulating water. Aluminum or copper tubes, painted black, are efficient absorbers of solar energy. Water is pumped through the tubes to heat a large body of water in a storage tank. A cover of glass or clear plastic is used to obtain higher temperatures and to reduce heat loss to the atmosphere. Panels of solar collectors are usually placed on the roof of the building.

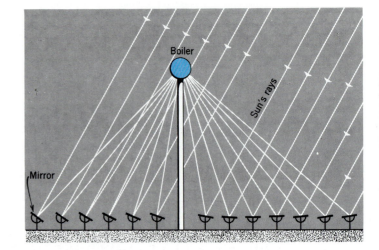

FIGURE 2.19 Design for a solar power plant using an array of movable mirrors (heliostats) to concentrate the sun's rays on a boiler. Each heliostat is controlled by a computer program so as to focus its rays on the tower as the sun travels its daily path across the sky.

Water is heated to a temperature of 65°C (150°F) and can be used to transfer heat to a hot-water supply, as well as to a conventional space-heating system.

Solar energy power plants have been designed to use reflecting mirror systems. A large number of movable mirrors, called heliostats, reflect solar rays to the top of a central tower where a boiler and electric generator are located (Figure 2.19). The extremely high temperatures and pressures produced in this way allow a number of kinds of gases and fluids to be used in the boiler. Hydrogen gas, which can be generated by this process, provides an ideal medium in which to store energy for later conversion to electricity when the solar input is cut off. A heliostat power plant constructed in 1981 near Barstow, California, delivers 10 megawatts of electric power (Figure 2.20).

We turn now to the technology of generating electricity directly from the impact of the sun's rays. The photovoltaic effect is known to amateur photographers through the light meter. As you move the glass window of the meter across a scene, a small hand wavers over a calibrated dial to tell the varying intensities of the incoming light. The sensing device in this meter is a

FIGURE 2.20 Solar One, a pilot plant for generating electricity with assembled heliostats, began operations in 1982. Its 1818 heliostats present a total reflecting area of 70,000 m². (Courtesy of Southern California Edison Company.)

photovoltaic cell; it transforms light energy into electricity. The hand on the meter is actually showing the amount of electric current generated by the cell.

As with direct heat-gathering solar panels, arrays of photovoltaic solar cells would occupy large areas of ground surface to produce important outputs of energy. Storage of the electricity is another problem, since the system does not work at night. Storage as hydrogen fuel is an attractive possibility for large-scale systems, while batteries can be used to store power in small systems, such as those of homes, farms, and ranches. One suggested small-scale use is to turn electric water pumps needed for irrigation of fields in the sunny dry season of tropical countries, for example, on wheat farms of Pakistan and northern India.

Indirect, or secondary, sources of solar energy, described in later chapters, make use of converted solar energy stored in different forms. For example, air in motion as wind and water in motion as ocean waves or flowing rivers are forms of kinetic energy of matter in motion within flow systems powered by solar energy. Quite different in character is solar energy that has been converted into chemical energy by plants and stored in plant tissues. These organic energy systems are described in Chapter 20. The fossil fuels—crude oil and coal—are derived from hydrocarbon compounds of organic origin that can be traced back to solar energy captured and stored in the distant geologic past.

The Human Impact on the Earth's Radiation Balance

Although our analysis of the earth's radiation balance is far from complete, it should be obvious by now that the balance is a sensitive one, involving as it does a number of variable factors that determine how energy is transmitted and absorbed. Has industrial activity already altered the components of the planetary radiation balance? An increase in carbon dioxide may be expected to increase the absorption of longwave radiation by the atmosphere. Will this change cause a rise in atmospheric temperature? Has such a change already occurred? If continued, where will it lead? Increase in atmospheric dusts high in the troposphere will increase scatter of incoming shortwave radiation and perhaps reduce the atmospheric temperature near the ground; but, on the other hand, increased dust content at low levels will act to absorb more longwave radiation and so raise the atmospheric temperature. Has either change occurred?

Humans have profoundly altered the

earth's land surfaces by cultivation and urbanization. Have these changes in surface albedo and in the capacity of the ground to absorb and to emit longwave radiation resulted in a changed energy balance?

Answers to such questions require further study of the processes of heating and cooling of the earth's atmosphere, lands, and oceans; these are the subject of the next chapter.

REMOTE SENSING FOR PHYSICAL GEOGRAPHY

In various branches of physical geography, as in other fields of the earth sciences, a new technical discipline called *remote sensing* has expanded rapidly within the past decade and is adding greatly to our ability to perceive and analyze the physical, chemical, biological, and cultural character of the earth's surface. In its broadest sense, remote sensing is the measurement of some property of an object using means other than direct contact. Hearing and seeing are remote-sensing activities of organisms and depend on reception of wave forms of energy transmitted from the object to the observer. In the more restricted meaning we use here, remote sensing refers to gathering information from great distances and over broad areas, usually through instruments mounted on aircraft or orbiting space vehicles. All substances, whether naturally occurring or synthetic, are capable of reflecting, absorbing, and emitting energy in forms that can be detected by instruments known collectively as remote sensors.

Two Kinds of Sensing Systems

There are two classes of electromagnetic sensor systems: passive systems and active systems. *Passive systems* measure radiant energy reflected or emitted by an object. Reflected energy falls mostly in the visible light and near-infrared regions, whereas emitted energy lies in the longer thermal infrared region. The most familiar instrument of the passive type is the camera, using film sensitive to reflected energy at wavelengths in the visible range. *Active systems* use a beam of wave energy as a source, sending the beam toward an object or surface. A part of the energy is reflected back to the source, where it is recorded by a detector. A simple analogy would be the use of a spotlight on a dark night to illuminate a target, which reflects light back to the eye.

Microwaves and Radar

At longer wavelengths beyond the visible to infrared portions of the electromagnetic spectrum lies a form of radiation known as *microwaves*. Most persons are familiar with microwaves as the form of energy used in the microwave oven for the rapid heating or cooking of foods. Microwaves are also used in direct-line transmission of messages from one tower to another across country. Within the microwave region is the *radar* region. (The word "radar" was originally an acronym for "*RA*dio *D*etection *A*nd *R*anging," but now is recognized as a word in its own right.) Radar systems are active microwave sensor systems. The radar system emits a short pulse of microwave radiation and then "listens" for a returning microwave echo.

Many radar systems can penetrate clouds to provide images of the earth's surface in any weather. At short wavelengths, however, microwaves can be scattered by water droplets and produce a return signal sensed by the radar apparatus. This effect is the basis for weather radars, which can detect rain and hail and are used in local weather forecasting.

Aerial Photography

Of the passive sensor systems, camera photography in the visible portion of the spectrum is most familiar to the average person. Black-and-white aerial photographs, taken by cameras from aircraft, have been in wide use by geographers and other environmental scientists since before World War II. Commonly, the field of one photograph overlaps the next along the plane's flight path, so that the photographs can be viewed stereoscopically for a three-dimensional effect. Color film can be used to increase the level of information on the aerial photographs. Because of its high resolution (degree of sharpness), aerial photography remains one of the most valuable of the older remote-sensing techniques.

Photography has been extended to greater distances through the use of cameras operated by astronauts in orbiting space vehicles. Most persons are familiar with the striking color photographs obtained during the Gemini missions of the early 1960s. More recently, a Space Shuttle flight included a specially constructed large-format camera, designed to produce very large, very detailed transparencies of the earth's surface suitable for precise topographic mapping. An excellent example is shown in Figure 2.21.

Infrared Photography

Photography using reflected electromagnetic radiation also extends into the ultraviolet and infrared wavelengths. Within the infrared spectrum, there is a region in the near-infrared, immediately adjacent to the visible red region, in which reflected rays can be recorded by cameras with suitable film and filter combinations. Because the atmosphere is very clear in this portion of the spectrum, direct infrared photographs taken from high-flying aircraft are extremely sharp and render a great deal of information on vegetation, soil surface conditions, and land use.

Color infrared film is often used in aerial photography (Figures 2.22 and 2.23). In this type of film, the red color is produced as a response to infrared light, the green color is produced by red light, and the blue color by green light. Because healthy, growing vegetation reflects strongly in the infrared band, vegetation has a characteristic red appearance on color infrared film. Thus, agricultural crops appear as color shades ranging from pink to orange-red to deep red. Mature crops and dried vegetation (dormant grasses) appear yellow or brown. Urbanized areas typically appear in tones of blue and gray. Shallow water areas appear blue; deep water appears dark blue to blue-black. Color infrared photography is particularly useful in geographic interpretation and land-use planning, and thus has found wide application since it was first developed during World War II.

Digital Images

Within the decade 1975–1985, the science of remote sensing made great strides. This progress was largely due to the use of computers to process remotely sensed data. Computer processing of pictorial data, however, requires that images be in digital format, i.e., as collections of numbers which a computer can process. Figure 2.24 illustrates the concept of a *digital image.* The picture can be thought of as consisting of a very large number of grid cells, each of which records a brightness value. The cells are referred to as *pixels,* a term that arises as a contraction of the term "picture element." Normally, low numbers code for darkness (low reflectance) and high numbers code for light (high reflectance). The numbers are typically stored on magnetic media (disks or tapes) by the computer, and so the image does not exist in a viewable form. To create an image that is visible—as, for example, on a television screen—the brightness values are fed to a special electronic device that generates a corresponding television signal. Or, the digital image may be sent to a film-writing device that exposes film to light, a tiny spot at a time, in proportion to the brightness values within the digital image. The result is a film negative or transparency that can be printed or viewed directly.

The importance of the digital image is that it allows the numbers that constitute the image to be modified by the computer. This activity is referred to as *image processing.* By manipulating these numbers in the computer, it is easy to enhance the image—for example, to modify the contrast selectively within certain areas of the image, or to emphasize edges or boundaries within the image (Figure 2.25).

FIGURE 2.21 This infrared color photograph was taken with a large-format camera carried on the NASA Space Shuttle. It gives an extremely detailed picture of the terrain and is suitable for precision mapping. The area shown, about 125 km in width, includes the Sulaiman Range in West Pakistan (upper left). The Indus and Sutlej rivers, fed by snowmelt from the distant Hindu Kush and Himalaya ranges, cross the scene flowing from northeast to southwest. The larger of the two, the Indus, is crossed by a diversion dam (upper right) that feeds water into irrigation canals. A mottled pattern of green fields (red) interspersed with barren patches of saline soil (white) covers much of the lower fourth of the area. Small streams arising in deeply eroded anticlines of sedimentary rock (upper left) have built a series of alluvial fans (white and pale blue) that bear clusters of irrigated fields.
(NASA No. 57-20380-4107-1783.)

FIGURE 2.22 High-altitude infrared photograph of an area near Bakersfield, California, in the southern San Joaquin Valley, taken by a NASA U-2 aircraft flying at approximately 18 km (60,000 ft). The original photo scale was about 1:120,000. Photos such as this can be used to study problems associated with agriculture. Problems affecting crop yields arise in (A) perched ground water areas, which appear dark, and (B) areas of high soil salinity, which appear light. The various red tones are associated with the different types of crops that are being grown. By knowing which crops are typically grown in this area and their growing cycle, the total areal extent of individual crops being cultivated can be determined. The areal estimates can then be combined with other data on weather conditions and soils to predict crop yields. (Photo by NASA, compiled and annotated by John E. Estes and Leslie W. Singer.)

A. Mt. Etna on the eastern coast of Sicily. The highest volcano in Europe, it is still active, as evident from the thin plume of smoke coming from its crater (top center). On Etna's flanks recent lava flows appear blue-black, while older flows and volcanic debris, which now support vegetation, appear in shades of red.

B. Flagstaff, Arizona, and vicinity in winter. San Francisco Peak, a forested extinct volcano is at far left, the Sunset Crater cinder cone field above it, and the city of Flagstaff at lower left. Meteor Crater, its circular rim seen in upper right, is approximately 1.6 km (1 mi) in diameter. Snow mapping from photos such as this allow the seasonal water resource capacity to be estimated.

FIGURE 2.23 These Skylab color infrared photos were taken with the Earth Resources Experiment Package terrain camera. (NASA)

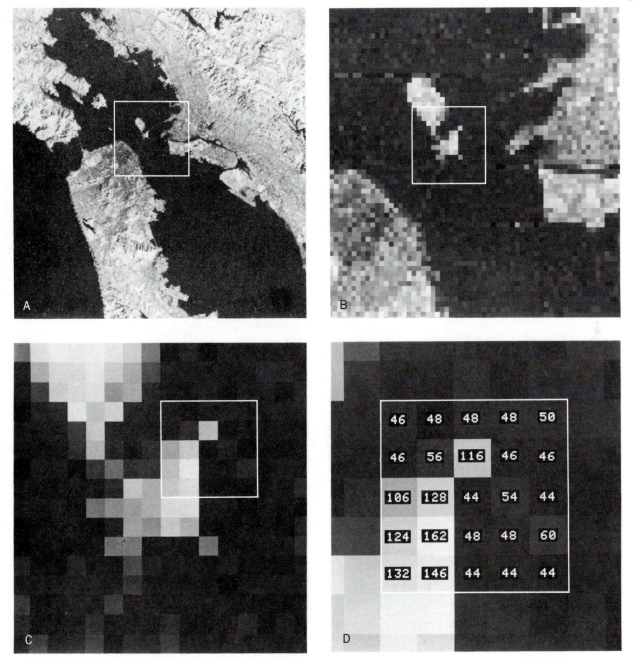

FIGURE 2.24 Four Landsat images of San Francisco (infrared band) illustrating the concept of the digital image. (A–C) Progressively smaller subimages zooming in on the Bay Bridge and Yerba Buena Island in San Francisco Bay. (D) Actual brightness values for a small array of 25 pixels, scaled to range from 0 (darkest) to 255 (lightest). (Alan H. Strahler.)

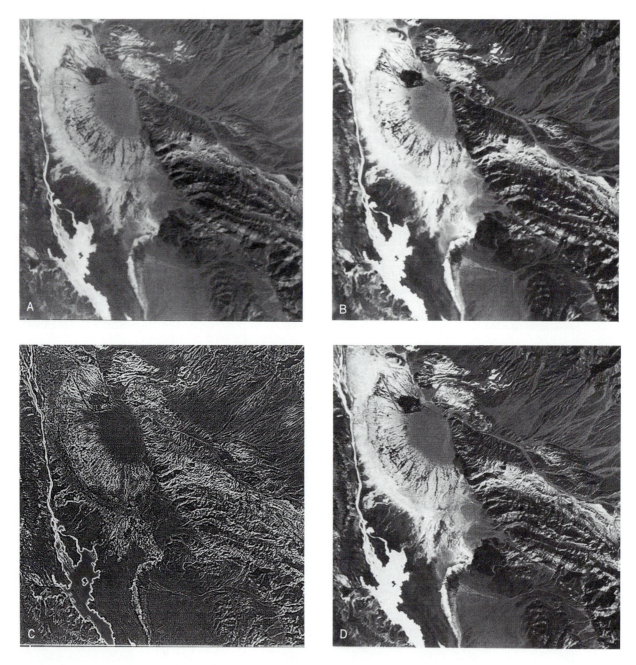

FIGURE 2.25 Thematic Mapper image of Death Valley (red band) showing how digital images can be enhanced. (A) Original image. (B) Contrast-enhanced image. (C) Image produced by applying edge detection procedure. (D) Edge-enhanced image created by adding images B and C together. (Alan H. Strahler.)

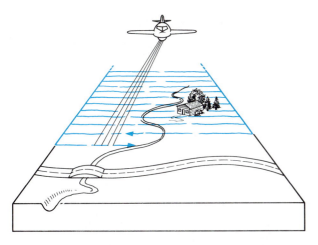

FIGURE 2.26 Multispectral scanning from aircraft. As the aircraft flies forward, the scanner sweeps side to side. The result is a digital image covering the overflight area. (Alan H. Strahler.)

Scanning Systems

Digital images are usually generated by *scanning systems* mounted in aircraft or orbiting space vehicles. Scanning is the process of receiving information instantaneously from only a very small portion of the area under surveillance (Figure 2.26). The scanning instrument senses a very small field of view that runs rapidly across the ground scene.

Light from the field of view is focused on a detector that measures its intensity, and a string of digital brightness values is the product. As the scans are repeated, information along a set of closely spaced parallel lines is obtained. In this way, a full digital image is built up by the scanning system.

Most scanning systems in common use are *multispectral scanners*. These devices measure brightness in several wavelength regions simultaneously. An example is the Multispectral Scanning System (MSS) used aboard the Landsat series of earth-observing satellites. This instrument simultaneously collects reflectance data in four spectral bands. Two are bands in the visible light spectrum (one green, one red) and two in the infrared region. A successor to this system, the Landsat Thematic Mapper (TM), collects data in seven spectral bands.

Multispectral Imagery

Although it is possible to produce black-and-white photos or TV signals from an image within a single spectral band, multispectral data are normally viewed as *multispectral images*. Because the human eye can respond to three primary colors at once, it is possible to assign one spectral band to each color and view the result as a color image. The colors produced will then not resemble true colors unless, of course, the primary colors of red, green, and blue are assigned to the red, green, and blue bands of the sensor. For example, Landsat MSS data are often presented as a color composite image with the red color intensity controlled by one of the infrared bands, the green color controlled by the red band, and the blue color controlled by the green band. Several examples of multispectral color imagery are found in later chapters. (See Figures 13.24, 15.24, 16.20, 16.29, 17.24, 17.28, 17.29, 17.32, 17.42, 17.46, 18.3, and 18.13.)

Thermal Infrared Sensing

Electromagnetic radiation is emitted by all objects because they possess sensible heat. At the range of temperatures encountered on the earth's surface, this radiation is within the *thermal infrared* portion of the spectrum. Thermal infrared radiation is usually sampled by a scanning system. However, the detector that absorbs the radiation so that it can be measured and digitized is composed of a material sensitive to infrared wavelengths rather than to light. Since warmer objects emit more infrared radiation than do cooler ones, the former will appear lighter on thermal infrared imagery. Note also that objects will emit thermal infrared radiation as long as they possess any sensible heat, and therefore thermal infrared imagery can be collected during both day and night.

An example of infrared imagery acquired at night is shown in Figure 2.27. Keep in mind while examining this image that the differences in tone by means of which objects are delineated are caused by differences in the level of emission of infrared energy. Differences in temperature

FIGURE 2.27 This infrared imagery of Brawley, a small town in the Imperial Valley of California, looks like an air photograph. It was taken in darkness between 2 and 4 A.M. Special lenses and films can record the infrared rays coming up from the surface. (Environmental Analysis Department, HRB—Singer, Inc.)

are the most important variable in causing the differences in tone, and in general the lighter tones indicate warmer temperatures. In Figure 2.27, an image taken during the early morning hours (between 2 and 4 A.M.), pavement and water emit infrared radiation more strongly and appear light in tone. Trees lining many of the roadways also appear light in tone. Agricultural areas, moist soil surfaces, and buildings are "cooler" and appear darker.

Radar-Sensing Systems

We turn next to the active mode of remote sensing within the radar portion of the electromagnetic spectrum. As noted earlier, the active sensing system uses pulses of energy emitted by transmitters mounted on aircraft or spacecraft. A beam of such pulses is directed at the ground, typically at an angle of 30 to 60 degrees from the horizon. A portion of the energy is returned as an echo signal. The strength of the return signal will depend partly on the nature of the surface. A smooth surface will

act like a mirror, scattering the pulse forward and away from the sensor; it will therefore look relatively dark on a radar image. A rough surface will contain many facets or projections that scatter part of the pulse back toward the sensor, and thus will appear brighter.

Radar-sensing systems are used effectively on aircraft or spacecraft. The type most often used to produce imagery is known as *side-looking airborne radar* (SLAR). A SLAR system sends its impulses toward either side of the vehicle, producing a long swath of imagery as the aircraft or spacecraft flies forward. SLAR images show terrain features with remarkable sharpness and contrast (Figure 2.28). Surfaces oriented most nearly at right angles to the radar beam will return the strongest echo and therefore appear lightest in tone. In contrast, those surfaces facing away from the beam will appear darkest. The effect is to produce an image resembling a relief map using plastic shading. Various types of surfaces, such as forest, rangeland, and agricultural fields, can also be identified by variations in image tone and pattern.

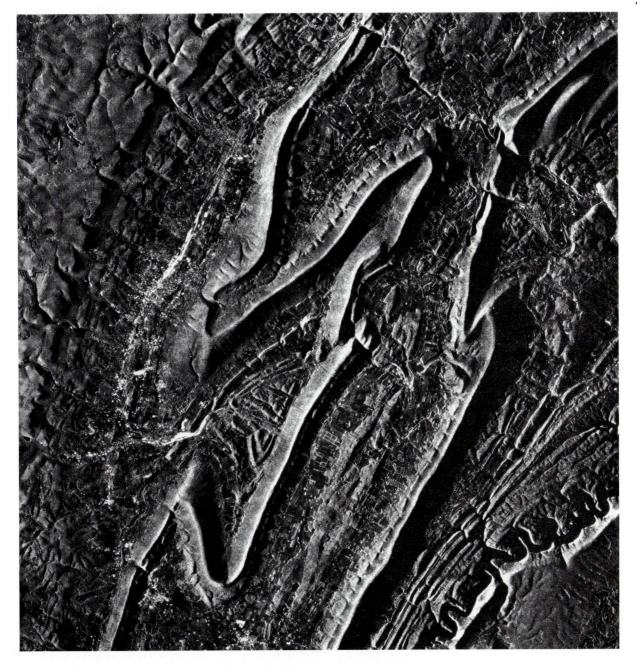

FIGURE 2.28 Radar image of a portion of the Folded Appalachians in south-central Pennsylvania. Zigzag ridges in the vicinity of Hollidaysburg include two plunging anticlines and two plunging synclines. At the lower right are the intrenched meanders of the Raystown Branch of the Juniata River. The area shown is about 40 km (25 mi) wide. (SAR image by courtesy of Intera Technologies Corporation, Calgary, Alberta, Canada.)

Orbiting Earth Satellites

It is only since the advent of orbiting earth satellites that remote sensing has burgeoned into a major branch of geographic research, going far beyond the limitations of conventional aerial photography. One of the reasons for this development has already been mentioned—the fact that most satellite remote sensors provide digital images that can be processed and enhanced by computers. The other reason is the ability of orbiting satellites to monitor nearly all the earth's surface.

In determining the orbit of a satellite, two factors are important. First, the orbit must allow the satellite to image as much as possible of the globe. Second, the orbit should be chosen so that the plane of the satellite's orbit remains fixed with respect to the sun. The latter ensures that the solar illumination of the earth under the satellite's track will remain constant in direction and intensity. The only orbit that allows a satellite to image the entire globe is a polar orbit, in which the satellite crosses both poles. However, a perfectly polar orbit will hold the plane of the satellite's motion fixed with respect to the stars, not to the sun, and thus illumination conditions will change with the seasons.

The solution to this problem is a *sun-synchronous orbit*, illustrated in Figure 2.29. The plane of this orbit makes an angle of 80 degrees with the plane of the earth's equator, and the satellite's earth track makes a tangent contact with the 80th parallels of latitude north and south. With this orbit, the torque exerted by the earth's equatorial bulge upon the satellite's motion is just sufficient to shift the orbit westward at a rate matching the shift in angle between the sun and the stars. Therefore, the satellite will always pass overhead at the same time of solar day, no matter what the season.

In the example illustrated in Figure 2.29, the satellite completes one orbit in about 1 hour and 54 minutes (114 min). In a single day the satellite will image about thirteen swaths that are about 3000 km apart at the equator.

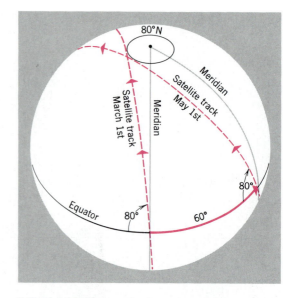

FIGURE 2.29 Earth-track of a sun-synchronous satellite. In the period from March 1 to May 1 the orbit has shifted eastward about 60° with respect to space coordinates. (Copyright © 1971 by Arthur N. Strahler.)

The Landsat Program

A major step forward in remote sensing of the environment was taken by the National Aeronautic and Space Administration (NASA) in July 1972 with the launching of the first of the earth-observing satellites known as Landsats. Since then, five Landsats have been launched. All use a sun-synchronous orbit. The satellite repeats the same ground track every 16 or 18 days, depending on its orbiting height.

The Landsat orbit requires the satellite to descend across the daylight portion of the globe between about 9 and 10 A.M. local time. This time of day was chosen for two reasons. First, by 10 A.M. the sun has not reached the zenith and thus casts strong shadows that serve to emphasize three-dimensional relief features. This timing aids geologic interpretation. Second, by that time of the morning, cloud cover created by convection from unequal solar heating of the surface has not usually begun to develop. Thus images acquired at this time of the day will tend to have fewer clouds.

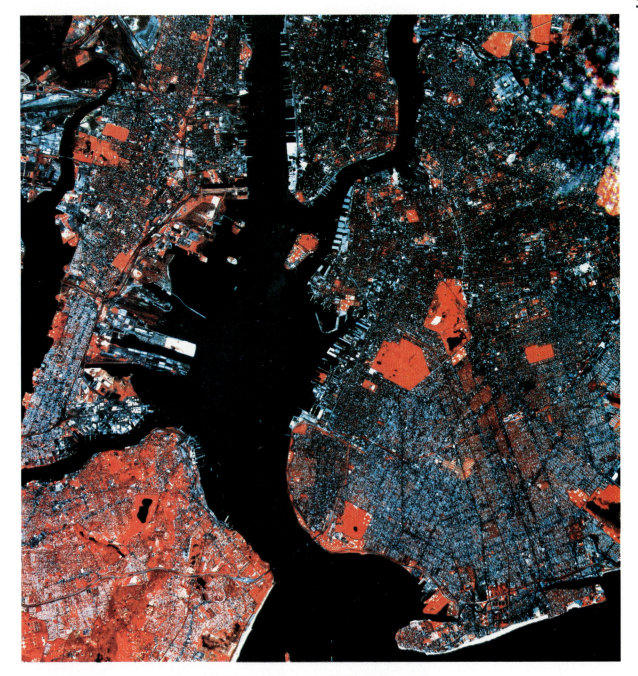

FIGURE 2.30 Digital image of New York Harbor, taken by the SPOT satellite on May 1, 1986. The satellite was built and launched by CNES, the French national aerospace consortium. In this image, the data are presented as a false-color picture. The tip of Manhattan is at the top center; the blocky, dark structures at the tip are Wall Street skyscrapers and their shadows. Just below the tip is Governor's Island, a Coast Guard military reservation, and immediately to the west (left) is small Liberty Island, home of the Statue of Liberty. The long container-loading piers of Port Newark are visible on the left, as are piers on the Brooklyn shoreline to the right. At the center bottom, the Verrazano Narrows bridge connects Brooklyn with Staten Island. The fine detail of the street grid pattern, and even of the wakes of ships in the harbor, show this satellite to have the finest resolution of any unclassified satellite system. (Copyright © 1986 by CNES, provided by SPOT Image Corporation, Reston, VA.)

Receiving and Storing Landsat Imagery

Imagery obtained by the multispectral imaging instruments is converted to digital form by electronics on board the satellite. The digital information is then transmitted by radio to receiving stations operating in the United States, Canada, Brazil, China, Italy, Japan, India, and Australia. The earlier Landsats also carried on-board tape recorders to store digital images when the satellite was out of range of receiving stations. Transmissions of the Thematic Mapper are relayed through NASA's geostationary communications satellite network, which is designed to handle data from these types of satellites as well as data generated by flights of the Space Shuttle. After relay, the radio signals are beamed to a receiving station at White Sands, New Mexico, and then further relayed by domestic communications satellites to Goddard Space Flight Center in Maryland, where the data are processed into a computer-readable format. They are then archived at a data center in Sioux Falls, South Dakota. Users may order digital data or may have the data converted to various photographic formats.

The SPOT Satellite

Although NASA has dominated the field of earth-observing satellites in the past, new satellites have been developed and launched by other nations. First on the line was the French satellite, SPOT (an acronym for the French title *Systeme Probatoire d'Observation de la Terre*), launched in 1986. Figure 2.30 shows a SPOT color image of the New York harbor area. The resolution and color are sufficient to reveal many fine details.

Many technical advances will continue to be made in remote sensing. Several nations are making important contributions through the launching of new satellites bearing even more sophisticated remote-sensing systems. Research in physical geography will be extended into the most remote areas of the earth, never previously mapped in detail and only rarely visited by Western scientists.

CHAPTER

3

HEAT AND COLD IN THE LIFE LAYER

THE THERMAL ENVIRONMENT   Daily and seasonal rhythms characterize the thermal environment of life on earth.

AIR TEMPERATURE   The daily cycle of air temperature is a rhythmic response to the changing radiation balance.

TEMPERATURE INVERSIONS   Close to the ground, the daily temperature cycle is strongly intensified. Strong chilling of a basal air layer—an inversion—can cause a killing frost.

THE ANNUAL CYCLE   Closely linked to the radiation balance is the annual cycle of air temperature; it is strongly influenced by latitude and the seasonal range of insolation.

LAND AND WATER CONTRASTS   Oceans and continents show remarkable temperature contrasts, linked to the unlike thermal properties of land and water surfaces.

CLIMATE CHANGE   Cycles of rise and fall of average air temperature are an expression of global climate change, possibly caused by changes in carbon dioxide and dust content of the atmosphere. The extent to which human activity will cause such climate change remains uncertain.

TEMPERATURE of the air and the soil layer is of major interest to the physical geographer. Temperature is a measure of heat energy available in the air and the soil. Organisms respond directly to heating and cooling of the environmental substance that surrounds them. We can refer to this influence as the *thermal environment*.

Our study of the radiation balance has shown that temperature changes result from the gain or loss of energy by the absorption or emission of radiant energy. When a substance absorbs radiant energy, the surface temperature of that substance is raised. This process represents a transformation of radiant energy into the energy of sensible heat, which is the physical property measured by the thermometer. Heat can also enter or leave a substance by conduction or can be lost as latent heat during evaporation. Many of the biochemical processes taking place within organisms as well as many common inorganic chemical reactions are intensified by an increase in temperature of the solutions in which these reactions are occurring. Severe cold, which is simply the lack of heat energy within matter, may greatly reduce or even completely stop biochemical and inorganic reactions. This is why the vital environmental ingredient of heat—heat in the air, water, and soil—needs to be understood.

We are all familiar with natural cycles of temperature change. There is a daily rhythm of rise and fall of air temperature as well as a seasonal rhythm. There are also systematic average changes in air temperatures from equatorial to polar latitudes and from oceanic to continental surfaces. These temperature changes require that the lower atmosphere and the surfaces of the lands and oceans receive and give up heat in daily and seasonal cycles. There must also be great differences in the quantities of heat received and given up annually in low latitudes as compared to high latitudes. Temperature cycles—daily and seasonal—and the influence of latitude on temperature will be dominant themes of this chapter.

The Troposphere

The lowermost atmospheric layer, the *troposphere,* is of most direct importance to humans in their environment at the bottom of the atmosphere. Almost all phenomena of weather and climate that physically affect us take place within the troposphere, which has a thickness of 12 to 15 km (7 to 9 mi).

In addition to pure, dry air, the troposphere contains *water vapor,* a colorless, odorless gaseous form of water that mixes perfectly with the other gases of the air. The quantity of water vapor present in the atmosphere is of great importance in weather phenomena. Water vapor can condense into clouds and fog. When condensation is rapid, rain, snow, hail, or sleet–collectively termed precipitation—are produced and fall to earth. Where water vapor is present only in small proportions, extremely dry deserts are found.

In addition, a most important function is performed by water vapor. Like carbon dioxide, it is a gas capable of absorbing heat in the radiant form coming from the sun and from the earth's surface. Water vapor helps to give the troposphere the qualities of an insulating blanket, which inhibits the

escape of heat from the earth's surface. In terms of the great earth realms, water vapor is part of the hydrosphere. In this respect, the atmosphere holds a very small part of the hydrosphere.

The troposphere contains myriads of tiny dust particles, so small and light that the slightest movements of the air keep them aloft. They have been swept into the air from dry desert plains, lake beds and beaches, or explosive volcanoes. Strong winds blowing over the ocean lift droplets of spray into the air. These may dry out, leaving as residues tiny crystals of salt that are carried high into the air. Forest and brush fires are another important source of atmospheric dust particles. Countless meteors, vaporizing from the heat of friction as they enter the upper layers of air, contribute dust particles. Industrial processes involving combustion of fuels are also a major source of atmospheric dust.

In terms of earth realms, mineral solids suspended as atmospheric dust are very small bits of the lithosphere, temporarily detached and held in the atmosphere.

Dust in the troposphere contributes to the occurrence of twilight and the red colors of sunrise and sunset, but the most important function of dust particles cannot be seen and is rarely appreciated. Certain types of dust particles serve as nuclei, or centers, around which water vapor condenses to produce cloud particles.

Above the troposphere lies the *stratosphere*; it extends upward to a height of roughly 50 km (30 mi). The stratosphere is almost entirely free of water vapor and dust. Clouds are rare in the stratosphere, but there are high-speed winds in narrow zones.

Measurement of Air Temperatures

Air temperature is one of the most familiar bits of weather information we hear and read daily through the news media. This information comes from numerous National Weather Service observing stations and is taken following a carefully standardized procedure. Thermometers are mounted in a standard instrument shelter, shown in Figure 3.1. The shelter shades the instruments from sunlight, but louvers allow

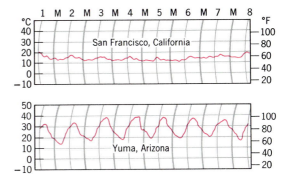

FIGURE 3.3 The recording thermometer made these continuous records of the rise and fall of air temperature over a period of one week. The daily cycle is strongly developed at Yuma, a desert station. In contrast, the record for San Francisco shows a very weak daily cycle.

FIGURE 3.1 This instrument shelter houses thermometers and an instrument for measuring relative humidity. (Arthur N. Strahler.)

air to circulate freely past the thermometers. The instruments are mounted from 1.2 to 1.8 m (4 to 6 ft) above ground level, at a height easy to read.

The Fahrenheit temperature scale is still widely used in the United States for public weather reports issued by the National Weather Service and the news media. As Figure 3.2 shows, the freezing point of water on the Fahrenheit scale is 32°F and the boiling point is 212°F. We prefer the Celsius temperature scale, which is the

scientific standard throughout the world. On the Celsius scale, the freezing point of water is 0°C and the boiling point is 100°C. Thus, 100 Celsius degrees are the equivalent of 180 Fahrenheit degrees (1 C° = 1.8 F°; 1 F° = 0.56 C°). Conversion formulas are given in Figure 3.2. We give Fahrenheit degrees in parentheses so that you can relate temperatures to sensations of bodily comfort or discomfort.

At most stations, only the highest and lowest temperatures of the day are recorded. To save the observer's time, the *maximum–minimum thermometer* is used. This instrument uses two thermometers; one to show the highest temperature, since it was last reset; and the other to show the lowest. Another labor-saving device is the recording thermometer, which draws a continuous record on graph paper attached to a slowly moving drum. Figure 3.3 shows typical traces for several days of record.

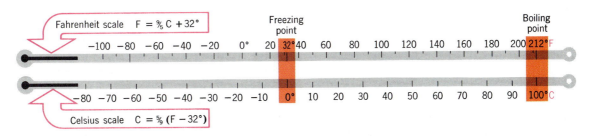

FIGURE 3.2 A comparison of the Celsius and Fahrenheit temperature scales.

64

When the maximum and minimum temperatures of a given day are added together and divided by two, we obtain the *mean daily temperature*. The mean daily temperatures of an entire month can be averaged to give the *mean monthly temperature*. Averaging the daily means (or monthly means) for a whole year gives the *mean annual temperature*. Usually such averages are compiled for records of many years' duration at a given observing station. These averages are used in describing the climate of the station and its surrounding area.

The Daily Cycle of Air Temperature

Because the earth turns on its axis, the net radiation undergoes a daily cycle of change. This cycle determines the daily cycle of rising and falling air temperatures with which we are all familiar. Let us see how net radiation and air temperature are linked in this cycle.

Three graphs in Figure 3.4 show average curves of daily insolation, net radiation, and air temperature, greatly generalized for a typical observing station at lat. 40 to 45° N in the interior United States.

Graph A shows insolation; units are langleys per minute (ly/min). At equinox, insolation begins about at sunrise (6 A.M. by local time), rises to a peak value at noon, and declines to zero at sunset (6 P.M.). At June solstice, insolation begins about 2 hours earlier (4 A.M.) and ends 2 hours later (8 P.M.). The June peak is much greater than at equinox and the total insolation for the day is also much greater. At December solstice, insolation begins about 2 hours later (8 A.M.) and ends 2 hours earlier (4 P.M.). Both the peak intensity and daily total insolation are greatly reduced in the winter solstice season.

Graph B shows net radiation in the same units as insolation. Net radiation shows a positive value—a surplus—shortly after sunrise and rises sharply to a peak at noon. The afternoon decline reaches zero shortly before sunset, and becomes a negative quantity, or deficit. The deficit continues through the hours of darkness with a more-or-less constant value. At June

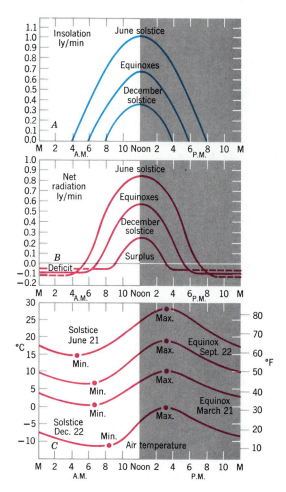

FIGURE 3.4 Idealized cycles of insolation, net radiation, and air temperature for a midlatitude station in the interior United States.

solstice, a radiation surplus begins earlier and ceases later than at equinox, generating a much larger daily surplus. At December solstice, the surplus period is greatly shortened and the surplus is small. Now, because the nocturnal deficit period is long, the total deficit exceeds the surplus, so that the net daily total is a small negative quantity. (We will investigate the annual cycle of net radiation in later paragraphs.)

Graph C shows the typical, or average, daily temperature cycle. The minimum daily temperature usually occurs just after sunrise, corresponding with the onset of a radiation surplus. Heat now begins to flow upward from the ground surface to warm the lower air layer. Temperature rises sharply in the morning hours, and continues to rise long after the noon peak of net radiation. We should expect the air temperature to rise as long as a radiation surplus is in effect, and in theory this should produce a temperature maximum just before sunset. Another atmospheric process, however, begins to take effect in the early afternoon. Mixing of the lower air by eddies distributes sensible heat upward, offsetting the temperature rise and setting back the temperature peak to about 3 P.M. (Time of daily maximum varies usually between 2 and 4 P.M., according to local climatic conditions.) By sunset, air temperature is falling rapidly and continues to fall, but at decreasing rate, through the entire night.

In midlatitudes and high latitudes, there is a very strong seasonal change in the air temperature curve. Upward displacement of the curve to warmer temperatures occurs in summer; downward displacement to colder temperatures occurs in winter. These changes constitute the annual temperature cycle, which we will examine in later paragraphs. At this point, you should note that at summer solstice, the time of minimum daily temperature occurs about 2 hours earlier than at equinox; at winter solstice the minimum occurs about 2 hours later than at equinox. The time of daily maximum temperature is not greatly affected by the seasonal cycle. For simplicity, in Graph C the maximum is shown as 3 P.M. throughout the year.

The daily cycle of air temperature can be strongly intensified close to the ground surface, as Figure 3.5 shows. Under the direct sun's rays, a bare soil surface or pavement heats to much higher temperatures than one reads in the thermometer shelter. At night, this surface layer cools to temperatures much lower than in the shelter. These effects are quite weak under a forest cover where the ground is shaded and moist. On the dry desert floor and on the pavements of city streets and parking lots the surface temperature extremes are large.

Soil temperatures play a major role in the seasonal rhythms of plant physiology and in biological activity generally within the soil. Soil scientists regard soil temperature as a major factor in determining many soil properties. As Figure 3.5 suggests, the daily cycle of soil temperature will show its greatest range at the surface, whereas the daily cycle gradually dies out with depth.

Temperature Structure of the Atmosphere

The atmosphere has been subdivided into layers according to temperatures and zones of temperature change. If we sent up a sounding balloon carrying a recording thermometer and repeated this operation many times, we would obtain an average or

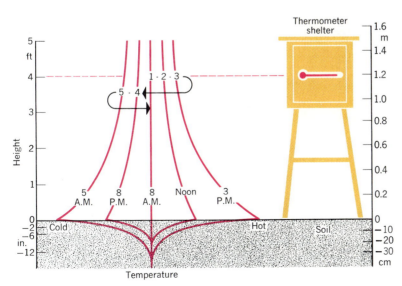

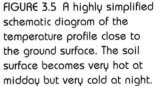

FIGURE 3.5 A highly simplified schematic diagram of the temperature profile close to the ground surface. The soil surface becomes very hot at midday but very cold at night.

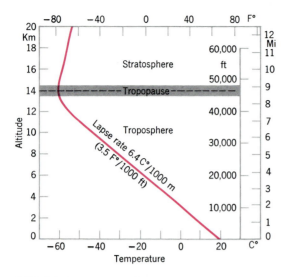

FIGURE 3.6 A typical environmental temperature lapse-rate curve for a summer day in midlatitudes.

representative profile of temperature.

Figure 3.6 shows a typical air sounding into the troposphere at midlatitudes (45° N) on a summer day. Altitude is plotted on the vertical axis and temperature on the horizontal axis. We find that air temperature falls rather steadily with increasing altitude through the entire troposphere, until a height of about 12 to

15 km (7 to 9 mi) is reached (Figure 3.6). For example, on a typical summer day in middle latitudes, when the air temperature near the surface is a pleasant 21°C (70°F), the air near the top of the troposphere at an altitude of about 14 km (46,000 ft) will be a chilly −60°C (−76°F).

At the top of the troposphere, a level called the *tropopause*, the air temperature no longer continues to fall, but remains nearly constant with increasing altitude. Then, a gradual temperature increase sets in within the next overlying layer, or stratosphere. These changes are shown in Figure 3.6.

Within the troposphere, the average rate of temperature decrease is about 6.4 C°/ 1000 m (3½ F°/1000 ft). This rate is known as the *environmental temperature lapse rate*. When used repeatedly, we shorten this ponderous term to *lapse rate*. Keep in mind that at a given time and place the lapse rate may be quite different from the average value we have given.

Air Temperatures on High Mountains

Because air density decreases upward, an increase in altitude of the ground surface brings important changes in the balances of radiation and heat. At a high altitude, the

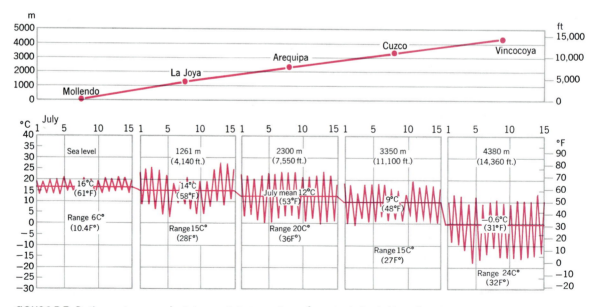

FIGURE 3.7 Daily maximum and minimum air temperatures for mountain stations in Peru, lat. 15° S. All data cover the same 15-day observation period in July. (Data from Mark Jefferson.)

thinner atmosphere has relatively less carbon dioxide, water vapor, and dust. This thinner air absorbs and reflects less incoming solar energy, so that insolation at the ground surface is of higher intensity than at low altitudes.

Increasing intensity of insolation at higher altitude has a profound influence on daily extremes of air temperature. Surfaces exposed to sunlight heat rapidly and intensely; shaded surfaces are quickly and severely cooled. This same effect results in rapid air heating during the day and rapid cooling at night at high-mountain locations.

This principle shows up in the set of daily-temperature graphs of Figure 3.7. The data cover two weeks in July at several stations on the west side of South America at lat. 15° S. The stations range in altitude from sea level to a maximum of over 4300 m (14,000 ft). Not only does the average temperature of the month decrease with higher altitude, but the daily temperature range becomes much larger at higher altitude.

The visible effects of decreasing average air temperature with increasing altitude are strikingly displayed in a landscape seen in the high Andes of Bolivia (see Figure 8.25). The profound influences of altitude zonation on plants and human food resources in the low-latitude environment are discussed in greater length in Chapter 8 (see Figure 8.26).

Temperature Inversion and Frost

During nights when the sky is clear and the air calm, the ground surface rapidly radiates longwave energy into the atmosphere above it. As we have explained, the soil surface temperature drops rapidly, and the overlying air layer becomes colder. When temperature is plotted against altitude, as in Figure 3.8, the straight, slanting line of the normal environmental lapse rate becomes bent to the left in a "J" hook. In the case shown, the air temperature at the surface, point A, has dropped to $-1°C$ (30°F). This value is the same as at point B, some 750 m (2500 ft) aloft. As we move up from ground level, temperatures become warmer up to about 300 m (1000 ft). Here the curve reverses

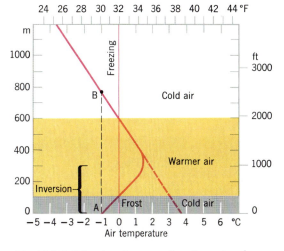

FIGURE 3.8 A low-level temperature inversion. In this case a layer of air close to the ground has dropped below the freezing temperature, while the air above remains warmer.

itself and the normal environmental lapse rate takes over.

The lower, reversed portion of the lapse rate curve is called a *low-level temperature inversion*. In the case shown, temperature of the lowermost air has fallen below the freezing point, 0°C (32°F). For sensitive plants this condition is a *killing frost* when it occurs during the growing season. Killing frost can be prevented in citrus groves by setting up an air circulation that mixes the cold basal air with warmer air above. One method is the use of oil-burning heaters; another is to operate powerful motor-driven propellers to circulate the air.

Low-level temperature inversion is also prevalent over snow-covered surfaces in winter. Inversions of this type are very intense and often extend a few thousand meters into the air.

The Annual Cycle of Air Temperature

As the earth revolves about the sun, the tilt of the earth's axis causes an annual cycle of incoming solar radiation, as we have already learned. This cycle is also felt in an annual cycle of the net radiation, which, in turn, causes an annual cycle in the mean daily and monthly air temperatures. In this way the climatic seasons are generated.

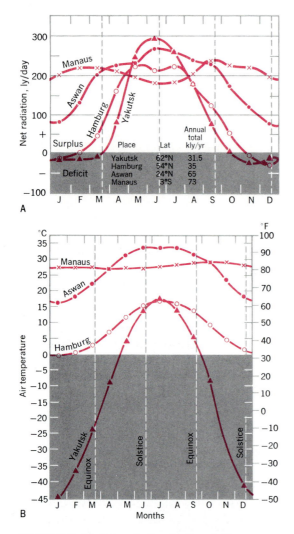

Place	Lat	Annual total kly/yr
Yakutsk	62°N	31.5
Hamburg	54°N	35
Aswan	24°N	65
Manaus	3°S	73

FIGURE 3.9 (*A*) Net radiation for four stations ranging from the equatorial zone to the arctic zone. (*B*) The annual cycles of monthly mean temperatures for the same stations. (Data courtesy of David H. Miller.)

Graph *A* of Figure 3.9 shows the yearly cycle of net radiation for four stations, ranging in latitude from the equator almost to the arctic circle. Graph *B* shows mean monthly air temperatures for these same stations. Starting with Manaus, a city on the Amazon River in Brazil, let us compare the net radiation graph with the air temperature graph. At Manaus, almost on the equator, the net radiation shows a large surplus in every month. The average surplus is about 200 ly/day, but there are two minor maxima, approximately coinciding with the equinox, when the sun

is nearly straight overhead. A look at the temperature graph of Manaus shows monotonously uniform air temperatures, averaging about 27°C (81°F) for the year. The *annual temperature range,* or difference between the highest mean monthly temperature and lowest mean monthly temperature, is only 1.7 C° (3 F°). In other words, near the equator one month is about like the next, thermally speaking. There are no temperature seasons.

We go next to Aswan, Egypt, on the Nile River at lat. 24° N. Here we are in a very dry desert. The net radiation curve has a strong annual cycle, and the surplus is large in every month. The surplus rises to more than 250 ly/day in June and July, but falls to less than 100 ly/day in December and January. The temperature graph shows a corresponding annual cycle, with an annual range of about 17 C° (30 F°). June, July, and August are terribly hot, averaging over 32°C (90°F).

Moving farther north, we come to Hamburg, Germany, lat. 54° N. The net radiation cycle is strongly developed. The surplus lasts for nine months, and there is a deficit for three winter months. The temperature cycle reflects the reduced total insolation at this latitude. Summer months reach a maximum of just over 16°C (60°F); winter months reach a minimum of just about freezing (0°C; 32°F). The annual range is 17 C° (30°F).

Finally, we travel to Yakutsk, Siberia, lat. 62° N. During the long, dark winters, there is an energy deficit; it lasts about six months. During this time air temperatures drop to extremely low levels. For three of the winter months monthly mean temperatures are between −35 and −45°C (−30 and −50°F). Actually, this is one of the coldest places on earth. In summer, when daylight lasts most of the 24 hours, the energy surplus rises to a strong peak, reaching 300 ly/day. This is a value higher than any of the other three stations. As a result, air temperatures show a phenomenal spring rise to summer-month values of over 13°C (55°F). In July the temperature is about the same as for Hamburg. The annual range at Yakutsk is enormous—over 60 C° (110 F°). No other region on earth, even the south pole, has so great an annual range.

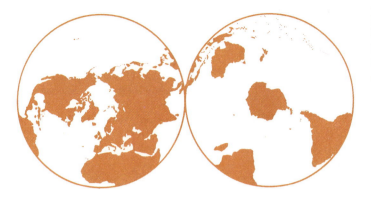

FIGURE 3.10 Northern and southern hemispheres—a study in contrasts in the distribution of lands and oceans. One pole bears a deep ocean and the other a great continental landmass.

Land and Water Contrasts

The odd distribution of continents and ocean basins gives our planet its great variety of climates. Look at the two global hemispheres—northern and southern—outlined in Figure 3.10. The northern hemisphere displays a small polar sea surrounded by massive continents; the southern hemisphere shows the very opposite—a pole-centered continent surrounded by a vast ocean. The Americas form a north–south barrier between two oceans—Atlantic and Pacific. The continents of Eurasia and Africa together form another great north–south barrier. Oceans and continents have quite different properties when it comes to absorbing and radiating energy.

Land surfaces behave quite differently from water surfaces. The important principle is this: the surface of any extensive deep body of water heats more slowly and cools more slowly than the surface of a large body of land when both are subject to the same intensity of insolation.

The slower rise of water-surface temperature can be attributed to four causes (Figure 3.11): (1) solar radiation penetrates water, distributing the absorbed heat throughout a substantial water layer; (2) the specific heat of water is large (a gram of water heats up much more slowly than a gram of rock); (3) water is mixed through eddy motions, which carry the heat to lower depths; and (4) evaporation cools the water surface.

In contrast, the more rapid rise of land-surface temperature can be attributed to

these causes: (1) soil or rock is opaque, concentrating the heat in a shallow layer, thus there is little transmission of heat downward; (2) specific heat of mineral matter is much lower than that of water; (3) soil, if it is dry, is a poor conductor of heat; and (4) no mixing occurs in soil and rock.

The effect of land and water contrasts is seen in two sets of daily air temperature curves shown in Figure 3.12. El Paso, Texas, exemplifies the temperature environment of an interior desert in midlatitudes. Soil moisture content is low, vegetation sparse, and cloud cover generally light. Responding to intense heating and cooling of the ground surface, air temperatures show an average daily range of 11 to 14 C° (20 to 25 F°). North Head, Washington, is a coastal station strongly influenced by air brought from the adjacent Pacific Ocean by prevailing westerly winds. Consequently, North Head exemplifies a

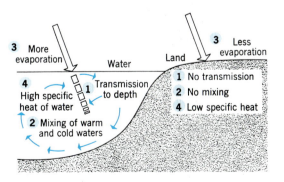

FIGURE 3.11 Four reasons why a land surface heats more rapidly and more intensely than the surface of a deep ocean body.

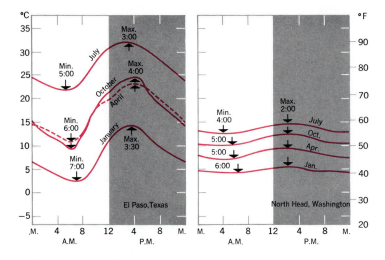

FIGURE 3.12 The average daily cycle of air temperature for four different months shows the effect of the seasons. Daily and seasonal ranges are great at El Paso, a station in the continental interior, but only weakly developed at North Head, Washington, close to the Pacific Ocean.

maritime temperature environment. The average daily range at North Head is a mere 3 C° (5 F°) or less. Persistent fogs and cloud cover also contribute to the small daily range. Refer also to Figure 3.3, which shows the same environmental contrast when the record of Yuma is compared with that of San Francisco.

The principle of contrasts in heating and cooling of water and land surfaces also explains the differences in the annual temperature cycles of coastal and interior stations. Let us examine the seasonal temperature curves for two places at approximately the same latitude, where

insolation is about the same for both. Two such places are Winnipeg, Manitoba, located in the heart of the continent, and the Scilly Islands, England, exposed to the Atlantic Ocean. Figure 3.13 shows the average daily temperatures through an entire year as well as the insolation curve for lat. 50° N.

Although insolation reaches a maximum at summer solstice, the hottest part of the year for inland regions is about a month later, since heat energy continues to flow into the ground well into August. The air temperature maximum, closely corresponding with maximum ground output of longwave radiation, is

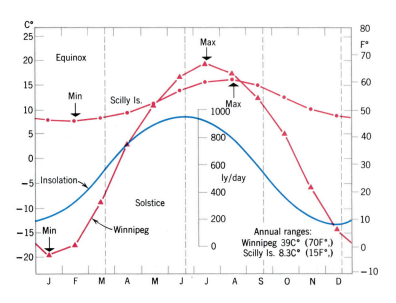

FIGURE 3.13 Annual cycles of monthly mean air temperature for two stations at lat. 50° N: Winnipeg, Manitoba, Canada, and Scilly Islands, England.

correspondingly delayed. (Bear in mind that this cycle applies to middle and high latitudes, but not to the region between the tropics of cancer and capricorn.) Similarly, the coldest time of year for large land areas is January, about a month after winter solstice, because the ground continues to lose heat even after insolation begins to rise.

Over the oceans there are two differences: (1) maximum and minimum temperatures are reached about a month later than on land—in August and February, respectively—because water bodies heat or cool much more slowly than land areas; and (2) the yearly range is less than that over land, following the law of temperature differences between land and water surfaces. Coastal regions are usually influenced by the oceans to the extent that maximum and minimum temperatures occur later than in the interior. This principle shows nicely for monthly temperatures of the Scilly Islands. February is slightly colder than January (Figure 3.13).

Figure 3.12 reinforces the evidence of Figure 3.13. The annual temperature range at El Paso is about 20 C° (35 F°), whereas that at North Head is only about 8 C° (15 F°).

Air Temperature Maps

The distribution of air temperatures over large areas can best be shown by a map composed of *isotherms*, lines drawn to

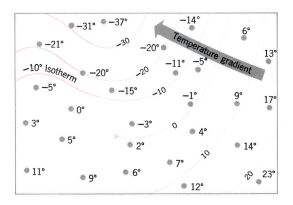

FIGURE 3.14 Isotherms are used to make temperature maps. Each line connects those points having the same temperature.

connect all points having the same temperature. Figure 3.14 shows a map on which the observed air temperatures have been recorded in the correct places. These may represent single readings taken at the same time everywhere, or they may represent the averages of many years of records for a particular day or month of a year, depending on the purposes of the map.

Usually, isotherms representing 5 or 10° differences are chosen, but they can be drawn for any selected temperatures. The value of isothermal maps is that they make clearly visible the important features of the prevailing temperatures. Centers of high or low temperatures are clearly outlined.

World Patterns of Air Temperature

World maps of air temperature (Figure 3.15) allow us to compare thermal conditions in the two months of greatest extremes—January and July. Isotherms are shown for Celsius degrees.

Patterns taken by the isotherms are largely explained by three controls: (1) latitude; (2) continent–ocean contrasts; and (3) altitude. (Ocean currents play an important, but secondary role.) Here are the important features you should identify when interpreting these maps.

1. The trend of isotherms is east–west, with temperatures decreasing from equatorial zone to polar zones. This feature shows best in the southern hemisphere, because the great Southern Ocean girdles the globe as a uniform expanse of water, and the continent of Antarctica is squarely centered on the south pole. The parallel pattern of isotherms is explained, of course, by the general decrease in intensity of insolation from equator to poles.

2. Large landmasses located in the subarctic and arctic zones develop centers of extremely low temperatures in winter. The two large landmasses we have in mind are North America and Eurasia. Find these centers on the January map.

3. Isotherms change very little in position from January to July over the equatorial zone, particularly over the oceans. This feature reflects the uniformity

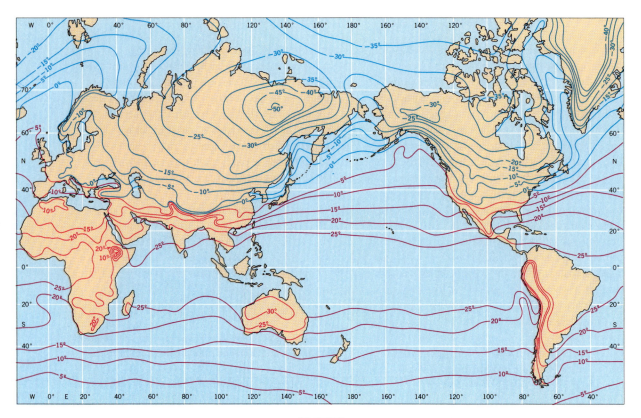

JANUARY

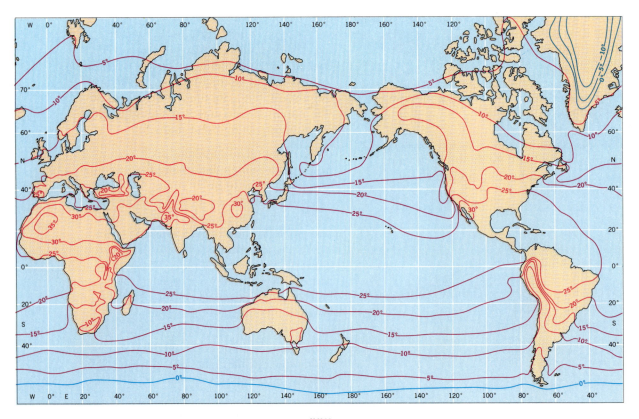

JULY

FIGURE 3.15 Mean monthly air temperatures (°C) for January and July, Mercator and stereographic projections. (Compiled by John E. Oliver from station data by World Climatology Branch, Meteorological Office, *Tables of Temperature*, 1958, Her Majesty's Stationery Office, London; U.S. Navy, 1955, *Marine Climate Atlas*, Washington, D.C.; and P. C. Dalrymple, 1966, American Geophysical Union.)

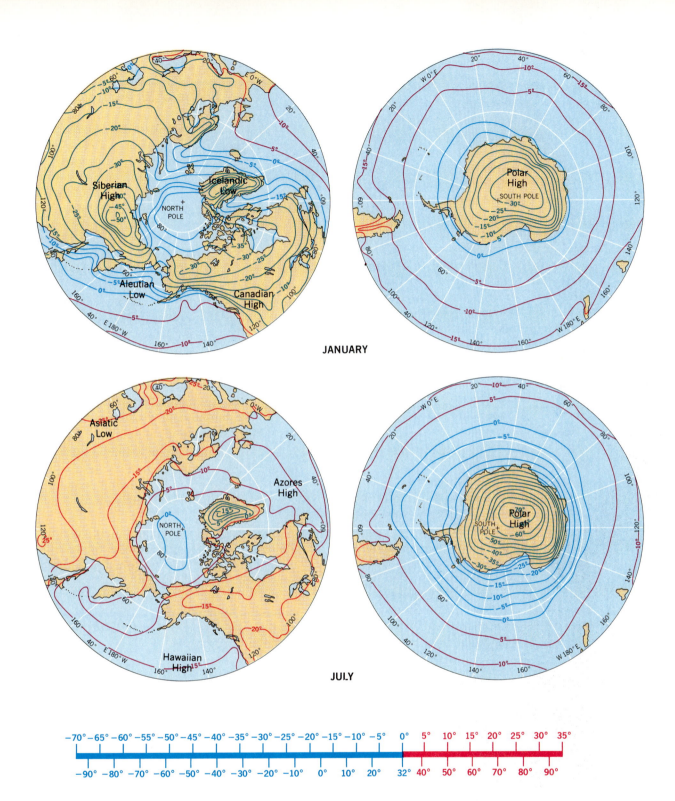

JANUARY

JULY

−70° −65° −60° −55° −50° −45° −40° −35° −30° −25° −20° −15° −10° −5° 0° 5° 10° 15° 20° 25° 30° 35°

−90° −80° −70° −60° −50° −40° −30° −20° −10° 0° 10° 20° 32° 40° 50° 60° 70° 80° 90°

of insolation throughout the year near the equator.

4. Isotherms make a large north–south shift from January to July over continents in the midlatitude and subarctic zones. Figure 3.16 shows this principle. Check this principle for North America. In January the 15°C isotherm lies over central Florida; in July this same isotherm has moved far north, cutting the southern shore of Hudson Bay and then looping far up into northwestern Canada. The 15°C isotherm on the Eurasian continent shows this same effect. We can explain the large north shift in position over these continents by the law of land and water contrasts.

5. Highlands are always colder than surrounding lowlands. Look for the Andes range, running along the western side of South America. The isotherms loop equatorward in long fingers over this lofty mountain chain.

6. Areas of perpetual ice and snow are always intensely cold. Greenland and Antarctica are the two great ice sheets. Notice how they stand out as cold centers in both January and July. Not only do these ice sheets have high surfaces, rising to over 3000 m (10,000 ft) in their centers, but the white snow surfaces have a high albedo and reflect much of the insolation. The Arctic Ocean, bearing a cover of floating ice, also maintains its cold throughout the year, but the cold is much less intense in July than over the Greenland Ice Sheet.

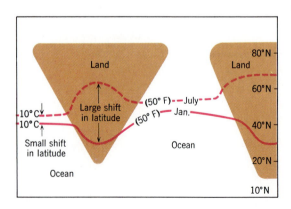

FIGURE 3.16 The seasonal migration of isotherms is much greater over the continents than over the oceans.

The Annual Range of Air Temperatures

Figure 3.17 is a world map showing the annual range of air temperatures. The lines resembling isotherms tell the difference between the January and July monthly means. This map serves as a summary of concepts we have previously emphasized. Note these features.

1. The annual range is extremely great in the subarctic and arctic zones of both Asia and North America. The effect is shown well in the annual temperature graph for Yakutsk (Figure 3.9).

2. The annual range is moderately large on land areas in the tropical zone, near the tropics of cancer and capricorn. North Africa, southern Africa, and Australia are examples. Here the annual ranges are substantially greater than those over adjacent oceans.

3. The annual range is very small over oceans in the equatorial zone. At all latitudes the annual range over all oceans is generally less than that over lands.

This review of air temperatures over the globe will form the groundwork for your understanding of climates, which will be developed in later chapters.

Carbon Dioxide, Dust, and Global Climate Change

Atmospheric changes induced by human activity fall into four categories with respect to basic causes: (1) changes in concentrations of the natural component gases of the lower atmosphere; (2) changes in the water-vapor content of the troposphere and stratosphere; (3) alteration of surface characteristics of the lands and oceans in such a way as to change the interaction between the atmosphere and those surfaces; and (4) introduction of finely divided solid substances into the lower atmosphere, along with gases not normally found in substantial amounts in the unpolluted atmosphere.

Under preindustrial conditions of recent centuries, the atmospheric content of carbon dioxide (CO_2) was maintained at a

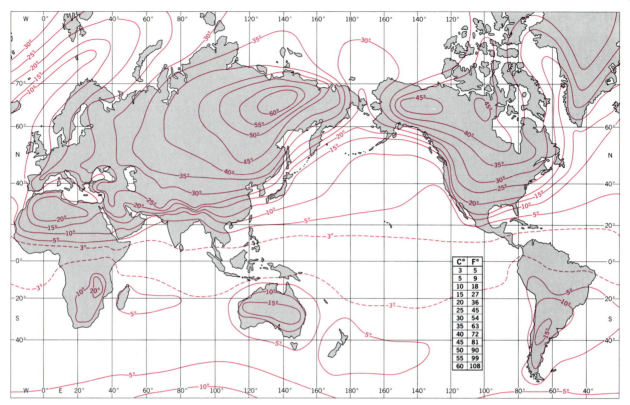

FIGURE 3.17 Annual range of air temperature in Celsius degrees. Data show differences between January and July means. (Same data sources as Figure 3.15.)

level of about 0.0294 percent by volume, or 294 parts per million (ppm). The environmental problem is that industries then began to extract and burn fossil fuels (coal, petroleum, natural gas) that had previously been locked in the earth's crust. Combustion of these fuels releases into the atmosphere both water and CO_2, and also a great deal of heat.

During the past 120 years (1860–1980) atmospheric CO_2 has increased about 13 percent by volume; it reached a level of 334 ppm in 1979. As Figure 3.18 shows, the rate of increase in this period, though slow

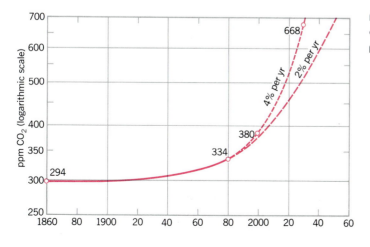

FIGURE 3.18 Increase in atmospheric carbon dioxide, observed to 1980 and predicted into the 21st century.

at first, became much more rapid toward the end of the period. Projection of the present curve of increase into the future suggests a CO_2 value of about 380 ppm by the year 2000. At that time, if the prediction is correct, the content of CO_2 will have increased by about 35 percent over the 1860 value. Doubling of the present level is predicted by about 2030 if the combustion of fossil fuels continues to increase at the present annual rate of 4 percent. Doubling time can be delayed, however, until about 2050 with a fuel combustion rate reduced to half of the present value, and to well into the 22nd century if the combustion rate remains at the present level.

Consider now the environmental effects to be anticipated from an increase of atmospheric CO_2. Because CO_2 is an absorber and emitter of longwave radiation, its presence in larger proportions will tend to raise the level of absorption of outgoing longwave radiation, changing the energy balance and resulting in a rise of air temperature in the troposphere. Scientists have estimated that a doubling of atmospheric CO_2 will cause an average global warming of about 3 C° (5.5 F°), with a probable error of 1.5 C° (3 F°) greater or less than that amount. It is generally agreed that this warming will lead to significant

changes in regional climatic patterns, such as the distribution of precipitation. We will look next at available temperature data to see if a global temperature increase has already occurred.

Figure 3.19 shows the mean global air temperature change for approximately the past century, as calculated in 1981 by scientists of the NASA Institute for Space Studies. From 1880 to 1940, when fuel consumption was rising rapidly, the average temperature increased by about 0.5 C° (0.9 F°). This rise follows the predicted pattern. After 1940, however, a decline set in, despite the rising rate of fuel combustion. Assuming that atmospheric warming because of increased CO_2 is a valid effect in principle, some other factor working in the opposite direction entered the picture and its cooling effect outweighed that of warming by CO_2. Since about 1965, however, the trend has reversed to one of increase.

Downturn in world and northern hemisphere air temperature after 1940 may have been the result of a greatly increased input of dust into the upper atmosphere by a number of major volcanic eruptions, the first of which was the 1947 eruption of the Icelandic volcano Hekla. An increase in volcanic dust at high altitudes increases the losses of insolation to space through diffuse reflection; the result is that less energy arrives at lower atmosphere levels. A reduction in level of sensible heat of the lower atmosphere is the predicted effect.

The prediction of an eventual rise in global air temperature continues to be supported by recent evaluations. On the other hand, the schedule on which these changes will operate is highly uncertain at this time because many variables are involved. One major uncertainty lies in predicting how quickly the great world ocean will respond to increased CO_2. It seems likely that a considerable time lag is involved and many researchers are concerned that the warming effect has not yet been felt.

In an attempt to get a better picture of climate changes over the past two centuries, scientists of the Lamont–Doherty Geological Observatory of Columbia set up a tree-ring laboratory to measure the temperature response of old-aged trees in North

FIGURE 3.19 Changes in the mean global air temperature from 1880 to 1985. The graph shows the five-year running average based on data of more than 1,000 stations. The single point in the upper right corner is the average for 1987. (From J. Hansen and S. Lebedeff. Copyright © 1988 by the American Geophysical Union. Used by permission.)

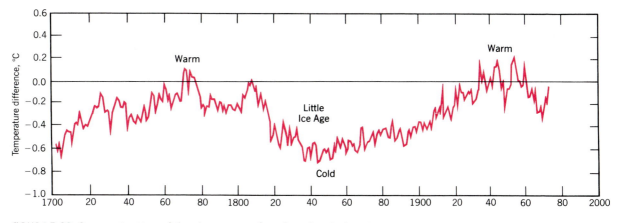

FIGURE 3.20 A reconstruction of the departures of northern hemisphere temperatures from the 1950–1965 mean, based on analyses of tree rings sampled along the northern tree limit of North America. (Courtesy of Gordon C. Jacoby of the Tree-Ring Laboratory of the Lamont–Doherty Geological Observatory of Columbia University.)

America as revealed in their annual growth rings. The resulting graph, based on hundreds of trees from dozens of sites along the northern tree line (subarctic limit of tree growth), is shown in Figure 3.20. From 1880 to the present, the rising trend and its reversal are quite similar to that in Figure 3.19. Looking further back in time, however, another cycle of rise and decline is revealed. The low point around 1840 is an expression of a well-known event, the "Little Ice Age," which is documented by the advance of European alpine glaciers (see Chapter 18). Similar cycles of warming and cooling, each cycle lasting on the order of 150 to 200 years, appear in a record going back over 800 years. The message of the long-term temperature records seems to be that the predicted warming because of increased CO_2 from industrial sources may not have yet made a measurable impact.

Another complicating aspect of the problem of a human-induced climate change is that several other gases are known to be effective in implementing the greenhouse effect. They are called "trace gases" and their potency is great, even in small amounts. The list includes methane, nitrous oxide, ozone, and the halocarbons. Regarding methane, its rapid increase in the past 200 years has now been documented in the record of Antarctic and Greenland ice cores. The possibility exists

that increases in the trace gases may have a reinforcing effect that could more than double the temperature rise predicted on the basis of CO_2 alone. Many years of research may yet lie ahead before a consensus can be reached as to when the global environment will receive the full impact of human-induced warming that seems inevitable.

The Ozone Layer—A Shield to Life

Of vital concern to humans and all other life-forms on earth is the presence of an *ozone layer* occurring within the stratosphere. The ozone layer begins at an altitude of about 15 km (9 mi) and extends upward to about 55 km (35 mi). The ozone layer is a region of concentration of the form of oxygen molecule known as *ozone* (O_3) in which three oxygen atoms are combined instead of the usual two atoms (O_2). Ozone is produced by the action of ultraviolet rays on ordinary oxygen atoms.

The ozone layer serves as a shield, protecting the troposphere and earth's surface from most of the ultraviolet radiation found in the sun's rays. If these ultraviolet rays were to reach the earth's surface in full intensity, all exposed bacteria would be destroyed and animal tissues severely damaged. In this protective role the presence of the ozone layer is an essential factor in maintaining a habitable earth.

A serious threat to the ozone layer is posed by the release into the atmosphere of Freons, synthetic compounds containing carbon, fluorine, and chlorine atoms. Compounds of this class are also called *halocarbons*. Prior to a ban issued in 1976 in the United States by the Environmental Protection Agency, many aerosol spray cans used in the household were charged with halocarbons. They are also widely used as the cooling fluid in refrigerators and air conditioning units, a practice that continues to contribute to the release of halocarbons.

Molecules of halocarbons drift upward through the troposphere and eventually reach the stratosphere. As these compounds absorb ultraviolet radiation, they are decomposed and chlorine is released. The chlorine in turn attacks molecules of ozone, converting them in large numbers by a chain reaction into ordinary oxygen molecules. In this way the ozone concentration within the stratospheric ozone layer can be reduced, and the intensity of ultraviolet radiation reaching the earth's surface can be increased. A marked increase in the incidence of skin cancer in humans is one of the predicted effects. Other possible effects are reduction of crop yields of various plants and the killing of certain forms of aquatic life found in the surface layer of the oceans and in streams and lakes.

The effects of halocarbons have been further analyzed by using improved atmospheric measurements and computer modeling of complex atmospheric chemistry. These studies have yielded the conclusion that the rate of depletion of stratospheric ozone on the basis of continued emission of halocarbons at the 1977 rate will result in a decrease of 2 to 4 percent in stratospheric ozone by the 21st century—a much smaller decrease than previously predicted. An interesting new twist in a 1984 prediction is that although the smaller rate of decrease of ozone in the stratosphere is welcome news in terms of reduced future risk, a substantial increase in ozone is predicted at a height of 10 km (6 mi), which is well within the troposphere. Here, the added ozone may have a

significant impact on weather processes and may influence the predicted warming by the greenhouse effect anticipated from an increase in carbon dioxide.

In 1985 scientists discovered that in the stratosphere over Antarctica there occurs each year a dramatic thinning of the stratospheric ozone layer. The effect starts in the antarctic spring, which begins in September as the sun reappears over the horizon, and lasts through October. Centered over the south pole, the depleted zone is popularly called the "Ozone Hole." Because of concern about the observed deepening of the hole, it has been the subject of intense scientific research and a number of different causes have been proposed for its occurrence. Halocarbons have been linked to the phenomenon and fears have been expressed that it may worsen.

The Thermal Environment in Review

In this chapter we have covered a vital environmental factor in the life layer. Heat, as recorded by the thermometer, is an essential ingredient of climate near the ground. All organisms respond to changes in temperatures of the medium that surrounds them, whether it is air, soil, or water.

Cycles of air temperature change are particularly important in the thermal environment. Both daily and seasonal changes in air temperature can be explained by daily and seasonal cycles of insolation and the energy balance. Superimposed on these astronomical rhythms is the powerful effect of latitude. A zoning of thermal environments from the equatorial zone to the polar zones is one of the most striking features of the global climate. But equally important is the way in which large land areas, especially the continents of North America and Eurasia, subvert the simple latitude zones to cause great seasonal extremes of temperature. In contrast, the southern hemisphere is strongly dominated by the simple effects of latitude. From the long-range standpoint, combustion of fuels and continued injection of halocarbons into the atmosphere have some serious implications.

CHAPTER 4

GLOBAL CIRCULATION OF THE ATMOSPHERE AND OCEANS

ATMOSPHERIC PRESSURE   Our atmosphere exerts barometric pressure, which is a vital and pervasive environmental factor.

ATMOSPHERIC MOTION   Ceaseless motion of the atmosphere is also a strong factor in the environment of the life layer.

WINDS   A pressure gradient force, generated by place-to-place differences in barometric pressure, causes the wind to blow.

CORIOLIS AND WINDS   Caused by earth rotation, the Coriolis effect turns the direction of the wind toward the right or left, depending on which hemisphere we refer to.

CYCLONES AND ANTICYCLONES   Winds form spiraling patterns of inflow into cyclones and outflow from anticyclones.

PREVAILING WINDS   Global patterns of barometric pressure determine the earth's prevailing winds, both at the surface and aloft.

OCEAN CURRENTS   Geared to the global system of surface winds are ocean currents, taking the form of enormous gyres of continual water motion.

THE physical environment at the earth's surface depends for its quality as much on atmospheric motions as it does on the flow of heat energy by radiation. In the form of very strong winds—those in hurricanes and tornadoes—air in motion is a severe environmental hazard. Winds also transfer energy to the surface of the sea, as wind-driven waves. Wave energy, in turn, travels to the shores of continents, where it is transformed into vigorous surf and coastal currents capable of reshaping the coastline.

But air in motion has another, more basic role to play in the planetary environment. Large-scale air circulation transports heat, both as sensible heat and as latent heat present in water vapor. Because of the global radiation imbalance—a surplus in low latitudes and a deficit in high latitudes—atmospheric circulation must transport heat across the parallels of latitude from the

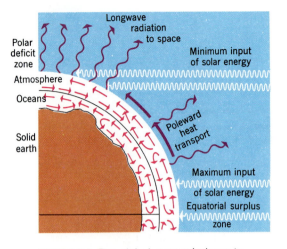

FIGURE 4.1 The global energy balance is maintained by transport of heat from low to high latitudes.

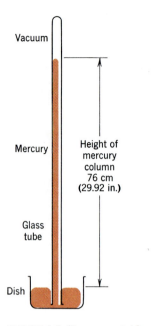

FIGURE 4.2 The mercurial barometer of Torricelli.

region of surplus to the regions of deficit. Figure 4.1 shows this meridional transport in schematic form. Notice that circulation of ocean waters is also involved in transport of sensible heat.

Atmospheric Pressure

Although we are not constantly aware of it, air is a tangible, material substance, exerting *atmospheric pressure* on every solid or liquid surface exposed to it. At sea level, this pressure is about 1 kilogram per square centimeter (1 kg/sq cm) or about 15 pounds per square inch. Because this pressure is exactly counterbalanced by the pressure of air within liquids, hollow objects, or porous substances, its ever-present weight goes unnoticed. The pressure on 1 sq cm of surface can be thought of as the actual weight of a column of air 1 cm in cross section extending upward to the outer limits of the atmosphere. Air is readily compressible. That which lies lowest is most greatly compressed and is, therefore, densest. Upward, both density and pressure of the air fall off rapidly.

Atmospheric science uses another method of stating the pressure of the atmosphere, based on a classic experiment of physics first performed by Torricelli in the year 1643. A glass tube about 1 m (3 ft)

long, sealed at one end, is completely filled with mercury. The open end is temporarily held closed. Then the tube is inverted and the end is immersed into a dish of mercury. When the opening is uncovered, the mercury in the tube falls a few centimeters, but then remains fixed at a level about 76 cm (30 in.) above the surface of the mercury in the dish (Figure 4.2). Atmospheric pressure now balances the weight of the mercury column. When the air pressure increases or decreases, the mercury level rises or falls correspondingly. Here, then, is a device for measuring air pressure and its variations.

Any instrument that measures atmospheric pressure is a *barometer*. The type devised by Torricelli is known as the *mercurial barometer*. With various refinements over the original simple device it has become the standard instrument. Pressure is read in centimeters or inches of mercury, the true measure of the height of the mercury column. Atmospheric pressure at sea level averages about 76 cm (30 in.).

Atmospheric pressure is an important environmental factor affecting the life processes of plants and animals. Air-breathing animals depend on air pressure to force air into the lungs. They rely on the abundant supply of oxygen present in air of

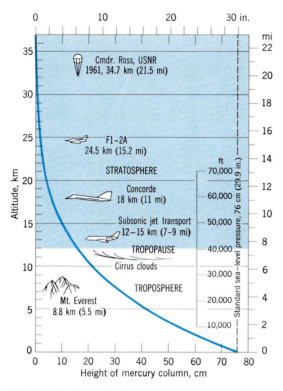

FIGURE 4.3 Atmospheric pressure decreases with increasing height above the earth's surface, but the rate of decrease falls off rapidly.

Supplementary oxygen is needed for full physical activity above 5400 m (18,000 ft) in unpressurized aircraft. Lightweight oxygen containers enabled climbers to reach the summit of Mount Everest (8800 m; 29,000 ft), a feat first accomplished in 1953. At the summit of Everest, the height of the mercury column is about 25 cm (10 in.)—only a third that at sea level. The cabins of commercial subsonic jet aircraft are pressurized so as to be equivalent to an altitude of about 1500 m (4500 ft) with a pressure of 64 cm (25 in.), whereas the outside pressure at cruising altitude may be only 15 to 20 cm (6 to 8 in.).

As atmospheric pressure becomes less, the boiling point of water becomes lower. For example, at 3000 m (10,000 ft), the boiling point, which is 100°C (212°F) at sea level, is reduced to 90°C (194°F); at 5000 m (16,000 ft), to 84°C (183°F). As the boiling point decreases, the time required to cook foods by boiling increases. As many experienced campers know, boiling potatoes is a very slow process at altitudes above 2000 m (6600 ft). Thus, a pressure cooker is a most useful device for persons living at high altitudes.

normal density for exchange with waste carbon dioxide through the lung tissues.

Figure 4.3 shows the rate of decrease of atmospheric pressure with altitude, using height of the mercury column as the horizontal scale. For every 275 m (900 ft) of rise in altitude, pressure is diminished by one-thirtieth of itself. Steepening of the curve as it rises shows that the rate of decrease in pressure is rapid at first, but becomes less rapid with increasing altitude.

The physiological effects on humans of a pressure decrease are well known from the experiences of flying and mountain climbing. Lowered pressure decreases the amount of oxygen entering the blood through the lungs. At altitudes of 3000 to 4500 m (10,000 to 15,000 ft) mountain sickness (altitude sickness) can occur, characterized by weakness, headache, nosebleed, or nausea. Persons who remain at these altitudes for a day or two normally adjust to the conditions, but physical exertion is always accompanied by shortness of breath.

Winds and the Pressure Gradient Force

Wind is air motion with respect to the earth's surface, and it is dominantly horizontal. (Dominantly vertical air motions are referred to by other terms, such as updrafts or downdrafts.) To explain winds, we must first consider barometric pressure and its variations from place to place.

Standard barometric pressure at sea level is 76 cm (29.92 in.) of mercury. In atmospheric science another pressure unit is used—the *millibar* (mb). One centimeter of mercury is equivalent to 13.3 mb (1 in. equals about 34 mb). Standard sea-level pressure is 1013.2 mb.

For an atmosphere at rest the barometric pressure will be the same within a given horizontal surface at any chosen altitude above sea level. In that case the surfaces of equal barometric pressure, called *isobaric surfaces,* are horizontal. In a cross section of a small portion of the atmosphere at rest, isobaric surfaces appear as horizontal lines, as shown in Figure 4.4A. (For convenience,

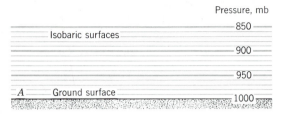

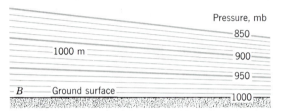

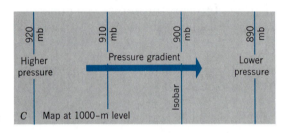

FIGURE 4.4 Isobaric surfaces and the pressure gradient. Diagrams A and B are vertical cross sections through the atmosphere. Diagram C is a map.

higher pressure (at the left) to lower pressure (at the right). You can think of the sloping pressure surface as a sloping hillside; the downward slope of the ground surface is analogous to the pressure gradient.

Where a pressure gradient exists, air molecules tend to drift in the same direction as that gradient. This tendency for mass movement of the air is referred to as the *pressure gradient force*. The magnitude of the force is directly proportional to the steepness of the gradient. That is, a steep gradient is associated with a strong force. Wind is the horizontal motion of air in response to the pressure gradient force.

Sea and Land Breezes

Perhaps the simplest example of the relationship of wind to the pressure gradient force is a common phenomenon of coasts. It is the sea breeze and land breeze effect shown in Figure 4.5.

On a clear day, the sun heats the land surface rapidly and a shallow air layer near

we have shown surface pressure as 1000 mb.)

Suppose now that the rate of upward pressure decrease is more rapid in one place than another, as shown in Figure 4.4B. As we proceed from left to right across the diagram, the upward rate of pressure decrease is more rapid. The isobaric surfaces now slope down toward the right. At a selected altitude, say 1000 m (horizontal line), barometric pressure declines from left to right. Figure 4.4C is a type of map; it shows that the 1000-m horizontal surface cuts across successive pressure surfaces. The trace of each pressure surface is a line on the map; the line is known as an *isobar*. The isobar is thus a line showing the location on a map of all points having the same barometric pressure.

The change in barometric pressure across the horizontal surface of a map constitutes a *pressure gradient;* its direction is indicated by the broad arrow in Figure 4.4C. The gradient is in the direction from

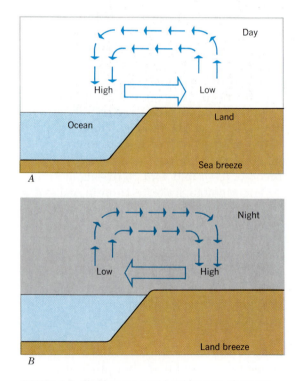

FIGURE 4.5 Sea breeze and land breeze alternate in direction from day to night.

the ground is strongly warmed (Figure 4.5A). The warm air expands. Because it has become less dense than the cool air offshore, low pressure is formed over the coastal belt, while pressure remains higher over the water. Now there is a pressure gradient from ocean to land, and a *sea breeze* is set in motion. The sea breeze can become very strong in a shallow layer close to shore and is pleasantly cooling on a hot summer day.

At night, the land surface cools more rapidly than the ocean surface and a cool air layer of higher pressure develops (Figure 4.5B.). Now the pressure gradient is from land to ocean, setting up a *land breeze*. These circulation systems are completed by rising and sinking motions of the air and by a weak return flow at higher levels, as indicated by the small arrows in the diagrams.

This illustration shows that a pressure gradient can be developed through unequal heating or cooling of a layer of the atmosphere. Air that is warmed expands and becomes less dense. Air that is cooled contracts and becomes denser. Heat energy drives this kind of circulation system by changing the air densities and setting up barometric pressure gradients. The entire mechanism is often described as a *heat engine,* because energy of air in motion is derived from the input of heat. The heat engine system of air circulation is also important on a large scale in the global atmosphere.

Measurement of Surface Winds

A description of winds requires measurement of two quantities: direction and speed. Direction is easily determined by a *wind vane,* commonest of the weather instruments. Wind direction is stated in terms of the direction from which the wind is coming (Figure 4.6). Thus an east wind comes from the east, but the direction of air movement is toward the west. The direction of movement of low clouds is an excellent indicator of wind direction and can be observed without the aid of instruments.

Speed of wind is measured by an *anemometer.* There are several types. The commonest one seen at weather stations is

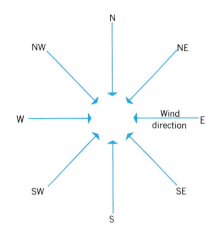

FIGURE 4.6 Winds are designated according to the compass point from which the wind comes. An east wind comes from the east, but the air is moving westward.

the cup anemometer. It consists of three hemispherical cups mounted as if at the ends of spokes of a horizontal wheel. The cups travel with a speed proportional to that of the wind. One type of anemometer turns a small electric generator, and the current it produces can be transmitted to a meter calibrated in units of wind speed. Units are meters per second or miles per hour.

For wind speeds at higher levels a small hydrogen-filled balloon is released into the air and observed through a telescope. The rate of climb of the balloon is known in advance. Knowing the balloon's vertical position by measuring the elapsed time, an observer can calculate the horizontal drift of the balloon downwind. For upper-air measurements of wind speed and direction the balloon carries a target that reflects radar waves and can be followed when the sky is overcast.

The Coriolis Effect and Winds

If the earth did not rotate on its axis, winds would follow the direction of pressure gradient. As we mentioned in Chapter 1, earth rotation produces the *Coriolis effect,* which is the tendency of air in motion to be turned toward the right or left. Although in strict terms of physics the Coriolis effect is

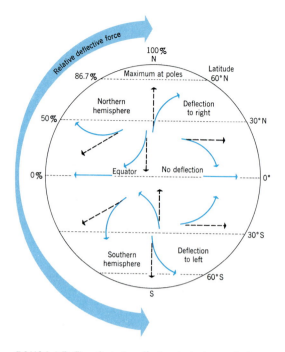

FIGURE 4.7 The Coriolis effect acts to turn winds or ocean currents to the right in the northern hemisphere and to the left in the southern hemisphere.

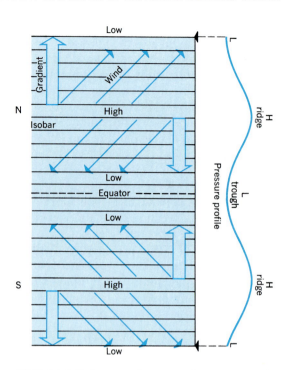

FIGURE 4.8 Surface winds cross isobars at an angle as the air moves from higher to lower pressure. Turn the figure sideways to view the pressure profile.

not a true force, it can be visualized as if it were a force acting on air in motion. The direction of action of this turning effect can be stated thus: any object or fluid moving horizontally in the northern hemisphere tends to be deflected to the right of its path of motion, regardless of the compass direction of the path. In the southern hemisphere a similar deflection is toward the left of the path of motion. The Coriolis effect is absent at the equator but increases in strength toward the poles.

In Figure 4.7 the small arrows show how an initial straight line of motion is modified by the Coriolis effect. Note especially that the compass direction is not of any consequence. If we face down the direction of motion, turning will always be toward the right hand in the northern hemisphere. Because the turning effect is very weak, its action is conspicuous only in freely moving fluids, such as air or water. Consequently, both winds and ocean current patterns are greatly modified by the Coriolis effect.

Our next step is to apply the Coriolis principle to winds close to the ground surface. Figure 4.8 shows isobars running east–west, forming a ridge of high pressure in each hemisphere. From each ridge pressure decreases to both the north and south toward belts of low pressure. Broad arrows show the pressure gradient. The Coriolis effect turns the wind so that it crosses the isobars at an angle. For surface winds, the angle of turning is limited by the force of friction of the air with the ground. The diagram shows the wind making an angle of 45° with the isobars. The angle is subject to some variation, depending on the character of the ground surface.

Looking first at the northern hemisphere case, the deflection is to the right. The northward pressure gradient gives a southwest wind. The southward gradient gives a northeast wind. In the southern hemisphere winds are deflected to the left and the pattern is the mirror image of that in the northern hemisphere. We turn next to the case of concentric isobars, using the same rules.

Cyclones and Anticyclones

In the language of meteorology, rotation of air around a center of low pressure is called

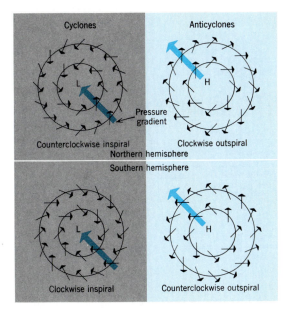

FIGURE 4.9 Surface winds spiral inward toward the center of a cyclone, but outward from the center of an anticyclone.

Global Pressure Systems

To understand the earth's surface wind systems, we must first study the global system of barometric pressure distribution. Once we grasp the patterns of isobars and pressure gradients, the prevailing or average winds can be predicted.

World isobaric maps are usually constructed to show average pressures for the two months of seasonal temperature extremes over large landmasses—January and July (Figure 4.10). Because observing stations lie at various altitudes above sea level, their barometric readings must first be reduced to sea-level equivalents, using the standard rate of pressure change with altitude. When this has been done, and the daily readings are averaged over long periods of time, small but distinct pressure differences remain.

A reading of 1013 mb is taken as standard sea-level pressure. Readings higher than this value will frequently be observed in midlatitudes, occasionally up to 1040 mb or higher. These pressures are designated as "high." Pressures ranging down to 982 mb or below are "low."

Over the equatorial zone is a belt of somewhat lower than normal pressure, between 1011 and 1008 mb, which is known as the *equatorial trough*. Lower pressure is conspicuous by contrast with belts of higher pressure lying to the north and south and centered at about lat. 30° N and S. These are the *subtropical belts of high pressure* in which pressures exceed 1020 mb. In the southern hemisphere this belt is clearly defined but contains centers of high pressure, called *high-pressure cells*.

In the southern hemisphere, south of the subtropical high-pressure belt, is a broad belt of low pressure, extending roughly from the midlatitude zone to the arctic zone. The axis of low pressure is centered at about lat. 65° S. This pressure trough is called the *subantarctic low-pressure belt*. Lying over the continuous expanse of the Southern Ocean, this trough has average pressures as low as 984 mb. On the surface of the continent of Antarctica is a permanent center of high pressure known as the *polar high*. It contrasts strongly with the encircling subantarctic low.

The pressure belts shift annually through

a *cyclone;* rotation around a center of high pressure is an *anticyclone*. Cyclones and anticyclones may remain in one location, or they may be rapidly moving pressure centers such as those that create the weather disturbances described in Chapter 6.

For surface winds, which move obliquely across the isobars, the systems for cyclones and anticyclones in both hemispheres are shown in Figure 4.9. Winds in a cyclone in the northern hemisphere show a counterclockwise inspiral. In an anticyclone there is a clockwise outspiral. Note the reversal between the labels "counterclockwise" and "clockwise" in the southern hemisphere.

In both hemispheres the surface winds spiral inward on the center of the cyclone, so that the air is converging on the center and must also rise to be disposed of at higher levels. For the anticyclone, by contrast, surface winds spiral out from the center. This motion represents a diverging of air flow and must be accompanied by a sinking (subsidence) of air in the center of the anticyclone to replace the outmoving air.

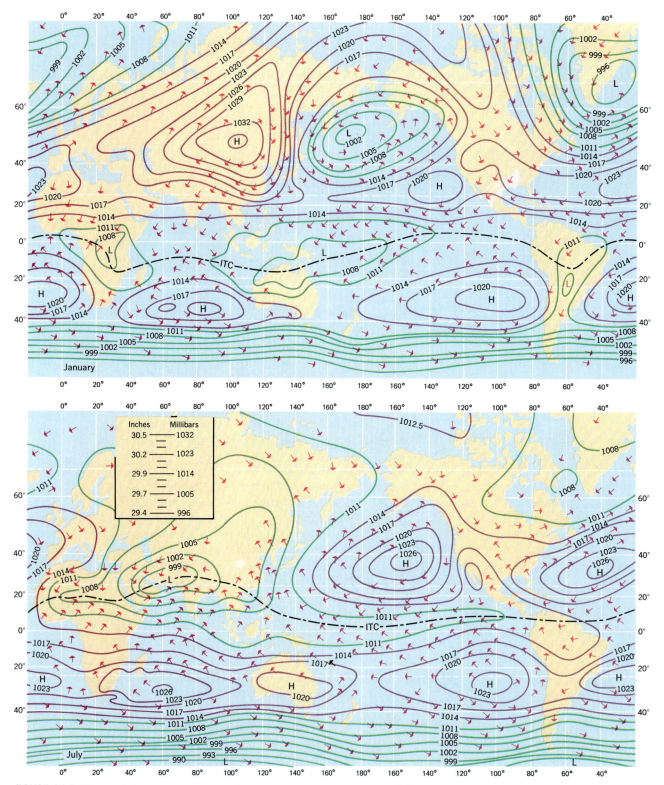

FIGURE 4.10 Mean monthly atmospheric pressure and prevailing surface winds for January and July. Pressure units are millibars reduced to sea level. Many of the wind arrows are inferred from isobars. (Data compiled by John E. Oliver.)

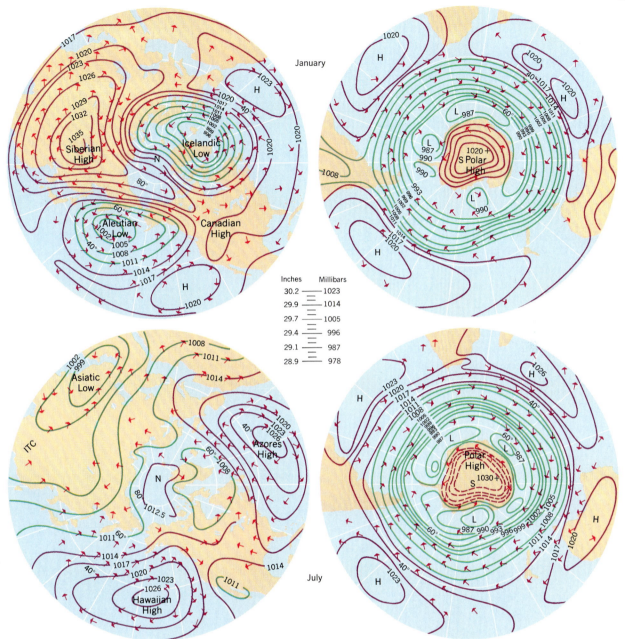

January

July

Inches   Millibars

30.2 —— 1023
29.9 —— 1014
29.7 —— 1005
29.4 —— 996
29.1 —— 987
28.9 —— 978

Mb	948	952	956	960	964	968	972	976	980	984	988	992	996	1000	1004	1008
In	28.0	28.1	28.2	28.3	28.4	28.5	28.6	28.7	28.8	28.9	29.0	29.1	29.2	29.3	29.4	29.5	29.6	29.7	29.8

Mb	996	1000	1004	1008	1012	1016	1020	1024	1028	1032	1036	1040	1044	1048	1052	1056
In	29.4	29.5	29.6	29.7	29.8	29.9	30.0	30.1	30.2	30.3	30.4	30.5	30.6	30.7	30.8	30.9	31.0	31.1	31.2

several degrees of latitude, along with the isotherm belts. These annual pressure belt migrations are important in causing seasonal climate changes. We will have several occasions to refer to these effects in analyzing world climates.

Northern Hemisphere Pressure Centers

The vast continents of North America and Eurasia and the intervening North Atlantic and North Pacific oceans exert a powerful control over pressure conditions in the northern hemisphere. As a result, the belted arrangement typical of the southern hemisphere is absent.

In winter the large, very cold land areas develop high-pressure centers. At the same time, intense low-pressure centers form over the warmer oceans. Over north central Asia in winter we find the Siberian high, with pressure exceeding 1030 mb (Figure 4.10). Over central North America there is

a clearly defined, but much less intense, ridge of high pressure, called the Canadian high. Over the oceans are the Aleutian low and the Icelandic low, named after the localities over which they are centered. These two low-pressure areas have much cloudy, stormy weather in winter.

In summer, pressure conditions are essentially the opposite of winter conditions. In summer the land areas develop low-pressure centers because at this season land-surface temperatures rise sharply above temperatures over the adjoining oceans. At the same time, the ocean areas develop strong centers of high pressure. This system of pressure opposites is striking on both the January and July isobaric maps (Figure 4.10). The low in Asia is intense: it is centered in southern Asia. Over the Atlantic and Pacific oceans are two large, strong cells of the subtropical belt of high pressure. They have shifted northward of their winter position and are considerably expanded. They are called the Azores high

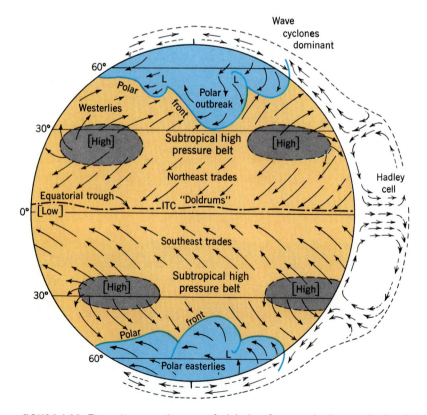

FIGURE 4.11 This schematic diagram of global surface winds disregards the disrupting effect of large continents in the northern hemisphere.

(or Bermuda high) and the Hawaiian high (Figure 4.10).

The Global Pattern of Surface Winds

Prevailing surface winds during January and July are shown by arrows on the pressure maps of Figure 4.10. The basic wind patterns are stressed in Figure 4.11, which is a highly diagrammatic representation showing the earth as if no land areas existed to modify the belted arrangement of pressure zones.

Let us begin with winds of the tropical zones. From the two subtropical high-pressure belts the pressure gradient is equatorward, leading down to the equatorial trough of low pressure. Following the simple model shown in Figure 4.8, air moving from high to low pressure is deflected by the Coriolis effect. As a result, two belts of *trade winds*, or *trades*, are produced. These are labeled in Figure 4.11 as the northeast trades and southeast trades. The trades are very persistent winds, deviating very little from a single compass direction. Sailing vessels traveling westward made good use of the trades.

The pattern of the trades suggests that they must converge somewhere near the equator. Meeting of the trades takes place within a narrow zone called the *intertropical convergence zone*. This long term is usually abbreviated to *ITC*. The position of the ITC is marked on the January and July maps (Figure 4.10). Converging winds require a rise of air to dispose of the incoming volume of air. This rise takes the form of stalklike flow columns carrying the air toward the top of the troposphere.

Along parts of the equatorial trough of low pressure at certain times of year the trades do not come together in convergence. Instead, there forms a belt of calms and variable winds, called the *doldrums*. Mariners on sailing ships knew that crossing the doldrums was hazardous because of the likelihood of lying becalmed for long periods of time.

The trades, doldrums, and ITC all shift seasonally north and south, along with the shifting of pressure belts and isotherms. The ITC migrates north and south only a

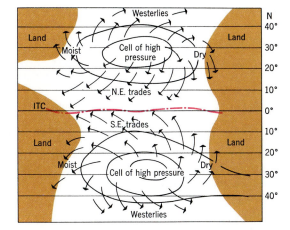

FIGURE 4.12 Over the oceans, surface winds spiral outward from the dominant cells of high pressure, feeding the trades and the westerlies.

few degrees of latitude over the Pacific and Atlantic oceans, but covers as much as 20 to 30° of latitude over South America, Africa, and the large region of Southeast Asia and the Indian Ocean. Important seasonal changes in winds, cloudiness, and rainfall accompany these migrations of the ITC and the trades.

We now return to the subtropical high-pressure belt, which ranges between lat. 25° and 40° N and S. Here we encounter large, stagnant, high-pressure cells (anticyclones). In the centers of the cells, winds are weak and are distributed around a wide range of compass directions; calms prevail as much as one quarter of the time. Because of the high frequency of calms, mariners named this belt the *horse latitudes*. It is said that the name originated in colonial times from the experiences of New England traders carrying cargoes of horses to the West Indies. When their ships were becalmed for long periods of time, the freshwater supplies ran low and the horses had to be thrown overboard.

Figure 4.12 is a schematic map of large anticyclonic cells centered over the oceans in two hemispheres. Winds make an outspiraling pattern that feeds into the converging trades. On the western sides of the cells air flows poleward; on the eastern sides the flow is equatorward. These flows have a strong influence on the climates of adjacent continental margins. Dryness of climate is a dominant general characteristic of the subtropical high-pressure belt and its cells, and we will emphasize this feature again.

Between lat. 35° and 60° N and S is the belt of *prevailing westerly winds,* or *westerlies.* These surface winds are shown in Figure 4.11 as blowing from a southwesterly quarter in the northern hemisphere and from a northwesterly quarter in the southern hemisphere. This generalization is somewhat misleading, however, because winds from polar directions are frequent and strong. It is more nearly accurate to say that within the westerlies, winds blow from all directions of the compass, but that the westerly components are definitely predominant. Rapidly moving cyclonic storms are common in this belt.

In the northern hemisphere, landmasses disrupt the westerlies, but in the southern hemisphere, between lat. 40° and 60° S, there is an almost unbroken belt of ocean. Here the westerlies gather great strength and persistence. Sailors on the great clipper ships called these latitudes "roaring forties,"

"furious fifties," and "screaming sixties." This belt was extensively used by sailing vessels traveling eastward from the South Atlantic Ocean to Australia, Tasmania, New Zealand, and the southern Pacific Islands. From these places it was then easier to continue eastward around the world to return to European ports. Rounding Cape Horn was relatively easy on an eastward voyage but, in the opposite direction, in the face of prevailing stormy westerly winds, it was a most dangerous operation.

A wind system called the *polar easterlies* is often said to be characteristic of the arctic and polar zones (Figure 4.11). The concept is at best greatly oversimplified and is certainly misleading when applied to the northern hemisphere. Winds in these regions take a variety of directions, dictated by local weather disturbances. On the other hand, Antarctica is an ice-capped landmass resting squarely on the pole and surrounded by a vast expanse of ocean. Here the outward spiraling flow of polar easterlies seems to be a dominant feature of the circulation. The polar maps in Figure 4.10 show these easterly winds.

Monsoon Winds of Southeast Asia

The powerful control exerted by the great landmass of Asia on air temperatures and pressures extends to the surface wind systems as well. In summer, southern Asia develops a cyclone into which there is a strong flow of air (July map, Figure 4.13). From the Indian Ocean and the southwestern Pacific, warm, humid air moves northward and northwestward into Asia, passing over India, Indochina, and China. This air flow constitutes the *summer monsoon* and is accompanied by heavy rainfall in southeastern Asia.

In winter, Asia is dominated by a strong center of high pressure from which there is an outward flow of air reversing that of the summer monsoon (January map, Figure 4.13). Blowing southward and southeastward toward the equatorial oceans, this *winter monsoon* brings dry weather for a period of several months.

North America does not have the remarkable extremes of monsoon winds experienced by southeastern Asia but, even

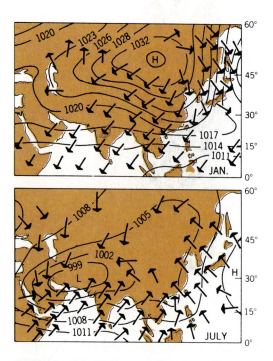

FIGURE 4.13 The Asiatic monsoon winds alternate in direction from January to July in response to reversals of barometric pressure over the large continent.

so, there is a distinct alternation of average temperature and pressure conditions between winter and summer. Wind records show that in summer there is a prevailing tendency for air originating in the Gulf of Mexico to move northward across the central and eastern part of the United States, whereas in winter there is a prevailing tendency for air to move southward from high-pressure sources in Canada. Wind arrows in Figure 4.10 show this seasonal alternation in the airflow pattern.

Local Winds

In certain localities, *local winds* are generated by immediate influences of the surrounding terrain rather than by the large-scale pressure systems that produce global winds and large traveling storms. Local winds are of environmental importance in various ways. They can exert a powerful stress on animals and plants when the air is dry and extremely hot or cold. Local winds are also important in affecting the movement of atmospheric pollutants.

One class of local winds—sea breezes and land breezes—was explained earlier in this chapter. The cooling sea breeze (or lake breeze) of summer is an important environmental resource of coastal communities, since it adds to the attraction of the shore zone as a recreation site.

Mountain winds and *valley winds* are local winds following a daily alternation of direction in a manner similar to the land and sea breezes. During the day, air moves from the valleys, upward over rising mountain slopes, toward the summits. At this time hill slopes are intensely heated by the sun. At night the air then moves valleyward, down the hill slopes, which have been cooled at night by radiation of heat from ground to air. These winds are responding to local pressure gradients set up by heating or cooling of the lower air.

Still another group of local winds are known as *drainage winds,* in which cold air flows under the influence of gravity from higher to lower regions. Such cold, dense air may accumulate in winter over a high plateau or high interior valley. When

general weather conditions are favorable, some of this cold air spills over low divides or through passes to flow out on adjacent lowlands as a strong, cold wind.

Drainage winds occur in many mountainous regions of the world and go by various local names. The *mistral* of the Rhone valley in southern France is a well-known example; it is a cold, dry local wind. On the ice sheets of Greenland and Antarctica, powerful drainage winds move down the gradient of the ice surface and are funneled through coastal valleys to produce powerful blizzards lasting for days at a time.

Another type of local wind occurs when the outward flow of dry air from a strong high-pressure center (an anticyclone) is combined with local effects of mountainous terrain. An example is the *Santa Ana,* a hot, dry easterly wind that, on occasion, blows from the interior desert region of southern California across coastal mountain ranges to reach the Pacific coast. Locally, this wind is funneled through narrow mountain gaps or canyon floors where it gains great force. At times the Santa Ana wind carries large amounts of dust. It is greatly feared for its ability to fan intense brush fires out of control.

Still another type of local wind, sometimes bearing the name *chinook,* results when strong regional winds passing over a mountain range are forced to descend on the lee side with the result that the air is heated and dried. This class of wind is explained in Chapter 5.

Winds Aloft

The surface wind systems we have examined represent only a shallow basal air layer a few hundred meters deep, whereas the troposphere is several kilometers deep. How does air move at these higher levels? Large, slowly moving high-pressure and low-pressure systems are found aloft, but these are generally simple in pattern, with smoothly curved isobars.

Winds high above the earth's surface are not affected by friction with the ground or water over which they move. The Coriolis effect turns the flow of air until it becomes parallel with the isobars, as Figure 4.14

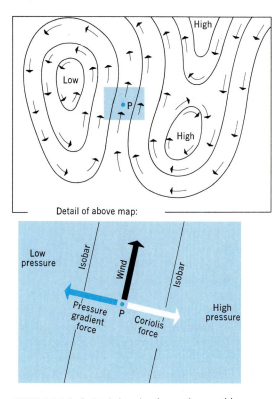

FIGURE 4.14 At high levels above the earth's surface, the wind blows parallel with the isobars. For the northern hemisphere, shown here, lower pressure is always toward the left of the direction of motion, whereas higher pressure is toward the right.

shows. (The Coriolis effect is shown as if it were a true force.) In this position the pressure gradient force and Coriolis force are exactly opposed and are exactly balanced.

The upper part of Figure 4.14 is a simplified map of pressure and winds high in the troposphere. Notice how the wind arrows run parallel with the isobars, forming circular flow patterns around lows and highs. Our rules for winds in cyclones and anticyclones need to be slightly modified, as compared with surface winds. Figure 4.15 shows the flow patterns for both hemispheres. For an airline pilot flying in the northern hemisphere and keeping a tailwind at all times, the rule would be "keep the highs on your right and the lows on your left."

The Global Circulation at Upper Levels

The general pattern of upper-air flow is sketched in Figure 4.16. Two systems dominate. One is the system of *upper-air westerlies* blowing in a complete circuit about the earth from about lat. 25° almost to the poles. At high latitudes these westerlies form a huge circumpolar vortex, coinciding with a great polar low-pressure center.

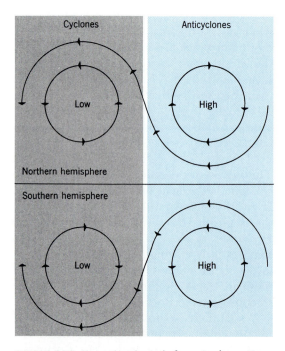

FIGURE 4.15 Upper-level winds form circular patterns around cyclones and anticyclones, but directions are reversed in the two hemispheres.

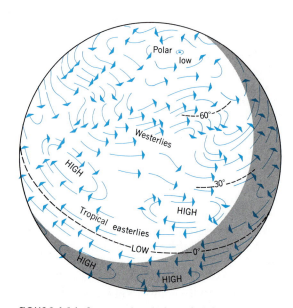

FIGURE 4.16 A generalized plan of global winds high in the troposphere.

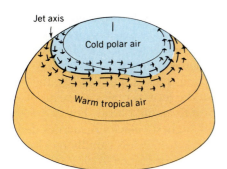

(A) The jet stream begins to undulate.

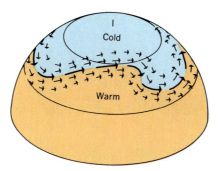

(B) Rossby waves begin to form.

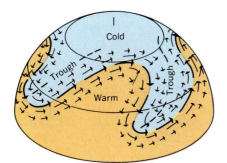

(C) Waves are strongly developed. The cold air occupies troughs of low pressure.

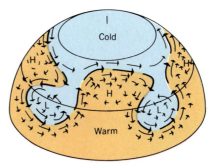

(D) When the waves are pinched off, they form cyclones of cold air.

FIGURE 4.17 Development of upper-air waves in the westerlies. (Data from J. Namias, NOAA, National Weather Service; from A. N. Strahler, *The Earth Sciences*, 2nd ed., Harper & Row, New York. Copyright © 1971 by Arthur N. Strahler.)

Toward lower latitudes the pressure rises steadily at a given altitude, forming high-pressure ridges at lat. 15° to 20° N and S. These are the high-altitude parts of the surface subtropical highs, but they are shifted somewhat equatorward.

Between the high-pressure ridges is a trough of weak low pressure in which the winds are easterly. These winds comprise the second major circulation system of the troposphere. They are called the *tropical easterlies*.

Seen in meridional cross section (Figure 4.11), circulation in equatorial and tropical latitude zones resolves itself into two circuits, one in each hemisphere. Heated air rises over the equatorial zone but subsides in the subtropical cells, forming the *Hadley cell*. Some of the subsiding air escapes poleward into the westerlies. This cellular circulation of low latitudes is basically a heat engine of the type described earlier in our explanation of sea and land breezes.

Rossby Waves and the Jet Stream

The uniform flow of the upper-air westerlies is frequently disturbed by the formation of large undulations, called *Rossby waves*. As shown in detail in Figure 4.17, these waves grow in amplitude and finally are cut off. The waves develop in a zone of contact between cold, polar air and warm, tropical air.

It is by means of the upper-air waves that warm air of low latitudes is carried far north at the same time that cold air of polar regions is brought equatorward. In this way horizontal mixing develops on a vast scale

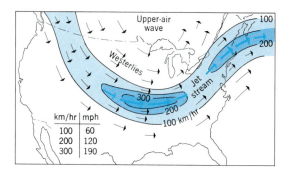

FIGURE 4.18 The jet stream is shown on this map by lines of equal wind speed. (National Weather Service.)

and provides heat exchange between the equatorial region of energy surplus and the polar regions of energy deficit.

Associated with the development of such upper-air waves at altitudes of 10 to 12 km (30,000 to 40,000 ft) are narrow zones in

which wind streams attain velocities up to 350 to 450 km (200 to 250 mi) per hour. This phenomenon is a *jet stream.* It consists of pulselike movements of air following broadly curving tracks (Figure 4.18). In cross section the jet resembles a stream of water moving through a hose. The centerline of highest velocity is surrounded by concentric zones of less rapidly moving fluid, as pictured in Figure 4.19.

The jet stream we have shown in Figure 4.17 as located along the fluctuating southern boundary of the great body of cold polar air is designated the *polar-front jet stream* (or, simply, the "polar jet stream"). In winter, it occupies a position generally between about 35° and 65° N. The polar-front jet stream is an important factor in the operation of jet aircraft in the range of their normal cruising altitudes. In addition to strongly increasing or decreasing the ground speed of the aircraft, the jet stream carries a form of air turbulence that at times reaches hazardous levels. This is clear air turbulence (CAT); it is avoided when known to be severe.

A second important jet stream forms in the subtropical latitude zone. Called the *subtropical jet stream,* it too flows from west to east. Its winter position fluctuates between 20° and 35° N. Cloud bands of the subtropical jet stream are pictured in Figure 4.20.

Temperature Layers of the Oceans

As with the atmosphere, the ocean has a layered structure. Ocean layers are recognized in terms of temperature. In the troposphere, air temperatures are generally highest at ground level and diminish upward. In the oceans, temperatures are generally highest at the sea surface and decline with depth. This trend is to be expected, since the source of heat is from the sun's rays and from heat supplied by the overlying atmosphere.

With respect to temperature, the ocean presents a layered structure in cross section, as shown in Figure 4.21. At low latitudes throughout the year and in middle latitudes in the summer, a warm surface layer develops. Here wave action mixes heated surface water with the water below it to give

FIGURE 4.19 A pictorial representation of the jet stream. (National Weather Service.)

FIGURE 4.20 A strong subtropical jet stream is made visible in this photo through a narrow band of cirrus clouds generated by the turbulent air stream. This jet stream is moving from west to east at an altitude of about 12 km (40,000 ft). The cloud band lies at about lat. 25° N. In this view, astronauts aimed their camera toward the southeast, taking in the Nile River valley and the Red Sea. At the left we see the tip of the Sinai Peninsula. (NASA.)

a warm layer that may be as thick as 500 m (1600 ft) with a temperature of 20° to 25°C (70° to 80°F) in oceans of the equatorial belt. Below the warm layer temperatures drop rapidly in a zone known as the *thermocline*. Below the thermocline is a layer of very cold water extending to the deep ocean floor. Temperatures near the base of the deep layer range from 0° to 5°C (32° to 40°F). In arctic and antarctic regions, the layer system is replaced by a single layer of cold water, as Figure 4.21 shows.

Surface currents of the oceans are broadly divided into warm currents, within the warm layer above the thermocline, and cold currents of the cold surface layer in arctic and polar latitudes.

Ocean Currents

An *ocean current* is any persistent, dominantly horizontal flow of ocean water. Ocean currents are important regulators of

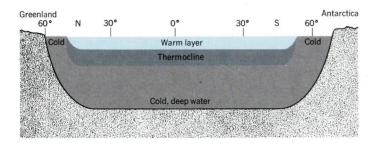

FIGURE 4.21 A schematic north–south cross section of the world ocean shows that the warm surface water layer disappears in arctic latitudes, where very cold water lies at the surface.

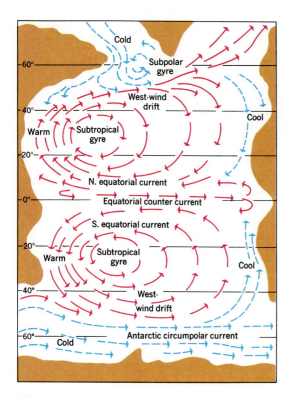

FIGURE 4.22 Two great gyres, one in each
hemisphere, dominate the circulation of shallow
ocean waters.

thermal environments at the earth's surface.
On a global scale, the vast current systems
aid in exchange of heat between the low
and high latitudes and are essential in
sustaining the global energy balance. On a
local scale warm water currents bring a
moderating influence to coasts in arctic
latitudes; cool currents greatly alleviate the
heat of tropical deserts along narrow coastal
belts.

Practically all the important surface
currents of the oceans are set in motion by
prevailing surface winds. Energy is
transferred from wind to water by the
frictional drag of the air blowing over the
water surface. Because of the Coriolis
effect, the water drift is impelled toward the
right of its path of motion (northern
hemisphere). Therefore, the current at the
water surface is in a direction about 45° to
the right of the wind direction.

To illustrate global surface water
circulation, we can refer to an idealized
ocean extending across the equator to

latitudes of 60° or 70° on either side (Figure
4.22). Perhaps the most outstanding
features are the circular movements, called
gyres, around the subtropical highs. The
gyres are centered about lat. 25° to 30° N
and S. An *equatorial current* with westward
flow marks the belt of the trades. Although
the trades blow to the southwest and
northwest, obliquely across the parallels of
latitude, the water movement follows the
parallels. The equatorial currents are
separated by an equatorial countercurrent.
A slow, eastward movement of water over
the zone of the westerlies is named the *west-
wind drift.* It covers a broad belt between lat.
35° and 45° in the northern hemisphere
and between lat. 30° and 60° in the
southern hemisphere. A world map, Figure
4.23 shows these ocean currents in greater
detail.

Along the west sides of the oceans in low
latitudes the equatorial current turns
poleward, forming a warm current
paralleling the coast. Examples are the Gulf
Stream (Florida stream or Caribbean
stream) and the Japan current (Kuroshio).
The currents bring higher than average
temperatures along these coasts.

The west-wind drift, upon approaching
the east side of the ocean, is deflected both
south and north along the coast. The
equatorward flow is a cool current,
accompanied by upwelling of colder water
from greater depths. It is well illustrated by
the Humboldt current (Peru current) off
the coast of Chile and Peru; by the
Benguela current off the southwest African
coast; by the California current off the west
coast of the United States; and by the
Canaries current off the Spanish and North
African coast.

In the northeastern Atlantic Ocean, the
west-wind drift is deflected poleward as a
relatively warm current. This is the North
Atlantic current, which spreads around the
British Isles, into the North Sea, and along
the Norwegian coast. The Russian port of
Murmansk, on the arctic circle, has year-
round navigability by way of this coast.

In the northern hemisphere, where the
polar sea is largely landlocked, cold water
flows equatorward along the west side of
the large straits connecting the Arctic Ocean
with the Atlantic basin. One example of a
cold current is the Labrador current,

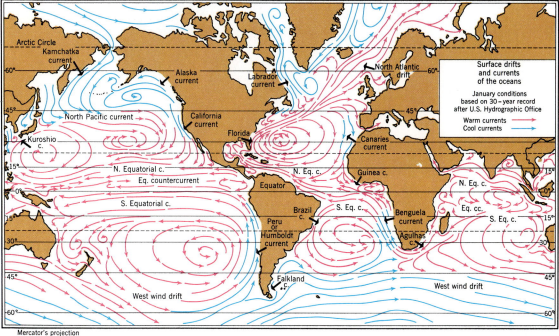

FIGURE 4.23 Surface drifts and currents of the oceans in January. (U.S. Navy Oceanographic Office.)

moving south from the Baffin Bay area through Davis Strait to reach the coasts of Newfoundland, Nova Scotia, and New England.

In both the North Atlantic and North Pacific oceans the Icelandic and Aleutian lows coincide in a very rough way with centers of counterclockwise circulation involving the cold arctic currents and the west-wind drifts. This type of center is labeled "subpolar gyre" in Figure 4.22.

The antarctic region has a relatively simple current scheme consisting of a single antarctic circumpolar current moving eastward around the antarctic continent in lat. 50° to 65° S, in a continuous expanse of open ocean.

Wind Power, Wave Power, and Current Power

Wind power is an indirect form of solar energy that has been used for centuries. The windmill of the Low Countries of Europe played a major role in pumping water from the polders as they were reclaimed from tidal land. The windmill was also used to grind grain in low, flat areas where streams could not be adapted to waterpower.

The design of new forms of windmills has intrigued inventors for many decades. The total supply of wind energy is enormous. The World Meteorological Organization has estimated that the combined electrical-generating wind power of favorable sites throughout the world comes to about 20 million megawatts, a figure about 100 times greater than the total electrical generating capacity of the United States. Many problems must be solved, however, in developing wind power as a major resource.

Small wind turbines are a promising source of supplementary electric power for individual farms, ranches, and homes. The Darrieus rotor, with circular blades turning on a vertical axis, is well adapted to small generators—those with less than 50 kilowatts output. Wind turbines presently in operation and being developed are adjusted

FIGURE 4.24 A windfarm in San Gorgonio Pass, near Palm Springs, California. Wind speeds here average 27 km (17 mi) per hour. (A. N. Strahler.)

in scale according to the purposes they serve. Turbines capable of producing power in the range of 50 to 200 kilowatts are already in operation to serve small communities.

Wind turbines with a generating capacity in the range from 50 to 100 kilowatts have been assembled in large numbers at favorable locations to form "windfarms." Arranged in rows along ridge crests, the turbines intercept local winds of exceptional frequency and strength. One such locality is the Tehachapi Pass in California, where winds moving southeastward in the San Joaquin Valley are funneled through a narrow mountain gap before entering the Mohave Desert. The San Gorgonio Pass, near Palm Springs, presents a similar situation favorable to the development of windfarms (Figure 4.24). Another important windfarm locality lies about 80 km (50 mi) east of San Francisco in the Altamont Pass area of Alameda and Contra Costa counties. Here daytime westerly winds of great persistence develop in response to the buildup of surface low pressure in the Great Valley to the east. A group of windfarms built here with a total of about 3000 turbines provides the Pacific Gas and Electric Company with a yield of over 500 million kilowatt-hours of electricity per year. Larger wind turbines, each capable of generating from 2 to 4 megawatts, have been constructed at a number of sites.

Wave energy is another indirect form of solar energy. Nearly all ocean waves are produced by the stress of wind blowing over the sea surface. Water waves are characterized by motion of water particles in vertical orbits. There is only a very slow downwind motion of the water, but a large flow of kinetic energy travels rapidly with the troughs and crests of the moving wave forms. Energy is extracted from wave motion by use of a floating object tethered to the seafloor. As the floating mass rises and falls, a mechanism is operated and drives a generator. Pneumatic systems use the principle of the bellows, operated by the changing pressure of the surrounding water as the water level rises and falls. A number of such devices are in the planning and testing stages.

Harnessing the vast power of a great current stream, such as the Gulf Stream or the Kuroshio, is a prospect that has not gone unnoticed. One rather grandiose plan, called the Coriolis Program after the effect that concentrates the Gulf Stream flow against the continental margins, proposes a large number of current-driven turbines to be tethered to the ocean floor so as to operate below the ocean surface. Each turbine would be 170 m (550 ft) in diameter and capable of generating 83 megawatts of power. An array of 242 such turbines could deliver 10,000 megawatts—a large part of Florida's power requirement and equivalent to the use of about 130 million barrels of crude oil per year.

Global Circulation and the Human Environment

We have outlined the major mechanical circuits within which sensible heat is transported from low latitudes to high latitudes. In low latitudes, the Hadley cell operates like a simple heat engine to transport heat from the equatorial zone to the subtropical zone. Upper-air waves take up the transport and move warm air poleward in exchange for cold air. Ocean currents perform a similar function through the turning of the great gyres.

The global atmospheric circulation also transports heat in the latent form held by water vapor. This heat is released by condensation, a process we will examine in the next chapter. The movement of water vapor also represents a mass transport of water and is a part of the world water balance.

Wind systems of the lower troposphere have direct environmental significance. Arriving at a mountainous coastal zone after a long travel path over a great ocean, these winds carry a large amount of water vapor, which is deposited as precipitation on the coast. In this way the distribution of our water resources is partly determined by the atmospheric circulation patterns. Winds also transport atmospheric pollutants, carrying them tens and hundreds of kilometers from the sources of pollution. These are environmental topics related to winds; we will investigate them in the next two chapters.

ATMOSPHERIC MOISTURE AND PRECIPITATION

ATMOSPHERIC MOISTURE   Our fresh water comes from the troposphere, where water in the vapor state passes into the liquid or solid state.

HUMIDITY   As air temperature changes, the capacity of the air to hold water vapor also changes. If sufficiently cooled, the air becomes saturated and condensation of the water vapor can result.

THE ADIABATIC PROCESS   A drop in temperature of air as it rises—the adiabatic effect—produces clouds and precipitation on a large scale.

CLOUDS   They come in different shapes—stratiform and cumuliform—each revealing a special pattern of atmospheric motion.

PRECIPITATION   Atmospheric water reaching the earth's surface takes various forms—rain, snow, sleet, hail.

THUNDERSTORMS   They are powerful updrafts in convection cells. Hail, lightning, and squall winds are environmental hazards associated with thunderstorms.

OROGRAPHIC PRECIPITATION   A mountain barrier generates precipitation by forcing moist air to rise. Deserts form on the opposite side of the barrier.

WE have stressed that heat and water are vital ingredients of the environment of the biosphere, or life layer. Plant and animal life of the lands, on which humans depend for much of their food, require fresh water. People use fresh water in many ways. The only basic source of fresh water is from the atmosphere through condensation of water vapor. In this chapter we are concerned mostly with water in the vapor state in the atmosphere and the processes by which it passes into the liquid or solid state and ultimately arrives at the surface of the ocean and the lands through the process of precipitation.

Water also leaves the land and ocean surfaces by evaporation and so returns to the atmosphere. Evidently, the global pathways of movement of water form a complex network. There is a global water balance, just as there is an energy balance; the water balance deals with flow of matter and so complements the energy balance.

At this point, you may wish to review the explanation of the three states of matter, presented in the Appendix.

Humidity

The amount of water vapor that may be present in the air at a given time varies widely from place to place. It ranges from almost nothing in the cold, dry air of arctic regions in winter to as much as 4 or 5 percent of a given volume of the atmosphere in the humid equatorial zone.

The general term *humidity* refers to the amount of water vapor present in the air. At any specified temperature, the quantity of moisture that can be held by the air has a definite limit. This limit is the *saturation point.* The proportion of water vapor present relative to the maximum quantity is the *relative humidity;* it is expressed as a percentage. At the saturation point, relative humidity is 100 percent; when half of the total possible quantity of vapor is present, relative humidity is 50 percent, and so on.

A change in relative humidity of the atmosphere can be caused in one of two ways. If an exposed water surface is present, the humidity can be increased by evaporation. The process is slow because it requires that the water vapor diffuse upward through the air. The second way is through a change of temperature. Even though no water vapor is added, a lowering of temperature results in a rise of relative humidity. This change is automatic because the capacity of the air to hold water vapor is lowered by cooling. After cooling, the existing amount of vapor represents a higher percentage of the total capacity of the air. Similarly, a rise of air temperature results in decreased relative humidity, even though no water vapor has been taken away.

The principle of relative humidity change caused by temperature change is illustrated by a graph of these two properties throughout the day (Figure 5.1). As air temperature rises, relative humidity falls, and vice versa.

A simple example may help to illustrate these principles (Figure 5.2). At 10 A.M., the air temperature is 16°C (60°F), and the relative humidity is 50 percent. By 3 P.M., the air has become warmed by the sun to 32°C (90°F). The relative humidity has automatically dropped to 20 percent, which is very dry air. Next, the air becomes chilled

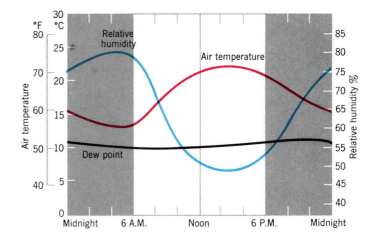

FIGURE 5.1 Relative humidity, air temperature, and dew-point temperature throughout the day at Washington, D.C. The curves show average values for the month of May. (Data from National Weather Service.)

during the night and, by 4 A.M., its temperature has fallen to 5°C (40°F). Now the relative humidity has automatically risen to 100 percent, and the air is saturated. Any further cooling will cause *condensation* of the excess vapor into liquid or solid form. As the air temperature continues to fall, the humidity remains at 100 percent, but condensation continues. This may take the form of minute droplets of dew or fog. If the temperature falls below freezing, condensation occurs as frost on exposed surfaces.

Dew point is the critical temperature at which air becomes saturated during cooling.

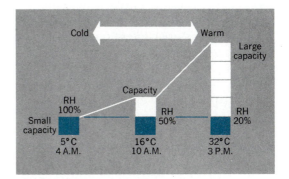

FIGURE 5.2 Relative humidity changes with temperature because capacity of warm air is greater than that of cold air.

Below the dew point, condensation usually sets in. An excellent illustration of condensation caused by cooling is seen in the summer, when beads of moisture form on the outside surface of a pitcher or glass filled with ice water. Air immediately adjacent to the cold glass or metal surface is chilled enough to fall below the dew-point temperature, causing moisture to condense on the surface of the glass.

Relative humidity is measured by an instrument called a *hygrometer.* Figure 5.3 illustrates a simple, homemade hygrometer in which a strand of human hairs is attached to the end of a pointer. The hairs change length in response to changes in relative humidity and, in this way, move the pointer up or down. In a more sophisticated instrument, the hygrograph, the strand of hairs or other fiber operates a pen point and makes a continuous record on graph paper wrapped around a slowly turning drum.

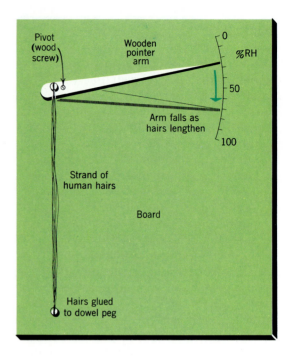

FIGURE 5.3 A simple hygrometer.

Specific Humidity

Although relative humidity is an important indicator of the state of water vapor in the air, it is a statement only of the relative quantity present compared to a saturation quantity. The actual quantity of moisture present is denoted by *specific humidity,* defined as the mass of water vapor contained in a given mass of air. Mass of water is given in grams; the unit mass of air is the kilogram. Specific humidity is stated in terms of grams per kilogram (gm/kg). For any specified air temperature, there is a maximum mass of water vapor that a kilogram of air can hold (the saturation quantity). Figure 5.4 is a graph showing this maximum moisture content of air for a wide range of temperatures.

Specific humidity is often used to describe the moisture characteristics of a large mass of air. For example, extremely cold, dry air over arctic regions in winter may have a specific humidity of as low as 0.2 gm/kg, whereas extremely warm moist air of equatorial regions often holds as much as 18 gm/kg. The total natural range on a worldwide basis is such that the largest values of specific humidity are from 100 to 200 times as great as the least. Figure 5.5 is

a graph showing how relative humidity and specific humidity vary with latitude. Notice that the relative humidity curve has two saddles, one over each of the subtropical high-pressure belts where the world's tropical deserts are found. Relative humidity is high in equatorial and arctic

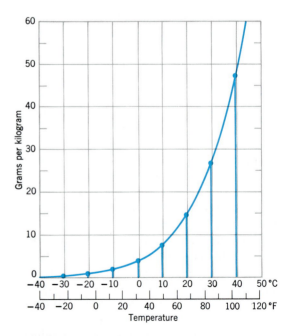

FIGURE 5.4 Maximum specific humidity of a mass of air increases sharply with rising temperature.

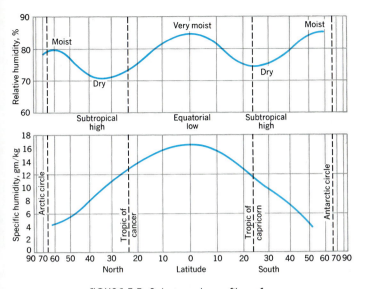

FIGURE 5.5 Pole-to-pole profiles of average humidity (above) and of specific humidity (below). (Data of Haurwitz and Austin.)

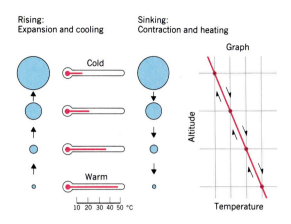

FIGURE 5.6 A schematic diagram of adiabatic cooling and heating accompanying the rising and sinking of a mass of air. (A. N. Strahler.)

zones. In contrast, the specific humidity curve has a single peak, near the equator, and declines toward high latitudes.

In a real sense, specific humidity is a geographer's yardstick of a basic natural resource—water—to be applied from equatorial to polar regions. It is a measure of the quantity of water that can be extracted from the atmosphere as precipitation. Cold air can supply only a small quantity of rain or snow; warm air is capable of supplying very large quantities.

The Adiabatic Process

Falling rain, snow, sleet, or hail are referred to collectively as *precipitation*. Only where large masses of air are experiencing a steady drop in temperature below the dew point can precipitation occur in appreciable amounts. Precipitation cannot be brought about by the simple process of chilling of the air through loss of heat by longwave radiation during the night. Precipitation requires that a large mass of air be rising to higher elevations.

One of the most important laws of meteorology is that rising air experiences a drop in temperature, even though no heat energy is lost to the outside. The drop of temperature is a result of the decrease in air pressure at higher elevations, permitting the rising air to expand. Because individual molecules of the gas are more widely diffused and do not strike one another so frequently, the sensible temperature of the expanding gas is lowered. The temperature drop is described as an *adiabatic process*, which simply means "occurring without any gain or loss of heat." In the adiabatic process heat energy as well as matter remain within the system. The process is thus completely reversible. Expansion always results in cooling. Compression, occurring when air descends to a lower altitude, always results in warming (Figure 5.6).

Within a rising body of air the rate of drop of temperature, termed the *dry adiabatic lapse rate*, is about 10 C° per 1000 m of vertical rise. In English units the rate is 5½ F° per 1000 ft. The dry rate applies only when no condensation is taking place. The dew point also declines gradually with rise of air: the rate is 2 C° per 1000 m (1 F° per 1000 ft).

Adiabatic cooling rate should not be confused with the environmental lapse rate, explained in Chapter 3. The environmental lapse rate applies only to nonrising air in which the temperature is measured at successively higher levels.

Lapse rates can be shown on a simple graph in which altitude is plotted on the vertical scale and temperature on the

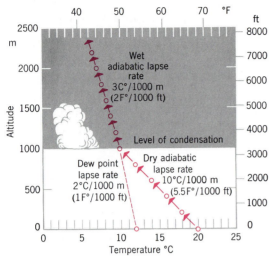

FIGURE 5.7 Adiabatic decrease of temperature in a rising mass of air leads to condensation of water vapor and the formation of a cloud.

energy of rapid motions of the gas molecules of water vapor. This heat amounts to about 600 calories for each gram of water. Originally, this stored energy was obtained during the process of *evaporation* at the time the water vapor entered the atmosphere.

Evaporation of water cools the liquid surface from which evaporation is taking place. In this manner, evaporation of perspiration cools the skin. The desert water bag also works on this principle. Evaporation of water seeping through the coarse flax cloth of the bag cools the remaining water. The heat drawn off by evaporation enters the atmosphere along with the water vapor in the form of the rapid motion of the gaseous water molecules. When these same molecules come to rest on a liquid surface as water, the energy of their motion turns into sensible heat.

Figure 5.8 shows the concept of change in state of water and the release or absorption of heat energy accompanying each change. Freezing of liquid water also results in the release of latent heat—about 80 calories for each gram of water. The left side of the diagram shows the process of *sublimation,* in which water vapor passes directly into the solid state as ice and the

horizontal scale (Figure 5.7). The chain of circles connected by arrows represents air rising as if it were a bubble. Suppose that a body of air near the ground has a temperature of 20°C (68°F) and that its dew-point temperature is 12°C (54°F); these are the conditions shown in Figure 5.7. If, now, a bubble of air undergoes a steady ascent, its air temperature will decrease much faster than its dew-point temperature. Consequently, the two lines on the graph are rapidly converging. At an altitude of 1000 m (3300 ft), the air temperature, now 10°C (50°F), will have met the dew-point temperature. The air bubble has now reached the saturation point. Further rise results in condensation of water vapor into minute liquid particles, and so a cloud is produced. The flat base of the cloud is a common visual indicator of the level of condensation.

Condensation and Latent Heat

As the bubble of saturated air continues to rise, further condensation takes place; but now a new principle comes into effect. When water vapor condenses, heat in the latent form is transformed into sensible heat, which is added to the existing heat content of the air. *Latent heat* represents

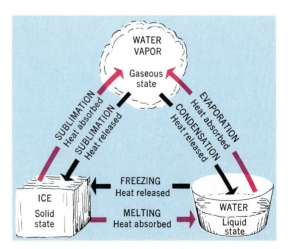

FIGURE 5.8 A schematic diagram of the three states of water. Arrows show the ways that any one state can change into either of the other two states. Heat energy is absorbed or released, depending on the direction of change.

reverse process, from ice directly into water vapor. The gradual disappearance of ice cubes stored in a frost-free freezer illustrates sublimation. Latent heat is also released or absorbed during sublimation, the amount being approximately equal to the sum of the latent heats released by condensation and freezing.

As condensation continues within a rising mass of air, the latent heat liberated by that condensation partly offsets the temperature drop by adiabatic cooling. As a result, the adiabatic rate is substantially reduced. The reduced rate, which ranges between 3 and 6 C° per 1000 m (2 and 3 F° per 1000 ft), is termed the *wet adiabatic lapse rate*. On the graph, this reduced rate is expressed by the more steeply inclined section of line above the level of condensation.

Cloud Particles

A *cloud* is a dense mass of suspended water or ice particles in the diameter range of 20 to 50 microns. Each cloud particle has formed on a *nucleus* of solid matter, which has a diameter in the range from $\frac{1}{10}$ to 1 micron. Nuclei of condensation must be present in large numbers, and they must be of such a composition as to attract water vapor molecules. In Chapter 3 we referred to these minute suspended particles collectively as atmospheric dust and noted that one source is the surface of the sea. Droplets of spray from the crests of waves are carried rapidly upward in turbulent air. Evaporation of the water leaves a solid residue of crystalline salt, which strongly attracts water molecules. Everyone is familiar with the way in which ordinary table salt becomes moist when exposed to warm humid air. Another example is calcium chloride, a commercial salt, widely used to control dust on dirt roads; it attracts atmospheric water vapor and keeps the soil damp, even in dry weather.

Although "clean air" is an environmental goal, the term is only relative, since all air of the troposphere is charged with dust. As a result, there is no lack of suitable condensation nuclei. As we will find in the discussion of air pollution, the heavy load of dust carried by polluted air over cities substantially aids in condensation and the formation of clouds and fog.

We are accustomed to finding that liquid water turns to ice when the surrounding

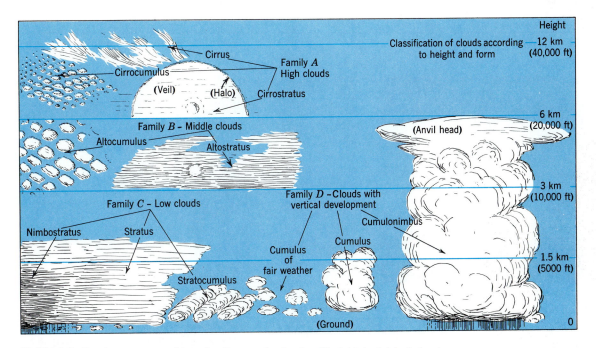

FIGURE 5.9 Clouds are grouped into families on the basis of height. Individual cloud types are named according to their form.

(A) Fibrous cirrus, or mares' tails, above; cumulus of fair weather, below. (Arthur N. Strahler.)

(B) Cirrus in broad, parallel bands, indicating a jet stream aloft. (G.R. Roberts.)

(C) Cirrocumulus forming a mackerel sky. (Jerome Wyckoff.)

(D) Altocumulus. (Jerome Wyckoff.)

(E) An altostratus layer, feathering out in the distance. (Arthur N. Strahler.)

FIGURE 5.10 Common cloud forms.

(F) A stratocumulus cloud deck seen from above. In the distance, a single active cumulus cloud is rising rapidly and, in time, may form a cumulonimbus cloud. (John S. Shelton.)

temperature falls to the freezing point 0°C (32°F) or below. Instead, water in such minute particles as those comprising clouds remains in the liquid state at temperatures far below freezing. Such water is described as *supercooled*. Clouds consist entirely of water droplets at temperatures down to about −12°C (10°F). Between −12° and −30°C (10° and −20°F), the cloud is a mixture of water droplets and ice crystals. Below −30°C (−20°F) the cloud consists predominantly of ice crystals; below −40°C (−40°F) all the cloud particles are ice crystals. Very high, thin clouds, formed at altitudes of 6 to 12 km (20,000 to 40,000 ft), are formed of ice particles.

Cloud Forms

Common cloud forms are pictured in Figures 5.9 and 5.10. Clouds are classified into families, arranged by height; these are named in Figure 5.9 along with individual types within each family. On the basis of form there are two major classes of clouds: *stratiform*, or layered clouds, and *cumuliform*, or globular clouds.

The stratiform clouds are blanketlike and cover large areas. A common type is *stratus*, which covers the entire sky. The important point about stratiform clouds is that they represent air layers being forced to rise gradually over stable underlying air layers of greater density. As forced rise continues, the rising layer of air is adiabatically cooled, and condensation is sustained over a large area. Dense, thick stratiform clouds can yield substantial amounts of rain or snow.

The cumuliform clouds are globular masses representing bubblelike bodies of warmer air spontaneously rising because they are less dense than the surrounding air. These clouds go by the name *cumulus* (Figure 5.9). Some cumuliform clouds are small masses no higher than wide; others develop tall, stalklike shapes and penetrate high into the troposphere. The tall, dense form is called *cumulonimbus* (see Figure 5.18). It is the thunderstorm, a weather disturbance with heavy rain and sometimes hail, to the accompaniment of lightning and thunder.

In later paragraphs we will explain the conditions under which both stratiform and cumuliform clouds are produced and the ways in which they yield their precipitation.

Fog

Fog is simply a cloud layer in contact with the land or sea surface or lying very close to the surface. Fog is a major environmental hazard in our industrialized world. Dense fog on high-speed highways is a cause of terrifying chain-reaction accidents, sometimes involving dozens of vehicles and often taking a heavy toll in injuries and deaths. Landing delays and shutdowns at major airports because of fog bring economic losses to airlines and inconvenience to thousands of travelers in a single day. For centuries fog at sea has been a navigational hazard. Today, with huge supertankers carrying oil, fog adds to the probability of ship collisions capable of creating enormous oil spills. Polluted fogs are a health hazard to urban dwellers and have at times taken a heavy toll in lives.

One type of fog, known as a *radiation fog*, is formed at night when temperature of the stagnant basal air falls below the dew point. This kind of fog is associated with low-level temperature inversion (Figure 3.8). Another fog type, *advection fog*, results from the movement of warm, moist air over a cold or snow-covered ground surface. As it loses heat to the ground, the air layer undergoes a drop of temperature below the dew point, and condensation sets in. A similar type of advection fog is formed over oceans where air from over a warm current blows across the cold surface of an adjacent cold current. Fogs of the Grand Banks off Newfoundland are largely of this origin because here the cold Labrador current comes in contact with warm waters of Gulf Stream origin.

A special type of fog is prevalent along the California coast. It forms within a cool marine air layer in direct contact with cold water of the California current (Figure 5.11). Similar fog conditions prevail close to all continental west coasts in the tropical latitude zone, where cool equatorward currents parallel the shoreline.

FIGURE 5.11 Coastal fog bank enveloping the Bay Bridge, San Francisco. (Frank Lambrecht/Berg & Associates.)

FIGURE 5.12 These individual snow crystals, greatly magnified, were selected for their variety and beauty. They were photographed long ago by W. A. Bentley, a Vermont farmer who devoted much of his life to snowflake photography. (National Weather Service.)

Precipitation Forms

During the rapid ascent of a mass of air in the saturated state, cloud particles grow rapidly and attain a diameter of 50 to 100 microns. They then coalesce through collisions and grow quickly into droplets of about 500 microns in diameter (about ¹⁄₅₀ in.). Droplets of this size reaching the ground constitute a drizzle, one of the recognized forms of precipitation. Further coalescence increases drop size and yields *rain*. Average raindrops have diameters of about 1000 to 2000 microns (¹⁄₂₅ to ¹⁄₁₀ in.), but they can reach a maximum diameter of about 7000 microns (¼ in.). Above this value they become unstable and break into smaller drops while falling.

One kind of rain forms directly by liquid condensation and droplet coalescence in warm clouds typical of the equatorial and tropical zones. However, much of the rain of middle and high latitudes is the product of the melting of snow as it makes its way to lower, warmer levels.

Snow is produced in clouds that are a mixture of ice crystals and supercooled water droplets. The falling crystals serve as nuclei to intercept water droplets. As these adhere, the water film freezes and is added to the crystalline structure (Figure 5.12). The crystals readily clot together to form larger snowflakes, and these fall more rapidly from the cloud. When the underlying air layer is below the freezing temperature, snow reaches the ground as a solid form of precipitation; otherwise, it will melt and arrive as rain. A reverse process, the fall of raindrops through a cold air layer, results in freezing of rain and produces pellets or grains of ice. These are commonly referred to in North America as *sleet*. (Among the British, sleet refers to a mixture of snow and rain.)

Hail, another form of precipitation, consists of large pellets or spheres of ice. The formation of hail will be explained in our discussion of the thunderstorm.

When rain falls on a frozen ground surface that is covered by an air layer of below-freezing temperature, the water freezes into clear ice after striking the ground or other surfaces, such as trees, houses, or wires (Figure 5.13). The coating of ice that results is called a *glaze*, and an *ice storm* is said to have occurred. Actually, no ice falls, so that ice glaze is not a form of precipitation. Ice storms cause great damage, especially to telephone and power wires and to tree limbs. Roads and sidewalks are made extremely hazardous.

FIGURE 5.13 Heavily coated wires and branches caused heavy damage in eastern New York State as a result of this icing storm (National Weather Service.)

FIGURE 5.14 This 10-cm (4-in.) rain gauge is made of clear plastic, allowing the amount of rain caught in the inner cylinder to be read from the outside. Rain exceeding the capacity of the inner cylinder overflows into the outer cylinder and can be measured after the rain period has ended. (Arthur N. Strahler.)

How Precipitation Is Measured

Precipitation is measured in units of depth of fall per unit of time; for example, centimeters or inches per hour or per day. One centimeter of rainfall is a quantity sufficient to cover the ground to a depth of 1 cm, provided that none is lost by runoff, evaporation, or sinking into the ground. A simple form of rain gauge can be operated merely by setting out a straight-sided, flat-bottomed pan and measuring the depth to which water accumulates during a particular period. Unless this period is short, however, evaporation seriously upsets the results.

A very small amount of rainfall, such as 2 mm (0.1 in.), makes too thin a layer to be accurately measured. To avoid this difficulty as well as to reduce evaporation loss, the *rain gauge* is made in the form of a cylinder whose base is a funnel leading into a narrow tube (Figure 5.14). A small amount of rainfall will fill the narrow pipe to a considerable height, thus making it easy to read accurately, once a simple scale has been provided for the pipe. This gauge requires frequent emptying unless it is equipped with automatic devices for this purpose.

Snowfall is measured by melting a sample column of snow and reducing it to an equivalent in water. In this way rainfall and snowfall records may be combined for purposes of comparison. Ordinarily, a 10-cm layer of snow is assumed to be equivalent to 1 cm of rainfall, but this ratio may range from 30 to 1 in very loose snow to 2 to 1 in old, partly melted snow.

Basic Causes of Precipitation

The most important basic cause of precipitation is the horizontal convergence of moist air toward a center or line of low barometric pressure. In Chapter 4 we discussed the convergence of air in a cyclone, pointing out that converging air at low levels can escape only by rising to higher altitude in the central part of the cyclone (see Figure 4.9). We also noted that outward flow (divergence) of air in an anticyclone must be accompanied by downsinking (subsidence) of air from higher altitudes over the center of the anticyclone. Thus convergence, accompanied by rise of air and its cooling by the adiabatic process, results in the formation of clouds and the production of precipitation. Divergence, accompanied by air subsidence, tends to dissipate cloudiness through the process of adiabatic warming, and thus prevents the occurrence of precipitation.

Convergence also takes place along a trough of low pressure (as distinct from a cyclone with its single center). An example on a global scope is convergence of the trade winds, or tropical easterlies, along the intertropical convergence zone (ITC) (see Figure 4.12). Here convergence leads to a general rise of air and results in a major zone of precipitation. Another global example is the subantarctic low, a trough of low pressure that encircles the globe over the Southern Ocean surrounding Antarctica. Here we find a belt of persistent cloudiness and storminess. The same principle can be applied to the Aleutian low and the Icelandic low in winter, when the lowest average pressures occur.

In Chapter 6 we will explain how moving cyclones and moving troughs of low pressure induce cloud formation and precipitation. These systems usually involve the forced lifting of a layer of warm, moist air over a denser layer of cold air.

A second basic cause of precipitation is the forced lift of moist air over a mountain barrier. This is a topic appropriate for study here, because we are examining the mechanism of precipitation in terms of the process of adiabatic cooling of a rising mass of air.

Orographic Precipitation

Precipitation generated by the forced ascent of moist air over a mountain barrier is referred to as *orographic precipitation*. The adjective "orographic" means "related to mountains." Figure 5.15 shows the steps associated with production of orographic precipitation. Moist air arrives at the coast after passing over a large ocean surface. As the air rises on the windward side of the range, it is cooled at the dry adiabatic rate. When cooling is sufficient, precipitation sets in. After passing over the mountain summit, the air begins to descend the lee side of the range. Now it undergoes compressional warming through the adiabatic process and, having no source from which to draw up moisture, the air becomes very dry. Air temperature increases at the dry adiabatic rate and, upon

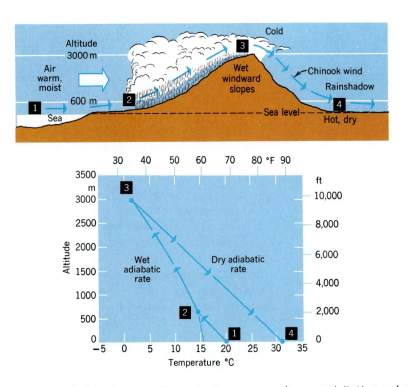

FIGURE 5.15 Forced ascent of oceanic air masses produces precipitation and a rainshadow desert. (From A. N. Strahler, *The Earth Sciences*, 2nd ed., Harper & Row, New York. Copyright © 1971 by Arthur N. Strahler.)

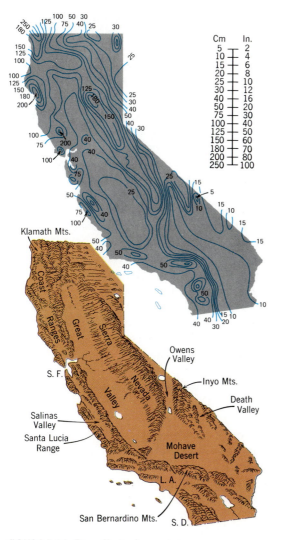

FIGURE 5.16 The effect of mountain ranges on precipitation is strong in the state of California. Isohyets in centimeters show the mean annual precipitation.

An excellent illustration of orographic precipitation and rainshadow occurs in the far west of the United States. A map of California, Figure 5.16, shows mean annual precipitation using lines of equal precipitation, called *isohyets*. Prevailing westerly winds bring moist air from the Pacific Ocean over the Coast Ranges of central and northern California and the great Sierra Nevada, whose summits rise to 4000 m (14,000 ft) above sea level. Heavy rainfall on the windward slopes of these ranges nourishes rich forests. Passing down the steep eastern face of the Sierra Nevada, air must descend nearly to sea level, even below sea level in Death Valley. The resulting adiabatic heating lowers the humidity, producing part of America's great desert zone, covering a strip of eastern California and all of Nevada.

Convectional Precipitation

When a body of warm, moist air is forced to rise because of convergence in a cyclone or along a trough of low pressure, or because of the orographic effect, a different mechanism often comes into play, radically changing the way in which the air rises to higher altitudes. This new mechanism is a spontaneous rise of air known as *convection* (or *convective activity*); it consists of strong updrafts taking place within a *convection cell*. Air rises in the cell because it is less dense than the surrounding air. Perhaps a fair analogy is the updraft of heated air in a chimney but, unlike the steady airflow in a chimney, air motion in a convection cell takes place in pulses as bubblelike masses of air rise in succession.

To illustrate the convection process, let us suppose that on a clear, warm summer morning the sun is shining on a landscape consisting of patches of open fields and woodlands. Certain of these types of surfaces, such as the bare ground, heat more rapidly and transmit radiant heat to the overlying air. Air over a warmer patch becomes warmed more than adjacent air and begins to rise as a bubble, much as a hot-air balloon rises after being released. Vertical movements of this type are often called "thermals" by sailplane pilots, who use them to obtain lift.

reaching sea level, it is much warmer than at the start. A belt of dry climate, often called a *rainshadow*, exists on the lee side of the range. Several great deserts of the earth are of this type.

Dry, warm *chinook winds* often occur on the lee side of a mountain range in the western United States. These winds cause extremely rapid evaporation of snow or soil moisture. Chinook winds result from the turbulent mixing of lower and upper air in the lee of the range. The upper air, which has little moisture to begin with, is greatly dried and heated when swept down to low levels.

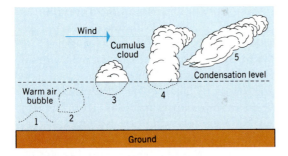

FIGURE 5.17 Rise of a bubble of heated air to form a cumulus cloud. (From A. N. Strahler, *The Earth Sciences*, 2nd ed., Harper & Row, New York. Copyright © 1971 by Arthur N. Strahler.)

As the air rises, it is cooled adiabatically so that eventually it is cooled below the dew point. At once condensation begins, and the rising air column appears as a cumulus cloud. The flat base shows the critical level above which condensation is occurring (Figure 5.17). The bulging "cauliflower" top of the cloud represents the top of the rising warm air column, pushing into higher levels of the atmosphere. Usually, the small cumulus cloud dissolves after drifting some distance downwind. Under a different set of atmospheric conditions, convection continues to develop and the cloud grows to a dense cumulonimbus mass, or thunderstorm, which yields heavy rain.

Why, you ask, does such spontaneous cloud growth take place and continue beyond the initial cumulus stage, long after the original input of heat energy is gone? Actually, the unequal heating of the ground served only as a trigger effect to release a spontaneous updraft, fed by latent heat energy liberated from the condensing water vapor. Recall that for every gram of water formed by condensation, 600 calories of heat are released. This heat acts like fuel in a bonfire.

Air capable of rising spontaneously during condensation is described as *unstable air*. In such air the updraft tends to increase in intensity as time goes on, much as a bonfire blazes with increasing ferocity as the updraft draws in greater supplies of oxygen. Of course, at very high altitudes, the bulk of the water vapor has condensed and fallen as precipitation, so that the

energy source is gone. When this happens the convection cell weakens and air rise finally ceases.

Unstable air, given to spontaneous convection in the form of heavy showers and thunderstorms, is most likely to be found in warm, humid areas, such as the equatorial and tropical zones throughout the year, and the midlatitude regions during the summer season.

Much orographic rainfall at low latitudes is actually of the convectional type in that it takes the form of heavy showers and thunderstorms. The convectional storms are set off by the forced ascent of unstable air as it passes over the mountain barrier. The torrential monsoon rains of the Asiatic and East Indian mountain ranges are largely of this type. For example, Cherrapunji, a hill station facing the summer monsoon air drift in northeast India, averages 1082 cm (426 in.) of rainfall annually.

Thunderstorms

Convection activity manifests itself in the *thunderstorm*, an intense local storm associated with a tall, dense cumulonimbus cloud in which there are very strong updrafts of air. Thunder and lightning normally accompany the storm, and rainfall is heavy, often of cloudburst intensity, for a short period. Violent surface winds may occur at the onset of the storm (Figure 5.18).

A single thunderstorm consists of individual convection cells. Air rises within each cell as a succession of bubblelike air bodies, instead of as a continuous updraft from bottom to top (Figure 5.19). As each bubble rises, air in its wake is brought in from the surrounding region. Precipitation in the thunderstorm cell can be in the form of rain in the lower levels, mixed water and snow at intermediate levels, and snow at high levels.

Upon reaching high levels, which may be 6 to 12 km (20,000 to 40,000 ft) or even higher, the rising rate diminishes, and the cloud top is dragged downwind to form an anvil top. Ice particles falling from the cloud top act as nuclei for condensation at lower levels, a process called *cloud seeding*. The rapid fall of raindrops adjacent to the

(A) Active thunderstorm cells producing cumulonimbus clouds over the southern Rocky Mountains. (Arthur N. Strahler.)

(B) Torrential rain beneath the flat base of a cumulonimbus cloud, North Rim of Canyon, Arizona. (Arthur N. Strahler.)

(C) Rain shower falling from the flat base of a thunderstorm cloud, Colorado. (Daniel Zirinsky/ Photo Researchers.)

(D) A succession of lightning strokes recorded by time exposure. (Deeks & Pribe/Science Source/ Photo Researchers.)

FIGURE 5.18 Thunderstorms, cumulonimbus clouds, and lightning.

rising air bubbles exerts a frictional drag on the air and sets in motion a downdraft. This downdraft strikes the ground where precipitation is heaviest and forms a local squall wind, which is sometimes strong enough to fell trees and do severe structural damage to buildings.

Modern principles of meteorology have been used in efforts to modify weather phenomena in various ways. One goal has been to increase precipitation in areas experiencing drought; another is to lessen the severity of storms. One method of inducing convectional precipitation is by artificial cloud seeding, the introduction of minute particles into dense cumulus clouds. The particles, which may be of silver iodide smoke, serve as nuclei that induce greater intensity of condensation, increase the size and height of cumulonimbus clouds, and perhaps result thereby in increased rainfall. Carefully conducted tests over Florida in 1978 and 1981 failed to confirm the ability of cloud seeding to increase rainfall.

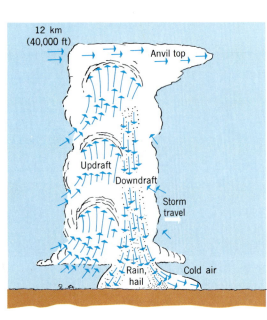

FIGURE 5.19 Schematic diagram of the interior of a thunderstorm cell.

FIGURE 5.20 Hailstones produced by a summer thunderstorm. (Runk and Schoenberger/Grant Heilman.)

Opinion among meteorologists remains divided as to the effectiveness of cloud seeding.

Hail and Lightning— Environmental Hazards

In addition to powerful wind gusts and torrential falls of rain, important environmental hazards connected with the thunderstorm are hail, lightning, and tornadoes. Hailstones are formed by the accumulation of ice layers on an ice pellet suspended in the strong thunderstorm updrafts. The phenomenon is much like that of the icing of aircraft flying through a cloud of supercooled water droplets. After the hailstones have grown to diameters that often reach 3 to 5 cm (1 to 2 in.), they escape from the updraft and fall to the ground (Figure 5.20).

An important aspect of planned weather modification is that of reducing the severity of hailstorms. Annual losses from crop destruction by hailstorms amount to several hundred million dollars (Figure 5.21). Damage to wheat crops is particularly severe in a north–south belt of the High Plains, running through Nebraska, Kansas, and Oklahoma. A much larger region of somewhat less hailstorm frequency extends eastward generally from the Rockies to the Ohio Valley. Corn is a major crop of this area. Scientists are studying the isolated cumulonimbus clouds that produce hail. Research may lead to development of cloud seeding techniques by means of which the severity of hailstorms can be reduced.

Another effect of convection cell activity is to generate lightning, one of the

FIGURE 5.21 These stalks of full-grown corn have been severely damaged by a hailstorm. (Courtesy of the Illinois State Water Survey.)

environmental hazards that results annually in the death of many persons and livestock and in the setting of many forest and building fires. Lightning is a great electric arc—a gigantic spark—passing between cloud and ground, or between parts of a cloud mass (see Figure 5.18D). During lightning discharge a current of as much as 60,000 to 100,000 amperes may develop. Rapid heating and expansion of the air in the path of the lightning stroke sends out intense sound waves, which we recognize as a thunderclap. In the United States lightning causes a yearly average of about 150 human deaths and property damage of over $100 million, including loss by fires set by lightning.

Latent Heat and the Global Balances of Energy and Water

Let us now take another look at the global energy balance and include the mechanism of heat transport in latent form. Figure 5.22 is a schematic diagram showing how evaporation and condensation are involved

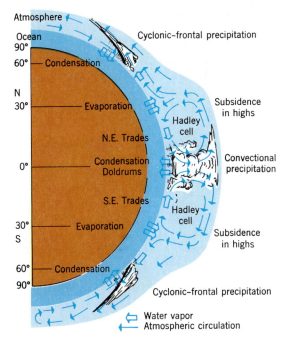

FIGURE 5.22 This schematic diagram summarizes the exchanges of water vapor from pole to pole. (From A. N. Strahler, The life layer, *Jour. of Geography*, vol. 69, no. 2, p. 74. Copyright © 1970 by *The Journal of Geography*. Used by permission.)

in energy exchange in each of the major latitude zones. We can now also visualize the pattern of transport of water vapor across the parallels of latitude; this is a transport of matter and is part of the global water balance.

The equatorial zone is characterized by a rise of moist, warm air in innumerable convection cells reaching to the upper limits of the troposphere. As condensation and precipitation occur here, enormous amounts of latent heat energy are liberated. This zone has been called the "fire box" of the globe in recognition of this intense production of sensible heat by condensation. Moving equatorward to replace the rising air are the trade winds, part of the Hadley cell circulation. Water evaporates from the ocean surface over the subtropical highs and is carried equatorward in vapor form.

Next, we move into the midlatitudes, where upper-air waves constantly form and dissolve in the westerlies. Here, cyclones and anticyclones exchange cold, polar air for warm, tropical air across the parallels of latitude. Water vapor is also transported poleward, carrying with it much latent heat. Condensation in cyclonic storms removes water vapor from the troposphere toward high latitudes, so that the movement of water vapor declines and reaches zero at the poles.

Air Pollution

We can make good use of an understanding of the processes of condensation and precipitation when probing into changes in the atmospheric environment caused by human activity. These are inadvertent changes, largely the result of urbanization and the growth of industrial processes, that have enormously increased the rate of combustion of hydrocarbon (fossil) fuels in the past half century or so. We have already considered the thermal effects of injecting carbon dioxide of combustion into the atmosphere. Now we turn to consider the kinds of foreign matter, or *pollutants*, injected into the lower atmosphere, and the effects of that pollution on air quality and urban climates.

We recognize two classes of atmospheric

pollutants. First, there are solid and liquid particles, which are designated collectively as *particulate matter*. Dusts found in smoke of combustion, as well as droplets naturally occurring as cloud and fog, fall into the category of particles. Second, there are compounds in the gaseous state, included under the general term *chemical pollutants* in that they are not normally present in measurable quantities in clean air remote from densely populated, industrialized regions. (Excess carbon dioxide produced by combustion is not usually classed as a pollutant.)

One group of chemical pollutants of industrial and urban areas is of primary origin, for example, produced directly from a source on the ground. Gases included in this group are carbon monoxide (CO), sulfur dioxide (SO_2), oxides of nitrogen (NO, NO_2, NO_3), and hydrocarbon compounds. These chemical pollutants cannot always be treated separately from particulate matter, since they are often combined within a single suspended particle. We have already seen that certain dusts, particularly the sea salts, have an affinity for water and easily take on a covering water film. The water film in turn absorbs the chemical pollutants.

When particles and chemical pollutants are present in considerable density over an urban area, the resultant mixture is known as *smog*. Almost everyone living in large cities is familiar with smog through its irritating effects on the eyes and respiratory system and its ability to obscure distant objects. When concentrations of suspended matter are less dense, obscuring visibility of very distant objects but not otherwise objectionable, the atmospheric condition is referred to as *haze*. Atmospheric haze builds up quite naturally in stagnant air layers as a result of the infusion of various surface materials. Haze is normally present whenever air reaches high relative humidity, because water films grow on suspended nuclei. Nuclei of natural atmospheric haze particles consist of mineral dusts from the soil, crystals of salt blown from the sea surface, hydrocarbon compounds (pollens and terpenes) exuded by plant foliage, and smoke from forest and grass fires. Dusts from volcanoes may, on occasion, add to atmospheric haze.

It is evident at this point that what we are calling atmospheric pollutants are of both natural and industrial origin, and that human activities can supplement the quantities of natural pollutants present. Table 5.1 illustrates this complexity by listing the primary pollutants according to sources.

Not all air pollution comes from the cities. Isolated industrial activities can produce pollutants far from urban areas. Particularly important are smelters and manufacturing plants in small towns and rural areas. Sulfide ores (metals in combination with sulfur compounds) are processed by heating in smelters close to the mine. Here, sulfur compounds are sent into the air in enormous concentrations from smokestacks. Fallout over the surrounding area is destructive to vegetation.

Mining and quarrying operations send mineral dusts into the air. For example, asbestos mines (together with asbestos processing and manufacturing plants) send into the air countless threadlike mineral particles, some of which are so small that they can be seen only with the electron microscope. These particles travel widely and are inhaled by humans, lodging permanently in the lung tissue. Nuclear test explosions inject into the atmosphere a wide range of particles, including many radioactive substances capable of traveling thousands of kilometers in the atmospheric circulation.

TABLE 5.1 Sources of primary atmospheric pollutants

Pollutants from Natural Sources	Pollutants Generated by Human Activities
Volcanic dusts	Fuel combustion (CO_2, SO_2, lead)
Sea salts from breaking waves	Chemical processes
Pollens and terpenes from plants (Aggravated by human activities)	Nuclear fusion and fission
(Aggravated by human activities): Smoke of forest and grass fires	Smelting and refining of ores
	Mining, quarrying
Blowing dust	Farming
Bacteria, viruses	

Data from Assn. of Amer. Geographers (1968), *Air Pollution*, Commission on College Geography, Resource Paper No. 2, Figure 3, p. 9.

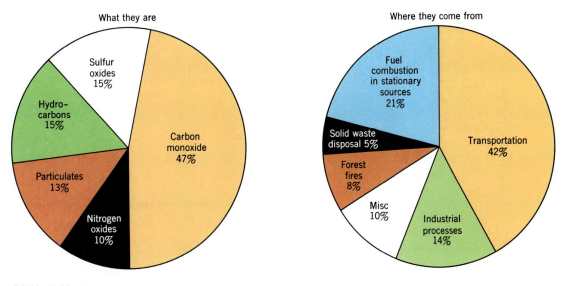

FIGURE 5.23 Air pollution emissions in the United States in percentage by weight. (Data from National Air Pollution Control Administration, HEW.)

Forest and grass fires, sometimes set by humans, add greatly to smoke palls in certain seasons of the year. Plowing, grazing, and vehicular traffic raise large amounts of mineral dusts from dry soil surfaces. Bacteria and viruses, which we have not yet mentioned, are borne aloft in air turbulence when winds blow over contaminated surfaces, such as farmlands, grazing lands, city streets, and waste disposal sites.

Figure 5.23 shows major atmospheric pollutants regarding both component matter and source in recent years. Much of the carbon monoxide, half of the hydrocarbons, and about a third of the oxides of nitrogen come from exhausts of gasoline and diesel engines in vehicular traffic. Generating of electricity and various industrial processes contribute most of the sulfur oxides, because the coal and lower-grade fuel oil used for these purposes are comparatively rich in sulfur. These same sources also supply most of the particulate matter. Fly ash consists of the coarser grades of soot particles emitted from smokestacks of generating plants. These particles settle out quite quickly within close range of the source. Combustion used for heating of buildings is a comparatively minor contributor to pollution, because the higher grades of fuel oil are low in sulfur and are usually efficiently burned. Very

finely divided carbon comprises much of the smoke of combustion and is capable of remaining in suspension almost indefinitely because of its colloidal size. Forest fires comprise a secondary contributor of particles. Burning of refuse is a minor contributor in all categories of pollutants.

In the smog of cities there are, in addition to the ingredients mentioned, certain chemical elements contained in particles contributed by exhausts of automobiles and trucks. Included are particles that contain lead, chlorine, and bromine.

Primary pollutants are conducted upward from the emission sources by rising air currents that are a part of the normal convectional process. The larger particles settle under gravity and return to the surface as *fallout*. Particles too small to settle out are later swept down to earth by precipitation, a process called *washout*. By a combination of fallout and washout the atmosphere tends to be cleaned of pollutants. In the long run a balance is achieved between input and output of pollutants, but there are large fluctuations in the quantities stored in the air at a given time. Pollutants are also eliminated from the air over their source areas by winds that disperse the particles into large volumes of cleaner air in the downwind direction. Strong winds can quickly sweep away most

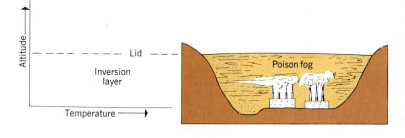

FIGURE 5.24 A low-level inversion was the predisposing condition for poison fog accumulation at Donora, Pennsylvania, in October 1948. (From A. N. Strahler, *Planet Earth*, Harper & Row, New York. Copyright © 1972 by Arthur N. Strahler.)

pollutants from an urban area but, during periods when a stagnant anticyclone is present, the concentrations rise to high values.

In polluted air certain chemical reactions take place among the components injected into the atmosphere, generating a secondary group of pollutants. For example, sulfur dioxide may combine with oxygen and then react with water of suspended droplets to yield sulfuric acid. This acid is irritating to organic tissues and corrosive to many inorganic materials. In another typical reaction, the action of sunlight on nitrogen oxides and organic compounds produces ozone, a toxic and destructive gas. Reactions brought about by the presence of sunlight are described as *photochemical reactions*. One toxic product of photochemical action is ethylene, produced from hydrocarbon compounds.

Ozone in urban smog has a harmful effect on plant tissues and, in some cases, has caused the death or severe damage of ornamental trees and shrubs. Sulfur dioxide is injurious to certain plants and is a cause of loss of productivity in truck gardens and orchards in polluted air. Atmospheric sulfuric acid in cities has in some places largely wiped out lichen growth.

Inversion and Smog

Concentration of pollutants over a source area rises to its highest levels when the vertical mixing (convection) of the air is inhibited by a stable configuration of the lower air layers. Recall from Chapter 3 that at night, when the air is calm and the sky clear, rapid cooling of the ground surface typically produces a low-level temperature inversion (see Figure 3.8). In cold air the reversal of the temperature gradient may extend several hundred meters into the air.

A low-level temperature inversion represents an unusually stable air structure. When this type of inversion develops over an urban area, conditions are particularly favorable for entrapment of pollutants to the degree that heavy smog or highly toxic fog can develop, as shown in Figure 5.24. The upper limit of the inversion layer coincides with the cap, or lid, below which pollutants are held. The lid may be situated 150 to 300 m (500 to 1000 ft) above the ground.

Although situations dangerous to health during prolonged low-level inversion have occurred a number of times over European cities since the Industrial Revolution began, the first major tragedy of this kind in the

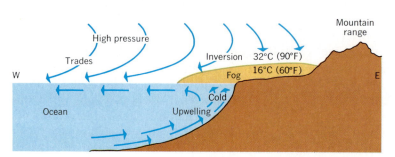

FIGURE 5.25 Subsiding air over a western continental coast produces a persistent temperature inversion and traps cool air and fog in a surface layer near the shore.

United States occurred at Donora, Pennsylvania, in late October 1948. The city occupies a valley floor hemmed in by steeply rising valley walls that prevent the free mixing of the lower air layer with that of the surrounding region (Figure 5.24). Industrial smoke and gases from factories poured into the inversion layer for five days, increasing the pollution level. Because humidity was high, a poisonous fog formed and began to take its toll. Twenty persons died and several thousand persons were made ill before a change in weather pattern dispersed the smog layer.

For the Los Angeles basin of southern California and, to a lesser degree, the San Francisco Bay area and over other west coasts in these latitudes generally, special climatic conditions produce prolonged inversions favorable to persistent smog accumulation (Figure 5.25). The Los Angeles basin is a low, sloping plain lying between the Pacific Ocean and a massive mountain barrier on the north and east sides. Cool air is carried inland over the basin on weak winds from the south and southwest, but it cannot move farther inland because of the mountain barrier. It is a characteristic of these latitudes that the air on the eastern sides of the subtropical high-pressure cells is continually subsiding, creating conditions of clear stable air for long periods at a time. The effect is particularly marked in the summer season, when the subtropical high is large and strong. The subsiding air over the Los Angeles basin is warmed adiabatically as well as heated by direct absorption of solar radiation during the day, so that it is markedly warmer than the cool stagnant air below. An inversion forms at the contact between the upper warm air and the lower cool air layer; the inversion acts as a lid to prevent mixing of the two layers. Pollutants accumulate in the cool air layer and produce the characteristic smog. The upper limit of the smog stands out sharply in contrast to the clear air above it, filling the basin like a lake and extending into valleys in the bordering mountains (Figure 5.26).

FIGURE 5.26 A dense layer of smog lying over the Los Angeles Basin. The view is from a point over the San Gabriel Mountains, looking southwest. (John S. Shelton.)

Modification of Urban Climate

By applying principles of insolation and temperature we can anticipate the human impact as cities spread, replacing a richly vegetated countryside with blacktop and concrete. In the urban environment the absorption of solar radiation causes higher ground temperatures for two reasons. First, foliage of plants is absent, so that the full quantity of solar energy falls on the bare ground. Absence of foliage also means absence of transpiration (evaporation from leaves), which elsewhere produces a cooling of the lower air layer. A second factor is that roofs and pavements of concrete and asphalt hold no moisture, so that evaporative cooling cannot occur as it would from a moist soil.

The thermal effect is that of converting the city into a hot desert. The summer temperature cycle close to the pavement of a city may be almost as extreme as that of the desert floor. This surface heat is conducted into the ground and stored there. The thermal effects within a city are actually more intense than on a sandy desert floor because the capacity of solid concrete, stone, or asphalt to conduct and hold heat is greater than that of loose, sandy soil. An additional thermal factor is that vertical masonry surfaces absorb insolation or reflect it to the ground and to other vertical surfaces. The absorbed heat is then radiated back into the air between buildings.

As a result of these changes in the energy balance, the central region of a city typically shows summer air temperatures a few degrees higher than for the surrounding suburbs and countryside. Figure 5.27 is a map of the Washington, D.C. area, showing air temperatures for a typical August afternoon. The lines of equal air temperature (isotherms) delineate a *heat island*. The heat island persists through the night because of the availability of a large quantity of heat stored in the ground during the daytime hours. In winter additional heat is radiated by walls and roofs of buildings, which conduct heat from the inside. Even in summer, city heat output is increased through use of air conditioners, which expend enormous amounts of energy at a time when the outside air is at its warmest.

Over large cities in the continental interior and east coast of North America pollutants are trapped beneath an inversion lid that takes the form of a broad *pollution dome* centered over the city when winds are light or a calm prevails (Figure 5.28). When there is a regional wind caused by a pressure gradient, the pollutants are carried downwind to form a *pollution plume*. Pollution plumes can extend far downwind to reach other urban centers.

An important physical effect of urban air pollution is that it reduces visibility and illumination. A smog layer can cut illumination by 10 percent in summer and 20 percent in winter. Ultraviolet radiation is absorbed by smog, which, at times,

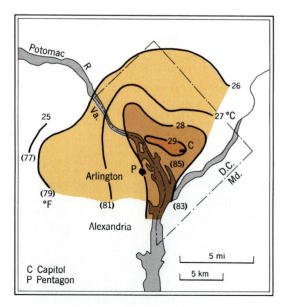

FIGURE 5.27 A heat island over Washington, D.C. Air temperatures were taken at 10 P.M. on a day in early August. (Data of H. E. Landsberg.)

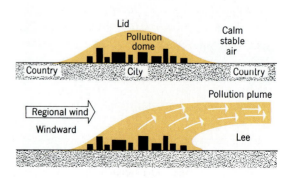

FIGURE 5.28 Pollution dome and pollution plume.

completely prevents these wavelengths from reaching the ground. Reduced ultraviolet radiation may prove to be important in permitting increased bacterial activity at ground level. City smog cuts horizontal visibility to one-fifth to one-tenth of the normal distance for clean air. Over cities, winter fogs are much more frequent than over the surrounding countryside. Coastal airports, such as those of New York City, Newark, and Boston, suffer severely from a high incidence of fogs augmented by urban air pollution.

A related effect of the urban heat island is the general increase in cloudiness and precipitation over a city as compared with the surrounding countryside. This increase results from intensified convection generated by heating of the lower air. For example, it has been found that, prior to the present period of strict pollution control, thunderstorms over London, England, produced 30 percent more rainfall over the city than over the surrounding country. Increased precipitation over an urban area is estimated to average from 5 to 10 percent over the normal for the region in which it lies.

Acid Deposition and Its Effects

Washout of sulfuric acid by precipitation results in rainwater with an abnormally high content of the sulfate ion, a condition known as *acid rain*. Nitric acid, formed by reactions involving pollutant nitrogen oxides, also contributes to the acidity of rainwater. Because fallout of solid pollutants in periods of dry weather also contributes these acids to the surfaces of plants, soil, streams, and lakes, the phenomenon is now described in broader terms as *acid deposition*. In earlier paragraphs on air pollution and its effects we covered the basic principles of acid deposition and made mention of its corrosive effects.

When tested for degree of acidity, measured in terms of the pH of the rainwater, acid rain shows pH values well below a pH of 5 to 6 typical of rainwater falling in unpolluted regions. (The pH scale is explained in Chapter 19. Rainwater is normally slightly acid because of the presence of carbon dioxide in solution,

which forms a weak concentration of carbonic acid.)

In the 1960s water chemists noted a significant lowering of the pH of rain in northwestern Europe. Values have been reduced to between pH 3 and 5. Because pH numbers are on a logarithmic scale, these values mean that rain in these areas is now often 100 to 1000 times more acid than previously.

American scientists studying the chemical quality of rainwater reported that since 1975 or thereabouts rainwater over a large area of the northeastern United States has had an average pH of about 4; they have observed pH values as low as 2.1 in rainwater of individual storms at certain localities. Other observations show that values less than pH 4 occur at times over many heavily industrialized United States cities, among them Boston, New York, Philadelphia, Birmingham, Chicago, Los Angeles, and San Francisco. Values between 4 and 5 have been observed near such urbanized areas as Tucson, Arizona; Helena, Montana; and Duluth, Minnesota— localities we do not usually associate with heavy air pollution.

Since about 1975 the level of acidity of rainwater seems to have leveled off in the heavily industrial northeastern United States and western Europe. At the same time, the contribution from nitrate has increased while that of sulfate has decreased, reflecting the change in proportion of these pollutants being injected into the atmosphere. Figure 5.29 shows the 1980 distribution of sulfur dioxide and nitrogen oxide in the United States. In recent years rainwater with pH below 5.0 has been identified locally in the southeastern and southwestern United States.

Some of the possible undesirable environmental effects of acid deposition upon natural systems are these: acidification of lakes and streams; excessive leaching of nutrients from plant foliage and the soil; various metabolic disturbances to organisms; and upsetting of the balances of predators and prey in aquatic ecosystems.

One possible instance of the effect of acidification of stream water, observed in Norway, has been the virtual elimination of salmon runs by inhibition of egg

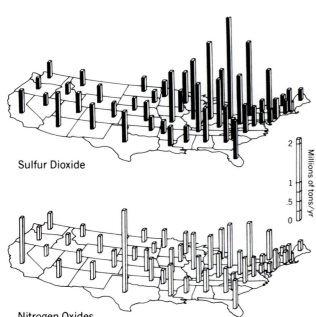

Sulfur Dioxide

Nitrogen Oxides

Millions of tons/yr

2

1

.5

0

FIGURE 5.29 Emissions of sulfur dioxide and nitrogen oxides by states for the year 1980. (Office of Technology Assessment, U.S. Congress, U.S. Government Printing Office, Washington, D.C.)

development. Even though the acidity of Norwegian streams has leveled off, salmon kills continue to occur, especially when rainstorms carry large amounts of sulfate into streams. Increased fish mortality, observed in Canadian lakes, has also been attributed to acidification. In 1980 it was reported by Canada's Department of Environment that 140 Ontario lakes had no fish and that thousands of other lakes of that province were threatened with similar fate. Forests, too, appear to have experienced damage from acid rain. In West Germany the impact has been especially severe in the Harz Mountains and the Black Forest. It was reported in 1983 that about one-third of West German forests showed visible damage. In the eastern United States, pine and spruce trees have apparently experienced damage in recent years.

By 1983, it had become abundantly clear to committees of the U.S. National Research Council and other research bodies that reductions of emissions of air pollutants will substantially reduce the level of acidity in precipitation. Their estimate was that a 50 percent reduction in sulfate and nitrate emissions would bring about a 50 percent reduction in acidity downwind of the emission sources. Debate continues within the scientific community as to the causal relationships between acid deposition and pollutant sources.

CHAPTER 6

AIR MASSES AND CYCLONIC STORMS

CYCLONIC STORMS Intense traveling cyclones bring much-needed precipitation, but they can also be environmental hazards because of the high winds and river floods that go with them.

WEATHER FRONTS Like armies on a great battlefield, unlike air masses meet in conflict along weather fronts.

WAVE CYCLONES Great spiraling whorls within the troposphere intensify to produce wave cyclones along weather fronts.

CYCLONE TRACKS Traveling wave cyclones of the middle and high latitudes use well-known tracks, one storm following the next.

THE TORNADO Its deadly funnel cloud appears near turbulent weather fronts.

LOW-LATITUDE WEATHER In the tropical and equatorial zones, weather disturbances range from mild rain-producers to violent tropical cyclones, known as hurricanes and typhoons.

THE atmosphere exerts stress, often severe on humans and other life-forms through weather disturbances involving extremes of high winds, cold, and precipitation. Under the category of severe storms, these phenomena generate environmental hazards directly by the impact of winds and precipitation and indirectly through their attendant phenomena—storm waves and storm surges on the seas and river floods, mudflows, and landslides on the lands.

Weather disturbances of lesser magnitude are among the beneficial environmental phenomena, because they bring precipitation to the land surfaces and so recharge the vital supplies of fresh water on which all terrestrial life-forms depend.

An understanding of weather disturbances of all intensities enables us to predict their times and places of occurrence and therefore give warnings and allow protective measures to be taken. This function of the atmospheric scientist can be placed under the heading of environmental protection. It may also be possible to a limited degree for us to modify atmospheric processes in a deliberate way in order to lessen the wind speeds of storms. These activities come under the heading of planned weather modification.

Traveling Cyclones

Much of the unsettled, cloudy weather we experience in middle and high latitudes is associated with traveling cyclones. Convergence of masses of air toward the cyclone centers is accompanied by lift of air and adiabatic cooling, which in turn produce cloudiness and precipitation. By contrast, much of our fair sunny weather is associated with traveling anticyclones in which the air subsides and spreads outward, causing adiabatic warming. This process produces stable air, unfavorable to the development of clouds and precipitation.

Many cyclones are mild in intensity and pass with little more than a period of cloud cover and light rain or snow. On the other hand, when pressure gradients are strong, winds ranging in strength from moderate to gale force accompany the cyclone. In such a case, the disturbance can be called a *cyclonic storm*.

Moving cyclones fall into three general classes: (1) The wave cyclone of midlatitude and arctic zones ranges in severity from a weak disturbance to a powerful storm. (2) The tropical cyclone of tropical and subtropical zones over ocean areas ranges from a mild disturbance to the highly destructive hurricane, or typhoon. (3) Although a very small storm, the tornado is an intense cyclonic vortex of enormously powerful winds; it is on a very much smaller scale of size than other types of cyclones and is related to severe convectional activity.

Air Masses

Cyclones of middle and high latitudes depend for their development on the coming together of large bodies of air of contrasting physical properties. A body of air in which the upward gradients of temperature and moisture are fairly uniform over a large area is known as an *air mass*. In horizontal extent a single air mass may be as large as a part of a continent; in vertical dimension it may extend through the troposphere. A given air mass is characterized by a distinctive combination of temperature, environmental lapse rate, and specific humidity. Air masses differ widely in temperature—from very warm to very cold—and in moisture content—from very dry to very moist.

A given air mass usually has a sharply defined boundary between itself and a neighboring air mass. This discontinuity is termed a *front*. We found an example of a front in the contact between polar and tropical air masses below the axis of the jet stream in upper-air waves, as shown in Figure 4.18. This feature we call the *polar front;* it represents the highest degree of global generalization. Fronts may be nearly vertical, as in the case of air masses having little motion relative to one another. Fronts may be inclined at an angle not far from the horizontal in cases where one air mass is sliding over another. A front may be almost stationary with respect to the earth's surface but, nevertheless, the adjacent air masses may be moving rapidly with respect to each other along the front.

The properties of an air mass are derived partly from the regions over which it passes. Because the entire troposphere is in more or less continuous motion, the particular air-mass properties at a given place may reflect the composite influence of a travel path covering thousands of kilometers and passing alternately over oceans and continents. This complexity of influences is particularly important in middle and high latitudes in the northern hemisphere, within the flow of the global westerlies.

On the other hand, over vast tropical and equatorial areas, an air mass reflects quite simply the properties of an ocean or a land surface over which it moves slowly or tends to stagnate. Over a warm equatorial ocean surface the lower levels of the overlying air mass develop a high water vapor content. Over a large tropical desert, slowly subsiding air forms a warm air mass with low relative humidities. Over cold, snow-covered land surfaces in the arctic zone in winter, the lower layer of the air mass remains very cold with a very low water-vapor content. Meteorologists have designated as *source regions* those land or ocean surfaces that strongly impress their temperature and moisture characteristics on overlying air masses.

Air masses move from one region to another following the patterns of barometric pressure. During these migrations, lower levels of the air mass undergo gradual modification, taking up or losing heat to the surface beneath, and perhaps taking up or losing water vapor as well. Air masses are classified according to two categories of source regions: (1) the latitudinal position on the globe, which primarily determines thermal properties; and (2) the underlying surface, whether continent or ocean, determining the moisture content. With respect to latitudinal position, five types of air masses are as follows:

Air Mass	Symbol	Source Region
Arctic	A	Arctic ocean and fringing lands
Antarctic	AA	Antarctica
Polar	P	Continents and oceans, lat. 50–60° N and S
Tropical	T	Continents and oceans, lat. 20–35° N and S
Equatorial	E	Oceans close to equator

With respect to type of underlying surface, two further subdivisions are imposed on the preceding types:

Air Mass	Symbol	Source Region
Maritime	m	Oceans
Continental	c	Continents

By combining types based on latitudinal position with those based on underlying

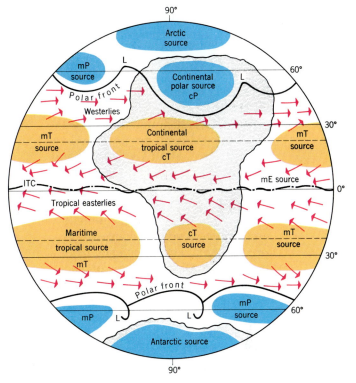

FIGURE 6.1 A schematic global diagram showing the source regions of air masses in relation to the polar front and the intertropical convergence zone (ITC).

surface, a list of six important air masses results (Table 6.1). Figure 6.1 shows the global distribution of source regions of these air masses. Table 6.1 also gives some typical values of temperature and specific humidity at the surface, although a wide range in these properties may be expected, depending on season.

Note that the polar air masses (mP, cP) originate in the subarctic latitude zone, not in the polar latitude zone. The

TABLE 6.1 Properties of typical air masses

| Air Mass | Symbol | Properties | Temperature | | Specific Humidity (gm/kg) |
			°C	(°F)	
Continental-arctic (and continental antarctic)	cA (cAA)	Very cold, very dry (winter)	−46°	(−50°)	0.1
Continental polar	cP	Cold, dry (winter)	−11°	(12°)	1.4
Maritime polar	mP	Cool, moist (winter)	4°	(39°)	4.4
Continental tropical	cT	Warm, dry	24°	(75°)	11
Maritime tropical	mT	Warm, moist	24°	(75°)	17
Maritime equatorial	mE	Warm, very moist	27°	(80°)	19

meteorologist's definition of the word "polar" for air masses has long been in use and has international acceptance; we cannot change this usage to conform with the geographer's system of latitude zones defined in Chapter 2.

The equatorial air mass (mE) holds about 200 times as much water vapor as the extremely cold and dry arctic and antarctic air masses, cA and cAA. The maritime tropical air mass (mT) and maritime equatorial air mass (mE) are quite similar in temperature and water vapor content. With very high values of specific humidity, both are capable of very heavy yields of precipitation. The continental tropical air mass (cT) has its source region over subtropical deserts of the continents. Although it may have a substantial water vapor content, it tends to be stable and has low relative humidity when highly heated during the daytime. The polar maritime air mass (mP) originates over midlatitude oceans. Although the quantity of water vapor it holds is not large compared with the tropical air masses, the mP air mass can yield heavy precipitation. Much of this precipitation is of the orographic type, over mountain ranges on the western coasts of continents. The continental polar air mass (cP) originates over North America and Eurasia in the subarctic zone. It has low specific humidity and is very cold in winter.

Cold and Warm Fronts

Figure 6.2 shows the structure of a front along which cold air is invading the warm-air zone. A front of this type is called a *cold front*. The colder air mass, being the denser, remains in contact with the ground and forces the warmer air mass to rise over it. The slope of the cold front surface is greatly exaggerated in the figure, being actually of the order of slope of 1 in 40 (meaning that the slope rises 1 km vertically for every 40 km of horizontal distance). Cold fronts are associated with strong atmospheric disturbance. As the unstable warm air is lifted, it may break out in severe thunderstorms. Thunderstorms often form a long line of massive clouds stretching for tens of kilometers (Figure 6.3).

Figure 6.4 illustrates a *warm front* in

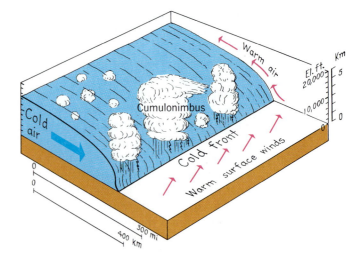

FIGURE 6.2 A cold front, along which a cold air mass is lifting a warm air mass, setting off a line of thunderstorms. (Drawn by A. N. Strahler.)

which warm air is moving into a region of colder air. Here, again, the cold air mass remains in contact with the ground and the warm air mass is forced to rise, as if it were ascending a long ramp. Warm fronts have lower slopes than cold fronts—on the order of 1 in 80, to as low as 1 in 200. Moreover, warm fronts commonly represent stable atmospheric conditions and lack the turbulent air motions of the cold front. Of course, if the warm air is unstable, it will develop convection cells and there will be heavy showers or thunderstorms.

Cold fronts normally move along the ground at a faster rate than warm fronts. So, when both types are in the same neighborhood, the cold front overtakes the warm front. An *occluded front* then results (Figure 6.5). The colder air of the fast-moving cold front remains next to the ground, forcing both the warm air and the less cold air to rise over it. The warm air mass is lifted completely free off the ground.

Wave Cyclones

The dominant type of weather disturbance of middle and high latitudes is the *wave cyclone*, a vortex that repeatedly forms, intensifies, and dissolves along the frontal zone between cold and warm air masses. At the time of World War I, the Norwegian meteorologist Jakob Bjerknes recognized

FIGURE 6.3 A line of cumulonimbus clouds marks the position of a cold front moving across the High Plains of eastern Colorado. The plane from which the photo was taken was flying in the cold air mass behind the front. (John S. Shelton.)

the existence of atmospheric fronts and developed his *wave theory* of cyclones.

The term "front," used by Bjerknes, was particularly apt because of the resemblance of this feature to the fighting fronts in western Europe that were then active. Just as vast armies met along a sharply defined front that moved back and forth, so masses of cold polar air meet in conflict with warm, moist tropical air. Instead of mixing freely, these unlike air masses remain clearly defined, but they interact along the polar front in great spiraling whorls.

A situation favorable to the formation of a wave cyclone is shown by a surface weather map (Figure 6.6). A trough of low pressure lies between two large anticyclones (highs); one is made up of a cold dry polar air mass, the other of a warm moist maritime air mass. Air flow is converging from opposite directions on the two sides of the front, setting up an unstable situation. The wave cyclone will begin to form along this low-pressure trough.

A series of individual blocks (Figure 6.7) shows the sequence of stages in the life

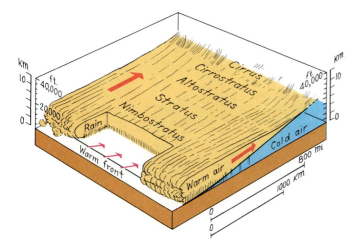

FIGURE 6.4 A warm front. Rain is falling from a dense stratiform cloud layer. (Drawn by A. N. Strahler.)

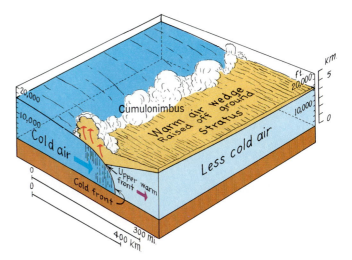

FIGURE 6.5 An occluded front. A warm front has been completely separated from contact with the ground. Abrupt lifting by the denser cold air has set off convectional activity. (Drawn by A. N. Strahler.)

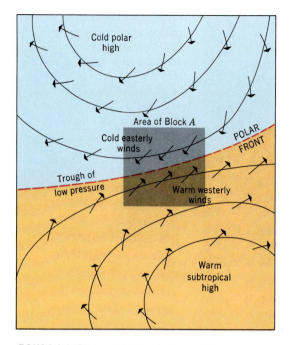

FIGURE 6.6 A trough between two high-pressure regions.

history of a wave cyclone. At the start of the cycle, the polar front is a smooth boundary along which air is moving in opposite directions. In Block *A* of Figure 6.7, the polar front shows a wave beginning to form. Cold air is turned in a southerly direction and warm air in a northerly direction so that each invades the domain of the other.

In Block *B* the wave disturbance along the polar front has deepened and intensified. Cold air is now actively pushing southward along a cold front; warm air is actively moving northeastward along a warm front. Each front is convex in the direction of motion. The zone of precipitation is now considerable, but wider along the warm front than along the cold front. In a still later stage the more rapidly moving cold front has reduced the zone of warm air to a narrow sector.

In Block *C*, the cold front has overtaken the warm front, producing an occluded

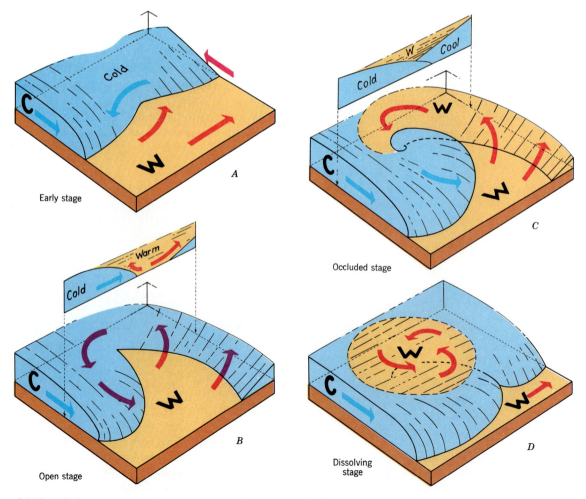

FIGURE 6.7 Development of a wave cyclone. (Drawn by A. N. Strahler.)

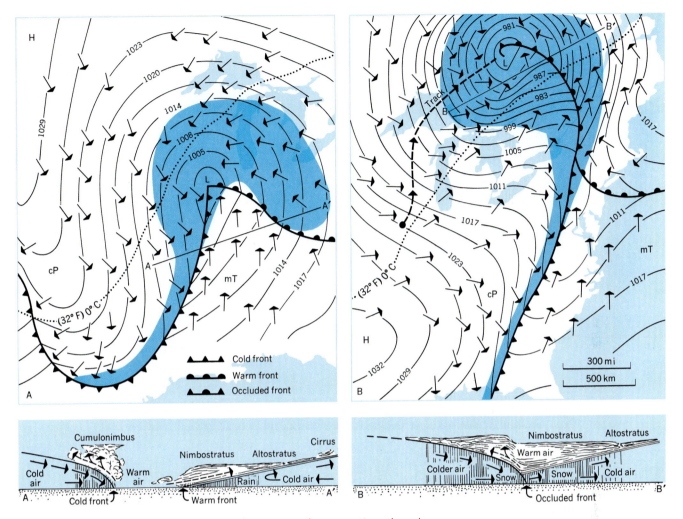

FIGURE 6.8 Simplified surface weather maps and cross sections through a wave cyclone. (A) Open stage. (B) Occluded stage.

front. The warm air mass has been forced off the ground and is now isolated from the parent region of warm air to the south. Because the source of moisture and energy has been cut off, the cyclonic storm gradually dies out. The polar front is now reestablished as originally (Block D).

Weather Changes within a Wave Cyclone

We can predict the changing weather conditions accompanying the passage of a wave cyclone in midlatitudes by examining some details of the internal structure of the storm. Figure 6.8 shows simplified surface weather maps depicting conditions on successive days. The structure of the storm is defined by the isobars, labeled in

millibars. Three kinds of fronts are shown by special line symbols. Areas in which precipitation is occurring are shown in color.

Map A shows the cyclone in an open stage, similar to that in Figure 6.7B. Note the following points. (1) Isobars of the low are closed to form an oval-shaped pattern. (2) Isobars make a sharp V where crossing cold and warm fronts. (3) Wind directions, indicated by arrows, are at an angle to the trend of the isobars and form a pattern of counterclockwise inspiraling. (4) In the warm-air sector there is northward flow of warm, moist tropical air toward the direction of the warm front. (5) There is a sudden shift of wind direction accompanying the passage of the cold front. This fact is indicated by the widely different wind directions at places close to the cold

FIGURE 6.9 An occluded cyclone over the eastern Pacific Ocean shows on this satellite image as a tight cloud spiral. The cold front makes a dense, narrow cloud band sweeping to the south and southwest of the cyclone center. View these frames with a stereoscope for strong three-dimensional effect. (NOAA-2 satellite image, courtesy of National Environmental Satellite Service.)

front, but on opposite sides. There is also a sharp drop in temperature accompanying the passage of the cold front. (6) Precipitation is occurring over a broad zone near the warm front and in the central area of the cyclone, but extends as a thin band down the length of the cold front. Cloudiness prevails generally over the entire cyclone. (7) The low is followed on the west by a high (anticyclone) in which low temperatures and clear skies prevail. (8) The 0°C (32°F) isotherm crosses the cyclone diagonally from northeast to southwest, showing that the southeastern part is warmer than the northwestern part.

A cross section through Map A along the line AA' shows how the fronts and clouds are related. Along the warm front is a broad layer of stratiform clouds. These take the form of a wedge with a thin leading

edge of cirrus. (See Figures 5.9 and 5.10 for cloud types.) Westward this wedge thickens to altostratus, then to stratus, and finally to nimbostratus with steady rain. Within the warm air mass sector the sky may partially clear with scattered cumulus. Along the cold front are cumulonimbus clouds associated with thunderstorms. These yield heavy rains, but only along a narrow belt.

The second weather map, Map B, shows conditions 24 hours later. The cyclone has moved rapidly northeastward, its track shown by a dashed line. The center has moved about 1600 km (1000 mi) in 24 hours, a speed of just over 65 km (40 mi) per hour. The cyclone has occluded. An occluded front replaces the separate warm and cold fronts in the central part of the disturbance. The high-pressure area, or tongue of cold polar air, has moved in to

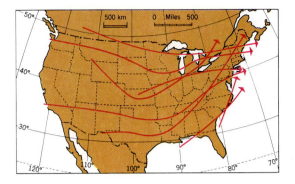

FIGURE 6.10 Common tracks taken by wave cyclones passing across the United States.

Cyclone Tracks and Cyclone Families

Long observation of the movements of cyclones and anticyclones has shown that certain tracks are most commonly followed. Figure 6.10 is a map of the United States showing the common cyclone paths. Notice that some cyclonic storms travel across the entire United States from a place of origin in the North Pacific; others originate in the Rocky Mountain region, the central states, or the Gulf Coast. Most tracks converge toward the northeastern United States and pass out into the North Atlantic, where they tend to concentrate in the region of the Icelandic low. (See also Figure 6.19).

In the northern hemisphere, wave cyclones are heavily concentrated in the neighborhood of the Aleutian and Icelandic lows. These cyclones commonly form in succession to travel in a chain across the North Atlantic and North Pacific oceans. Figure 6.11, a world weather map, shows several such *cyclone families*. As each cyclone moves northeastward, it deepens and occludes, becoming an intense upper-air vortex. For this reason, intense cyclones arriving at the western coasts of North America and Europe are usually occluded.

In the southern hemisphere, storm tracks are more nearly along a single lane,

the west and south of the cyclone, and the cold front has pushed far south and east. Within the anticyclone the skies are clear. A cross section below the map shows conditions along line *BB'*, cutting through the occluded part of the storm. Notice that the warm air mass is being lifted higher off the ground and is giving heavy precipitation.

Wave cyclones in all stages of development are seen in weather satellite photographs. A good example is pictured in Figure 6.9. The two photographs have been placed side by side in such a way that a strong three-dimensional effect results when you view them together with a stereoscope. Try it—the effect is quite stunning.

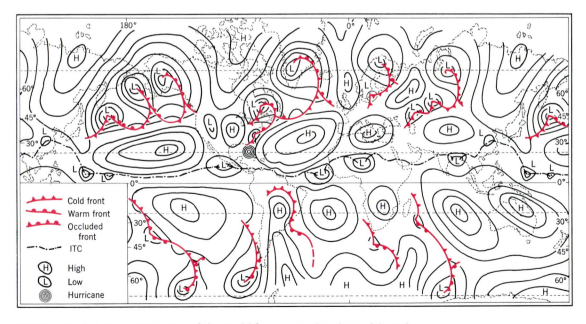

FIGURE 6.11 A daily weather map of the world for a given day during July or August might look like this map, which is a composite of typical weather conditions. (After M. A. Garbell.)

FIGURE 6.12 A tornado funnel cloud. (Howie Bluestein/Science Source/Photo Researchers.)

following the parallels of latitude. Three such cyclones are shown in Figure 6.11. This track is the result of uniform ocean surface circling the globe at these latitudes; only the southern tip of South America breaks the monotonous oceanic expanse. Also, the polar-centered ice sheet of Antarctica provides a centralized source of very cold air. (See Figure 6.19).

The Tornado—An Environmental Hazard

Associated with convectional activity along fast-moving fronts in middle latitudes is the *tornado*, a small but intense cyclonic vortex in which air is spiraling at tremendous speed. The tornado is a typically North American phenomenon, since most of the known violent tornadoes have occurred in the United States and Canada. Tornadoes are reported occasionally in a number of other localities in middle latitudes and are also observed in Australia.

The tornado appears as a dark *funnel cloud* hanging from the base of a dense cumulonimbus cloud (Figure 6.12). At its lower end, the funnel may be 100 to 450 m (300 to 1500 ft) in diameter. The base of the funnel appears dark because of the density of condensing moisture, dust, and debris swept up by the wind.

Wind speeds in a tornado exceed anything known in other storms. Estimates of wind speed run as high as 400 km (250 mi) per hour. As the tornado moves across

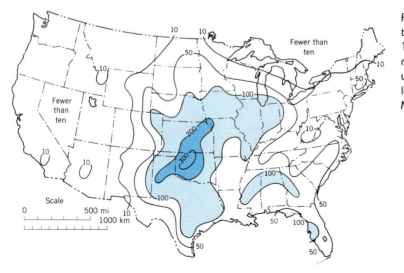

FIGURE 6.13 Distribution of tornadoes in the United States, 1955 to 1967. Figures tell number of tornadoes observed within two-degree squares of latitude and longitude. (Data of M. E. Pautz, 1969, ESSA.)

FIGURE 6.14 Swath of destruction left by a tornado that swept through Louisville, Kentucky. Undamaged houses at the extreme right and left delimit the tornado path. (George Hall/Woodfin Camp.)

the country, the funnel writhes and twists. The end of the funnel cloud may alternately sweep the ground, causing complete destruction of anything in its path, and rise in the air to leave the ground below unharmed.

Tornadoes occur as parts of cumulonimbus clouds traveling in advance of a cold front. They seem to originate where turbulence is greatest. They are most common in the spring and summer, but can occur in any month. Where maritime polar air lifts warm, moist tropical air on a cold front, conditions may become favorable for tornadoes. They occur in greatest numbers in the central and southeastern states and are rare over mountainous and forested regions. They are almost unknown from the Rocky Mountains westward and are relatively fewer on the eastern seaboard (Figure 6.13).

Devastation from a tornado is often completed within the narrow limits of its path (Figure 6.14). Only the strongest buildings constructed of concrete and steel can resist major structural damage. A tornado can often be seen or heard approaching, but those that come during the night may give no warning. The National Weather Service maintains a tornado forecasting and warning system. Whenever weather conditions conspire to favor tornado development, the danger area is alerted, and systems for observing and reporting a tornado are set in readiness.

Tropical and Equatorial Weather Disturbances

Weather systems of the tropical and equatorial zones show some basic differences from those of midlatitudes. The Coriolis effect is weak close to the equator, and there is a lack of strong contrast between air masses. Consequently, clearly defined fronts and large, intense wave cyclones are missing. On the other hand, there is intense atmospheric activity in the

FIGURE 6.15 Viewed from the Apollo 9 spacecraft, the tops of great cumulonimbus clouds over the Amazon jungle of South America show concentric patterns. As each thunderstorm cell rises and attains the top of the troposphere the ice cloud it produces spreads horizontally. Storms of this form are nearly stationary. (NASA AS9–19–3026.)

the airflow is almost directly from east to west in the form of persistent tropical easterlies, described in Chapter 4.

One of the simplest forms of weather disturbance is an *easterly wave*, a slowly moving trough of low pressure within the belt of tropical easterlies (trades). These waves occur in lat. 5° to 30° N and S over oceans, but not over the equator itself. Figure 6.16 is a simplified weather map of an easterly wave showing isobars, winds, and the zone of rain. The wave is simply a series of indentations in the isobars to form a shallow pressure trough. The wave travels westward at a rate of 300 to 500 km (200 to 300 mi) per day. Air flow converges on the eastern, or rear, side of the wave axis. This convergence causes the moist air to be lifted and to break out into scattered showers and thunderstorms. The rainy period may last for a day or two.

Another related disturbance is the *weak equatorial low*, a disturbance that forms near the center of the equatorial trough. Moist equatorial air masses converge on the center of the low, causing rainfall from many individual convectional storms. Several such weak lows are shown on the world weather map (Figure 6.11) lying along the ITC. Because the map is for a day in July or

form of numerous convection cells because of the high moisture content of the maritime air masses in these low latitudes (Figure 6.15). In other words, there is an enormous reservoir of energy in the latent form. This same energy source powers the most formidable of all storms—the tropical cyclone.

The world map (Figure 6.11) shows a typical set of weather conditions in the subtropical, tropical, and equatorial zones. Fundamentally the arrangement consists of two rows of high-pressure cells, one or two cells to each land or ocean body. The northern row lies approximately along the tropic of cancer; the southern row lies along the tropic of capricorn. Between the subtropical highs lies the equatorial trough of low pressure. Toward this trough the northeast and southeast trades converge along the intertropical convergence zone (ITC). At higher levels in the troposphere,

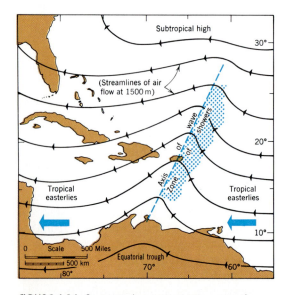

FIGURE 6.16 An easterly wave passing over the West Indies. (Data from H. Riehl, 1954, *Tropical Meteorology*, p. 213, Figure 9.3. McGraw–Hill, New York.)

August, the ITC is shifted well north of the equator. At this season the rainy monsoon is in progress in Southeast Asia.

Another distinctive feature of low-latitude weather is the occasional penetration of powerful tongues of cold polar air from the midlatitudes into very low latitudes. These tongues are known as *polar outbreaks;* they bring unusually cool, clear weather with strong, steady winds moving behind a cold front with squalls. The polar outbreak is best developed in the Americas. Outbreaks that move southward from the United States over the Caribbean Sea and Central America are called "northers" or "nortes"; those that move north from Patagonia into tropical South America are called "pamperos." One such outbreak is shown over South America on the world weather map (Figure 6.11). A severe polar outbreak may bring subfreezing air temperatures to highlands of South America and severely damage the coffee crop.

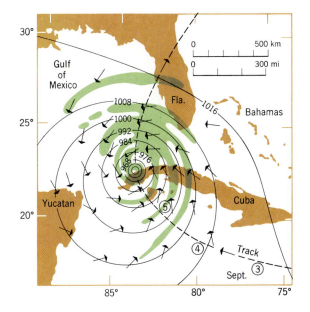

FIGURE 6.17 A hurricane passing over the western tip of Cuba. The recurving path will take the storm over Florida. Shaded areas show dense rain clouds as seen in satellite image.

The Tropical Cyclone—An Environmental Hazard

One of the most powerful and destructive types of cyclonic storms is the *tropical cyclone,* otherwise known as the *hurricane* or *typhoon.* The storm develops over oceans in lat. 8° to 15° N and S, but not close to the equator, where the Coriolis effect is extremely weak. It originates as a weak low, which deepens and intensifies, growing into a deep, circular low. High sea-surface temperatures that are over 27°C (80°F) in these latitudes are important in the environment of storm origin because warming of air at low level creates an unstable air mass. Once formed, the storm moves westward through the trade-wind belt. It may then curve northwest and north, finally penetrating well into the belt of westerlies.

The tropical cyclone is an almost circular storm center of extremely low pressure into which winds are spiraling at high speed, accompanied by very heavy rainfall (Figure 6.17). Storm diameter may be 150 to 500 km (100 to 300 mi). Wind speeds range from 120 to 200 km (75 to 125 mi) per hour, and sometimes much higher. Barometric pressure in the storm center commonly falls to 950 mb or lower.

A characteristic feature of the tropical cyclone is its *central eye,* in which calm prevails (Figures 6.18 and 6.22). The eye is

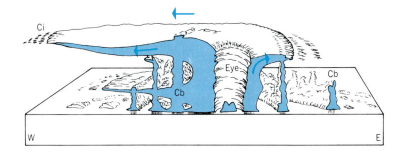

FIGURE 6.18 Schematic diagram of a hurricane. Cumulonimbus (Cb) clouds in concentric rings rise through dense stratiform clouds. Width of diagram represents about 1000 km (600 mi). (Redrawn from NOAA, National Weather Service, R. C. Gentry, 1964.)

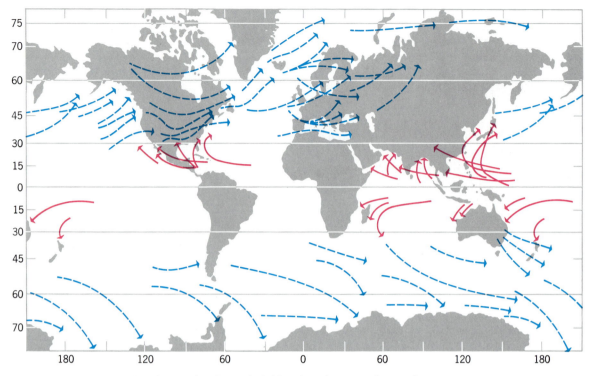

FIGURE 6.19 Typical paths of tropical cyclones (solid lines) and wave cyclones of middle latitudes (dashed lines). (Based on data of S. Pettersen, B. Haurwitz and N. M. Austin, J. Namias, M. J. Rubin, and J-H. Chang.)

a cloud-free vortex produced by the intense spiraling of the storm. In the eye, air descends from high altitude and is adiabatically warmed. Passage of the central eye may take about half an hour, after which the storm strikes with renewed ferocity, but with winds in the opposite direction.

World distribution of tropical cyclones is limited to seven regions, all of them over tropical and subtropical oceans (Figure 6.19): (1) West Indies, Gulf of Mexico, and Caribbean Sea; (2) eastern Pacific coastal region off Mexico and Central America; (3) western North Pacific, including the Philippine Islands, China Sea, and Japanese Islands; (4) Arabian Sea and Bay of Bengal; (5) south Indian Ocean, off Madagascar; (6) western South Pacific, in the region of Samoa and Fiji Islands and the east coast of Australia; and (7) Indian Ocean off the northwest coast of Australia. Curiously, these storms are unknown in the South Atlantic. Tropical cyclones never originate over land, although they often penetrate well into the margins of continents.

Tracks of tropical cyclones of the North

Atlantic are shown in Figure 6.20. Most of the storms originate at 10° to 20° latitude, travel westward and northwestward through the trades, then turn northeast at about 30° to 35° latitude into the zone of the westerlies. Here the intensity lessens and the storms change into typical midlatitude wave cyclones. In the trade-wind belt the cyclones travel 10 to 20 km (6 to 12 mi) per hour.

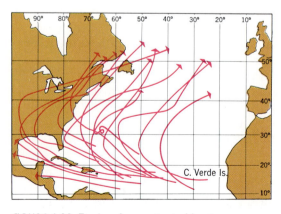

FIGURE 6.20 Tracks of some typical hurricanes occurring during August.

FIGURE 6.21 Only the foundations remain from a building leveled by winds and storm swash during the hurricane of September 1960 at Plantation Key, Florida. (Leslie F. Conover/Photo Researchers.)

The occurrence of tropical cyclones is restricted to certain seasons of the year, and these vary according to the global location of the storm region. For hurricanes of the North Atlantic the season runs from May through November, with maximum frequency in late summer or early autumn. The general rule is that tropical cyclones of the northern hemisphere occur in the season during which the ITC has moved north; those of the southern hemisphere occur when it has moved south.

The environmental importance of tropical cyclones lies in their tremendously destructive effect on inhabited islands and coasts (Figure 6.21). Wholesale destruction of cities and their inhabitants has been reported on several occasions. A terrible hurricane that struck Barbados in the West Indies in 1780 is reported to have torn stone buildings from their foundations, destroyed forts, and carried cannons more than 30 m (100 ft) from their locations. Trees were torn up and stripped of their bark. More than 6000 persons perished there.

Coastal destruction by storm waves and greatly raised sea level is perhaps the most serious effect of tropical cyclones. Where water level is raised by strong wind pressure, great storm surf attacks ground ordinarily far inland of the limits of wave action. A sudden rise of water level, known as *storm surge*, may take place as the hurricane moves over a coastline. Ships are lifted bodily and carried inland to become stranded. If high tide accompanies the storm the limits reached by inundation are even higher. The terrible hurricane disaster at Galveston, Texas, in 1900 was wrought largely by a sudden storm surge, inundating the low coastal city and drowning about 6000 persons. At the mouth of the Hooghly River on the Bay of Bengal, 300,000 persons died as a result of inundation by a 12 m (40 ft) storm surge that accompanied a severe tropical cyclone in 1737. In a similar disaster along the same coast in 1970, the death toll was estimated at 300,000 persons. Low-lying coral atolls of the western Pacific may be entirely swept over by wind-driven seawater, washing away palm trees and houses and drowning the inhabitants.

Important, too, is the large quantity of rainfall produced by tropical cyclones. A considerable part of the summer rainfall of some coastal regions can be traced to a few such storms. Although this rainfall is a valuable water source, it may prove a menace as a producer of unwanted river floods and, in areas of steep mountain slopes, gives rise to disastrous mudslides and landslides.

Attempts to reduce the severity of hurricanes by cloud seeding were made during the 1960s by scientists of the National Oceanic and Atmospheric Administration (NOAA) under an experimental program known as Project Stormfury. Seeding of four hurricanes on eight different days may have produced an observed reduction in wind speeds of between 10 and 30 percent on four of the days. Subsequent research led to the conclusion that hurricane clouds contain too much natural ice and too little supercooled water to respond satisfactorily to seeding. The project was abandoned in 1983, but much was learned about the physics of condensation processes.

Remote Sensing of Weather Phenomena

Remote sensing methods, described in Chapter 2, have been intensively developed for weather observation through the use of specialized orbiting satellites. The first of these, Nimbus-1, was launched into a sun-synchronous polar orbit in 1964. It

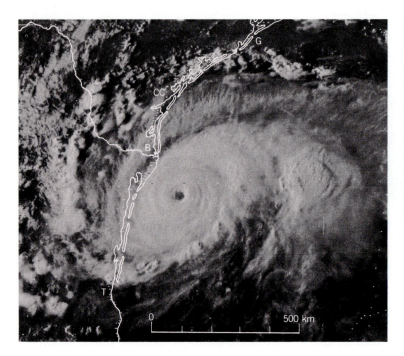

FIGURE 6.22 Hurricane Anita, over the western Gulf of Mexico, as observed by GOES-2 on September 1, 1977. The eye of the storm was about 175 km (100 mi) southeast of Brownsville, Texas. The hurricane was moving slowly in a southwesterly direction and later crossed the coast 230 km (140 mi) south of Brownsville. (Satellite Services Division, Environmental Data and Information Service, NOAA.)

transmitted TV photographs as well as high-resolution infrared images, the latter taken in darkness. Later Nimbus satellites and the subsequent NOAA TIROS series, all polar orbiters, have been continuously improved in their data-collecting ability in many categories. Besides transmitting images, these weather satellites record the vertical profile of temperatures in the atmosphere. They also provide data on ozone, water vapor, cloud and snow cover, precipitation, and sea ice.

NOAA's TIROS series has included eleven spacecraft in a comprehensive meteorological and environmental data collection program. Data are received, processed, and distributed at a central facility in Suitland, Maryland. The satellites collect data from several hundred stations, including not only fixed and floating platforms, but balloons as well. Both the U.S. National Weather Service and the World Weather Program use the TIROS data for forecasting purposes. The satellite images are particularly useful in tracking hurricanes and typhoons, a service that has helped to save countless lives and forestall a great amount of property damage.

Another class of weather satellites uses a *geostationary orbit*, in which the satellite holds a fixed position over a selected point on the earth's equator. These earth-synchronous satellites are in an equatorial orbit at a height of 36,200 km (22,300 mi); there the satellite speed exactly matches the speed of eastward rotation of the globe.

The first of the geostationary satellites, called the Synchronous Meteorological Satellite (SMS), was launched in 1974 and was followed by several more of the same type. Five geostationary weather satellites, all launched by NASA since 1980, are in simultaneous operation: GOES-East and GOES-West, operated by NOAA; INSAT, by India; Meteostat, by the European Space Agency; and GMS Sunflower, by Japan.

The scanning systems of these satellites sweep over the globe from north to south, yielding an image equal to about one hemisphere. Figure 2.3 is such an image. Notice the succession of cyclonic storms clearly shown in the southern hemisphere. Tropical storms, as well as wave cyclones and fronts of midlatitudes, are continuously monitored by the earth-synchronous satellites, supplementing imagery returned by the polar-orbiting satellites (Figure 6.22). In television broadcasts, up-to-date images of the changing regional weather phenomena are made available to millions of viewers as a regular part of the weather forecast segment of network news programs.

Weather Disturbances and the Global Environment

Reviewing the major concepts covered in this chapter, we can recognize first the important role that cyclonic storms and their fronts play in the global balances of heat and water. Along with convectional activity, cyclonic storms liberate enormous quantities of latent heat carried as water vapor. Moist maritime air masses migrate poleward from the low-latitude belt of surplus energy. These air masses carry heat as well as water, and this heat is released in higher latitudes.

Cyclonic storms supply precipitation vital to native and cultivated plants of the midlatitude and tropical zones. But these same storms also produce environmental hazards to life. As yet, humans have been unable to exert far-reaching or lasting control over the enormous releases of energy in cyclonic storms. However, looking into the future, we can perhaps anticipate more effective methods of achieving weather modification than those now being tried. With such success may come side effects that could be unwanted and undesirable, and these will require close watching.

EL NIÑO—A GLOBAL WEATHER DISTURBANCE

At intervals of some three to eight years there occurs a remarkable disturbance of ocean and atmosphere that begins in the eastern Pacific ocean and spreads its effects widely over the globe for more than a year's time, bringing with it unseasonal weather patterns with abnormalities in the form of droughts, heavy rainfalls, severe spells of heat and cold, or a high incidence of cyclonic storms. This phenomenon is called *El Niño*. Back in the 1980s it was reported that Peruvian fishermen were using the expression "Corriente del Niño" ("Current of the Christ Child") to describe an invasion of warm surface water that occurred once every few years around Christmas time and greatly depleted their catch of fish.

The occurrences of El Niño have been at irregular intervals and with varying degrees of intensity, the most notable ones being in 1891, 1925, 1940–1941, 1965, 1972–1973, and 1982–1983.

As we mentioned in Chapter 4, the cool Humboldt (Peru) current flows northward off the South American coast, then about at the equator it turns westward across the Pacific as the south equatorial current. The Humboldt current is characterized by upwelling of cold deep water, bringing with it nutrients that serve as food for plankton.

Fish feed on these high concentrations of plankton. The anchoveta, a small fish used commercially to produce fish meal for animal feed, thrives in great numbers and is harvested by the Peruvian fishermen. With the onset of El Niño, upwelling ceases, the cool water is replaced by warm water from the west, and the plankton and their anchoveta predators disappear. Vast numbers of birds that feed on the anchoveta die of starvation. Only since the event of 1982–1983 has the full story of El Niño been uncovered, although certain far-reaching weather phenomena that accompany El Niño have long been linked with it.

Cessation of the Humboldt current upwelling is linked to a major change in barometric pressures across the entire stretch of the equatorial zone as far west as southeastern Asia. Normally, as the world map of pressure and winds for January clearly shows (Figure 4.10), low pressure prevails over northern Australia, the East Indies, and New Guinea. Abundant rainfall normally occurs in this area during this season, which is the high-sun period in the southern hemisphere. During the 1982–1983 El Niño, this low-pressure system was replaced by persistent high pressure, and

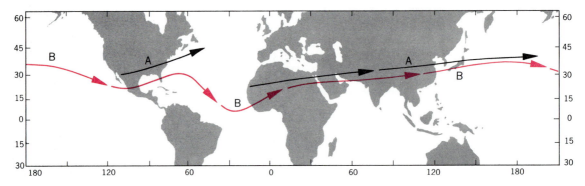

FIGURE 6.23 (A) The typical prevailing subtropical jet stream pattern for December through February. (B) Pattern during El Niño, December 1982 through February 1983. Wind speeds are in meters per second. (Eugene M. Rasmusson, NOAA.)

severe drought ensued. In contrast, in the equatorial zone of the eastern Pacific, pressures became lower than normal. This pressure reversal between west and east parts of the ocean resulted first in cessation of the usual easterly trade winds, then in the onset of low-level westerly winds.

Not only did this wind reversal cause the equatorial current to cease flowing, but actually set up an eastward movement of the warm surface water. As a result, sea-surface temperatures began to rise off the tropical western coasts of the Americas. Accompanying this temperature rise was a rise of sea level in what can be thought of as a kind of "sloshing" effect.

A consequence of the unusual warm water zone off the South American coast was the occurrence of torrential rainfalls, causing devastating floods in the Andean highlands of Peru, Bolivia, and Colombia. The warm surface layer spread northward along the Central American and North American coast, reaching latitudes as far north as Oregon. One effect of the rather sudden rise in water temperature and water level was the rapid heat-death in December 1982 of reef corals along more than 3000 km of reef-bound coasts. That event was regarded by some ecologists as a nonreversible ecological disaster leading to ultimate reef destruction.

Associated with the mature phase of El Niño was an important change in the pattern of the northern hemisphere subtropical jet stream, as shown in Figure 6.23. Notice that the El Niño jet swept

across Mexico and the Caribbean and continued strong across the Atlantic. It then entered North Africa and continued in strength across the Arabian Peninsula and northern India. One effect of the far southward position of the jet stream over North America was to bring frequent and intense invasions of cyclonic storms. These dumped heavy rains on coastal areas and heavy snows over the western mountain ranges, as well as heavy rains over the Gulf Coast and Florida. Precipitation in January and February 1983 exceeded 200 percent of normal in these locations.

Another probable effect of El Niño was the unusual occurrence between December 1982 and April 1983 of several tropical cyclones in the south-central Pacific, where normally few of these storms occur.

Early in 1983, the dry high-pressure area of the western Pacific moved slowly northward to cover the Philippines and then spread westward to include southern India and Sri Lanka, bringing drought conditions with it. Meantime, a severe drought was occurring in southeast Africa.

Various other kinds of disturbances of the atmosphere—collectively called "weather anomalies"—occurred in higher latitudes during the 1982–1983 El Niño. These inferred chains of worldwide effects belong to a class of meteorological events called "teleconnections." The atmospheric and oceanic interactions involved in such postulated teleconnections are obscure and are the object of continued research.

CHAPTER 7

THE GLOBAL SCOPE OF CLIMATE

CLIMATE AND HUMANS   As a species, we humans can adapt to a wide range of climates—cold, hot, wet, or dry—whereas a plant species limits its adaptation to a particular climate.

WHAT IS CLIMATE?   Climate is the characteristic condition of the atmosphere at a given place; it can be defined in various ways for different purposes.

INGREDIENTS OF CLIMATE   For physical geography, average values of both air temperature and precipitation are important ingredients of climate.

CLIMATE CLASSIFICATION   The global distribution of air-mass source regions and frontal zones provides a meaningful basis for a climate classification system.

CLIMATE GROUPS   Three major climate groups—low-latitude, midlatitude, and high-latitude—show the dominance of special combinations of air-mass source regions and frontal zones.

THE SOIL-WATER BALANCE   An accounting for the gains, losses, and storage changes of the soil water within a layer that supplies water to the roots of plants.

GLOBAL SOIL-WATER REGIMES   Soil-water budgets differ greatly from place to place over the globe. These differences are closely linked to climate.

A goal of physical geography is to recognize and describe environmental regions of significance to humans and to all life-forms generally. Which are the vital components of the environment of the life layer with respect to humans and to the biosphere in general? Keep in mind that the basic food supply of animals is organic matter synthesized by plants. Plants are the primary producers of organic matter that sustains other life.

In view of our dependence on the primary producers, it seems logical that we should emphasize those components of the environment that are vital to plant growth. Plants occupy both marine and terrestrial environments. Because plant life of the lands is most directly exposed to the atmosphere for exchanges of energy and matter, and because most of our food is derived from terrestrial plants, we focus on the land–atmosphere interface.

Two vital, atmospherically derived components vary greatly from place to place and from season to season: energy and water. Both must be in available forms. Plants require exposure to solar radiation in order to carry on photosynthesis.

(Photosynthesis is the process by which carbohydrate molecules are synthesized.) Plants need carbon dioxide as well as water to synthesize carbohydrate compounds. But carbon dioxide is nearly uniform in its atmospheric concentration over the globe at all seasons, so we can omit it as a variable factor in the environment.

Plants require the presence of sensible heat as measured by air and soil temperatures within specified limits. Plants also require water in the root zone of the soil. Although land animals are consumers they, too, require fresh water and a tolerable range of surrounding temperatures. Plants also release energy and water to the atmosphere through the process of respiration, in which the carbohydrate molecules are broken down into carbon dioxide and water. (Photosynthesis and respiration are discussed in Chapter 20).

Besides energy, carbon dioxide, and water, plants need nutrients, which they derive from the soil. The nutrients are released by the plants when the plant tissue dies. In this way the nutrients are returned, or recycled, to the soil.

To summarize, land plants require light energy, carbon dioxide, water, and nutrients. The input of these ingredients comes from two basic sources: (1) the adjacent atmosphere and (2) the soil. Input of light energy, heat energy, and water from the atmosphere is encompassed by the concept of climate. In short, plants depend on climate and soil. Of these two sources of energy and matter, the soil is the more strongly affected by the plants it serves through the recycling of matter.

Climate, when broadly defined as a source of energy and water, is an independent agent of control. Climate is determined by latitude and by large-scale air motions and air-mass interactions within the troposphere. If we are to establish a pyramid of priorities and interactions, climate occupies the apex as the independent control (Figure 7.1). Below it and forming the base of the pyramid are (1) the organic process of plants and (2) the soil process. Plants and soil interact with one another at the basal level.

Climate and Climate Classification

Climate has always been a keystone in physical geography and has formed the basis for defining physical regions of the globe. Let us first examine the content of traditional *climatology*, the science of climate. In the broadest sense, *climate* is the characteristic condition of the atmosphere near the earth's surface at a given place or over a given region. Components that enter into the description of climate are mostly the same as weather components used to describe the state of the atmosphere at a given instant. If we think of weather information as dealing with the specific event, then climate represents a generalization of weather. A statement of the climate of a given observing station, or of a designated region, is described through the medium of weather observations accumulated over many years' time. Not only are mean, or average, values taken into account, but the departures from those means and the probabilities that such departures will occur are also considered.

Earlier chapters contain much information falling within the definition of climate. For example, world maps showing average January and July barometric pressures, winds, and air temperatures are expressions of climate. The physical components of climate are many; they include measurable quantities, such as

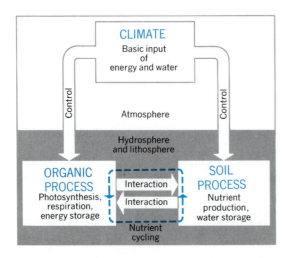

FIGURE 7.1 The role of climate is one of dominance over the environmental processes of the life layer.

radiation, sensible heat, barometric pressure, winds, relative and specific humidity, dew point, cloud cover and type, fog, precipitation type and intensity, evaporation and transpiration, incidence of cyclones and anticyclones, and frequency of frontal passages. Which of these components are really significant in our analysis of climate? Approaching climatology as geographers, we will take only what we need.

Our problem in devising a meaningful system of climate classes (i.e., a classification of climates) is to select categories of available information that correlate closely with the needs of life on the lands. We then use those categories to define and delineate regional classes of climates. If we have done our work well, the regional units within the climate system will reflect strongly the control of the atmosphere on terrestrial life. This same system will give some indication of the opportunities and constraints that the atmospheric environment imposes on humans as they seek to increase their food supplies and water resources at the same time that they extend the areas of urban and industrial land use. The climate classification we seek must have utility in guiding land-use planning and population growth as well as describing the natural environment. This role is particularly crucial in guiding the progress of the developing nations as they seek to augment inadequate food resources.

Let us now select those categories of information that will produce a meaningful system of climate classification.

Air Temperature as a Basis for Climate Classification

All species of organisms, whether plants or animals, are subject to limiting temperatures of the surrounding air, water, or soil; above or below these limits survival is not possible. Few growing plants can survive temperatures over 50°C (120°F) for more than a few minutes. Many species of plants native to the tropical and equatorial latitude zones die if they experience even a short period of below-freezing temperature (0°C; 32°F). Freezing of water in the plant tissue causes physical disruption in plants not

adapted to such conditions. Alternate freezing and thawing of the soil can have a disruptive effect on plant roots and is an important factor in limiting plant growth in the arctic and alpine zones.

Apart from extreme limits of temperature tolerance, plants react to an increase in air and water temperature through an increase in the intensity of physical and chemical activity. The optimum rate at which photosynthesis takes place increases with rising temperature in the range from near freezing to 20° to 25°C (70° to 80°F), after which the rate begins to decline. Knowing that plants are strongly influenced by the factor of sensible air and soil temperatures that surround them, we include air temperature as one of the essential climate factors.

Besides being an important environmental factor in plant physiology and reproduction, air temperature also enters into many activities of animal life, for example, hibernation and migration. For humans, air temperature is an important physiological factor and relates directly to the quantity of energy expended in space heating and air conditioning of buildings. Nevertheless, air temperature alone does not define meaningful climate classes, because the ingredient of water availability is missing.

Temperature of the lower air layer, as measured in the standard thermometer shelter (Chapter 3), has long provided one essential variable quantity in leading systems of climate classification. Monthly data based on daily readings of the maximum–minimum thermometer have been accumulated for many decades at thousands of observing stations the world over. Consequently, availability has been an important factor in favoring the use of air temperature in climate classification.

Thermal Regimes

In Chapter 3 we studied the annual air temperature cycle and its relationship to the annual energy balance cycle. Figure 3.9 showed annual temperature cycles for four stations, ranging from the equatorial zone to the subarctic zone. Let us now carry this investigation of annual temperature cycles a

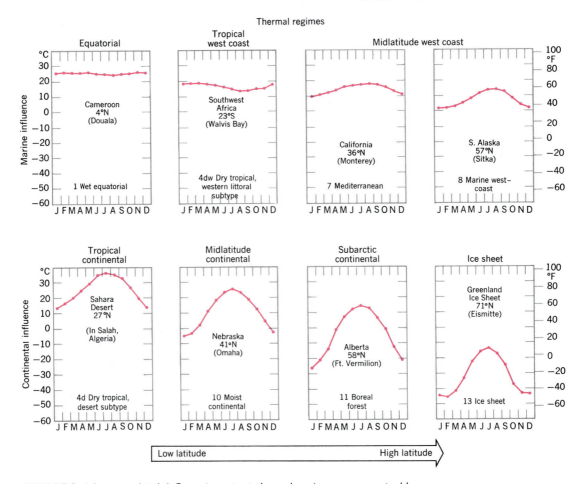

FIGURE 7.2 (above and right) Some important thermal regimes, represented by annual cycles of monthly mean air temperature. (Based on Goode Base Map.)

step further to explore the global range of annual cycles.

A comparison of annual air temperature cycles for many observing stations over the globe allows us to recognize a number of distinctive types, which can be called thermal regimes. A few of these are shown in Figure 7.2. Each regime has been labeled according to the latitude zone in which it lies: equatorial, tropical, midlatitude, and subarctic. Some labels also include words that describe the climate location in terms of position on a landmass. "Continental" refers to a continental interior location; "west coast" and "marine" refer to a location close to the ocean, usually on the western side of the continent.

The equatorial regime is uniformly very warm; temperatures are close to 27°C (80°F) year-round, and there are no temperature seasons. The tropical continental regime ranges from very hot when the sun is high,

near one solstice, to mild at the opposite solstice. However, close to the ocean we find the tropical west-coast regime with only a weak annual cycle and no extreme heat. The same weak annual cycle can be traced into the midlatitude west-coast regime, and it persists even much farther poleward. In the continental interiors the strong annual cycle prevails through the midlatitude continental regime into the subarctic continental regime, where the annual range is enormous. The ice sheet regime of Greenland is in a class by itself, with severe cold all year.

Other regimes can be identified and, because they grade into one another, the list could be expanded indefinitely. Instead of memorizing the eight regimes shown in Figure 7.2, you should simply get a feel for the concepts they illustrate. One concept is that of *continentality*, the tendency of a large landmass to impose a large annual range on

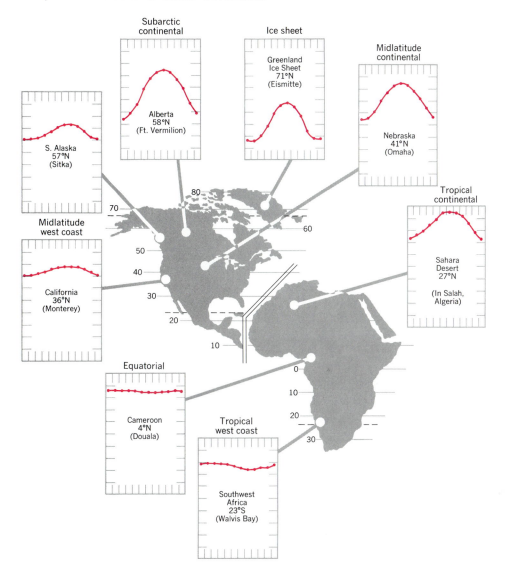

the annual cycle. Continentality becomes stronger with higher latitude, because the insolation cycle shows stronger seasonal extremes with increasing latitude. Another concept is that of the marine influence, which tends to weaken the annual cycle and to hold temperatures in a moderate range. This effect is due, of course, to the capacity of the oceans to hold a large amount of heat in storage and to take up and give off that heat very slowly, as compared with land areas.

Precipitation as a Basis for Climate Classification

Precipitation data, obtained by the simple rain gauge (Chapter 5), are abundantly available for long periods of record for thousands of observing stations widely distributed over the globe. Small wonder, then, that monthly and annual precipitation data form the cornerstone of most of the widely used climate classifications.

At this point it is important for you to obtain an overview of the global patterns of precipitation and their relationship to air-mass source regions and prevailing movements of air masses.

Average annual precipitation is shown on a world map (Figure 7.3). This map uses *isohyets,* lines drawn through all points having the same annual precipitation. For regions where all or most of the precipitation is rain, we use the word "rainfall"; for regions where snow is an important part of the annual total, we use the word "precipitation."

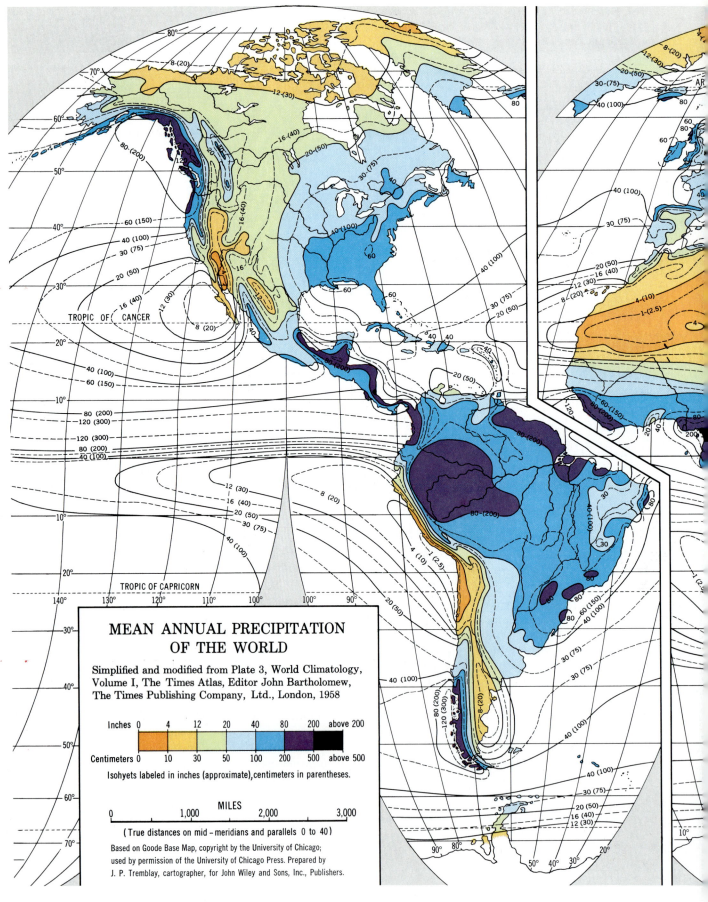

MEAN ANNUAL PRECIPITATION
OF THE WORLD

Simplified and modified from Plate 3, World Climatology,
Volume I, The Times Atlas, Editor John Bartholomew,
The Times Publishing Company, Ltd., London, 1958

Inches   0   4   12   20   40   80   200 above 200

Centimeters 0   10   30   50   100 200 500 above 500

Isohyets labeled in inches (approximate), centimeters in parentheses.

MILES

0   1,000   2,000   3,000

(True distances on mid-meridians and parallels 0 to 40)

Based on Goode Base Map, copyright by the University of Chicago;
used by permission of the University of Chicago Press. Prepared by
J. P. Tremblay, cartographer, for John Wiley and Sons, Inc., Publishers.

FIGURE 7.3 World precipitation.

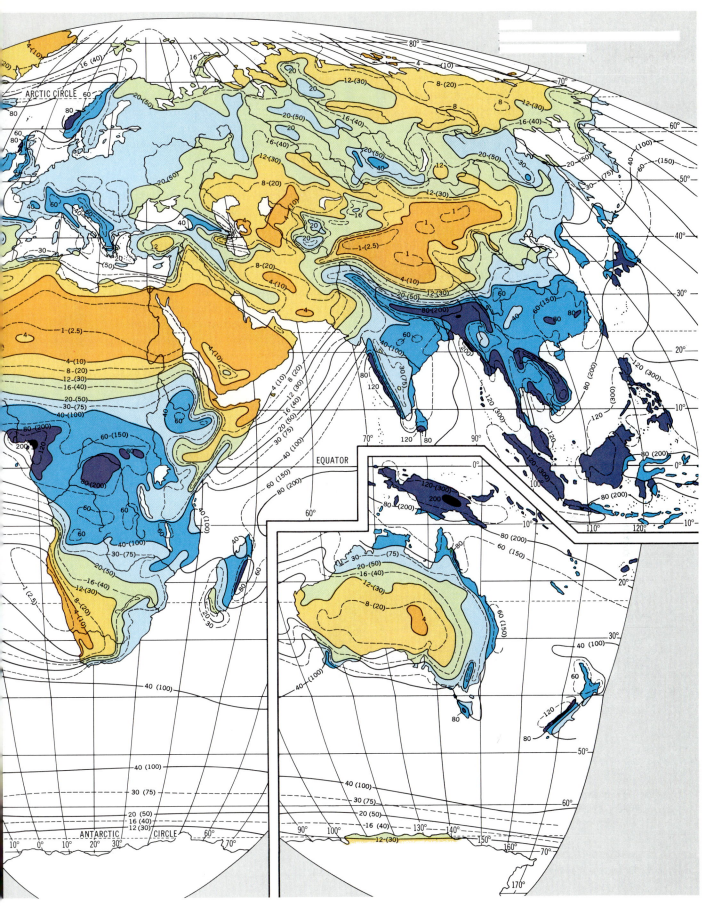

148

Seven precipitation regions can be recognized in terms of annual total in combination with location. Figure 7.4 is a schematic diagram of the global distribution of precipitation, simplifying the picture as much as possible in terms of an imaginary continent. Notice that the adjectives "wet," "humid," "subhumid," "semiarid," and "arid" have been applied to each of five levels of annual precipitation ranging from 200 cm (80 in.) down to zero.

1. *Wet equatorial belt.* This zone of heavy rainfall, over 200 cm (80 in.) annually, straddles the equator and includes the Amazon River basin in South America, the Congo River basin of equatorial Africa, much of the African coast from Nigeria west to Guinea, and the East Indies. Here the prevailing warm temperatures and high moisture content of the mE air masses favor abundant convectional rainfall. Thunderstorms are frequent year-round.

2. *Trade-wind coasts.* Narrow coastal belts of high rainfall, 150 to 200 cm (60 to 80 in.), and locally even more, extend from near the equator to latitudes of about 25° to 30° N and S on the eastern sides of every

continent or large island. Examples include the eastern coast of Brazil, Central America, Madagascar, and northeastern Australia. These are the trade-wind coasts, where moist mT air masses from warm oceans are brought over the land by the trades. As they encounter coastal hills and mountains, these air masses produce heavy orographic rainfall.

3. *Tropical deserts.* In striking contrast to the wet equatorial belt astride the equator are the two zones of vast tropical deserts lying approximately on the tropics of cancer and capricorn. These are hot, barren deserts, with less than 25 cm (10 in.) of rainfall annually and in many places with less than 5 cm (2 in.). They are located under and are caused by the subtropical cells of high pressure where the subsiding cT air mass is adiabatically warmed and dried. These deserts extend off the west coasts of the lands and out over the oceans. Rain here is largely convectional and extremely unreliable.

4. *Midlatitude deserts and steppes.* Farther northward, in the interiors of Asia and North America between lat. 30° and lat. 50°, are great continental midlatitude deserts

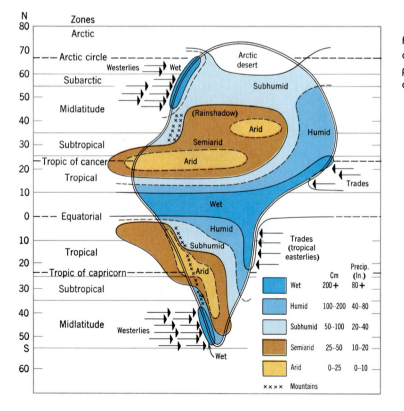

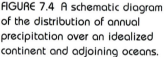

FIGURE 7.4 A schematic diagram of the distribution of annual precipitation over an idealized continent and adjoining oceans.

and expanses of semiarid grasslands known as steppes. Annual precipitation ranges from less than 10 cm (4 in.) in the driest areas to 50 cm (20 in.) in the moister steppes. Dryness here results from remoteness from ocean sources of moisture. Located in a region of prevailing westerly winds, these arid lands occupy the position of rainshadows in the lee of coastal mountains and highlands. For example, the Cordilleran Ranges of Oregon, Washington, British Columbia, and Alaska shield the interior of North America from moist mP air masses originating in the Pacific. Upon descending into the intermontane basins and interior plains, the mP air masses are warmed and dried.

Similarly, mountains of Europe and the Scandinavian peninsula obstruct the flow of moist mP air masses from the North Atlantic into western Asia. The great southern Asiatic ranges likewise prevent the entry of moist mT and mE air masses from the Indian Ocean.

The southern hemisphere has too little land in the midlatitudes to produce a true continental desert, but the dry steppes of Patagonia, lying on the lee side of the Andean chain, are roughly the counterpart of the North American deserts and steppes of Oregon and northern Nevada.

5. *Humid subtropical regions.* On the southeastern sides of the continents of North America and Asia, in lat. 25° to 45° N, are the humid subtropical regions, with 100 to 150 cm (40 to 60 in.) of rainfall annually. Smaller areas of the same kind are found in the southern hemisphere in Uruguay, Argentina, and southeastern Australia. These regions lie on the moist western sides of the subtropical high-pressure centers in such a position that moist mT air masses from the tropical ocean are carried poleward over the adjoining land. Commonly, too, these areas receive heavy rains from tropical cyclones.

6. *Midlatitude west coasts.* Another distinctive wet location is on midlatitude west coasts of all continents and large islands lying between about 35° and 65° in the region of prevailing westerly winds. These zones have been mentioned in Chapter 5 as good examples of coasts on which abundant orographic precipitation falls as a result of forced lift of mP air

masses. Where the coasts are mountainous, as in Alaska and British Columbia, southern Chile, Scotland, Norway, and South Island of New Zealand, the annual precipitation is over 200 cm (80 in.). These coasts formerly supported great valley glaciers that carved the deep bays (fiords) so typically a part of their scenery.

7. *Arctic and polar deserts.* A seventh precipitation region is formed by the arctic and polar deserts. Northward of the 60th parallel, annual precipitation is largely under 30 cm (12 in.), except for the west coast belts. Cold cP and cA air masses cannot hold much moisture; consequently, they do not yield large amounts of precipitation. At the same time, however, the relative humidity is high and evaporation rates are low.

As you would expect, zones of gradation lie between these precipitation regions. Moreover, this list does not recognize the fact that seasonal patterns of precipitation differ from region to region.

Seasonal Patterns of Precipitation

Although the annual total precipitation is a useful quantity in establishing the character of a climate type, it can be a misleading statistic because there may be strong seasonal cycles of precipitation. It makes a great deal of difference in terms of native plants and agricultural crops if there are alternately dry and wet seasons instead of a uniform distribution of precipitation throughout the year. It also makes a great deal of difference whether the wet season coincides with a season of higher temperatures or with a season of lower temperatures, because plants need both water and heat.

The seasonal aspects of precipitation can be largely covered by three major patterns: (1) uniformly distributed precipitation; (2) a precipitation maximum during the summer (or season of high sun) in which insolation is at its peak; and (3) a precipitation maximum during the winter or cooler season, when insolation is least. The first pattern can include a wide range of possibilities from little or no precipitation in any month to abundant precipitation in all months.

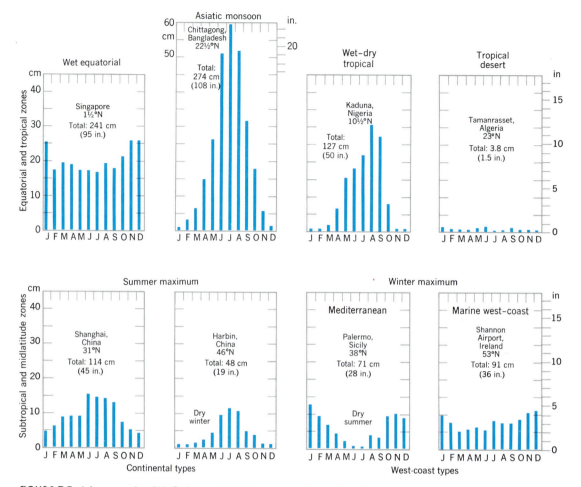

FIGURE 7.5 (above and right) Eight precipitation types selected to show various seasonal patterns. (Based on Goode Base Map.)

A study of monthly means of precipitation throughout the year at many observing stations over the globe shows a number of outstanding seasonal precipitation types. Some of these are illustrated in Figure 7.5. For each, the average monthly precipitation is shown by the height of the bar in the small graphs.

Dominant in very low latitudes is an equatorial rainy type, illustrated by Singapore, near the equator. Rain is abundant in all months, but some months have considerably more than others. The tropical desert type has so little rain in any month that it scarcely shows on the graph. The tropical wet–dry type has a very wet season at time of high sun (summer solstice) and a very dry season at time of low sun (winter solstice). This seasonal alternation is carried to its greatest extreme in the Asiatic monsoon type, with an extremely wet high-

sun monsoon season and a very dry period at low sun.

The summer precipitation maximum is carried into higher latitudes on the eastern sides of continents in a subtropical moist type. This same feature persists into the midlatitude continental type, which has a long, dry winter but a marked summer rain period.

A cycle with a winter precipitation maximum is typical of the west sides of continents in midlatitudes. A Mediterranean type, named for its prevalence in the Mediterranean region, has a very dry summer but a moist winter. This same cycle is carried into higher midlatitudes along narrow strips of west coasts. In the midlatitude west-coast type there is precipitation throughout the year, but with a distinct maximum in winter and a distinct minimum in summer.

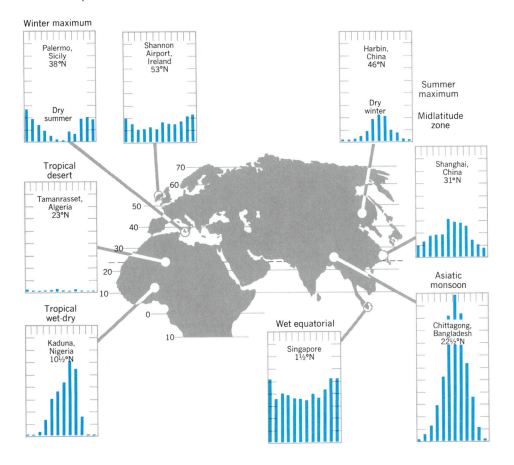

We do not intend that you memorize these precipitation types, but that you get a feel for the patterns illustrated. Each will appear in more detail in the descriptions of individual climates in later paragraphs. The reasons for annual precipitation cycles will need to be developed further in terms of shifting belts of pressure and winds and the seasonal changes in domination by different air masses.

A Climate System Based on Air Masses and Frontal Zones

Recall from Chapter 6 that air masses are classified according to the general latitude of their source regions and the surface qualities of the source region, whether land or ocean. This air-mass classification has built into it a recognition of the factors of temperature and precipitation, because (1) air-mass temperature decreases poleward and (2) precipitation capability of an air mass is related to sources of moisture.

Source regions of air masses change seasonally along with shifting belts of air temperature, pressure, and winds. Frontal zones also migrate seasonally. In this way seasonal cycles of both air temperature and precipitation can be described in terms of changing air-mass activity.

The climate system we shall develop here uses the concepts of thermal regimes and precipitation types, but explains them in terms of air-mass activity. This explanatory system makes use of all the global climatic information developed systematically in previous chapters. The system allows you to use that information to understand a given climate and to appreciate the reasons for its occurrence in a particular location.

Our climate system is based on the location of air-mass source regions and the nature and movement of air masses, fronts, and cyclonic storms. A schematic global diagram (Figure 7.6) shows the principles of the classification. Notice that this figure is the same as Figure 6.1. We have placed it here to show how three major climate groups can be superimposed.

Group I: Low-Latitude Climates

Group I includes the tropical air-mass source regions and the equatorial trough, or ITC, that lies between. Climates of Group I are controlled by the subtropical high-pressure cells, or anticyclones, which are regions of air subsidence and are basically dry, and by the great equatorial trough of convergence that lies between them. Although air of polar origin occasionally invades the tropical and equatorial zones, the climates of Group I are almost wholly dominated by tropical and equatorial air masses. Easterly waves and tropical cyclones are important in this climate group.

Group II: Midlatitude Climates

Climates of Group II are in a zone of intense interaction between unlike air masses: the *polar front zone*. Tropical air masses moving poleward and polar air masses moving equatorward are in conflict in this zone, which contains a procession of eastward-moving wave cyclones. Locally and seasonally, either tropical or polar air masses may dominate in these regions, but neither has exclusive control.

Group III: High-Latitude Climates

Climates of Group III are dominated by polar and arctic (including antarctic) air masses. The two polar continental air-mass source regions of northern Canada and Siberia fall into this group, but there is no southern hemisphere counterpart to these continental centers. In the arctic belt of the 60th to 70th parallels, air masses of arctic origin meet polar continental air masses along an *arctic front zone*, creating a series of eastward-moving wave cyclones.

Climate Types

Within each climate group are a number of *climate types* (or simply, *climates*): four in Group I, six in Group II, and three in Group III for a total of 13 climate types. For ease in identification on maps and diagrams, the climate types are numbered, but the numbers have no significance as a code. The order in which climates are taken within each group may be quite arbitrary, particularly for the climates of Group II. You need use only the climate names,

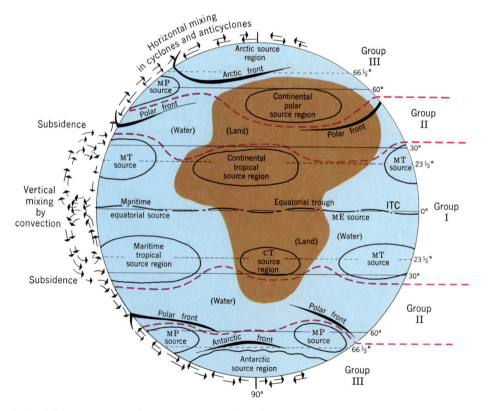

FIGURE 7.6 Three major climate groups are related to air-mass source regions and frontal zones.

because these describe climate quality and suggest global location.

In presenting the climate types, we will make use of a pictorial device called a *climograph*. It is a simple graph that shows simultaneously the cycles of monthly mean air temperature and monthly mean precipitation. The climograph combines temperature and precipitation graphs such as those used in Figure 7.2 and 7.5. Figure 7.7 is a set of typical climographs for the 13 climates. Reduced to small size, it is useful for comparing climates and for demonstrating the orderly changes in temperature and precipitation regimes that occur with changing latitude and continental location.

Another useful graphic device is a generalized diagram of the climate groups and climate types (Figure 7.8). Boundaries are drawn in smooth, simple lines on an imaginary supercontinent, as we did for precipitation (see Figure 7.4). Coastal highlands are shown because they have strong local effects in producing orographic precipitation belts and rainshadows. The diagram shows typical positions of each climate type and the common boundaries between climates. Boundaries between groups are labeled.

Our world map of climates, Figure 7.9, shows the actual distribution of climate types and subtypes on the continents. This map is based on available data of a large number of observing stations. It is a simplified map, because the climate boundaries are uncertain in many areas where observing stations are thinly distributed. Precise definitions of the climates and their boundaries are available, but our concern here is to analyze the 13 climates in purely descriptive terms, using actual figures of temperature and precipitation very broadly and only as examples.

Dry and Moist Climates

All but two of the 13 climate types are classified as either dry climates or moist climates. *Dry climates* are those in which total annual evaporation of moisture from the soil and from plant foliage exceeds the annual precipitation by a wide margin.

Generally speaking, the dry climates do not support permanent streams. The soil is dry much of the year and the land surface is clothed with sparse plant cover—scattered grasses or shrubs—or simply lacks a plant cover. *Moist climates* are those with sufficient rainfall to maintain the soil in a moist condition through much of the year and to sustain the year-round flow of the larger streams. Moist climates support forests or prairies of dense tall grasses.

Within the dry climates there is a wide range of degree of aridity, ranging from very dry deserts through transitional levels of aridity in the direction of adjacent moist climates. We will refer to three dry *climate subtypes:* (1) semiarid (or steppe); (2) semidesert; and (3) desert. The *semiarid subtype* (or steppe subtype), designated by the letter *s*, has enough precipitation to support sparse grasslands of a variety referred to by geographers as steppes, or steppe grasslands. The steppe climate subtype will be found adjacent to moist climates. The *desert subtype*, letter *d*, is extremely dry and has so little precipitation that only scattered hardy plants can grow. The *semidesert subtype*, letters *sd*, is transitional between semiarid and desert subtypes.

Within the moist climates there is a wide range of degree of wetness. Bordering the dry climates is a *subhumid subtype*, designated by the letters *sh*, in which the evaporation losses of moisture from soil and plant cover approximately balance the precipitation, on an annual average basis. Where precipitation is great enough to produce stream flow through most of the year and to support forests, the *humid subtype* prevails, designated by the letter *h*. Where precipitation is very heavy, with copious stream flow, the *perhumid subtype* prevails, letter *p*.

Two of our 13 climates cannot be accurately described as either dry or moist climate. Instead, they show a seasonal alteration between a very wet season and a very dry season. This striking contrast in seasons gives a special character to the two climates we have singled out for recognition as moist–dry climates. In Figure 7.5, they are designated as having the wet–dry tropical and the Mediterranean precipitation patterns.

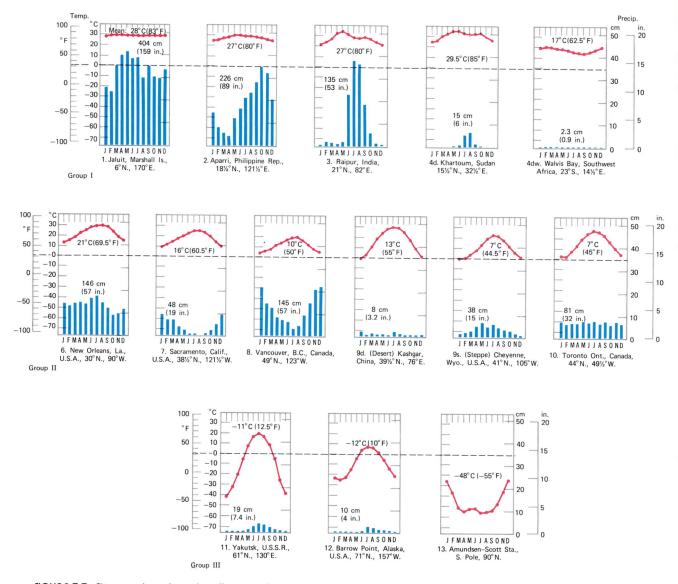

FIGURE 7.7 Climographs selected to illustrate climate types within each climate group.

Soil Moisture and Climate

A basic concept of physical geography is that the availability of water to plants is a more important factor in the environment than precipitation itself. Much of the water received as precipitation is lost in a variety of ways and is not usable for plants. Just as in a fiscal budget, when the monthly or yearly loss of moisture exceeds the precipitation, a budgetary deficit results; when precipitation exceeds the losses, a budgetary surplus results. An analysis of the water budget of a given area of the earth's

surface is actually arrived at in much the same way as the understanding of a fiscal budget. The accounting requires only additions and subtractions of amounts within fixed periods of time, such as the calendar month or year.

(Text continued on page 158.)

FIGURE 7.8 (Right) A schematic diagram of the placement of climates on an idealized continent. Compare with the world map, Figure 7.9.

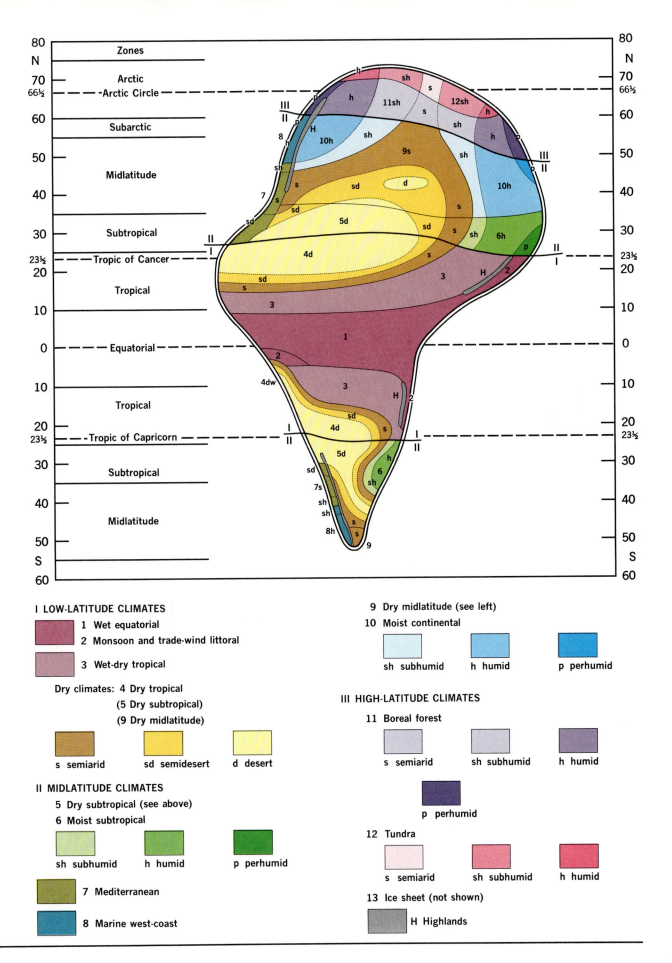

I LOW-LATITUDE CLIMATES

1 Wet equatorial
2 Monsoon and trade-wind littoral

3 Wet-dry tropical

Dry climates: 4 Dry tropical
(5 Dry subtropical)
(9 Dry midlatitude)

s semiarid   sd semidesert   d desert

II MIDLATITUDE CLIMATES

5 Dry subtropical (see above)
6 Moist subtropical

sh subhumid   h humid   p perhumid

7 Mediterranean

8 Marine west-coast

9 Dry midlatitude (see left)
10 Moist continental

sh subhumid   h humid   p perhumid

III HIGH-LATITUDE CLIMATES

11 Boreal forest

s semiarid   sh subhumid   h humid

p perhumid

12 Tundra

s semiarid   sh subhumid   h humid

13 Ice sheet (not shown)

H Highlands

155

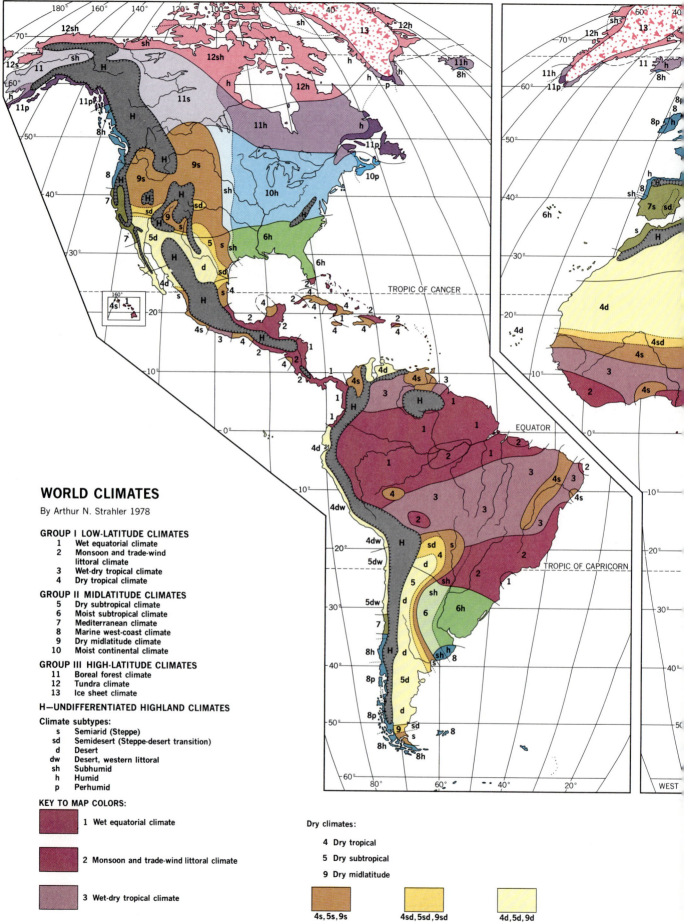

WORLD CLIMATES

By Arthur N. Strahler 1978

GROUP I LOW-LATITUDE CLIMATES
1. Wet equatorial climate
2. Monsoon and trade-wind littoral climate
3. Wet-dry tropical climate
4. Dry tropical climate

GROUP II MIDLATITUDE CLIMATES
5. Dry subtropical climate
6. Moist subtropical climate
7. Mediterranean climate
8. Marine west-coast climate
9. Dry midlatitude climate
10. Moist continental climate

GROUP III HIGH-LATITUDE CLIMATES
11. Boreal forest climate
12. Tundra climate
13. Ice sheet climate

H—UNDIFFERENTIATED HIGHLAND CLIMATES

Climate subtypes:
- s   Semiarid (Steppe)
- sd   Semidesert (Steppe-desert transition)
- d   Desert
- dw   Desert, western littoral
- sh   Subhumid
- h   Humid
- p   Perhumid

KEY TO MAP COLORS:

	1 Wet equatorial climate
	2 Monsoon and trade-wind littoral climate
	3 Wet-dry tropical climate

Dry climates:

4   Dry tropical
5   Dry subtropical
9   Dry midlatitude

| 4s, 5s, 9s | 4sd, 5sd, 9sd | 4d, 5d, 9d |

156

FIGURE 7.9 Climates of the world. (Based on Goode Base Map.)

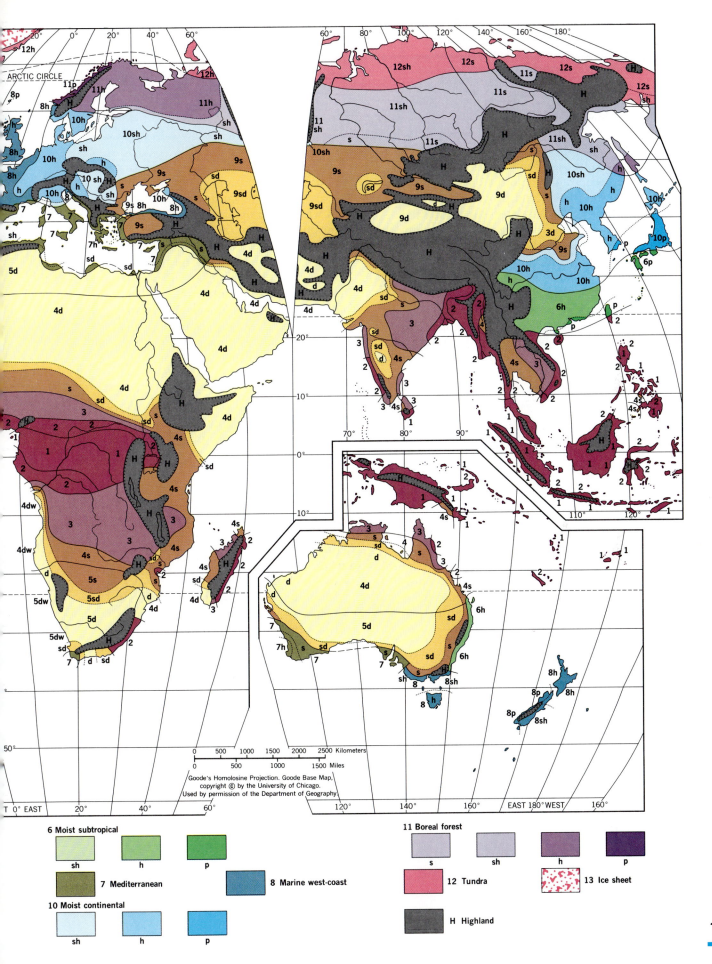

ARCTIC CIRCLE

Goode's Homolosine Projection. Goode Base Map,
copyright © by the University of Chicago.
Used by permission of the Department of Geography.

T 0° EAST   20°   40°   60°   120°   140°   160°   EAST 180° WEST   160°

6 Moist subtropical

sh   h   p

7 Mediterranean   **8 Marine west-coast**

10 Moist continental

sh   h   p

11 Boreal forest

s   sh   h   p

12 Tundra   **13 Ice sheet**

H Highland

Evaporation and Transpiration

Soil moisture, which we refer to here as *soil water,* is water held in pore spaces of the soil layer and available to plants. Between periods of rain the soil water is gradually given up by a twofold drying process. First, direct evaporation into the open air occurs at the soil surface and progresses downward. Air also enters the soil freely and may actually be forced alternately in and out of the soil by atmospheric pressure changes. Even if the soil did not "breathe" in this way, there would be a slow diffusion of water vapor surfaceward through the open soil pores. Ordinarily only the first 30 cm (1 ft) of soil is dried by evaporation in a single dry season. In the prolonged drought of deserts, a dry condition extends to depths of many meters.

Second, plants draw the soil water into their systems through vast networks of tiny rootlets. This water, after being carried upward through the trunk and branches into the leaves, is discharged through leaf pores into the atmosphere in the form of water vapor. The process is termed *transpiration.*

In studies of climatology it is convenient to use the term *evapotranspiration* to cover the combined water loss from direct evaporation and the transpiration of plants. The rate of evapotranspiration slows down as soil-water supply becomes depleted during a dry summer period because plants employ various devices to reduce transpiration. In general, the less moisture remaining, the slower is the loss through evapotranspiration.

Water in the Soil

Evapotranspiration is only one of the many flow paths of water in the *hydrologic cycle,* or total water cycle operating in the earth's atmosphere, oceans, and solid crust. We will investigate the hydrologic cycle in Chapter 10. Here, attention focuses on the shallow surface layer of the lands. As Figure 7.10 shows, precipitation enters the soil by *infiltration.* Precipitation can also escape by flow over the sloping ground surface (runoff by overland flow). Precipitation that has infiltrated the surface, and is temporarily held in the soil layer as soil water, occupies the *soil-water belt.* This belt gains water through precipitation and infiltration. As the minus signs show, the soil loses water through transpiration and evaporation, combined as evapotranspiration. Excess water also leaves the soil by downward *gravity percolation* to the ground water zone below. Between the soil-water belt and the ground water zone is an *intermediate belt.* Here water is held at a depth too great to be returned to the surface by evapotranspiration, since it is below the level of plant roots.

When infiltration occurs, the water is drawn downward through the soil pores, wetting successively lower layers. This

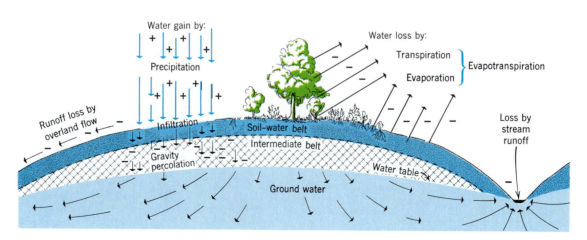

FIGURE 7.10 The soil-water belt occupies an important position in the hydrologic cycle.

activity is called *soil-water recharge.* Eventually the soil layer holds the maximum possible quantity of water, although the larger pores remain filled with air. Water movement then continues downward through the underlying intermediate belt.

Suppose now that the rain stops and several days of dry weather follow. The excess soil water continues to drain downward, but some water clings to the soil particles and effectively resists the pull of gravity because of the force of capillary tension. We are all familiar with the way a water droplet seems to be enclosed in a "skin" of surface molecules, drawing the droplet together into a spherical shape, so that it clings to the side of a glass indefinitely without flowing down. Similarly, tiny films of water adhere to the soil grains, particularly at the points of grain contacts, and will stay until disposed of by evaporation or by absorption into plant rootlets.

When a soil has first been saturated by water, then allowed to drain under gravity until no more water moves downward, the soil is said to be holding its *storage capacity* of water. Drainage takes no more than two or three days for most soils. Most excess water is drained out within one day. Storage capacity is measured in units of depth,

usually centimeters or inches, as with precipitation. For example, a storage capacity of 2 cm in the uppermost 10 cm of the soil would mean that for a given cube of soil, 10 cm on a side (1 cubic decimeter), all the extractable water would form a layer of water 2 cm deep in a 10 × 10-cm container with flat bottom and vertical sides. This depth would be equivalent to complete absorption of a 2-cm rainfall by a completely dry 10-cm layer of soil.

Storage capacity of a given soil depends largely on its texture. Sandy soil has a very small storage capacity; clay soil has a large capacity.

The Soil-Water Cycle

We can turn next to consider the annual water balance of the soil. Figure 7.11 shows the annual cycle of soil water for a single year at an agricultural experiment station in Ohio. This example can be considered generally representative of conditions in moist, midlatitude climates where there is a strong temperature contrast between winter and summer.

Let us start with the early spring (March). At this time the evaporation rate is low, because of low air temperatures. The abundance of melting snows and rains has

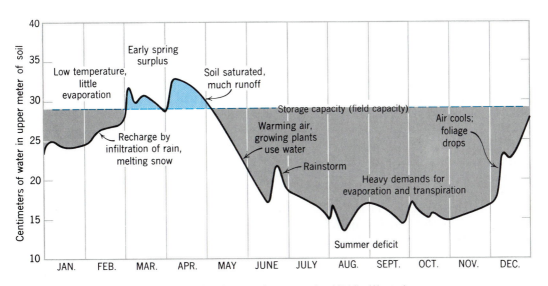

FIGURE 7.11 A typical annual cycle of soil-water change in the Middle West shows a short period of surplus in the spring and a long period of deficit during the summer and fall. (Data of Thornthwaite and Mather.)

restored the soil water to a surplus quantity. For two months the quantity of water percolating through the soil and entering the ground water keeps the soil pores nearly filled with water. This is the time of year when one encounters soft, muddy ground conditions, whether driving on dirt roads or walking across country. This, too, is the season when runoff is heavy, and major floods may be expected on larger streams and rivers. In terms of the soil-water budget, a *water surplus* exists.

By May the rising air temperatures, increasing evaporation, and full growth of plant foliage bring on intense evapotranspiration. Now the soil water falls below the storage capacity, although it may be restored temporarily by unusually heavy rains in some years. By midsummer a large *water deficit* exists in the water budget. Even the occasional heavy thunderstorm rains of summer cannot restore the water lost by heavy evapotranspiration. Small springs and streams dry up, and the soil becomes firm and dry. By November (and sometimes in September), however, the soil water again begins to increase. This is because the plants go into a dormant state, sharply reducing transpiration losses. At the same time, falling air temperatures reduce evaporation. By late winter, usually in February at this location, the storage capacity of the soil is again fully restored.

The Soil-Water Balance

From the example of soil-water change in a single year, we move forward to a more generalized concept. The gain, loss, and storage of soil water are accounted for in the *soil-water balance*. Figure 7.12 is a pictorial flow diagram that illustrates the components of the balance. Water held in storage in the soil-water zone is increased by recharge during precipitation, but decreased by use through evapotranspiration. Surplus water is disposed of by downward percolation to the ground water zone or by overland flow.

To proceed, we must recognize two ways to define evapotranspiration. First is *actual evapotranspiration*, which is the true or real rate of water vapor return to the atmosphere from the ground and its plant

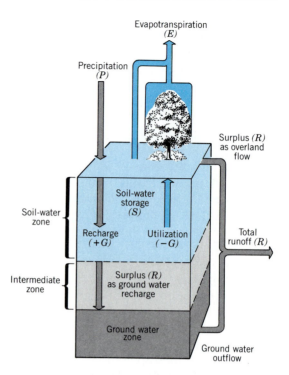

FIGURE 7.12 Schematic diagram of the soil-water balance in a soil column. (From A. N. Strahler, *The Earth Sciences*, 2nd ed., Harper & Row, New York. Copyright © 1971 by Arthur N. Strahler.)

cover. Second is *potential evapotranspiration*, representing the water vapor loss under an ideal set of conditions. One condition is that there be present a complete (or closed) cover of uniform vegetation consisting of fresh green leaves and no bare ground exposed through that cover. The leaf cover is assumed to have a uniform height above ground—whether the plants be trees, shrubs, or grasses. A second condition is that there be an adequate water supply, such that the storage capacity of the soil is maintained at all times. This condition can be fulfilled naturally by abundant and frequent precipitation, or artificially by irrigation. To simplify the ponderous terms we have just defined, they may be transformed as follows:

actual evapotranspiration is *water use*
potential evapotranspiration is *water need*

The word "need" signifies the quantity of soil water needed if plant growth is to be maximized for the given conditions of solar radiation and air temperature and the available supply of nutrients.

The difference between water use and water need is the *soil-water shortage*. This is the quantity of water that must be furnished by irrigation to achieve maximum crop growth within an agricultural system.

A Simple Soil-Water Budget

We can now make a simple accounting of the monthly and annual quantities of the components of the soil-water balance. The numerical accounting is called a *soil-water budget;* it involves only simple addition and subtraction of monthly mean values for a given observing station. All terms of the soil-water budget are stated in centimeters of water depth, the same as for precipitation.

A simplified soil-water budget is shown in Figure 7.13. The seven terms we need for a complete budget are listed as follows, along with abbreviations used on the graph:

> Precipitation, P
> Water need, Ep
> Water use, Ea
> Storage withdrawal, −G
> Storage recharge, +G
> Soil-water shortage, D
> Water surplus, R

Points on the graph represent average monthly values. They are connected by smooth curves to enhance annual cycles of change. In our example, precipitation (P) is much the same in all months, with no strong annual cycle. In contrast, water need (Ep) shows a strong seasonal cycle, with low values in winter and a high summer peak. For a moist climate in middle latitudes, this model is about right.

At the start of the year a large water surplus (R) exists; it is disposed of by runoff. By May, conditions have switched over to a water deficit. In this month plants begin to withdraw soil water from storage. *Storage withdrawal* (−G) is represented by the difference between the water-use curve and the precipitation curve. As storage withdrawal continues, the plants reduce their water use to less than the optimum quantity, so that without irrigation the water-use curve departs from the water-need curve. Storage withdrawal continues

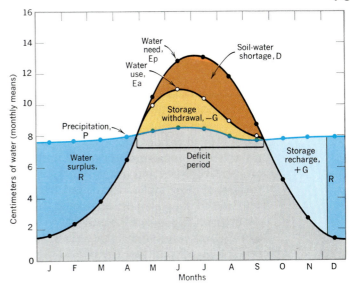

FIGURE 7.13 A simplified soil-water budget typical of a moist climate in middle latitudes.

throughout the summer. The deficit period lasts through September. The area labeled soil-water shortage (D) represents the total quantity of water needed by irrigation to ensure maximum growth throughout the deficit period.

In October, precipitation (P) again exceeds water need (Ep), but the soil must first absorb an amount equal to the summer storage withdrawal. So there follows a period of *storage recharge* (+G); it lasts through November. In December the soil has reached its full storage capacity, arbitrarily fixed at 30 cm (12 in.). Now, a water surplus (R) again sets in, lasting through the winter.

The soil-water budget was developed by C. Warren Thornthwaite, a distinguished climatologist and geographer. He was concerned with practical problems of crop irrigation. He developed the calculation of the soil-water budget in order to place crop irrigation on a precise, accurate basis. The Thornthwaite method was later adopted by soil scientists of the U.S. Department of Agriculture to identify classes of soils associated with various soil-water conditions.

In a world beset by severe and prolonged food shortages, the Thornthwaite concepts and calculations are of great value in assessing the benefits to be gained by increased irrigation. Only in a few parts of the tropical and midlatitude zones is precipitation ample to fulfill the water need during the growing season. In contrast, the equatorial zone generally has a large water surplus throughout the year.

The Global Range of Water Need

Figure 7.14 illustrates the effect of latitude on the annual cycle of water need. Both air temperature cycles and insolation cycles are reflected in these cycles. Each chart has been labeled as to climate, according to the system given in Figures 7.8 and 7.9.

In the wet equatorial climate (1) water need is high all year and the total is over 150 cm (60 in.). For the monsoon and trade-wind littoral climate (2) in the tropical latitude zone, there is a pronounced annual cycle, but a large annual total water need: almost 150 cm (60 in.). In the tropical desert climate (4d) the annual cycle is very strongly developed and the total water need remains high: nearly 127 cm (50 in.).

In midlatitudes on the west coast, the Mediterranean climate (7) shows a well-developed annual cycle, but monthly values of water need remain moderately great throughout the mild winter. In the moist continental climate (10) three winter months have no water need, but the summer peak is high; the total is 75 cm (30 in.). In the marine west-coast climate (8), however, close to the Pacific Ocean, water need remains substantial throughout the winter because of the mild temperatures.

Farther poleward in the boreal forest climate (11) the number of months of zero water need increases to six, and the annual total water need diminishes to 40 cm (16 in.). North of the arctic circle, the tundra climate station (12) shows nine consecutive winter months of zero water need and a very narrow summer peak. The total water need here is least of all the examples: 20 cm (8 in.).

The details of these examples are not important; the trends they show are. Total annual water need diminishes from a maximum in the equatorial zone to a minimum in the arctic zone. At the same time, the annual cycle becomes stronger and stronger in the poleward direction. As the cold of winter makes its effects more pronounced, the number of months of zero

water need increases. All soil water is solidly frozen in this period. Of course, plant growth is essentially zero when soil water is frozen, so the growing season is nicely reflected in the graphs of water need. Water need persists through the winter far poleward along narrow western coastal zones (littorals), in contrast to the continental interior regions.

The Soil-Water Balance and World Climates

Principles of the soil-water balance can be easily fitted into our system of global climates. For each climate type, a typical soil-water budget—similar to that shown in Figure 7.13—can be constructed. A soil-water graph of this kind can be matched to each of the climographs shown in Figure 7.7. In Chapters 8 and 9 corresponding graphs appear for many of the climate types described.

Figure 7.15 is a schematic diagram of world climates arranged according to water need and the degree of wetness or dryness. The centerline of the graph divides the graph into two halves. To the left lie the dry climates, graded according to the quantity of soil-water shortage. To the right lie the moist climates, graded according to the annual water surplus. The individual climates are arranged in tiers, the lowest tier being that with the greatest annual water need, and grading upward to climates of successively smaller water need.

Although 11 of the 13 climate types fit neatly into the two-way classification scheme shown in Figure 7.15, two types do not fit into boxes because they cannot be classified as being either moist or dry. Instead, these two types (3 and 7) have a very wet season alternating with a very dry season, with the result that each shows both a large soil-

FIGURE 7.14 (right) Some representative annual cycles of water need (potential evapotranspiration) over a wide latitude range. (Data of C. W. Thornthwaite Associates, Laboratory of Climatology, Centerton, N.J.)

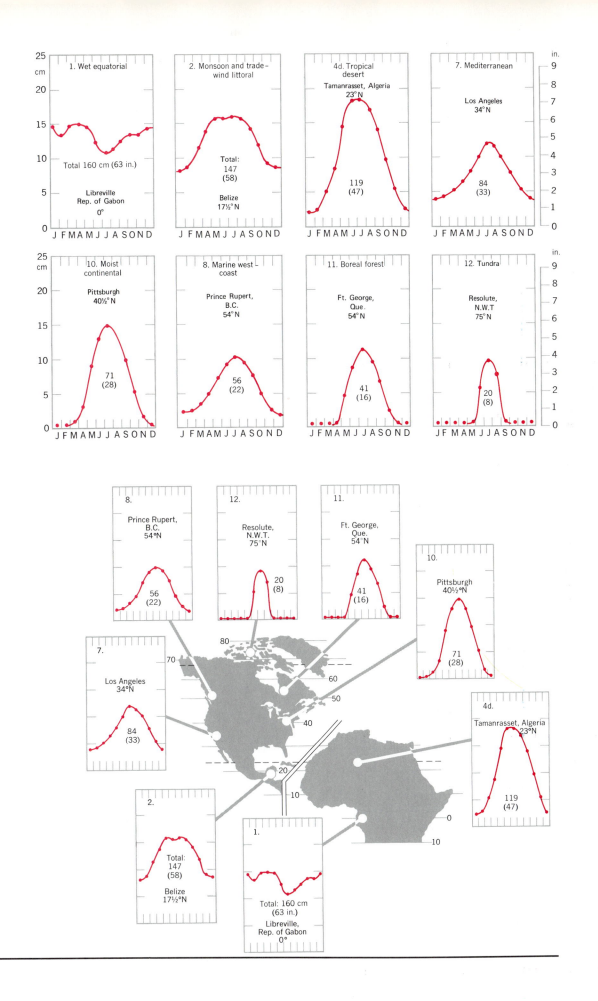

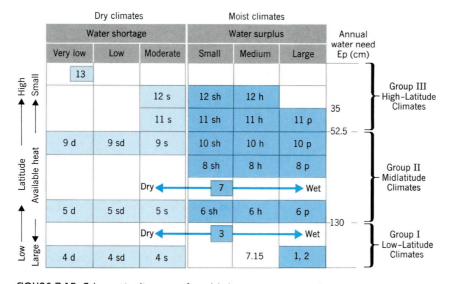

FIGURE 7.15 Schematic diagram of world climates in terms of water need and the degree of dryness and wetness.

water shortage and a large soil-water surplus. The two climates of this type are the wet–dry tropical climate (3) and the Mediterranean climate (7).

This chapter has laid the groundwork for a study of world climate types. With a classification system at our disposal, we turn in the next two chapters to an in-depth examination of each climate type, noting the geographical distribution of the climate, the controlling factors of latitude and position with respect to the great continental landmasses, and the role played by air masses and frontal zones in varying the precipitation throughout the annual seasonal cycle.

THE KÖPPEN CLIMATE SYSTEM

Air temperature and precipitation data have formed the basis for several climate classifications. One of the most important of these is the Köppen climate system, devised in 1918 by Dr. Vladimir Köppen of the University of Graz in Austria. For several decades this system, with various later revisions, was the most widely used climate classification among geographers. Köppen was both climatologist and plant geographer, so that his main interest lay in finding climate boundaries that coincided approximately with boundaries between major vegetation types.

Under the Köppen system each climate is defined according to assigned values of temperature and precipitation, computed in terms of annual or monthly values. Any given station can be assigned to its particular climate group and subgroup solely on the basis of the records of temperature and precipitation at that place provided, of course, that the period of record is long enough to yield meaningful averages.

The Köppen system features a shorthand code of letters designating major climate groups, subgroups within the major groups, and further subdivisions to distinguish particular seasonal characteristics of temperature and precipitation.

Five major climate groups are designated by capital letters as follows:

A *Tropical rainy climates.* Average temperature of every month is above

18°C (64.4°F). These climates have no winter season. Annual rainfall is large and exceeds annual evaporation.

B **Dry climates.** Evaporation exceeds precipitation on the average throughout the year. No water surplus; hence no permanent streams originate in *B* climate zones.

C **Mild, humid (mesothermal) climates.** Coldest month has an average temperature under 18°C (64.4°F), but above −3°C (26.6°F); at least one month has an average temperature above 10°C (50°F). The *C* climates have thus both a summer and a winter.

D **Snowy-forest (microthermal) climates.** Coldest month average temperature under −3°C (26.6°F). Average temperature of warmest month above 10°C (50°F), that isotherm coinciding approximately with poleward limit of forest growth.

E **Polar climates.** Average temperature of warmest month below 10°C (50°F). These climates have no true summer.

Note that four of these five groups (*A, C, D,* and *E*) are defined by temperature averages, whereas one (*B*) is defined by precipitation-to-evaporation ratios. Groups *A, C,* and *D* have sufficient heat and precipitation for the growth of forest and woodland vegetation. Figure 7.16 shows the boundaries of the five major climate groups. Figure 7.17 is a world map of Köppen climates.

Subgroups within the five major groups are designated by a second letter according to the following code.

S Semiarid (steppe).

W Arid (desert).

(The capital letters *S* and *W* are applied only to the dry *B* climates.)

f Moist. Adequate precipitation in all months. No dry season. This modifier is applied to *A, C,* and *D* groups.

w Dry season in winter of the respective hemisphere (low-sun season).

s Dry season in summer of the respective hemisphere (high-sun season).

M Rainforest climate despite short, dry season in monsoon type of precipitation cycle. Applies only to *A* climates.

From combinations of the two letter groups, 12 distinct climates emerge:

Af **Tropical rainforest climate.** Rainfall of the driest month is 6 cm (2.4 in.) or more.

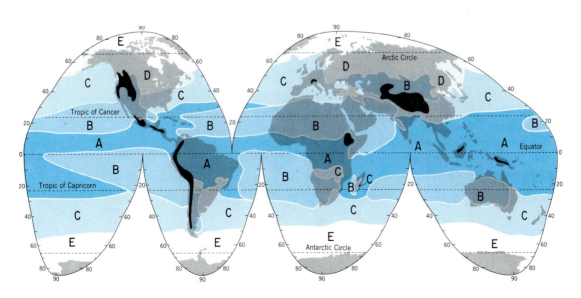

FIGURE 7.16 Highly generalized world map of major climatic regions according to the Köppen classification. Highland areas are in black. (Based on Goode Base Map.)

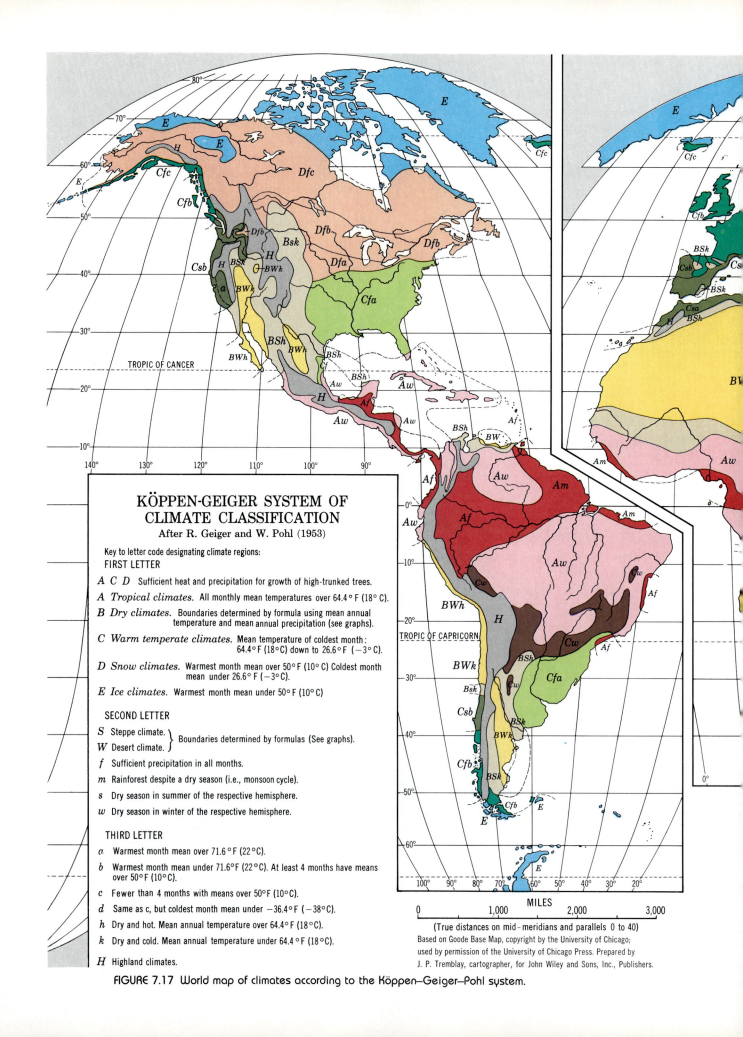

KÖPPEN-GEIGER SYSTEM OF CLIMATE CLASSIFICATION

After R. Geiger and W. Pohl (1953)

Key to letter code designating climate regions:

FIRST LETTER

A C D   Sufficient heat and precipitation for growth of high-trunked trees.

A   *Tropical climates.*   All monthly mean temperatures over 64.4°F (18°C).

B   *Dry climates.*   Boundaries determined by formula using mean annual temperature and mean annual precipitation (see graphs).

C   *Warm temperate climates.*   Mean temperature of coldest month: 64.4°F (18°C) down to 26.6°F (−3°C).

D   *Snow climates.*   Warmest month mean over 50°F (10°C) Coldest month mean under 26.6°F (−3°C).

E   *Ice climates.*   Warmest month mean under 50°F (10°C)

SECOND LETTER

S   Steppe climate.
W   Desert climate.   } Boundaries determined by formulas (See graphs).

f   Sufficient precipitation in all months.

m   Rainforest despite a dry season (i.e., monsoon cycle).

s   Dry season in summer of the respective hemisphere.

w   Dry season in winter of the respective hemisphere.

THIRD LETTER

a   Warmest month mean over 71.6°F (22°C).

b   Warmest month mean under 71.6°F (22°C). At least 4 months have means over 50°F (10°C).

c   Fewer than 4 months with means over 50°F (10°C).

d   Same as c, but coldest month mean under −36.4°F (−38°C).

h   Dry and hot. Mean annual temperature over 64.4°F (18°C).

k   Dry and cold. Mean annual temperature under 64.4°F (18°C).

H   Highland climates.

MILES
0   1,000   2,000   3,000

(True distances on mid-meridians and parallels 0 to 40)

Based on Goode Base Map, copyright by the University of Chicago; used by permission of the University of Chicago Press. Prepared by J. P. Tremblay, cartographer, for John Wiley and Sons, Inc., Publishers.

FIGURE 7.17 World map of climates according to the Köppen–Geiger–Pohl system.

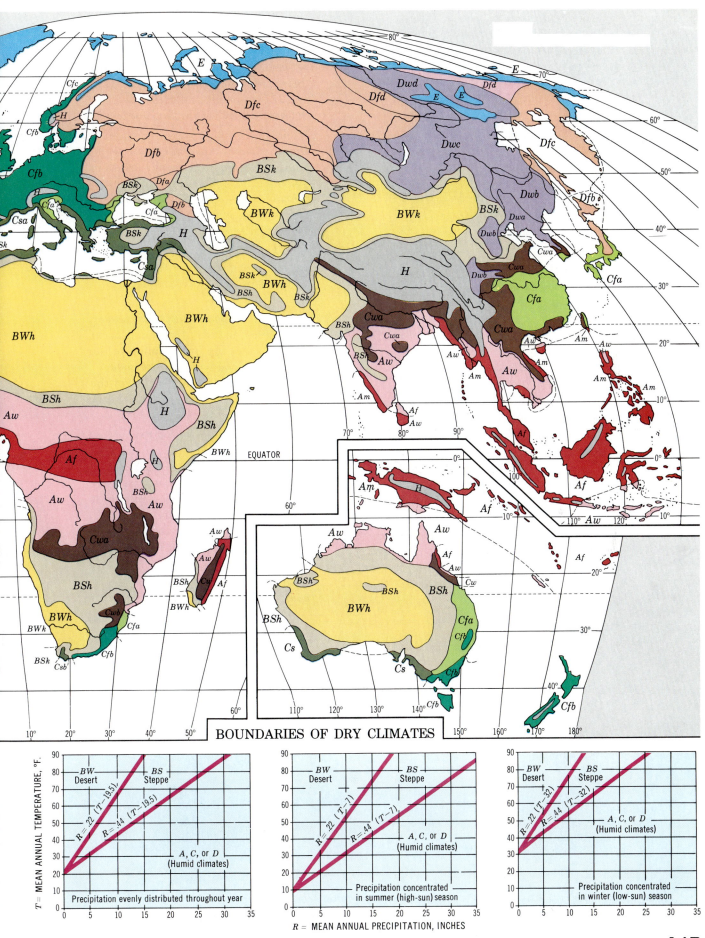

BOUNDARIES OF DRY CLIMATES

Graph 1 (left):

T = MEAN ANNUAL TEMPERATURE, °F.

BW Desert   BS Steppe

$R = 22 (T - 19.5)$

$R = 44 (T - 19.5)$

$A, C, or D$ (Humid climates)

Precipitation evenly distributed throughout year

Graph 2 (center):

BW Desert   BS Steppe

$R = 22 (T - 7)$

$R = 44 (T - 7)$

$A, C, or D$ (Humid climates)

Precipitation concentrated in summer (high-sun) season

Graph 3 (right):

BW Desert   BS Steppe

$R = 22 (T - 32)$

$R = 44 (T - 32)$

$A, C, or D$ (Humid climates)

Precipitation concentrated in winter (low-sun) season

R = MEAN ANNUAL PRECIPITATION, INCHES

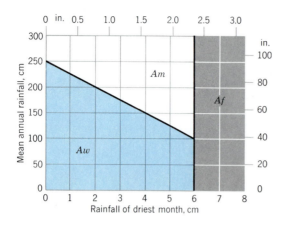

FIGURE 7.18 Boundaries of the A climates.

Am **Monsoon variety of Af.** Rainfall of the driest month is less than 6 cm (2.4 in.). The dry season is strongly developed.

Aw **Tropical savanna climate.** At least one month has rainfall less than 6 cm (2.4 in.). The dry season is strongly developed.

Figure 7.18 shows the boundaries between Af, Am, and Aw climates as determined by both annual rainfall and rainfall of the driest month.

BS **Steppe climate.** A semiarid climate characterized by grasslands. It

occupies an intermediate position between the desert climate (BW) and the more humid climates of the A, C, and D groups. Boundaries are determined by formulas given in Figure 7.19.

BW **Desert climate.** An arid climate with annual precipitation usually less than 40 cm (15 in.). The boundary with the adjacent steppe climate (BS) is determined by formulas given in Figure 7.19.

Cf **Mild humid climate with no dry season.** Precipitation of the driest month averages more than 3 cm (1.2 in.).

Cw **Mild humid climate with a dry winter.** The wettest month of summer has at least 10 times the precipitation of the driest month of winter. (Alternate definition: 70 percent or more of the mean annual precipitation falls in the warmer six months.)

Cs **Mild humid climate with a dry summer.** Precipitation of the driest month of summer is less than 3 cm (1.2 in.). Precipitation is at least three times as much as the driest month of summer. (Alternate definition: 70 percent or more of the mean annual precipitation falls in the six months of winter.)

Df **Snowy-forest climate with a moist winter.** No dry season.

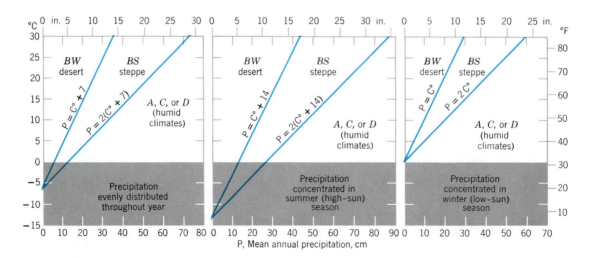

FIGURE 7.19 Boundaries of the B climates. (See Figure 7.17 for equivalent graphs in English units.)

Dw *Snowy-forest climate with a dry winter.*

ET *Tundra climate.* Mean temperature of the warmest month is above 0°C (32°F) but below 10°C (50°F).

EF *Perpetual frost climate.* Ice-sheet climate. Mean monthly temperatures of all months are below 0°C (32°F).

To denote further variations in climate, Köppen added a third letter to the code group. Meanings are as follows:

a With hot summer; warmest month is over 22°C (71.6°F); *C* and *D* climates.

b With warm summer; warmest month is below 22°C (71.6°F); *C* and *D* climates.

c With cool, short summer; less than four months are over 10°C (50°F); *C* and *D* climates.

d With very cold winter; coldest month is below −38°C (−36.4°F); *D* climates only.

h Dry-hot; mean annual temperature is over 18°C (64.4°F); *B* climates only.

k Dry-cold; mean annual temperature is under 18°C (64.4°F); *B* climates only.

As an example of a complete Köppen climate code, *BWk* refers to a *cool desert climate,* and *Dfc* refers to a *cold, snowy forest climate with cool, short summer.*

CHAPTER 8

LOW-LATITUDE CLIMATES

LOW-LATITUDE CLIMATES   They lie mostly between the tropics of cancer and capricorn, spanning the vast equatorial and tropical latitude zones.

WIND AND PRESSURE BELTS   Climates here are dominated by the equatorial belt of doldrums and intertropical convergence zone, the belt of tropical easterlies, or trades, and large parts of the oceanic subtropical highs.

DRY AND MOIST CLIMATES   Low-latitude climates include types ranging from extremely dry to extremely moist.

SEASONAL CLIMATES   Whereas the wet equatorial climate is monotonously uniform—wet throughout the year—the wet–dry tropical climate alternates seasonal drought with a rainy season.

HIGHLAND CLIMATES   Mountains of low latitudes rise as climate islands, moister and cooler than surrounding lowlands.

A DIVERSE GROUP   Diversity is the key word for low-latitude climates as a group. Agricultural practices adapted to these latitudes are correspondingly diverse, with many serious difficulties to be overcome.

CLIMATES of Group I, the low-latitude climates, lie for the most part between the tropics of cancer and capricorn, but with poleward extensions of their domain in North Africa and southwestern Asia reaching to latitudes as far north as about the 35th parallel. Similar but narrow poleward extensions of latitude occur in the southern hemisphere along east coasts windward to the trades. In terms of world latitude zones, defined in Figure 2.13, the low-latitude climates occupy all the equatorial zone (10° N to 10° S), most of the tropical zone (10–15° N and S), and part of the subtropical zone.

In terms of prevailing pressure and wind systems, the low-latitude climates occupy the equatorial belt of doldrums and intertropical convergence zone (ITC), the belt of tropical easterlies (northeast and southeast trades), and large portions of the oceanic subtropical highs.

As a group, the four low-latitude climates include types ranging from extremely dry to extremely moist, which is to say that the group includes climates with extremely large annual water deficits as well as those with extremely large water surpluses. Low-latitude climates span the range from a climate of extreme seasonality of precipitation—the wet–dry tropical climate (3)—to one with heavy precipitation throughout the year—the wet equatorial climate (1). Thermal regimes are likewise quite varied, from the monotonous uniformity (equability) found in the wet equatorial climate to a strong annual range in the interior tropical deserts. Cycles of water need (potential evapotranspiration) are correspondingly varied in range. Perhaps the only climatic criterion shared by all the low-latitude climates is that the total annual water need is always great and does not exceed 130 cm.

Europeans who colonized the regions of low-latitude climates found themselves in very strange surroundings, difficult to cope with in terms of conventional social and agricultural practices with which they were familiar. Soils were unlike any they cultivated effectively in Europe and, in many areas, proved to be either infertile or difficult to till. Abundant rainfall that might have seemed at first to be a great asset to crop cultivation proved detrimental to that end. Extremes of heat proved difficult to bear and many strange, lethal diseases were endemic. Even today, native inhabitants of the Third World nations of the low-latitude climates have not yet learned to adapt their borrowed agricultural practices and economic systems to the unrelenting limitations of climate, soils, and native vegetation—limitations their ancestors understood and respected.

The Wet Equatorial Climate (1)

This is a climate of the intertropical convergence zone (ITC), dominated by warm, moist maritime equatorial (mE) and maritime tropical (mT) air masses that yield heavy convectional rainfall. Rainfall is copious in all months; the annual total often exceeds 250 cm (100 in.). There are, however, strong seasonal differences in monthly rainfall; these can be attributed to changes in position of the ITC and to local orographic effects. Remarkably uniform temperatures prevail throughout the year.

Both monthly mean and mean annual temperatures are typically close to 27°C (80°F).

Latitude Range: 10° N to 10° S

Major Regions of Occurrence: Amazon lowland of South America; Congo Basin of equatorial Africa; East Indies, from Sumatra to New Guinea.

Example: Figure 8.1 is a climograph for Iquitos, Peru, a typical wet equatorial station located close to the equator in the broad, low basin of the upper Amazon River. Notice the very small annual range in temperature and the very large annual rainfall total.

Soil-Water Budget: The modern city of Singapore, located almost on the equator at the southern tip of the Malay Peninsula, illustrates the soil-water budget of a warm wet climate of the equatorial zone (Figure 8.2). Notice first that the water need (Ep) is highly uniform throughout the year,

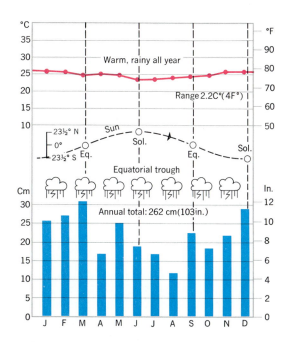

FIGURE 8.1 Wet equatorial climate (1). Iquitos, Peru, is located in the upper Amazon lowland, close to the equator. Temperatures differ very little from month to month, and there is copious rainfall throughout the year.

172

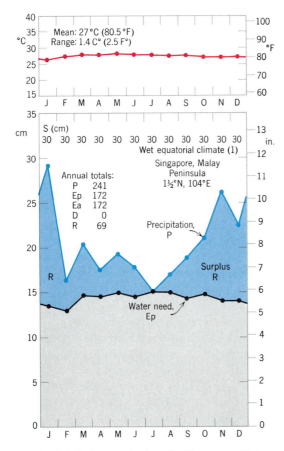

FIGURE 8.2 Soil-water budget for Singapore, Malay Peninsula, lat. 10° N. (Data from C. W. Thornthwaite Associates, Laboratory of Climatology, Centerton, NJ.)

corresponding with the remarkably uniform air temperature. Precipitation (P) is very large in every month, exceeding the water need in every month but one. Every month but one has a substantial water surplus (R). The total annual water surplus is therefore very large.

A row of figures at the top of the soil-water graph gives the amount of soil water in storage (S) for each month. At this locality the storage is at the maximum value of 30 cm in every month. For most stations in this climate, storage is over 25 cm in at least 10 months.

Daily Range in Air Temperature

A remarkable feature of the wet equatorial climate is the extreme uniformity of monthly mean air temperature, which is typically between 26° and 29°C (79° and 84°F) for stations at low elevation in the equatorial zone. The equatorial thermal regime is illustrated by graphs of minimum and maximum daily temperatures for two months in the city of Panama, lat. 9° N (Figure 8.3). The selected months—July and February—represent the extremes of the yearly temperature cycle, yet the means differ by only 0.16 C° (0.3 F°). The daily temperature range, in contrast, is usually from 8 to 11 C° (15 to 30 F°). Thus the daily range greatly exceeds the annual range. Geographers have a saying about this: "Night is the winter of the tropics."

The Monsoon and Trade-Wind Littoral Climate (2)

Trade winds bring moisture-laden maritime tropical (mT) air masses to narrow eastern littorals. ("Littoral" means a narrow coastal zone.) Orographic rainfall is produced by coastal hills and mountains. This shower activity is intensified by arrivals of easterly waves. Rainfall shows a strong annual cycle, peaking in the season of high sun, when the ITC is close by. A marked short season of reduced rainfall occurs just following the

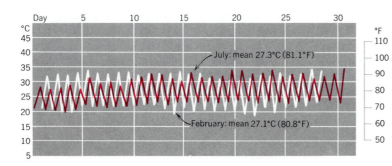

FIGURE 8.3 July and February temperatures at Panama, lat. 9° N. The sawtooth graph shows daily maximum and minimum readings for each day of the month. (Data from Mark Jefferson.)

season of low sun. Temperatures are warm throughout the year, but with a marked annual cycle. Minimum temperatures occur at time of low sun.

Latitude Range: 5° to 25° N and S

Major Regions of Occurrence: Trade-wind littorals occur along the east sides of Central and South America, the Caribbean Islands, Madagascar (Malagasy), Indochina, the Philippines, and northeast Australia. Monsoon west coasts are found in western India (Malabar Coast) and Burma. Large inland extensions of the monsoon climate occur in Bangladesh and Assam, central and western Africa, and southern Brazil.

Example: Figure 8.4 is a climograph for Belize, a Central American east-coast city at lat. 17° N, which is exposed to the tropical easterlies. Rainfall is copious from June through November, when the ITC is in this latitude zone. Easterly waves are common in this season, and an occasional tropical cyclone strikes the coast, bringing torrential rainfall. Following the winter solstice (end

of December), rainfall is greatly reduced, with minimum values in March and April. At this time the ITC lies farthest away and the climate is dominated by the subtropical high-pressure cell. Air temperatures show an annual range of 5 C° (9 F°) with maximum in the high-sun months.

Soil-Water Budget: Aparri, a station in Luzon Island in the Philippines at lat. 18° N, is in the tropical zone, on a coast exposed to the moist air masses of the trade-wind belt (Figure 8.5). The soil-water budget is strikingly different from that of Singapore in several respects. Water need (Ep) shows a well-developed annual cycle with a maximum in the high-sun period, May

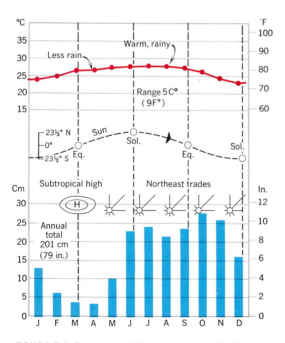

FIGURE 8.4 Trade-wind littoral climate (2). This climograph for Belize, a Central American east-coast city, at lat. 17° N, shows a marked season of low rainfall following the period of low sun.

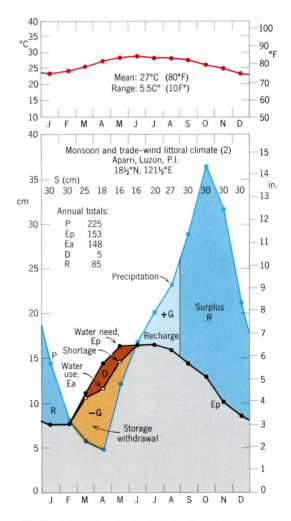

FIGURE 8.5 Soil-water budget for Aparri, Luzon Island, Philippine Islands, lat. 18° N. (Same data source as figure 8.3.)

FIGURE 8.6 Unsurpassed as an agricultural wonder is this rice terrace system at Banaue in northern Luzon, Philippine Islands. The bright green patches are young rice plants, ready to be uprooted and transplanted in rows in the flooded rice paddies. These remarkable terraces have been maintained for centuries. (S. Vidler/ Leo de Wys, Inc.)

through August. Monthly precipitation follows a strongly developed annual cycle with a minimum in April and a very strongly peaked maximum in the period from September through November. Thus, the curves of precipitation and water need cross each other twice each year, producing a very large water surplus (R) at one season and a modest soil-water shortage (D) in the short dry season. During the season of shortage, the quantity of soil water in storage declines considerably, but there is always a substantial supply of soil water available to plants. The climate here is ideal for rice, which needs very wet conditions for growth, as well as a dry harvest season (Figure 8.6).

Asiatic Monsoon Variety

A special monsoon variety or subtype of the monsoon and trade-wind littoral climate is associated with the Asiatic monsoon system.

It features an extreme peak of rainfall during the high-sun period and a well-developed dry season with two or three months of only small rainfall amounts.

The Asiatic monsoon variety is illustrated in Figure 8.7 by a climograph for Cochin, India, lat. 10° N. Located on the west coast of lower peninsular India, Cochin occupies a windward location with respect to the southwest winds of the rainy (summer) monsoon season. In the rainy season, monthly rainfall is extremely great in both June and July. A strongly pronounced season of low rainfall occurs at time of low sun—December through March. Air temperatures show only a very weak annual cycle, cooling a bit during the rains, but the annual range is small at this low latitude.

The Rainforest Environment

Our first two climates—wet equatorial (1) and monsoon and trade-wind littoral (2)—

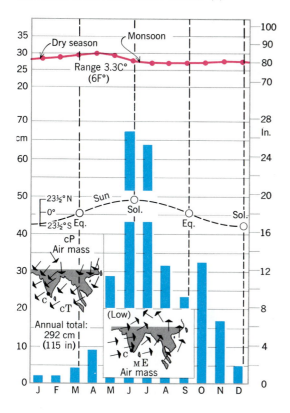

FIGURE 8.7 Monsoon climate (2). Cochin, India, on a windward coast at lat. 10° N, shows an extreme peak of rainfall during the rainy monsoon, contrasting with a short dry season at time of low sun.

have in common certain environmental properties of great interest and importance to geographers. First is the extreme uniformity of monthly mean air temperature. Second is the combination of a very large annual total rainfall, a great annual water surplus, and high soil-water storage throughout the year. Both of these factors create a special environment for the development of a unique combination of soil type and native vegetation. We can call this the *low-latitude rainforest environment.*

Streams flow copiously throughout most of the year and river channels are lined along the banks with dense forest vegetation (Figure 8.8). Natives of the rainforest traveled the rivers in dugout canoes, and today they enjoy easier travel by attaching an outboard motor. Over a century ago, larger shallow-draft river craft turned the major waterways into the main arteries of trade; towns and a few cities became situated on the river banks. Aircraft have added a new dimension in mobility, crisscrossing the almost trackless green sea of forest to find landings in clearings or on broad reaches of the larger rivers.

In the low-latitude rainforest environment, the large water surplus and prevailing warm soil temperatures promote

FIGURE 8.8 Rainforest of the western Amazon lowland, northwest of Iquitos, Peru. A petroleum exploration well is being drilled from a barge, floated 3700 km (2300 mi) up the Amazon River and its tributary, seen here, the Cuinico River. (Phillips Petroleum Company.)

the decay and decomposition of rock to great depths, so that a thick soil layer is usually present. The soil is typically rich in oxides of iron and has a deep red color; it belongs to a soil class called the Oxisols (Chapter 19). This kind of soil has largely lost its ability to hold nutrient substances (ions) needed by plants such as grasses and grain crops important in modern agriculture practiced in higher latitudes. There are, however, many kinds of plants capable of surviving on Oxisols. Most conspicuous of these are the great forest trees with broad leaves that do not undergo a seasonal shedding; these comprise the *broadleaf evergreen rainforest* (Figure 8.8). The key to the success of this kind of forest lies in the ability of the trees to quickly reuse (recycle) the essential plant nutrients that are released by the decay of fallen leaves and branches.

A major difference between the low-latitude rainforest and forests of higher latitudes is the great diversity of species that it possesses. In the low-latitude rainforest one can find as many as 3000 different tree species in an area of only a few square kilometers—midlatitude forests would possess perhaps only one-tenth that number.

The animal assemblage, or fauna, of the rainforest is also very rich. A 16-km^2 (6-mi^2) area in the Canal Zone, for example, contains about 20,000 species of insects, whereas there are only a few hundred in all of France. This very large number of species occurs because the rainforest environment is so uniform and free of physical stress.

Animal life of the rainforest is most abundant in the upper layers of the rainforest. Above the canopy, birds and bats are important carnivores and feed on insects above and within the topmost leaf canopy. Below this level live a wide variety of birds, mammals, reptiles, and invertebrates. These animals feed on the leaves, fruit, and nectar abundantly available in the main part of the canopy. Ranging between the canopy and ground are small climbing animals—monkeys, for example—that forage in both layers. At the surface are the large ground animals, including herbivores that graze the low leaves and fallen fruits, and carnivores that prey on the abundant animals.

FIGURE 8.9 Tuberous manioc roots are soaked in a river after being peeled. The roots may be prepared for eating by roasting or boiling, or they may be used to make a coarse meal. This scene is in the Amazon basin of Brazil, where manioc is an important starchy staple. (Francois Gohier/Photo Researchers.)

Plant Products and Food Resources of the Rainforest

Several forest products are of economic value. Rainforest lumber, such as mahogany, ebony, or balsawood, is an important export. Quinine, cocaine, and other drugs come from the bark and leaves of tropical plants; cocoa comes from the seed kernel of the cacao plant. Natural rubber is made from the sap of the rubber tree. The tree comes from South America, where it was first exploited. Rubber trees also are widely distributed through the rainforest of Africa. Today, the principal production is from plantations in Indonesia, Malaysia, Thailand, Vietnam, and Sri Lanka (Ceylon).

An important class of food plants native to the wet low-latitude environment are starchy staples; some are root structures and others are fruits. Manioc, also known as cassava, is one of these staples. The plant has a tuberous root—something like a sweet potato—that reaches a length over 0.3 m (1

FIGURE 8.10 A coastal grove of coconut palms, City of Refuge, Kona Coast, Island of Hawaii. (Arthur N. Strahler.)

ft) and may weigh several kilograms (Figure 8.9). When properly prepared to remove a poisonous cyanide compound, the manioc roots yield a starchy food with very little (1 percent or less) protein. Manioc is used as a food in the Amazon basin of Brazil; it was taken to Africa in the 16th century by the Portuguese. In the present century, its use in Africa has increased widely. Manioc is now also important in Indonesia. The food unfortunately contributes to malnutrition because it contains too little protein. Manioc is cultivated in small plantations placed in forest clearings.

Yams are another starchy staple of the wet low-latitude regions. Like the manioc, the yam is a large underground tuber. Yams are a major source of food in West Africa. The plant was introduced into the Caribbean region during the era of slavery and is an important food in the region today. The yam has a higher protein content than the manioc.

Taro is another starchy staple of wet low-latitude climates. The taro plant has large leaves, which are edible, but the food value (largely starch) lies mostly in an enlarged underground portion of the plant, called a "corm." Visitors to Hawaii know the taro plant through its transformation into poi, a fermented paste. Few mainland tourists eat poi a second time, but it has long been a favored food of native Hawaiians. Taro was imported into Africa from Southeast Asia and eventually reached the Caribbean region.

The banana and plantain are starchy staples familiar to everyone. The banana plant lacks woody tissue and is, in fact, a perennial herb; the fruit is classed as a berry by botanists. The banana was first cultivated for food in Southeast Asia, then spread to Africa. It was imported into the Americas in the 16th century and quickly became well established. The plantain is a coarse variety of the banana that is starchy, has little sugar, and requires cooking.

Perhaps the plant most important to humans in the low latitudes has been the coconut palm. Besides being a staple food, it provides a multitude of useful products in the form of fiber and structural materials. The coconut palm flourishes on islands and coastal fringes of the wet equatorial and tropical climates (Figure 8.10). Copra, the dried meat of the coconut, is a valuable source of vegetable oil. Copra and coconut oil are major products of Indonesia, the Philippines, and New Guinea. Palm oil and palm kernels of other palm species are important products of the equatorial zone of West Africa and the Congo River basin.

In Chapter 20, we present a special section on exploitation of the low-latitude rainforests, contrasting the primitive practice of shifting cultivation of small forest plots with the wholesale destruction of the rainforests in progress today over large areas in Central and South America, Africa, and the East Indies. Are these priceless ecological systems doomed to destruction?

The Wet–Dry Tropical Climate (3)

The wet–dry tropical climate results from a seasonal alternation between dominance by moist mT and mE air masses along the ITC and dry cT air masses of the subtropical high-pressure belt. As a result there is a very wet season at time of high sun and a very dry season at time of low sun. Cooler temperatures accompany the dry season, but give way to a very hot period before the rains begin.

Latitude Range: 5° to 20° N and S (Asia: 10° to 30° N)

Major Regions of Occurrence: India; Indochina; West Africa; southern Africa; South America, both north and south of the Amazon lowland; north coast of Australia.

Example: Figure 8.11 is a climograph for Timbo, Guinea, at lat. 10½° N in West Africa. Here, the rainy season begins just after vernal equinox, reaching a peak in July and August, or about two months following summer solstice. At this time the

ITC has migrated to its most northerly position and large infusions of moist mE air masses occur from the ocean lying to the south. Rainfall then declines as the low-sun season arrives. Three months—December through February—are practically rainless. At this season the subtropical high-pressure cell dominates the climate, with a stable subsiding cT air mass over the area. The temperature cycle is closely linked with precipitation. In February and March air temperature rises sharply, bringing on a brief hot season. As soon as the rains set in the effect of cloud cover and evaporation of rain causes the temperature to decline. By July, temperatures have resumed an even level.

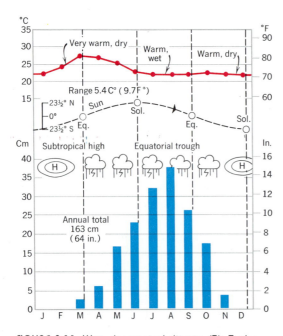

FIGURE 8.11 Wet–dry tropical climate (3). Timbo, Guinea, at lat. 10½° N, is in West Africa. A long wet season at time of high sun alternates with an almost rainless dry season at time of low sun.

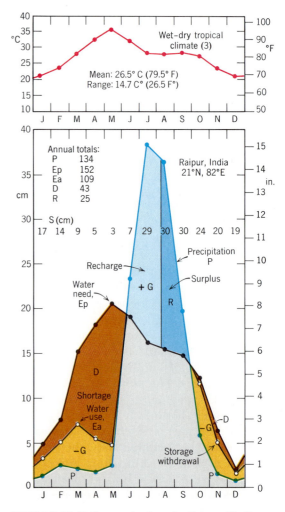

FIGURE 8.12 Soil-water budget for Raipur, Madhya Pradesh, India, lat. 21° N. (Same data source as figure 8.3.)

FIGURE 8.13 Savanna woodland of the Serengeti Plains, Tanzania, East Africa. Acacia trees with flattened crowns remain green, although the coarse grasses have turned to straw in the dry season. (H. Barad/Photo Researchers, Inc.)

Soil-Water Budget: The wet–dry climate of the monsoon lands of southern Asia is illustrated by the soil-water budget for Raipur, located well inland at lat. 21° N in the Indian province of Madhya Pradesh (Figure 8.12). At first glance, this graph seems to be quite similar to that for Aparri. The important difference is the more strongly developed dry season at Raipur. Although precipitation is very heavy in the brief rainy season, most of this rainfall goes into recharging the soil water; only a small surplus (R) is generated. During the long dry season a large soil-water shortage (D) accumulates, becoming very severe during the hot season in March, April, and May. Notice that soil-water storage (S) becomes very low in those months.

The Savanna Environment

Because of the great sweep of the annual alternations of wet and dry seasons in this climate, the native vegetation of the wet–dry tropical climate is characterized by plants capable of surviving through several consecutive months of drought, then bursting into leaf and bloom to grow rapidly in a rainy season that comes with the high-sun period of the year. For that reason the native plant cover can be described as *rain-green vegetation,* a name that covers a wide range in terms of the kinds of plants present and the ways in which they are distributed over the land surface.

Rain-green vegetation consists of two basic types. First is *savanna woodland.* By "woodland" we mean an open forest in which trees are widely spaced apart. In the savanna woodland, coarse grasses occupy the open space between the trees, which are typically coarse-barked and may bear large thorns (Figure 8.13). Large expanses of grassland may also be present (Figure 8.14). The grasses turn to straw in the dry season, and many of the tree species shed their leaves to cope with the drought. Second, in the more arid parts of the climate region, small thorny trees and large shrubs form dense patches; this is referred to as thorntree–tall-grass savanna. (Further details are given in Chapter 20.) Because of the prevalence of the savanna vegetation in the tropical wet–dry climate, it can be identified as the *savanna environment* when we wish to refer to the unique package of features of climate, soils, and vegetation that it presents.

In the savanna environment, most river channels are nearly or completely dry in the low-sun dry season. (An exception would be rivers fed from moist mountain regions.) In the rainy season, these river channels become filled to their banks with swiftly flowing, turbid water. The rains are not reliable and agriculture without irrigation is hazardous at best. When the rains fail, a devastating famine can ensue. In Chapter 20, we present as a special topic the problems of drought and famine in the

FIGURE 8.14 Vast herds of wildebeest and zebra graze the lush vegetation of the Serengeti plain in Tanzania, still green from the rains. These animals migrate long distances in search of food and water. (M. P. Kahl/Photo Researchers.)

border zone (the Sahel) of the savanna environment in Africa.

Soils of the savanna environment are similar in their physical characteristics and fertility to those of the rainforest environment, but this statement would apply only when we compare soils that occupy well-drained upland surfaces. Oxisols can be found in this environment along with a related soil group that shares the low fertility and red color of the Oxisols. In contrast, substantial areas of the savanna environment have fertile soils developed and sustained by the slow infall of windblown dusts from adjacent deserts. Equally important are highly fertile alluvial soils of major rivers that flow through the regions of tropical wet–dry climate. Annual flooding of these rivers leaves deposits of fertile silt carried down from distant mountain ranges.

Animal Life of the African Savanna

Closely adapted to the vegetation and climate is the natural animal life of the savanna grasslands and woodlands. These are the regions of the carnivorous game animals and a vast multitude of grazing animals on which they feed (Figure 8.14). The savanna of Africa is the natural home of herbivores, such as wildebeest, gazelle, deer, antelope, buffalo, rhinoceros, zebra, giraffe, and elephant. Their predators are the lion, leopard, hyena, wild dog, and jackal. Some of the herbivores depend on fleetness of foot to escape the predators. Others, such as the rhinoceros, buffalo, and elephant, defend themselves by their size, strength, or armor-thick hide. The giraffe is a peculiar adaptation to savanna woodlands; its long neck permits browsing on the higher foliage of scattered trees.

The dry season brings a severe struggle for existence to animals of the African savanna. As streams and hollows dry up, the few muddy waterholes must supply all drinking water. Danger of attack by carnivores is greatly increased.

The savanna ecosystem in Africa faces the prospect of widespread destruction. Parks set aside to preserve the ecosystem confine the grazing animals to a narrow range and prevent their seasonal migrations in search of food. In some instances the growth in animal populations has been phenomenal because they have been protected from hunting. In confined preserves they rapidly consume all available vegetation in futile attempts to survive. Rapidly growing human populations are bringing increasing pressure on management agencies to allow encroachment on game preserves in order to expand cattle grazing and agriculture.

Poaching on a large scale is now threatening the extinction of the elephant and rhinoceros in many areas.

Agricultural Resources of the Savanna Environment

In terms of agricultural resources and practices, no one sweeping statement can apply meaningfully to the entire savanna environment wherever it occurs. Instead, we achieve a truer picture by considering Southeast Asia separately from Africa and other world areas of wet–dry tropical climate. The Asiatic monsoon system imparts a special quality to agricultural patterns of Southeast Asia. There are, moreover, vastly larger human populations in the Asiatic monsoon nations than elsewhere in low latitudes, and these peoples have evolved their own unique culture patterns.

Of the staple foods grown widely in Southeast Asia, rice is dominant. About one-third of the human race subsists on rice, and most of this vast population is crowded into the arable lands of Southeast Asia. Intensive rice cultivation of this part of the world actually spans three climate types: monsoon and trade-wind littoral climate (2), wet–dry tropical climate (3), and moist subtropical climate (6). Rice requires flooding of the ground at the time the seedling plants are cultivated, and this activity has been traditionally timed to coincide with the peak of the rainy monsoon season (Figure 8.6). The crop matures and is harvested in the dry season. Sugarcane is another important crop that grows rapidly during the rainy season and is harvested in the dry season.

Sorghum, also called kaffir corn or guinea corn, is an important food crop of the savanna environment in Africa and India. It is a grain capable of survival under conditions of a short wet season and a long, hot dry season. The peanut is another major food crop of the savanna environment in India, and it has been introduced into a corresponding climate zone of West Africa.

Food and Woodland Resources of the African Savanna

Because the savanna environment in Africa is a transition zone between rainforest and desert environments, there is a corresponding gradation of plant resources and the ways in which they are used to support human life. In the moister zones, those of the savanna woodland with a long wet season, agriculture follows a pattern known as *bush-fallow farming*. This practice is related in some ways to the slash-and-burn agriculture of the rainforest. Trees are cut from a small area, piled up, and burned. The ash provides fertilizer for cultivated plots. After a few years, the land is allowed to revert to shrub and tree growth. Some of the same crops that are grown in the rainforest agriculture are grown in wetter areas of the bordering savanna. Crops include grains, yams, soybeans, and sugarcane. Tobacco and cotton are also grown.

In the drier savanna grassland and thorntree-savanna, where the wet season is short, the main subsistence crops of uplands are sorghum, millet (a kind of grain), peanuts, and corn. Cotton, along with peanuts, is an important cash crop for export. Besides the agricultural system of the permanent farmers, there exists a shifting cattle culture, in which large numbers of cattle are maintained as a display of wealth (Figure 8.15). The cattle provide food in the form of milk, butter, and blood. Following a nomadic pattern, the cattle are moved into the semidesert zone in the rainy season to graze on grasses, then returned to the savanna grassland zone in the dry season, where they graze on fallow croplands and rely on water holes for survival.

The savanna woodland belt of Africa provides a number of other plant resources besides cultivated food crops. Trees are cut for firewood, which is the only fuel for cooking available to most inhabitants. Trees also provide construction poles for dwellings. Among the savanna woodland trees that furnish important export products are the cashew-nut tree and the kapok tree. Gum arabic is taken from acacia trees. Most native trees of the savanna woodland have little commercial value as export lumber (an exception is black teak), but plantations of exotic tree species have been introduced in some areas. Both pine and eucalyptus have been successfully introduced as sources of pulpwood. These

FIGURE 8.15 Masai herdsmen in southwestern Kenya, East Africa. Cattle represent a form of wealth and prestige; they also supply food in the form of milk and blood. Quality of these animals is secondary in importance to their numbers. (Marvin E. Newman/Woodfin Camp.)

trees grow very rapidly, as compared with pulpwood trees in high latitudes.

The Dry Tropical Climate (4)

The tropical dry climate occupies source regions of the cT air mass in high-pressure cells centered over the tropics of cancer and capricorn. This subsiding air mass is stable and dry; it becomes highly heated at the surface under intense insolation. A strong annual temperature cycle follows the changing declination of the sun. The high-sun period brings extreme heat; the low-sun period, a comparatively cool season. Extremely dry areas, recognized as a desert subtype (4d), are over the tropics of cancer and capricorn. This dry zone grades on the equatorward side through a narrow zone of semidesert subtype (4sd), into a semiarid, or steppe climate subtype (4s). In this transition zone, a short rainy season occurs, grading into the long rainy season of the wet–dry tropical climate (3). Recognized as a special climate subtype, a narrow western littoral zone (4dw) has a cool, uniform thermal regime because of the presence of a cool marine air layer.

Latitude Range: 15° to 25° N and S

Major Regions of Occurrence: Sahara–Arabia–Iran–Thar desert belt of North Africa and southern Asia; a large part of Australia; small areas in Central America; South America; South Africa. Important areas of the steppe subtype (4s) are found in India and Thailand, with many small scattered dry areas to the lee of highlands in the trade-wind belt.

Examples: Figure 8.16 is a climograph for a tropical desert (4d) station in the heart of the North African desert. Wadi Halfa, Sudan, lies at lat. 22° N, almost on the tropic of cancer. The temperature record shows a strong annual cycle with a very hot period at the time of high sun, when three

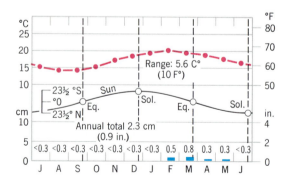

FIGURE 8.16 Dry tropical climate, desert subtype (4d). Wadi Halfa is a city on the Nile River in Sudan at lat. 22° N, close to the Egyptian border. Too little rain falls to be shown on the graph. Air temperatures are very high during the high-sun months.

consecutive months average 32°C (90°F). Daytime maximum air temperatures are frequently between 43° and 48°C (110° to 120°F) in the warmer months. There is a comparatively cool season at time of low sun, but the coolest month averages a mild 16°C (60°F) and freezing temperatures are very rarely recorded. Precipitation averages less than 0.25 cm (0.1 in.) in all months; over a 39-year period the maximum rainfall recorded in a 24-hour period was only 0.75 cm (0.3 in.).

Another example of a station in the tropical desert (4d) of North Africa is Bou-Bernous, Algeria, at lat. 27½° N. Figure 8.17 shows details of the monthly temperature cycle and includes the extreme values of the record. Notice that the annual range of the mean of daily means is substantially greater than that for Wadi Halfa, a result of the high-sun months being hotter than at Wadi Halfa. On rare occasions in January, temperatures fall close to the freezing mark.

Soil-Water Budget: We have selected the city of Khartoum, Sudan, lat. 15½° N, to represent the soil-water budget of the dry tropical climate (Figure 8.18). The desert subtype (4d) is illustrated here. Water need (Ep) follows the temperature trend, but with a much stronger annual cycle. Water need in this climate exceeds that of all other climates. The potential for evaporation of soil water is enormous, but there is rarely much soil water to evaporate. Because precipitation (same as water use, Ea) is always much less than water need, there is a water shortage in every month. Soil water in storage is so small that it is given a zero value in every month. In the two months (July and August) that show substantial precipitation, any soil water derived from brief rain showers quickly evaporates.

Semiarid Subtype (4s)

Figure 8.19 is a climograph for Kayes, Mali, located at lat. 14½° N in West Africa. This station represents the semiarid climate subtype (4s) and is in the transition zone between the dry Sahara Desert to the north and the wet–dry tropical climate (3) immediately to the south (see climograph of Timbo, Figure 8.11). The short wet season

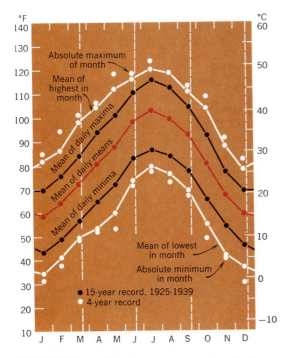

FIGURE 8.17 Monthly air temperature data for Bou-Bernous, Algeria, at lat. 27½° N in the heart of the Sahara Desert of North Africa.

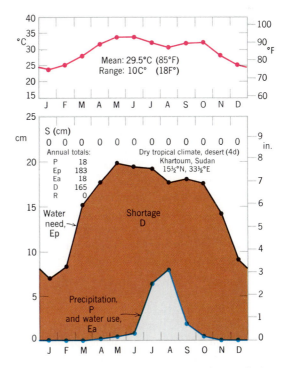

FIGURE 8.18 Soil-water budget for Khartoum, Sudan, lat. 15½° N. (Same data source as Figure 8.3.)

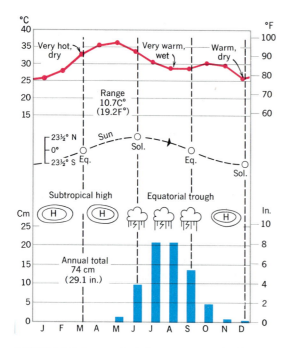

FIGURE 8.19 Dry tropical climate, semiarid subtype (4s). Kayes, Mali, at lat. 14½° N, lies in the Sahel of West Africa. In normal years there is a short wet season. The dry season is long, with a succession of rainless months.

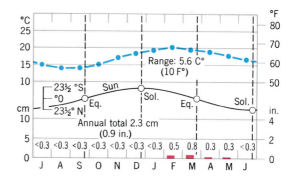

FIGURE 8.20 Dry tropical climate, western littoral desert subtype (4dw). Walvis Bay, Namibia (South-West Africa), is a desert station on the west coast of Africa at lat. 23° S. Air temperatures are cool and remarkably uniform throughout the year.

Another important occurrence of the western littoral desert subtype is along the western coast of South America in Peru and Chile. Figure 8.21 shows a stretch of this coastline in northern Chile, where it is known as the Atacama Desert.

brings a moderate amount of rainfall, mostly in four months. The long dry season is nearly rainless for four consecutive months. Extremely high temperatures characterize the hot season that precedes the rains.

Western Littoral Desert Subtype (4dw)

Figure 8.20 shows the desert climate typical of narrow western littorals (4dw). The station is Walvis Bay, a port city on the west coast of Namibia (South-West Africa), at lat. 23° S. (The yearly cycle begins with July, because this is a southern hemisphere station.) For a location nearly on the tropic of capricorn (23½° S) the monthly temperatures are remarkably cool, the warmest month mean being only 19°C (67°F). The coolest month is 14°C (57°F), making an annual range of only 5 C° (10 F°). The cool Benguela current, with upwelling of cold, deep water, chills the lower air layer. This condition explains both coolness and uniformity of the air temperatures. Coastal fog is a persistent feature of this littoral climate.

FIGURE 8.21 The desert coast of northern Chile. Known as the Atacama Desert, this barren coastal strip between the Andes Mountains and the Pacific Ocean comes close to being a truly rainless zone. (George Gerster/Photo Researchers.)

The Tropical Desert Environment

The tropical deserts and their bordering semiarid zones comprise a global environmental region sustained by subsiding air masses of the continental high-pressure cells. Figure 21.5 is a world map of all major desert and semidesert areas of the low and middle latitudes expressed in terms of the extent of vegetation classes.

Because desert rainfall is unreliable in its occurrence, even at those seasons when it is most likely to occur, river channels and the beds of smaller streams are dry most of the time. However, a sudden and intense desert downpour can cause local flooding of brief duration that transports large amounts of silt, sand, gravel, and boulders. These events can be called "flash floods." Major river channels often end in flat-floored basins having no outlet. Here clay and silt are deposited and accumulate, along with layers of soluble salts. Shallow salt lakes occupy some of these basins. The action of desert streams and the deposition of their salts by evaporation are covered in some detail in Chapters 14 and 15.

Large expanses of the very dry low-latitude desert appear to be entirely devoid of plant life except, perhaps, along the banks of some of the larger dry stream channels. Large barren areas bear a cover of loosely fitted rock particles (sometimes called "desert pavement"), whereas loose drifting sand completely mantles other large areas. These kinds of surfaces are described in Chapter 17.

Plants that are capable of survival in the true desert (4d) do so because they are quick to take advantage of a rare rainfall that may arrive only once in several years. Many of these plants are annuals. Some kinds of plants survive because they are hard-leaved or spiny shrubs equipped to resist water loss by transpiration. Other kinds are succulent plants—the cacti, for example—that store water in their spongy tissues. The tamarisk, a shrub or small tree of the Sahara Desert, thrives along the dry beds of major watercourses by sending its roots deep into the coarse alluvium beneath to reach stored water that infiltrated in time of a river flood.

In various low places in the dry desert, water can be reached by digging or drilling wells that tap the ground water zone, where porous rock material is saturated with fresh water. Where such water supplies are available, they are often used to irrigate agricultural plots in otherwise extremely hostile surroundings, thus creating an oasis. A particularly good example is the Algerian oasis of Souf, shown in Figure 8.22.

FIGURE 8.22 Looking down on the Algerian oasis of Souf, in the heart of the Sahara Desert, we see small groves of date palms planted on the floors of hollows in a sea of dune sand. (George Gerster/Rapho-Photo Researchers.)

FIGURE 8.23 Thorntree semidesert of Botswana, in the Kalahari region of southern Africa. In this dry-season scene, the larger tree has shed its leaves. The native inhabitants are Bushmen, whose temporary shelters are seen here. (S. Trevor/Bruce Coleman.)

Hollows maintained in sterile dune sand expose patches of arable soil in which groves of date palms can thrive and provide shelter to citrus trees, small plots of grain, and vegetable gardens.

Native vegetation of the tropical deserts can be divided into two general classes. First is the *dry desert,* described in the foregoing paragraphs and associated with the dry desert climate subtype (4d). Second is *semidesert,* associated with the semiarid climate subtype (4sd). A particularly important occurrence is the *thorntree semidesert* of Africa, characterized by thorny trees and shrubs that shed their leaves for the long dry season. Figure 8.23 shows this thorntree vegetation in the Kalahari Desert of southern Africa. Few if any native Bushmen survive in this natural setting today.

Highland Climates of Low Latitudes

Highland climates, shown by a distinctive pattern on the world climate map, are cool to cold, usually moist climates occupying mountains and high plateaus. High mountain ranges exhibit narrow belts of rapid climate change; these become colder with increasing altitude. The climate of a given highland area is usually closely related to the climate of the surrounding lowland in seasonal character, particularly the form of the annual temperature cycle and the times of occurrence of wet and dry seasons. In lowlands of arid climate, narrow mountain ranges tend to be islands of moist climate.

Highland climates are not usually included in the broad schemes of climate classification. Many small highland areas are simply not shown on a world map.

An example of the altitude effect in the tropical zone is shown by climographs for two stations in close geographical proximity (Figure 8.24). New Delhi, the capital city of India, lies in the Ganges lowland. Simla, a mountain refuge from the hot weather, is located about 2000 m (6500 ft) in altitude in the foothills of the Himalayas. When hot-season averages are over 32°C (90°F), Simla

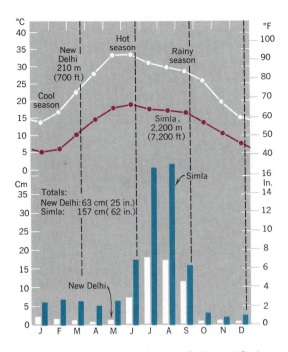

FIGURE 8.24 Climographs for New Delhi and Simla, both in northern India. Simla is a welcome refuge from the intense heat of the Gangetic Plain in May and June.

is enjoying a pleasant 18°C (65°F), which is a full 14 C° (25 F°) cooler. Notice, however, that the two temperature cycles are quite similar in phase, with the minimum month being January for both.

The general effect of increased altitude is first to bring an increase in precipitation, at least for the first few kilometers of altitude increase. This change is due to the production of orographic rainfall, generated by the forced ascent of air masses (see Chapter 5). The altitude effect shows nicely in the monthly rainfalls of New Delhi and Simla (Figure 8.24). New Delhi shows the typical rainfall pattern of the wet–dry tropical climate of Southeast Asia, with monsoon rains peaking in July and August.

FIGURE 8.25 Snowclad peaks of the Bolivian Andes rise to altitudes over 6000 m (20,000 ft). An arctic-type climate above the snowline nourishes living glaciers, even though the location is only a few degrees of latitude removed from the equator. The Indian village in the foreground lies in an altitude zone between 3600 and 4200 m (12,000 and 14,000 ft) in which wheat, barley, and potatoes can be grown. Here, the air temperature averages about 10°C (50°F) from one month to the next throughout the entire year. (Courtesy American University Field Staff.)

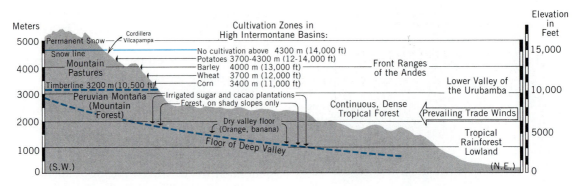

FIGURE 8.26 Altitude zoning of climates in the equatorial Andes of southern Peru, lat. 10° to 15° S. (After Isaiah Bowman.)

Simla has the same pattern, but the amounts are larger in every month, and the monsoon peak is very strong. Simla's annual total is well over twice that of New Delhi.

Mountain Agriculture in the High Equatorial Andes

In Chapter 3, we studied the effect of increasing altitude on air temperature for a series of mountain stations in Peru at latitude 15° N (Figure 3.7). As daily mean temperatures steadily declined to reach the freezing mark at about 8400 m (14,000 ft), the daily range increased greatly in comparison with the sea-level value.

Agriculture at high altitudes in the equatorial Andes of Peru and Bolivia has been practiced for centuries despite a hostile environment that includes cold and intense solar radiation (Figure 8.25). High intermontane basins of the Bolivian Plateau, or Altiplano, are in the altitude range 3200 to 4300 m (10,500 to 14,000 ft) and are above the upper limits of the rainforest. Here, in small plots, corn, wheat, barley, and potatoes can be cultivated on a limited

scale, and nearby mountain pastures provide grazing for domesticated animals (Figure 8.26).

The Low-Latitude Climates in Review

No generalization is meaningful for the four low-latitude climates considered together, because they include great extremes of the climatic spectrum. Anyone who makes a sweeping statement about the "tropics" is very poorly informed. How can the extreme aridity of the tropical deserts be equated with the extreme wetness of the equatorial zone in reference to the thermal environment, the soil-water balance, or the food resource? If diversity is the outstanding characteristic of the low-latitude climates, shall we then look to the midlatitudes for some sweeping simplicity of character? The midlatitude zone has for centuries been called the "temperate zone." How "temperate" is the midlatitude zone? Keep this question in mind as we turn next to the climates of the midlatitude zone.

CHAPTER 9

MIDLATITUDE AND HIGH-LATITUDE CLIMATES

MIDLATITUDE CLIMATES   They occupy large areas of both the midlatitude zone and the subtropical zone. The great preponderance of this area is in the northern hemisphere.

AIR MASSES AND FRONTAL ZONES   The midlatitude climates of the northern hemisphere lie mostly in a broad zone of intense interaction between tropical and polar air masses; this is the polar front zone.

A BELT OF TRAVELING STORMS   In the midlatitude zone, traveling cyclones and weather fronts move continuously from west to east, bringing with them much of the precipitation these climates experience.

A DIVERSE GROUP   The midlatitude climates, ranging from dry to very moist, exhibit great diversity, which strongly affects the natural plant cover and soil layer.

LANDS OF COLD AND SNOW   Climates of the high-latitude zone have prevailingly low air and soil temperatures, with frozen soil and snow accumulation through several months of the year.

CLIMATES of Group II, the midlatitude climates, almost fully occupy the land areas of the midlatitude zone and a large proportion of the subtropical latitude zone. Along the western fringe of Europe, they extend into the subarctic latitude zone as well, reaching to the 60th parallel. Unlike the low-latitude climates, which are about equally distributed in northern and southern hemispheres, the great preponderance of the midlatitude climate area is in the northern hemisphere. In the southern hemisphere, land area poleward of the 40th parallel south is so small that the climates are under dominance of the great Southern Ocean and lack the continentality of their northern hemisphere counterparts.

In terms of air masses and frontal zones, the midlatitude climates of the northern hemisphere lie in a broad zone of intense interaction between two groups of very unlike air masses. From the subtropical zone, tongues of maritime tropical (mT) air masses enter the midlatitude zone, where they meet in conflict with tongues of maritime polar (mP) and continental polar (cP) air masses along the discontinuous and shifting polar front zone.

In terms of prevailing pressure and wind systems, the midlatitude climates include the poleward halves of the great oceanic subtropical high pressure systems, and much of the belt of prevailing westerly winds. As a result, weather systems, such as traveling cyclones and their fronts, characteristically move from west to east. This dominant circumglobal eastward airflow gives a strong asymmetry to the distribution of climates from west to east across the great North American and Eurasian continental masses. In this respect, then, the midlatitude climates are the direct opposites of the low-latitude climates, which are dominated by the tropical easterlies (trades), maintaining a westward circumglobal airflow.

The six midlatitude climate types range from two that are very dry to three that are extremely moist, so that the group includes climates with large annual water deficits and those with large water surpluses. The

midlatitude climates span the range from a climate of extreme seasonality of precipitation—the Mediterranean climate—to those of more-or-less uniformly distributed precipitation. Thermal regimes are likewise quite varied, the low annual ranges seen along the windward west coasts contrasting with the great annual ranges in the continental interiors. Annual cycles of water need (potential evapotranspiration) are correspondingly varied, but not to the great degree seen in the low-latitude climates. Total annual water need lies between 130 cm in the warmer subtropical zone to about 50 cm in the colder northern fringe, and this is a rather strong gradient of change poleward across the entire midlatitude climate zone.

Clearly, the midlatitude climates exhibit a diversity so great that no single common characteristic stands out to unify them. Europeans who migrated to North America quickly discovered this remarkable diversity of climate and its effects on the natural plant cover they encountered and the soils they began to cultivate.

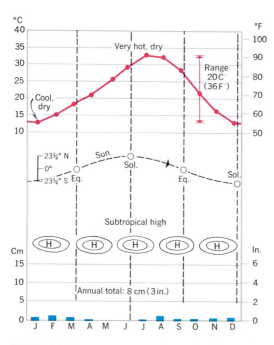

FIGURE 9.1 Dry subtropical climate, desert subtype (5d). Yuma, Arizona, lat. 33° N, has a strong seasonal temperature cycle. Compare with Wadi Halfa (Figure 8.16).

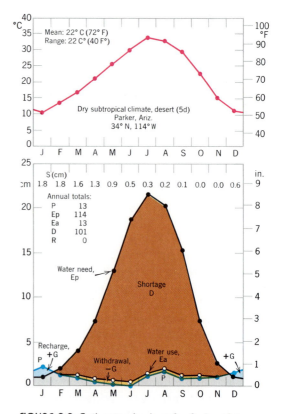

FIGURE 9.2 Soil-water budget for Parker, Arizona, lat. 34° N. (Same data source as Figure 8.3.)

The Dry Subtropical Climate (5)

The dry subtropical climate is simply a poleward extension of the dry tropical climate, caused by somewhat similar air-mass patterns. A point of difference is in the annual temperature range, which is greater for the dry subtropical climate. There is a distinct cool season in the lower-latitude portions and a cold season in the higher-latitude portions. The cold season, occurring at time of low sun, is due in part to invasions of cold cP air masses from higher latitudes. In this way the dry subtropical climate shows the influence of polar air masses. Precipitation that occurs in the low-sun season is produced by midlatitude cyclones that make incursions into the subtropical zone. As in the case of the dry tropical climate, subtypes include steppe subtype (5s), semidesert subtype (5sd), and desert subtype (5d).

Latitude Range: 25° to 35° N and S

Major Regions of Occurrence: North Africa; Near East (Jordan, Syria, Iraq); southwestern United States and northern Mexico; southern part of Australia; Argentina (Patagonia); South Africa.

Example: Figure 9.1 is a climograph for Yuma, Arizona, a city close to the Mexican border at lat. 33° N. The annual temperatures show a strong seasonal cycle with hot summers. A cold season brings monthly means as low as 13°C (55°F). Freezing temperatures—0°C (32°F) and below—can be expected at night in December and January. The annual range is 20 C° (36 F°). Precipitation, which totals about 8 cm (3 in.), is small in all months, but with two maxima. The August maximum is caused by invasion of mT air masses, which result in thunderstorm activity. Higher rainfalls from December through March are produced by midlatitude cyclones following a southerly path. Two months, May and June, are nearly rainless.

Soil-Water Budget: Parker, Arizona, is located on the lower Colorado River at lat. 34° N. The soil-water budget (Figure 9.2)

illustrates the differences between this dry subtropical desert climate (5d) and the tropical desert climate (4d) of Khartoum. Water need (Ep) at Parker is much lower in the cool season than at Khartoum, but rises to a sharp summer peak with values even greater than at Khartoum. This strong cycle reflects the large annual range of the air temperature cycle. Average monthly precipitation (P) at Parker shows a small quantity in every month except in May and June, when values are close to zero. In both December and January, precipitation exceeds water need, bringing on a brief period of soil-water recharge (+G), but this water is withdrawn over the remaining months. The annual soil-water shortage (D) is very large, though substantially less than that at Khartoum.

The Subtropical Desert Environment

Much of what we said in Chapter 8 about the tropical desert environment applies to the adjacent subtropical desert. The boundary between these two climate types is arbitrarily drawn and would be invisible to an observer traveling from one to the other. But if we were to travel northward in the subtropical climate zone of North America, arriving at about 34° N in the interior Mojave Desert of southeastern California, we would encounter environmental features significantly different from those of the low-latitude deserts of tropical Africa, Arabia, and northern Australia. Differences would be found in soils, native vegetation, and animal life.

Although the great summer heat of the low-elevation basin floors of the Mojave Desert is comparable with that experienced in the Sahara Desert, the low sun brings a clearly recognizable winter season not found in the tropical deserts. In the Mojave Desert cyclonic precipitation can occur in most months, including the cool low-sun months.

In the Mojave Desert (and adjacent Sonoran Desert), plants are often large and numerous, in some places giving the appearance of an open woodland. One example is the occurrence of forestlike stands of the tall, cylindrical saguaro cactus (Figure 20.26); another is the woodland of Joshua trees found in higher parts of the Mojave Desert. Other large shrubs or small trees include the prickly pear cactus (Figure

FIGURE 9.3 The many-branched ocotillo (*Fouquiera splendens*) is shown here with its thin, soft leaves, which appear only after rain has moistened the ground. Later, they will fall off. It will display flame-red flower clusters in March and April. The creosote bush is seen to left and right. Anza Borrega State Park, California. (A. N. Strahler.)

20.4), the ocotillo plant, the creosote bush, and the smoke tree (Figure 9.3).

Animals of the Mojave and Sonoran deserts show many interesting adaptations to the dry environment. Many of the invertebrates follow much the same life pattern as the ephemeral annual plants, which is to remain dormant in dry periods but to emerge when rain falls and take advantage of the event. For example, the tiny brine shrimp of the American desert may wait many years in dormancy until normally dry lake beds fill with water, an event that occurs perhaps three or four times per century. The shrimp then emerge and complete their life cycles before the lake evaporates.

The mammals are by nature poorly adapted to the desert environment, yet many survive there by employing a variety of mechanisms to avoid water loss. Many desert mammals do not sweat through skin glands; they rely instead on other methods of cooling. For example, the huge ears of the jackrabbit serve as efficient radiators of heat to the clear sky. Many desert mammals conserve water by excreting highly concentrated urine and relatively dry feces. The desert mammals also conserve water by limiting their physical activity to the night. In this respect, they are joined by most of the rest of the desert fauna in spending their days in cool burrows in the soil and their nights foraging for food.

Humans adapt to the American subtropical desert environment by importing the environment to which they

are accustomed. Irrigation projects allow water to be imported in abundance and lavishly applied to croplands, where most of it is lost by intense evapotranspiration. Pipelines and highways facilitate the importation of building materials, machinery, home appliances (the air conditioner), and fuels; transmission lines bring electricity from distant dams and coal-fired power plants. These human technological adaptations have run into serious environmental difficulties, one of which we will present in depth in Chapter 10. It is the deterioration of croplands by deposition of salts left by evaporation of irrigation waters and the accompanying waterlogging of the soil.

The Moist Subtropical Climate (6)

The moist subtropical climate occupies the subtropical continental margins under the domination of the maritime tropical (mT) air mass, flowing out from the moist western sides of the oceanic high-pressure cells. This air mass brings copious summer rainfall, much of it convectional. An occasional tropical storm brings heavy rainfall. Summers are warm, with persistent high humidity. Winter precipitation is also copious, produced in midlatitude cyclones. Invasions of the continental polar (cP) air mass are frequent in winter, bringing spells of subfreezing weather. No winter month has a mean temperature lower than 0°C (32°F). Subhumid, humid, and perhumid subtypes are identified. In Southeast Asia, this climate is characterized by a strong monsoon effect, with increased summer rainfall.

Latitude Range: 20° to 35° N and S

Major Regions of Occurrence: Southeastern United States; southern China; Taiwan (Formosa); southernmost Japan; Uruguay and adjoining parts of Brazil and Argentina; eastern coast of Australia.

Example: Figure 9.4 is a climograph for Charleston, South Carolina, located on the eastern seaboard at lat. 33° N. In this region, a marked summer maximum of precipitation is typical. Total annual rainfall is copious and ample precipitation falls in every month. The annual temperature cycle is strongly developed, with large annual range. Winters are mild, with the January mean temperature well above the freezing mark.

Soil-Water Budget: Baton Rouge, Louisiana, lat. 30½° N, illustrates the moist subtropical climate of the United States Gulf Coast (Figure 9.5). Precipitation (P) is substantial in all months, but shows a small peak in July and a sharp dip in October. The annual cycle of water need (Ep) has a large annual range, but in summer reaches maximum values that are slightly greater than precipitation. Consequently, there is a summer period of soil-water withdrawal (−G) but the shortage (D) is very small. A high level of soil-water storage (S) is maintained through the summer. Recharge (+G) in the early winter quickly restores the used soil water and a surplus (R) sets in by December. The total surplus is large, so the climate belongs in the humid subtype (6h).

Because air temperatures average well above the freezing mark throughout the winter, water need continues to be appreciable in the coolest months. This fact suggests that some kinds of plants—those

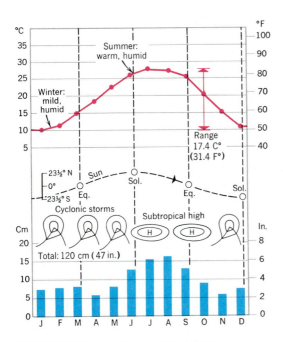

FIGURE 9.4 Moist subtropical climate (6). Charleston, South Carolina, lat. 33° N, has a mild winter and a warm summer. There is ample precipitation in all months, but a definite summer maximum.

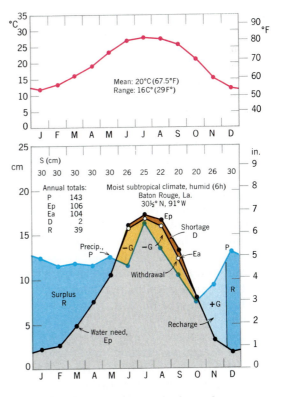

FIGURE 9.5 Soil-water budget for Baton Rouge, Louisiana, lat. 30½° N. (Same data source as Figure 8.3.)

FIGURE 9.6 Broadleaf evergreen forest of the Gulf Coast region is represented here by Evangeline oaks bearing Spanish "moss," an epiphyte that forms long beardlike streamers. The ground beneath is maintained as a lawn. Evangeline State Park, Bayou Teche, Louisiana. (Arthur N. Strahler.)

with evergreen foliage—can continue their growth throughout the mild winter.

The Moist Subtropical Forest Environment

With a very small summer water shortage and a large winter water surplus, soil water remains adequate for plant growth without irrigation in most years. Rivers and streams flow copiously through much of the year. Flooding can be severe from tropical cyclones (hurricanes in the Gulf Coast region; typhoons in southeastern China and southern Japan) that come inland and produce torrential rains in the high-sun months.

In such a moist regime, much of the natural vegetation present when explorers and colonists first visited this area in the New World consisted of broadleaf forest. In a narrow Gulf coastal zone of the United States and in a large part of southern China and the south island of Japan, the native

broadleaf forest was of the evergreen type, in which a leaf canopy remains green throughout the year. This kind of forest is called *broadleaf evergreen forest.* In Louisiana, the evergreen Evangeline oak and the magnolia flourished as representative trees of the evergreen broadleaf forest (Figure 9.6). Farther inland in the Gulf states and in Florida, the forest vegetation is of an entirely different kind—southern pine forest—adapted to sandy soils (see Figure 9.9). Near its colder northern limits, vegetation of the moist subtropical climate grades into broadleaf deciduous forest. (In a deciduous forest, leaf shedding occurs annually and the trees are bare throughout the winter season.) Today, large areas of these former forests have been replaced by agricultural croplands.

The comparatively warm climate and high rainfall of the moist subtropical environment favors the leaching out of nutrient elements (ions) from the soil layer. As in the soils of the moist low-latitude climates, iron oxides accumulate in the soil, giving it colors ranging from yellow to red. In terms of agricultural crops, especially the grains or cereals, the soils of the moist subtropical environment rate as low in fertility and require large applications of fertilizers. Another unfavorable factor is the susceptibility of these soils to severe erosion and gullying when exposed by forest removal and intensive cultivation. We will discuss these problems of soil fertility and erodibility in a special feature in Chapter 19.

Agricultural Resources of the Moist Subtropical Forest Environment

No simple statement can cover agricultural adaptation to the moist subtropical forest environment in both North America and Southeast Asia. Differences in these two widely separated areas are partly historical and cultural, but partly reflect the stronger monsoon effect in Asia, which causes a stronger concentration of precipitation in the summer. The human population is vastly denser in Southeast Asia than in the New World and subsists largely on rice, the dominant staple food crop. Two and even three rice crops are harvested annually in southern China. The rice crop is often followed by planting of wheat, winter

FIGURE 9.7 Tea leaves are carefully picked from new growth of a cultivated evergreen shrub (*Camellia*) on this terraced tea plantation on Honshu Island, Japan. (Thomas Hopker/Woodfin Camp.)

legumes, peas, or green fertilizer crops. Except in the Mississippi delta region, little rice is produced in the southern United States. Both American and Asiatic regions produce sugarcane, peanuts, tobacco, and cotton, although not on an equal intensity in both regions. One striking difference is that tea is widely cultivated in Southeast Asia, but not at all in the southern United States (Figure 9.7). Corn is an important crop in the southern United States but is not important in southern China.

The potential of the moist subtropical climate to produce more food rests in more intensive land use instead of expansion of the area now under cultivation. The best land is already in use and, in Southeast Asia, elaborate terrace systems have been used for centuries to allow farming of steep hill slopes. New genetic strains of rice and corn offer promise of increased yields when the necessary fertilizers are used. Japan and Taiwan apply very high levels of fertilizers and achieve high rice yields (Figure 9.8). The People's Republic of China is only now in the process of sharply increasing its

FIGURE 9.8 Japanese farmers planting rice seedlings in a flooded rice paddy. By the time this rice crop was harvested, a single acre may have consumed a thousand person-hours of hand labor. Today, mechanical paddy transplanters perform this operation. (W. H. Hodge/Peter Arnold.)

FIGURE 9.9 This plantation of longleaf pine grows on sandy soil of the Georgia coastal plain. The trees are used as pulpwood for paper production. The blackened trunks show that fires periodically sweep through the area, consuming the undergrowth. (Charles R. Belinky/Photo Researchers.)

production and use of fertilizers from comparatively low levels of the recent past. China is also developing independently some new high-yielding strains of rice and wheat.

In the southern United States, cattle production is another source of increased food production and makes use of soils too sandy for field crops. With soil water frequently replenished through the long, warm summer, pasture and range land can be continuously productive. Tree farming is also an important use of sandy soils. Pines are well adapted to rapid growth on sandy soils and thrive where nutrient bases are in short supply (Figure 9.9).

The large water surplus of the moist subtropical climate has important implications in terms of economic development. The large flows of rivers can furnish abundant freshwater resources for urbanization and industry without competition from irrigation demands. Evaporative losses from reservoirs are much less important than in arid lands. The maintenance of copious stream flows tends to reduce the dangers of severe water pollution and its adverse effects on ecosystems of streams and estuaries.

The Mediterranean Climate (7)

The wet-winter, dry-summer Mediterranean climate results from a seasonal alternation of conditions causing the dry subtropical climate (5), which lies at lower latitudes, and the moist marine west-coast climate (8), which lies on the poleward side. The moist mP air mass invades in winter with cyclonic storms and generates ample rainfall. In summer, subsiding cT and mT air masses are dominant, with extreme drought of several months' duration. In terms of total annual rainfall, the Mediterranean climate spans a wide range from arid to humid, depending on location. Temperature range is moderate, with warm to hot summers and mild winters. Coastal zones between lat. 30° and 35° N. and S. show a smaller annual range, with very mild winters.

Latitude Range: 30° to 45° N and S

Major Regions of Occurrence: Central and southern California; coastal zones bordering the Mediterranean Sea; coastal western Australia and South Australia; Chilean coast; Cape Town region of South Africa.

Example: Figure 9.10 is a climograph for Monterey, California, a Pacific Coast city at lat. 36½° N. The annual temperature cycle is very weak. The small annual range reflects strong control by the cold California current and its cool marine air layer. A cool summer is typical of the narrow western littorals. Winter temperatures are much milder than those of inland locations at this latitude. Rainfall drops to nearly zero for four consecutive summer months, but rises to substantial amounts in the rainy winter season.

Soil-Water Budget: The great metropolis of Los Angeles, lat. 34° N, serves as a North American representative of the Mediterranean climate (Figure 9.11). The special feature of this soil-water budget lies in the opposite curvature of the seasonal cycles of precipitation (P) and water need (Ep). The precipitation cycle has a rainy winter, but declines to a near-total drought in midsummer. In contrast, the water-need cycle reaches its low in winter and peaks in midsummer. This relationship intensifies a large summer soil-water shortage (D) lasting a full eight months. Recharge begins in December, but does not produce any water surplus. This climate can be characterized as a semiarid subtype of dry climate (7s). Notice that the soil-water storage (S) never rises to more than half the storage capacity of 30 cm; storage falls to very low levels toward the end of the dry summer.

The Mediterranean Climate Environment
Like the wet–dry tropical climate of the low latitudes, the Mediterranean climate offers a "feast-or-famine" environment for plants. Along with that incongruous linkage of a very dry climate with a wet one there arise a number of special environmental problems

for opportunistic humans who have been strongly attracted to it. The attraction lies in the benign thermal cycle, especially in the narrow coastal zones, or littorals. There, the mild winters with considerable sunshine (despite periods of substantial rainfall) are a most welcome refuge from the severe winters of the midlatitude continental interiors of Eurasia and North America. The catch lies in the scarcity of local freshwater supplies to support the heavy load of humanity that insists upon the same lavish use of water that is easily afforded in the moist soil-water regimes.

Soil fertility in valley and lowland areas of the Mediterranean climate is naturally high, as it usually is in semiarid climates of the midlatitudes. Aridity has permitted the retention in the soil layer of nutrients essential to forage grasses and grains, fruit trees, vegetables, and many other varieties of plants. Despite winter rains, these nutrients have not been leached away. Soils native to this environment belong to a special category of subclasses, explained in Chapter 19; they are not easily described in a few words.

The native vegetation of the Mediterranean climate environment is

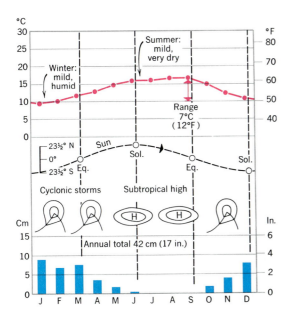

FIGURE 9.10 Mediterranean climate (7). Monterey, California, lat. 36½° N, has a very weak annual temperature cycle because of its closeness to the Pacific Ocean. The summer is very dry.

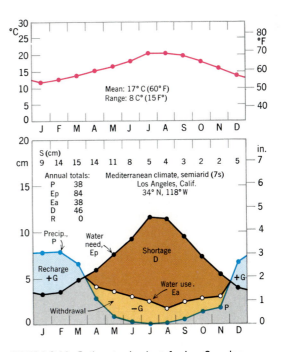

FIGURE 9.11 Soil-water budget for Los Angeles, California, lat. 34° N. (Same data source as Figure 8.3.)

FIGURE 9.12 This bark, stripped from the cork oak (*Quercus suber*), will be ground up and cemented into cork board and other structural products. Thick bark of choice quality is used for wine corks. Algeria, North Africa. (Arthur N. Strahler.)

FIGURE 9.14 Evergreen—oak woodland with grassland, Santa Ynez Valley, Santa Barbara County, California. (above) At the end of the long, dry summer the grasses are dormant, but the oak trees are green. (below) At the end of the cool, wet winter, grasses are a lush green whereas a few deciduous oaks in the foreground are nearly leafless. (Arthur N. Strahler.)

FIGURE 9.13 Olive trees on a steep, barren mountain slope. This scene is in the Atlas Mountains of Algeria. (Arthur N. Strahler.)

adapted to survival through the long summer drought. Shrubs and trees that can survive such drought are characteristically equipped with small, hard or thick leaves that resist water loss through transpiration. These plants are called *sclerophylls;* the prefix *scler,* from the Greek for "hard," is combined with *phyllo,* which is Greek for "leaf." (Compare with "atherosclerosis," the disease of hardening of the arteries.) Sclerophylls of the Mediterranean environment are typically evergreen and retain green leaves through the entire yearly cycle. Examples are the evergreen oaks, of which there are several common species in California, and the cork oak of the Mediterranean lands (Figure 9.12). Another is the olive tree, native to the Mediterranean lands (Figure 9.13). In Australia, the thick-leaved eucalyptus tree is the dominant sclerophyll. Oak woodland of California bears a ground cover of grasses that turn to straw in the summer (Figure 9.14). Another form of native vegetation in this environment is a cover of drought-resistant shrubs, including sclerophylls and spiny-leaved species. In the Mediterranean lands, this scrub vegetation goes under the name of *maquis* or *garrigue*. In California, where it is called *chaparral,* it clothes steep hill and mountain slopes too dry to support oak woodland or oak forest (see Figure 20.20). Figure 21.8 is a composite map showing the areas of sclerophyll vegetation associated with the Mediterranean climate.

Wildfire is an integral part of the Mediterranean environment of California. Chaparral is extremely flammable during the long fire season of summer. Brush fires rage through chaparral and oak forests and leave the soil surface bare and unprotected. When torrential rains occur in winter, large quantities of coarse mineral debris are swept downslope by overland flow and carried long distances by streams in flood. Mudflows and debris floods (usually called "mudslides" in the news media) are particularly destructive to human habitations on canyon floors and on piedmont fan surfaces that are often heavily urbanized.

Throughout the Mediterranean lands of Europe, North Africa, and the Near East, devastating soil erosion, induced by human activity over the past 2000 years or longer, has left its scars on the landscape. Many hillsides have been denuded of their soils and present a barren rocky aspect. Sediment, representing the displaced soil, has formed thick layers of sand and silt in adjacent valley floors. This human-induced change in the appearance of the Mediterranean landscape is the subject of a special topic in Chapter 15.

Agriculture in the Mediterranean Environment

Lands bordering the Mediterranean Sea produce cereals—wheat, oats, and barley—where arable soils are extensive enough to be cultivated. However, we usually think of that region as an important source of citrus fruits, grapes, and olives for European markets. Cork from the bark of the cork oak is also a product of economic value (Figure 9.12). In central and southern California, citrus, grapes, avocados, nuts (almond, walnut), and deciduous fruits are grown extensively (Figure 9.15). Irrigated alluvial soils are also highly productive of vegetable crops, such as carrots, lettuce, cauliflower, broccoli, artichokes, and strawberries, as well as sugar beets and forage crops (alfalfa). Cattle ranching and sheep grazing are of major importance on grassy hill slopes unsuited to field crops and orchards (see Figure 9.14) and on irrigated lowland pastures.

Because the Mediterranean environment is limited in global extent to comparatively small land areas, it offers little prospect for

FIGURE 9.15 Groves of lemon, orange, and avocado trees surround the homes and stables of the wealthy in Montecito, California. Chaparral covers the steep slopes of the Santa Ynez Mountains in the background. (A. N. Strahler.)

the expansion of croplands to provide major additions to the world's food supply. Irrigation is essential for high productivity, but there are hazards associated with heavy irrigation of lowland soils: salt accumulation and waterlogging. Urbanization and industrial development also face major problems of obtaining additional water supplies through importation over long distances by aqueduct. Nevertheless, the mild, sunny climate of southern California has proved a powerful population magnet, and water importation has already been developed on a mammoth scale. Expansion of suburban housing has, however, begun to take over rich, flat croplands, reducing the agricultural potential in a number of areas.

The Marine West-Coast Climate (8)

The marine west-coast climate occupies midlatitude west coasts, which receive the prevailing westerlies from over a large ocean and experience frequent cyclonic storms involving the cool, moist mP air mass. In this moist climate, precipitation is copious in all months, but with a distinct winter maximum. Where the coast is mountainous, the orographic effect causes a very large annual precipitation; this is the perhumid subtype (8p). The annual temperature range is comparatively small for midlatitudes. Winter temperatures are very mild compared with inland locations at equivalent latitudes.

Latitude Range: 35° to 60° N and S

Major Regions of Occurrence: Western coast of North America, spanning Oregon, Washington, and British Columbia; western Europe and the British Isles; Victoria and Tasmania; New Zealand; Chile, south of 35° S.

Example: Figure 9.16 is a climograph for Vancouver, British Columbia, just north of the United States–Canada border. The annual precipitation is very great and most of it falls during the winter months. Notice the greatly reduced rainfall in the summer months. The temperature cycle shows a remarkably small range for this latitude. Even the winter months have averages above the freezing mark.

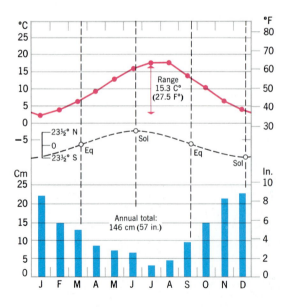

FIGURE 9.16 Marine west-coast climate (8). Vancouver, British Columbia, lat. 49° N, has a large annual total precipitation, but with greatly reduced amounts in the summer. The annual temperature range is small and winters are very mild for this latitude.

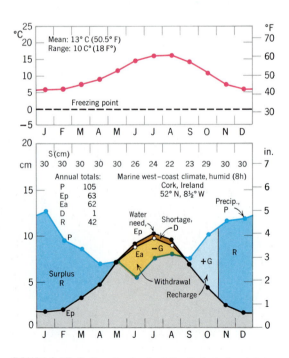

FIGURE 9.17 Soil-water budget for Cork, Ireland, lat. 52° N. (Same data source as Figure 8.3.)

Soil-Water Budget: The city of Cork is located at lat. 52° N on the south coast of Ireland, where it lies exposed to westerly winds and maritime air masses of the North Atlantic. Cork illustrates the humid subtype (8h) of the marine west-coast climate (Figure 9.17). Precipitation (P) shows a strong annual cycle with a winter maximum and a summer minimum, but the amounts are substantial in all months and the yearly total is large. Water need (Ep) also shows a strong annual cycle, continuing through the winter months because the winters are mild. (Notice that all winter months have temperature means well above freezing.) In summer, soil water is withdrawn from storage, but the soil-water shortage (D) is very small. Irrigation of crops would not be of benefit in an average year.

The Marine West-Coast Environment

Because of the copious precipitation, large soil-water surplus, and small water shortage, lowland soils of the marine west-coast climate regions show the effects of leaching out of nutrients, but in Europe have retained moderate fertility. Applications of fertilizers and lime are needed for bountiful crop production, and in Europe these soils have been successfully cultivated for centuries. Much of the land surface within the marine west-coast environment in northern Europe, British Columbia, southern Chile, and the South Island of New Zealand is on mountainous slopes that have been heavily scoured by the ice sheets and mountain glaciers of the recent Ice Age. Soils of these glaciated areas are extremely young and are poorly developed.

Forest is the native vegetation of this environment. In the perhumid mountainous areas of the northern Pacific coast there flourish dense needleleaf forests of redwood, fir, cedar, hemlock, and spruce (Figure 9.18). Under the lower precipitation regime of Ireland, southern England, France, and the Low Countries a broadleaf deciduous forest was the native vegetation, but much of it disappeared many centuries ago under cultivation, so that only scattered forest plots or groves remain (see Figure 9.19). Sometimes called "summergreen"

FIGURE 9.18 Needleleaf forest of Douglas fir and hemlock in Snoqualmie National Forest, Washington. The barren mountainside illustrates the practice of block cutting, in which all trees are removed. Severe soil erosion can follow, causing stream channels to be choked with debris. (Jay Lurie/Black Star.)

FIGURE 9.19 A rural scene in northern Scotland, showing diversified farming around the small village of Rhymie. The barren uplands are moors bearing heath, a cover of small plants. (Mark A. Melton.)

deciduous forest, it is dominated by tall broadleaf trees that provide a continuous and dense canopy in summer but shed their leaves completely in winter. Dominant tree species of this forest type in western Europe are oak and ash, with beech in the cooler and moister areas. Figure 21.9 is a composite map showing the distribution of needleleaf and broadleaf deciduous forests in the areas of marine west-coast climate.

Agricultural and Water Resources

The marine west-coast environment of western Europe and the British Isles has been intensively developed for centuries for such diverse uses as crop farming, dairying, orchards, and forest (Figure 9.19). It is an environment in most respects similar in agricultural character and productivity to the moist continental climate environment with which it merges on the east.

In North America the mountainous terrain of the coastal belt offers only limited valley floors for agriculture, which is generally diversified farming. Forests are the primary plant resource, and here they constitute perhaps the greatest structural and pulpwood timber resource on earth. Douglas fir, western cedar, and western hemlock are the principal lumber trees of the Pacific Northwest (Figure 9.18). The same mountainous terrain that limits agriculture in the Pacific Northwest is a producer of enormous water surpluses that run to the sea in rivers. Including now the ranges of the northern Rocky Mountains

with the Pacific coastal ranges, the potential for long-distance transfer of this excess water to dry regions of the western United States has not passed unnoticed.

The Dry Midlatitude Climate (9)

The dry midlatitude climate is limited almost exclusively to interior regions of North America and Eurasia. This dry climate occupies a rainshadow position with respect to mountain ranges on the west or south. Maritime air masses are blocked effectively much of the time, so that the continental polar cP air mass dominates the climate in winter. In summer a dry continental air mass of local origin is dominant. Summer rainfall is mostly convectional and is caused by sporadic invasions of maritime air masses. Steppe and semidesert subtypes (9s, 9sd) are extensive; true desert (9d) occurs only in basins of interior Asia. The annual temperature cycle is strongly developed, with a large annual range. Summers are warm to hot, but winters are very cold.

Latitude Range: 35° to 55° N

Major Regions of Occurrence: Western North America (Great Basin, Columbia Plateau, Great Plains); Eurasian interior, from steppes of eastern Europe to the Gobi Desert and northern China. (A small area is found in southern Patagonia.)

Example: Figure 9.20 is a climograph for Pueblo, Colorado, located at lat. 38° N, just east of the Rocky Mountains. The climate is of the semiarid subtype (9s), with a total annual precipitation of 31 cm (12 in.). Most of this precipitation is in the form of convectional summer rainfall, which occurs when moist mT air masses invade from the south and cause thunderstorms. In winter, snowfall is light and yields only small monthly precipitation averages. The temperature cycle has a large annual range, with warm summers and cold winters. January, the coldest winter month, has a mean temperature just below freezing.

Soil-Water Budget: Medicine Hat, Alberta, is located at lat. 50° N, near the northern limit of the Great Plains. It lies in a rainshadow to the lee of the Cordilleran Ranges and represents the semiarid subtype (9s) of the dry midlatitude climate (Figure 9.21). The annual cycle of water need (Ep) peaks strongly in summer, following five consecutive months in which water need is

zero because of subfreezing air and soil temperatures. Precipitation (P) shows a distinct annual cycle in which the summer months have about double the precipitation of the winter months. A substantial soil-water shortage (D) develops in summer. Storage recharge (+G) is not sufficient to raise the soil-water storage (S) to its full capacity of 30 cm. As a result, there is no water surplus. The recharge accumulates in the frozen state throughout the winter, to be released to the soil in the spring thaw.

The Dry Midlatitude Environment

Low annual precipitation combined with a large soil-water shortage under a strongly continental thermal regime has produced soils of high natural fertility that retain large supplies of the nutrient elements (positively charged ions), known to soil scientists as "bases." The principal nutrient bases are calcium, magnesium, potassium, and sodium. These soils are moderately to strongly alkaline, in contrast to soils of the humid and perhumid midlatitude climates

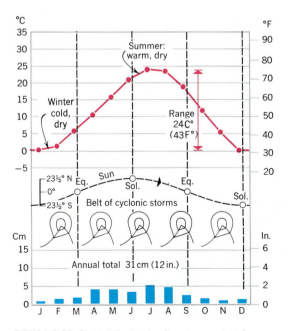

FIGURE 9.20 Dry midlatitude climate, semiarid subtype (9s). Pueblo, Colorado, lat. 38° N, shows a marked summer maximum of rainfall in the summer months.

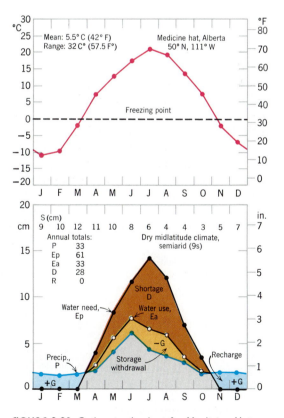

FIGURE 9.21 Soil-water budget for Medicine Hat, Alberta, lat. 50° N. (Same data source as Figure 8.3.)

FIGURE 9.22 A roadside soil exposure showing rocklike slabs of caliche (calcrete). Pecos Plains near Vaughan, New Mexico. (Arthur N. Strahler.)

that are acid in chemical balance. (See Chapter 19 for details.)

Grasses thrive on large supplies of these soil nutrients and a mildly alkaline

condition of the soil. Thus the native vegetation of the semiarid subtype (9s) consists principally of hardy perennial short grasses capable of enduring severe summer drought. We refer to this vegetation type as *short-grass prairie*. Among geographers, the Asiatic plains landscapes of these short grasses are known as *steppes*. (The singular form, *steppe*, is used as a general term for the total environment.) Refer to the composite map, Figure 21.11, showing the areas of short-grass prairie.

Soils of the short-grass prairie are dominantly of a major class known as *Mollisols*. The prefix of this recently coined word uses the Greet root *mollis*, meaning "soft." (Compare with the word "mollify," meaning "to soften.") This refers to the rather loose soil texture, consisting of small soil particles and giving the soil the property of being easily tilled. Soil color is brown to pale brown, and these soils are known variously as chestnut soils and brown soils.

FIGURE 9.23 Harvesting wheat on the rolling Palouse Hills of eastern Washington. The rich dark soil is formed on a thick layer of silt carried to this area by wind near the close of the Ice Age. As a natural prairie grassland, this area is ideally suited to wheat farming. (Grant Heilman.)

An important feature of Mollisols in the semiarid climate is that excess calcium carbonate (commonly known as "lime") accumulates in a lower layer of the soil, taking the form of rocklike nodules or plates. Pioneers who settled the rangelands of the American southwest adopted the Spanish word *caliche* for these white rocklike slabs and layers (Figure 9.22).

The dry midlatitude climate also comes in a semidesert subtype (9sd) and a desert subtype (9d). We shall not dwell on these subtypes. The semiarid subtype is transitional to the subtropical desert and, in the western United States, is characterized by sagebrush shrub vegetation. The desert subtype is found only in central Asia. This cold desert environment shares the major soil and vegetation characteristics of the interior desert subtype of the dry subtropical climate.

Agricultural Resources of the Short-Grass Prairie

Wheat is perhaps the most important single crop produced in unirrigated areas of short-grass prairie bordering the subhumid zone. One such important wheat-producing region lies in southern Alberta and Saskatchewan and in the northern border region of Montana. Another is the Palouse Hills region of southeastern Washington state and western Idaho (Figure 9.23). Here the crop is spring wheat, which is planted in the spring of the year. Using the soil water that has been recharged in early spring and the precipitation that falls in late spring and early summer, the crop is able to reach

FIGURE 9.24 The perfect circles of green laid out in orderly patterns are thriving plots of food and forage crops—corn, wheat, beans, alfalfa, or grass—irrigated by strange walking sprinkler systems. Called centerpivot irrigation, the system consists of a long pipe mounted on wheeled supports. The pipe is held fast to a center point where the water is injected and creeps slowly round and round in a huge circle emitting water from sprinkler heads along the pipe. Water is supplied from deep wells close by. (Courtesy of Valmont Industries, Incorporated, Valley, Nebraska.)

maturity for midsummer harvesting.

In Russia, the rich wheat region of the Ukraine continues in a narrow zone far eastward across the steppes of Kazakhstan. In northern China wheat is grown within a steppe region bordering the moist continental climate.

Wheat production of the midlatitude steppes is very much at the mercy of variations in seasonal rainfall. Good years and poor years follow cyclic variations. Soil water, not soil fertility, is the key to wheat production over these vast steppe lands lying beyond the practical limits of upland irrigation systems fed by major rivers. In recent years in the High Plains there has been a great increase in the use of ground water pumped to the surface and distributed by centerpivot irrigation systems (Figure 9.24). This ground water source is rapidly being depleted and will ultimately fail.

Semiarid steppes form the great sheep and cattle ranges of the world. The steppes of central Asia have for centuries supported a nomadic population whose sheep and goats find subsistence on the scanty grassland (Figure 9.25). On the vast expanses of the High Plains, the American bison lived in great numbers until being almost exterminated by hunters. The short-grass veldt of South Africa also supported much game at one time.

Steppe grasses do not form a complete sod cover; loose, bare soil is exposed between grass clumps. For this reason, overgrazing during a series of dry years can easily reduce the hold of grasses enough to permit destructive deflation (wind erosion), followed by water erosion and gullying.

On the Great Plains of Kansas, Oklahoma, and Texas, deflation and soil drifting reached disastrous proportions during a series of drought years in the

FIGURE 9.25 These nomads of northern Afghanistan are encamped on the floor of an arid valley, close to a snowfed stream. Camels will carry their tents and other possessions to a new location when necessary. (Victor Englebert/Photo Researchers.)

middle 1930s, following a great expansion of wheat cultivation. These former grasslands are underlain by friable Mollisols. During the drought a sequence of exceptionally intense dust storms occurred. Within their formidable black clouds visibility declined to nighttime darkness, even at noonday. The area affected became known as the Dust Bowl. (It included part of the adjacent subhumid belt of the moist continental climate.) Many centimeters of soil were removed from fields and transported out of the region as suspended dust, whereas the coarser silt and sand particles accumulated in drifts along fence lines and around buildings. The combination of environmental degradation and repeated crop failures caused widespread abandonment of farms and a general exodus of farm families.

Among geographers who have studied the Dust Bowl phenomenon, there is a difference of opinion as to how great a role soil cultivation and livestock grazing played in inducing deflation. The drought was a natural event over which humans had no control, but it seems reasonable that the natural grassland would have sustained far less soil loss and drifting if it had not been destroyed by the plow.

Although we cannot prevent cyclic occurrences of drought over the Great Plains, measures can be taken to minimize the deflation and soil drifting occurring in periods of dry soil conditions. Improved farming practices include use of listed furrows (deeply carved furrows) that act as traps to soil movement. Stubble mulching will reduce deflation when land is lying fallow, and tree belts may have significant effect in reducing the intensity of wind stress at ground level.

The Moist Continental Climate (10)

The moist continental climate is located in central and eastern parts of North America and Eurasia in the midlatitude zone. This climate is in the polar front zone—the battleground of polar and tropical air masses. Seasonal temperature contrasts are strong, and day-to-day weather is highly variable. Ample precipitation throughout the year is increased in summer by the invading mT air mass. Cold winters are dominated by cP and cA air masses from subarctic source regions.

In eastern Asia (China, Korea, Japan) the monsoon effect is strongly evident in a summer rainfall maximum and a relatively dry winter. In Europe, the moist continental climate lies in a higher latitude belt (45° to 60° N) and receives precipitation from the mP air mass coming from the North Atlantic.

Latitude Range: 30° to 55° N (Europe: 45° to 60° N)

Major Regions of Occurrence: Eastern parts of United States and southern Canada; northern China; Korea; Japan; central and eastern Europe.

Example: Figure 9.26 is a climograph for Madison, Wisconsin, lat. 43° N, in the American Midwest. The annual temperature range is very large. Summers are warm, but winters are cold, with three consecutive monthly means well below freezing. Precipitation is ample in all months and the annual total is large. There is a summer maximum of precipitation when the mT air mass invades and thunderstorms are formed along moving cold fronts and squall lines. Much of the winter precipitation is in the form of snow, which remains on the ground for long periods.

Soil-Water Budget: Located in the interior eastern region of the United States at lat. 40° N, Pittsburgh, Pennsylvania, represents the humid subtype (10h) of the moist continental climate (Figure 9.27). Precipitation (P) runs rather uniformly through the year but with a slight summer maximum; total annual precipitation is substantial. The annual cycle of water need (Ep) peaks strongly in summer, rising rapidly after three consecutive months of zero values when soil water is frozen and plants are dormant. A very small soil-water shortage (D) develops in summer. A substantial water surplus occurs in winter and early spring. Some of this surplus water is held in the frozen state in winter, to be released rapidly in the early spring thaw. Spring floods are highly probable. Larger

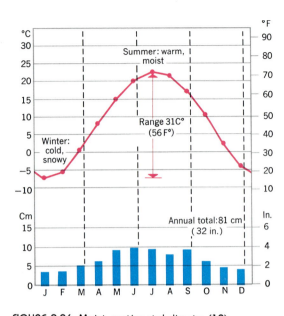

FIGURE 9.26 Moist continental climate (10). Madison, Wisconsin, lat. 43° N, has cold winters and warm summers, making the annual temperature range very large.

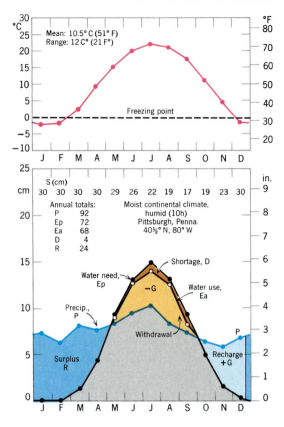

FIGURE 9.27 Soil-water budget for Pittsburgh, Pennsylvania, lat. 40° N. (Same data source as Figure 8.3.)

streams maintain their flow throughout the summer in most years.

The Moist Continental Forest and Prairie Environment

With ample precipitation throughout the year and only a small summer water shortage, the humid and perhumid subtypes of the moist continental climate support forests as the native vegetation. Soils beneath these forests show effects of the moist environment through the leaching out of soil bases and other soil components and a strong tendency to soil acidity.

These effects are most severe in the colder, more northerly parts of the climate zone, where strongly acidic soils are found on sandy surface layers. Here the evergreen needleleaf forest dominates. For example, pine forest was found in the Great Lakes region. Throughout much of the northeastern United States and southeastern Canada mixed coniferous and deciduous forest was the native type. It

graded southward into broadleaf deciduous forest, which was found in a large area of the eastern United States. Here, moderately leached forest soils are found, and these have retained a high level of natural fertility suitable for crop cultivation. The deciduous forests were also found in this climate in central and eastern Europe, and in a narrow belt penetrating far eastward into Siberia; they are also found in north-central China and in Korea.

An important environmental factor has been the activity of the great ice sheets that recently covered the northern parts of this climate region. The ice had a profound effect on the landforms of the region as well as on parent materials of the soils, which are of recent origin and poorly developed in many parts of the area. These effects are explained in Chapter 18.

A most important feature of the moist continental climate is that, when traced into the continental interior, it grades into a progressively less humid climate. In North

America the gradation to greater aridity is seen in a narrow north–south zone of a subhumid subtype (10sh) adjacent to the semiarid (steppe) climate (9s). The effect of this westward gradation is profound on both soils and native vegetation. There is a large region of the Middle West, starting about in Illinois and continuing west through Iowa and into Nebraska, where a variety of the Mollisols is the dominant soil type. Also called *prairie soil,* this soil is rich in nutrient bases and formerly supported a natural cover of tall, dense grasses; it was the *tall-grass prairie.* This kind of prairie once extended from about the United States–Canada border southward to the Gulf Coast, but today only a few small remnants can be found (see Figure 20.23).

Agricultural Resources of the Moist Continental Climate

Throughout Europe, large areas of the moist continental forest environment have been under field crops, pastures, and vineyards for centuries, while at the same time forests have been carefully cultivated over large areas (Figure 9.28). In this environment the potentials for both crop agriculture and forest culture (sylviculture) have reached a near-optimum adjustment in terms of the soils and terrain.

Because of the availability of soil water through a warm summer growing season, the moist continental environment has an enormous potential for food production. The cooler, more northerly sections in North America and Europe support dairy farming on a large scale. A combination of acid soils, known as Spodosols, and unfavorable glacial terrain in the form of bogs and lakes, rocky hills, and stony soils has deterred crop farming in many parts.

Farther south, plains formed on former lake floors and on undulating uplands are ideally suited to crop farming. Here, soils are of high fertility. Cereals grown extensively in North America and Europe include corn, wheat, rye (especially in Europe), oats, and barley. Corn is also an important crop in Hungary and Rumania. Beet sugar is an important product of this environmental region in Europe, but not in North America. On the other hand, soybeans are intensively cultivated in the

FIGURE 9.28 The Mosel River in its winding, entrenched meandering gorge through the Rhineland-Pfalz province of western Germany. On the steep, undercut valley wall at the right are vineyards and forested land. (Porterfield/Chickering Photo Researchers.)

midwestern United States and in northern China and Manchuria, but very little in Europe. Rice is a dominant crop in both South Korea and Japan, much farther poleward than elsewhere in Asia. The rice seedlings can be planted in paddies flooded during the brief but copious rains of midsummer, then harvested in the dry autumn. (Among geographers, this northern rice area is often included in the region called Monsoon Asia.) Agricultural productivity of the tall-grass prairie lands in the United States is now legendary under the name of the *corn belt*. Corn production is concentrated most heavily in the prairie plains of Illinois, Iowa, and eastern Nebraska. Wheat is also a major crop near the western limits of the tall-grass prairie in Kansas and Oklahoma.

The Midlatitude Climates in Review

The midlatitude climates run the gamut from very wet to very dry and from mild marine coastal climates to strongly seasonal continental climates. Soils and natural vegetation cover an equally great range of types befitting the spectrum of climate. No useful generalization is possible for such a diverse group of environments.

Production of food resources in the midlatitude regions spans as wide a range of intensities as the climates themselves. Here we have the richest food-producing regions of the world, fully developed by the Western nations through massive inputs of fertilizers, pesticides, and fuels and guided by the most advanced technology. But there are also unproductive deserts at the same latitudes. These midlatitude environments, taken as a whole, are neither temperate in climate nor uniform in plant resources. Perhaps this heterogeneous quality of the midlatitudes is just what we should expect of a global zone where polar and tropical air masses wage war incessantly over vast land areas that cut across the latitude zones and their prevailing westerly airflow.

High-Latitude Climates

Climates of Group III, the high-latitude climates, exclusive of the ice sheet climate,

lie almost entirely in northern hemisphere lands of North America and Eurasia. They occupy the northern subarctic and arctic latitude zones, but extend southward into the midlatitude zone as far south as about the 47th parallel in eastern North America and Asia. In terms of air masses and frontal zones, the high-latitude climates of the northern hemisphere lie in a zone of intense interaction between unlike air masses. Maritime polar (mP) air masses interact violently with continental polar (cP) and arctic (A) air masses in a discontinuous and constantly fluctuating arctic front zone. In summer, tongues of maritime tropical air masses (mT) reach the subarctic latitudes to interact with polar air masses and yield important precipitation.

In terms of prevailing pressure and wind systems, the high-latitude climates coincide closely with the belt of prevailing westerly winds that form the periphery of the circumpolar flow of the great upper-air polar vortex. Local reversals of surface airflow to east winds accompany traveling cyclones and extend upward to high levels in cutoff lows that are part of the air-mass exchange system capable of transporting water vapor into high latitudes.

The high-latitude climates have low annual total evapotranspiration, always less than about 50 cm, reflecting the prevailing low air and soil temperatures, and declining sharply poleward to extremely low values in the tundra climate and effectively to zero in the ice sheet climate. The frozen condition of the soil in several consecutive winter months causes plant growth to virtually cease, cutting off evapotranspiration. Snow that falls in this period is retained in surface storage until the spring thaw releases it for infiltration and runoff. Needless to say, the growing season for crops is short in the subarctic zone, but low air and soil temperatures are partly compensated for by the great increase in day length.

Europeans who came to settle the high latitudes of North America found familiar counterparts to their native climates in northern Europe. Experience in the boreal forest lands and tundra of Scandinavia, Finland, and northern Russia served them well in exploiting the natural environmental resources and they encountered few surprises.

The Boreal Forest Climate (11)

The boreal forest climate is a continental climate with long, bitterly cold winters and short, cool summers. It occupies the source region of the cP air mass, which is cold, dry, and stable in the winter. Invasions of the very cold cA air mass are common. The annual range of temperature is greater than that for any other climate, and is greatest in Siberia. Precipitation is substantially increased in summer, when maritime air masses penetrate the continent with traveling cyclones, but the total annual precipitation is small. Although much of the boreal forest climate is classed as humid, large areas in western Canada and Siberia have meager annual precipitation and fall into the subhumid or semiarid subtypes.

Latitude Range: 50° to 70° N

Major Regions of Occurrence: Central and western Alaska; Canada, from Yukon Territory to Labrador; Eurasia, from northern Europe across all of Siberia to the Pacific Coast.

Example: Figure 9.29 is a climograph for Fort Vermilion, Alberta, at lat. 58° N. The very great annual temperature range shown here is typical for North America. Monthly mean air temperatures are below freezing for seven consecutive months. The summers are short and cool. Precipitation shows a marked annual cycle with a strong summer maximum, but the total annual precipitation is small, and the climate can be characterized as subhumid. Although precipitation in winter is small, a snow cover remains over solidly frozen ground through the entire winter. On the same climograph, temperature data are shown for Yakutsk, U.S.S.R., a Siberian city at lat. 62° N. The enormous annual range is evident, as well as the extremely low winter-month means. January reaches a mean of about −42°C (−45°F), making this region the coldest on earth, except for the ice sheet interiors of Antarctica and Greenland. Precipitation is not shown for Yakutsk, but the annual total is very small.

Soil-Water Budget: Trout Lake, Ontario, lies in central Ontario in the heart of the

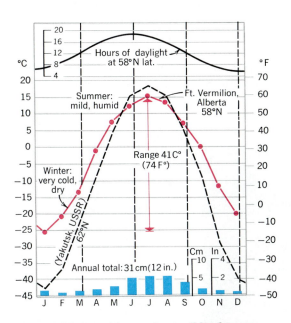

FIGURE 9.29 Boreal forest climate (11). Extreme winter cold and a very great annual range in temperature characterize the climates of Fort Vermilion, Alberta, and Yakutsk, U.S.S.R.

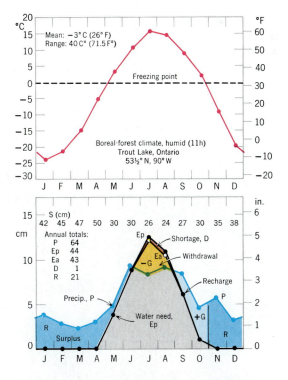

FIGURE 9.30 Soil-water budget for Trout Lake, Ontario, lat. 53½° N. (Same data source as Figure 8.3.)

Canadian Shield, not far south of Hudson Bay. This station illustrates the humid subtype (11h) of the boreal forest climate (Figure 9.30). Water need (Ep) rises to a sharp peak following six consecutive months of zero values when soil water is solidly frozen. Precipitation (P) also rises sharply during the summer. There is a brief period of soil-water use, but the soil-water shortage (D) is extremely small. Recharge is completed in October, after which precipitation accumulates as snow. Notice that the figures for soil-water storage (S) exceed 30 cm as this frozen water accumulates; it is released as runoff in May during the spring thaw. Streams flow copiously during the summer.

The Boreal Forest Environment

Land surface features of much of the region of boreal forest climate were shaped beneath the great Pleistocene ice sheets, which had their centers over the Hudson Bay–Labrador region, the northern Cordilleran Ranges, the Baltic region, and highland centers in Siberia. Severe ice erosion exposed hard bedrock over vast areas and created numerous shallow rock basins. Bouldery rock debris, called glacial till, mantles the rock surface in many places. Many of the shallower rock basins have been filled by organic bog materials forming a matlike layer called muskeg. Peat, a black substance consisting of partly decomposed plant matter, has accumulated in numerous bogs. These have provided a low-grade fuel in northern Europe (Figure 9.31). Peat and

those mineral soils poorly developed in the recent glacial materials are low in readily available plant nutrients and are acid in chemical balance.

The dominant upland vegetation of this climate region is boreal forest, consisting of needleleaf trees. In North America and Europe, these are evergreen needleleaf trees, mostly pine, spruce, and fir (Figure 9.32). Figure 21.13 is a composite map showing the extent of the boreal forest. In central and eastern Siberia, the boreal forest is dominated by the larch, which sheds its needles in winter and is thus a deciduous tree. Associated with the needleleaf trees are stands of aspen, balsam poplar, willow, and birch.

Along the northern fringe of boreal forest lies a zone of woodland in which low trees, such as black spruce, are spaced widely apart. The open areas are covered by a surface layer of lichens and mosses (Figure 9.33). This cold woodland is referred to by geographers as the *taiga*.

Crop farming in the continental subarctic environment is largely limited to lands surrounding the Baltic Sea in bordering Finland and Sweden. Cereals grown in this area include barley, oats, rye, and wheat. Along with dairying, these crops primarily provide food for subsistence. The principal nonmineral economic product throughout the subarctic lands of eastern Canada is pulpwood from the needleleaf forests. Logs are carried down the principal rivers to pulp mills and lumber mills. Forests of pine and fir in Sweden, Finland, and European

FIGURE 9.31 A peat bog in boreal forest near the border between Norway and Sweden. Blocks of peat are dried out on a crude rack of poles and will be used as household fuel. (John S. Shelton.)

FIGURE 9.32 A narrow road passes through dense boreal forest of spruce and fir in central Sweden. (John S. Shelton.)

FIGURE 9.33 Lichen woodland near Ft. McKenzie, lat. 57° N, in northern Quebec. The trees are black spruce. Between the trees is a carpet of lichen. (R. N. Drummond.)

Russia are the primary plant resource (Figure 9.32). The wood products are exported in the form of paper, pulp, cellulose, and construction lumber.

The Tundra Climate (12)

The tundra climate occupies arctic coastal fringes and is dominated by cP, mP, and cA air masses. Winters are long and severe. There is a very short mild season, which many climatologists do not recognize as a true summer. A moderating influence of the nearby ocean water prevents winter temperatures from falling to the extreme lows found in the continental interior.

Latitude Range: 60° to 75° N and S

Major Regions of Occurrence: Arctic zone of North America; Hudson Bay region and Baffin Island; Greenland coast; northern Siberia bordering the Arctic Ocean; Antarctic Peninsula.

Example: Figure 9.34 is a climograph for Upernivik, located on the west coast of Greenland at lat. 73° N. A short milder season is equivalent to a summer season in lower latitudes. The long winter is very cold, but the annual temperature range is not as large as for the boreal forest climate to the south. Total annual precipitation is small. Increased precipitation beginning in July is explained by the melting of the sea-ice cover and a warming of ocean water temperatures which increases the moisture content of the local air mass.

Soil-Water Budget: Hebron is a station on the Atlantic Coast of Labrador at lat. 58° N; it represents the humid subtype (12h) of the tundra climate (Figure 9.35). Because the winter is long and severe, water need (Ep) is zero for eight consecutive months. The total annual water need is small, but peaks sharply in the short warm season when the sun is in the sky for much of the 24-hour day. Precipitation is substantial, with a summer maximum. The soil-water shortage (D) is extremely small. A water surplus develops early in winter; this water is held in frozen storage until late spring.

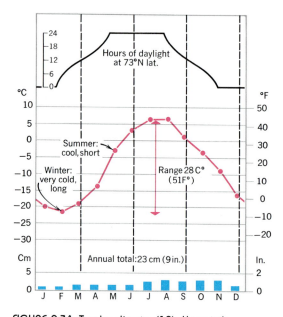

FIGURE 9.34 Tundra climate (12). Upernivik, Greenland, lat. 73° N, shows a smaller annual range than Fort Vermilion (Figure 9.29).

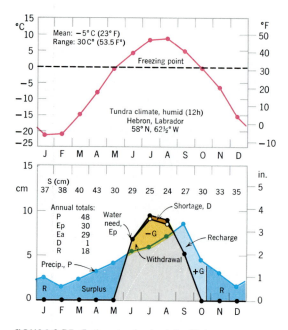

FIGURE 9.35 Soil-water budget for Hebron, Labrador, lat. 58° N. (Same data source as Figure 8.3.)

The Arctic Tundra Environment

The term *tundra* describes both an environmental region and a major class of vegetation. Figure 9.36 is a polar map showing the extent of the arctic tundra. (An equivalent climatic environment—called alpine tundra—prevails in many global locations in high mountains above the timberline.) Soils of the arctic tundra are poorly developed and consist of freshly broken mineral particles and varying amounts of humus (finely divided, partially decomposed plant matter). Peat bogs are numerous. Because soil water is solidly and permanently frozen not far below the surface, the summer thaw brings a condition of water saturation to the soil.

Trees exist in the tundra only as small,

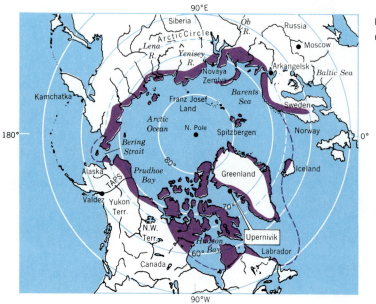

FIGURE 9.36 The tundra of the northern hemisphere.

shrublike features because of the seasonal damage to roots by freeze and thaw of the soil layer and to branches exposed to the abrading action of wind-driven snow. Vegetation of the treeless tundra consists of grasses, sedges, and lichens, along with shrubs of willow. Traced southward, the vegetation changes into birch–lichen woodland, then into the needleleaf boreal forest.

In some places a distinct tree line separates the forest and tundra. It coincides approximately with the 10°C (50°F) isotherm of the warmest month and has been used by geographers as a boundary between boreal forest and tundra.

Vegetation is scarce on dry, exposed slopes and summits—the rocky pavement of these areas gives them the name of "fell-field," the Danish term meaning "rock desert."

The number of species in the tundra ecosystem is small, but the abundance of individuals is high. Among the animals, vast herds of caribou in North America or reindeer (their Eurasian relatives) roam the tundra, lightly grazing the lichens and plants and moving constantly (Figure 9.37). A smaller number of musk-oxen are also consumers of the tundra vegetation. Wolves and wolverines, arctic foxes, and polar bears are predators. Among the smaller mammals, snowshoe rabbits and lemmings are important herbivores. Invertebrates are scarce in the tundra, except for a small number of insect species. Black flies, deerflies, mosquitoes, and "no-see-ums" (tiny biting midges) are all abundant and can make July on the tundra most uncomfortable for humans and animals. Reptiles and amphibians are also rare. The boggy tundra, however, offers an ideal summer environment for many migratory birds such as waterfowl, sandpipers, and plovers.

The food chain of the tundra ecosystem is simple and direct. The important producer is "reindeer moss," a lichen (Figure 9.38). In addition to the caribou and reindeer, lemmings, ptarmigan (arctic grouse), and snowshoe rabbits are important lichen grazers. The important predators are the fox, wolf, lynx, and bear, although all these animals may feed directly

FIGURE 9.37 Caribou migration across the arctic tundra of northern Alaska. (Warren Garst/Tom Stack and Assoc.)

FIGURE 9.38 Reindeer moss, a variety of lichen, seen here on rocky trundra of Alaska. (Steve McCutcheon.)

on plants as well. During the summer, the abundant insects help support the migratory waterfowl populations. The directness of the tundra food chain makes it particularly vulnerable to fluctuations in the populations of a few species.

Arctic Permafrost

Perennially frozen ground, or *permafrost*, prevails over the tundra region and a wide bordering area of boreal forest climate. The active layer of seasonal thaw is from 0.6 to 4 m (2 to 14 ft) thick, depending on latitude and the nature of the ground. Continuous permafrost, which extends without gaps or interruptions under all surface features, coincides largely with the tundra climate, but also includes a large part of the boreal forest climate in Siberia. Discontinuous permafrost, which occurs in patches separated by frost-free zones under lakes and rivers, occupies much of the boreal forest climate zone of North America and Eurasia. Sporadic occurrence of permafrost in small patches extends into the southern limits of the boreal forest climate.

Depth of permafrost reaches 300 to 450 m (1000 to 1500 ft) in the continuous zone near lat. 70° N. Much of this permanent frost is an inheritance from more severe conditions of the last ice age, but some permafrost bodies may be growing under existing climate conditions. Various surface features of the arctic permafrost are described in Chapter 14, including ice wedges and patterned ground (see Figures 14.5 and 14.6).

Environmental degradation of permafrost regions arises from surface changes made by humans. The undesirable consequences are usually related to the destruction or removal of an insulating surface cover, which may consist of a moss or peat layer in combination with living plants of the tundra or arctic forest. When this layer is scraped off, the summer thaw is extended to a greater depth, with the result that ice wedges and other ice bodies melt in the summer and waste downward. This activity is called thermal erosion. Meltwater mixes with silt to form mud, which is then eroded and transported by water streams.

The consequences of disturbance of permafrost terrain became evident in World War II, when military bases, airfields, and highways were hurriedly constructed without regard for maintenance of the natural protective surface insulation. In extreme cases, scraped areas turned into mud-filled depressions and even into small lakes that expanded in area with successive seasons of thaw, eventually engulfing nearby buildings. Engineering practices now call for placing buildings on piles with an insulating air space below or for the deposition of a thick insulating pad of coarse gravel over the surface prior to construction. Steam and hot-water lines are placed aboveground in a heated insulated tunnel called a "utilidor" to prevent thaw of the permafrost layer (Figure 9.39).

Another serious engineering problem of arctic regions involves the behavior of streams in winter. As the surfaces of

FIGURE 9.39 The elevated tunnel of sheet metal is a "utilidor" (utility corridor) in which steam and water lines are protected from freezing in the severe tundra winter. Notice that the buildings are mounted on posts and have small windows. Inuvik, Northwest Territories. (Mark A. Melton.)

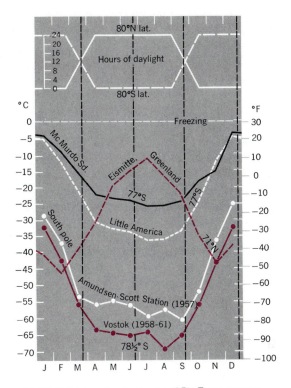

FIGURE 9.40 Ice-sheet climate (13). Temperature graphs for five ice-sheet stations.

streams and springs freeze over, the water beneath bursts out from place to place, freezing into huge accumulations of ice. Highways are thus made impassable.

The lessons of superimposing modern technology on a highly sensitive natural environment were learned the hard way— by encountering unpleasant and costly effects that were not anticipated. The threat of environmental destruction will persist as oil continues to flow through the Trans-Alaska Pipeline. This facility carries hot oil from the northern shores of Alaska, across a permafrost landscape, to the port of Valdez on the south coast. Effects of this pipeline on permafrost and other elements of the environment were hotly debated and were the subject of intensive investigation. In addition to the prospects of thaw of the permafrost layer by heat of the pipeline, there is the possibility of damage to the ecosystem from spills caused by pipe breakage.

The Ice-Sheet Climate (13)

The ice-sheet climate coincides with the source regions of arctic (A) and antarctic (AA) air masses situated on vast, high ice sheets and over polar sea ice of the Arctic Ocean. Mean annual temperature is much lower than that of any other climate, with

no above-freezing monthly mean. Strong temperature inversions develop over the ice sheets. The high surface altitude of the ice sheets intensifies the cold. Strong cyclones with blizzard winds are frequent. Precipitation, almost all occurring as snow, is very small, but accumulates because of the continuous cold.

Latitude Range: 65° to 90° N and S

Examples: Figure 9.40 shows temperature graphs for several representative ice-sheet stations. The graph for Eismitte, Greenland, shows the northern hemisphere temperature cycle, whereas the other four examples are all from Antarctica. Temperatures in the interior of Antarctica have proved to be far lower than at any other place on earth. The Russian meteorological station at Vostok, located about 1300 km (800 mi) from the south pole at an altitude of about 3500 m (11,400 ft), may be the world's coldest spot. Here a low of −88.3°C (−127°F) was observed. At the pole itself (Amundsen–Scott Station),

July, August, and September of 1957 had averages of about −60°C (−76°F). Temperatures run considerably higher, month for month, at Little America because it is located close to the Ross Sea and is at low altitude.

Soil-Water Budget: Water need of the ice-sheet climate (Ep) is effectively zero throughout the entire year, because no monthly temperature mean is above 0°C (32°F). Almost all the scarce precipitation is in the form of snow, which on land accumulates as glacial ice.

The Ice-Sheet Environment

Because of low monthly mean temperatures throughout the year over the ice sheets, this environment is devoid of vegetation and soils. The few species of animals found on the ice margins are associated with a marine habitat. In terms of habitation by humans, the ice-sheet environment is extremely hostile because of extreme cold, high winds, and a total lack of food and fuel resources. Enormous expenditures of energy are required to import these necessities of life and to provide shelter. These efforts are justified because of the need for scientific research, but in the foreseeable future there is little prospect that this icy environment will provide useful supplies of energy or minerals.

Global Climates in Review

In three chapters, we have surveyed the principal climates of the globe. Each climate type, together with its characteristic soils, natural vegetation, and landforms, comprises a unique natural environmental region.

We must apply this information to two grave questions facing the human race. "Can the developing nations increase their food production fast enough to stave off starvation?" Answers given by well-informed specialists span the range from extreme pessimism to extreme optimism. Another question we hear is: "Will there be enough fresh water to supply the rapidly increasing demands of the energy-consuming industrial nations?" Your appreciation of current problems of agriculture and freshwater supplies over the lands of the globe can be greatly increased by application of the soil-water budget to the climate of a problem area.

Perhaps the most important practical lesson of this chapter is that few soil-water budgets provide the full water need of food plants during the growing season, with neither too great a shortage nor too great a surplus. The soil-water budget sets stringent limitations on expansion of agricultural resources. Intelligent global planning for environmental management depends heavily on the soil-water balance.

CHAPTER 10

RUNOFF AND WATER RESOURCES

THE HYDROLOGIC CYCLE   A global system of pathways by which water moves through the atmosphere and oceans and upon or beneath the earth's land surfaces.

RUNOFF   Precipitation falling on the lands follows flow paths both beneath and above the surface; these are all forms of runoff.

SUBSURFACE WATER   Excess soil water that moves downward may reach a saturated zone to become ground water.

GROUND WATER FLOW   Moving very slowly through openings in rock, ground water eventually reaches streams, lakes, or the ocean shoreline, completing a flow path in the hydrologic cycle.

OVERLAND FLOW   Excess precipitation flows downhill over the land surface, eventually becoming concentrated in a stream.

DRAINAGE SYSTEMS   Streams converge and join to form rivers, which reach the sea to complete a part of the hydrologic cycle.

FLOODS   They are natural inundations of low-lying lands adjacent to rivers; a great hazard, too often aggravated by human activity.

THE primary concern of this chapter is with the science of *hydrology,* which is a study of water as a complex but unified system on the earth. We began our investigation of hydrology in Chapter 7 by covering that phase of hydrology in which soil water is recharged by precipitation and returned directly to the atmosphere by evapotranspiration, or to be disposed of as runoff. Our investigation now continues with surplus water and the paths it follows as subsurface water and surface water.

There are two basic paths of escape for surplus water. First, surplus water may percolate through the soil, traveling downward under the force of gravity to become part of the underlying ground water body. Following subterranean flow paths, this water emerges to become surface water, or it may emerge directly in the shore zone of the ocean. Second, surplus water may flow over the ground surface as runoff to lower levels. As it travels, the dispersed flow becomes collected into streams, which eventually conduct the runoff to the ocean. In this chapter, we trace both the subsurface and surface pathways of flow of surplus water.

Surplus water, as runoff, is a vital part of the environment of terrestrial life-forms and of humans in particular. Surface water in the form of streams, rivers, ponds, and lakes constitutes one of the distinctive environments of plants and animals.

Our heavily industrialized society requires enormous supplies of fresh water for its sustained operation. Urban dwellers consume water in their homes at rates of 150 to 400 liters (50 to 100 gallons) per person per day. Large quantities of water are used for cooling purposes in air-conditioning units and power plants.

In view of projections based on existing rates of increase in water demands, we will be hard put in the future to develop the needed supplies of pure fresh water. Water pollution also tends to increase as populations grow and urbanization advances over broader areas. A disconcerting concept is that the available resource of pure fresh water is shrinking while demands are rising. Knowledge of hydrologic processes enables us to evaluate the total water resource, to plan for its management, and to protect it from pollution.

Global Water in Storage

The hydrosphere can be thought of as an enormous storage pool of water in three states, but in greatly differing proportions, depending where and in what state it is stored. As Figure 10.1 shows, most of the hydrosphere—about 97 percent—consists of the salt water of the oceans. Next in bulk is

fresh water stored as ice in the world's ice sheets and mountain glaciers—a little over 2 percent.

Water in the liquid state is found both on and beneath the earth's land surfaces. Water occupying openings in soil and rock is called *subsurface water;* most of it is held in deep storage as ground water, where it makes up just over 0.6 percent of the hydrosphere. Water held in the soil, within reach of plant roots, comprises 0.005 percent. Water held in streams, lakes, marshes, and swamps is called *surface water;* it amounts to about 0.02 percent of the hydrosphere. As the right-hand circle in Figure 10.1 shows, most of this surface water is about evenly divided between freshwater lakes and salty lakes. An extremely small proportion is temporarily held in streams (rivers). Although the quantity of water held as vapor and cloud particles in the atmosphere is very small— 0.001 percent of the hydrosphere—its importance is enormous because this is the avenue of supply of all fresh water.

The Hydrologic Cycle

Water of oceans, atmosphere, and lands moves in a great series of continuous interchanges of both geographic position and physical state, known as the *hydrologic cycle* (Figure 10.2). A particular molecule of water might, if we could trace it continuously, travel through any one of a number of possible circuits involving alternately the water vapor state and the liquid or solid state.

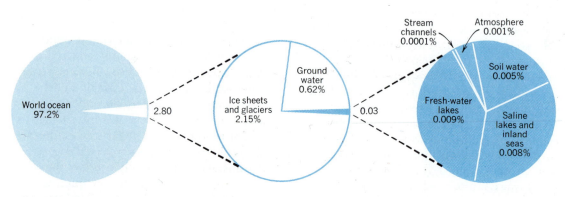

FIGURE 10.1 The total volume of global water in storage is largely held in the world ocean.

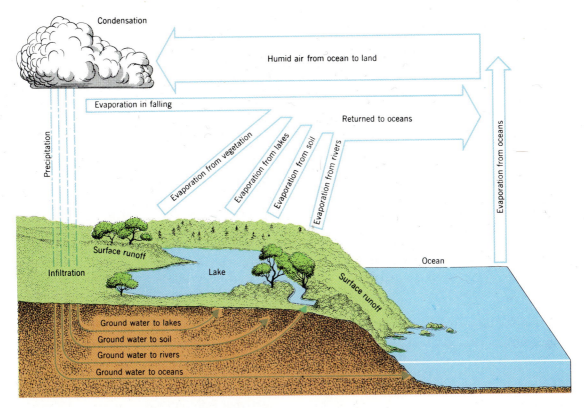

FIGURE 10.2 The hydrologic cycle traces the various paths of water from oceans, through atmosphere, to lands, and return to oceans.

The pictorial diagram of the hydrologic cycle given in Figure 10.2 can be quantified for the earth as a whole. Figure 10.3 is a mass-flow diagram relating the principal pathways of the water circuit. We can start with the oceans, which comprise the basic reservoir of free water. Evaporation from the ocean surfaces totals about 419,000 km³ per year. (English equivalents are shown in Figure 10.3). At the same time, evaporation from soil, plant foliage, and water surfaces of the continents totals about 69,000 km³. Thus, the total evaporation is 488,000 km³. Because this water vapor must eventually condense and return to the liquid or solid state, it is an amount equal to the total annual global precipitation.

Precipitation is unevenly divided between continents and oceans; 106,000 km³ are received by the land surfaces and 382,000 km³ by the ocean surfaces. Notice that the continents receive about 37,000 km³ more water as precipitation than they lose by evaporation. This excess quantity flows over or under the ground surface to reach the

sea; it is collectively termed *runoff*.

We can state the *global water balance* as

$$P = E + G + R$$

where P = precipitation
E = evaporation
G = net gain or loss of water in the system, a storage term
R = runoff (positive sign when running off the continents, negative sign when flowing into the oceans)

All terms are in units of cubic kilometers per year. When applied over the span of a year, and averaged over many years, the storage term G can be considered as zero, because the global system is essentially closed so far as matter is concerned. The quantities of water in storage in the atmosphere, on the lands, and in the oceans will remain about constant from year to year. The equation then simplifies to

$$P = E + R$$

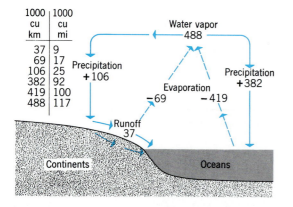

1000 cu km	1000 cu mi
37	9
69	17
106	25
382	92
419	100
488	117

FIGURE 10.3 The global water balance. Figures give average annual water flux in and out of world land areas and world oceans. (Based on data of John R. Mather.)

Using the figures given for the continents,

$$106,000 = 69,000 + 37,000$$

and for the oceans,

$$382,000 = 419,000 - 37,000$$

For the globe as a whole, combining continents and ocean basins, the runoff terms cancel out:

$$106,000 + 382,000 = 69,000 + 419,000$$

$$488,000 = 488,000$$

Infiltration and Runoff

Most soil surfaces in their undisturbed, natural states are capable of absorbing the water from light or moderate rains by infiltration. Most soils have natural passageways between poorly fitting soil particles, as well as larger openings, such as earth cracks, resulting from soil drying, borings of worms and animals, cavities left from decay of plant roots, or openings made by growth and melting of frost crystals. A mat of decaying leaves and stems breaks the force of falling drops and helps to keep these openings clear. If rain falls too rapidly to be passed downward through these soil openings, the excess amount flows as a surface water layer down the direction of ground slope. This surface runoff is called *overland flow*.

Gravity percolation carries excess water down to the ground water zone, in which all pore spaces are fully saturated (see Figure 7.10). Within this zone water moves slowly in deep paths, eventually emerging by seepage into streams, ponds, lakes, and oceans.

Excess water leaves the area by stream flow. Streams are fed directly by overland flow in periods of heavy, prolonged rain or rapid snowmelt. Streams that flow throughout the year—perennial streams— derive much of their water from ground water seepage. Streams fed only by overland flow run intermittently, and their channels are dry much of the time between rain periods.

Ground Water

Ground water is that part of the subsurface water that fully saturates the pore spaces of bedrock or soil material. The ground water occupies the *saturated zone* (Figure 10.4). Above it is the *unsaturated zone* in which water does not fully saturate the pores. Water is held in the unsaturated zone by capillary force in tiny films adhering to the mineral surfaces.

Ground water is extracted from wells dug or drilled to reach the ground water zone. In the ordinary shallow well, water rises to the same height as the *water table*, or upper boundary of the saturated zone (Figure 10.4).

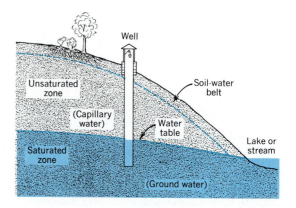

FIGURE 10.4 Zones of subsurface water.

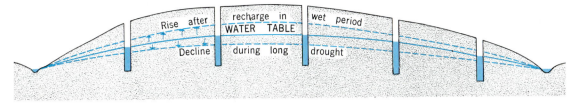

FIGURE 10.5 Configuration of the water-table surface conforms with the land surface above it.

Where wells are numerous in an area, the position of the water table can be mapped in detail by plotting the water heights and noting the trend of change in altitude from one well to the other. The water table is highest under the highest areas of land surface: hilltops and divides. The water table declines in altitude toward the valleys, where it appears at the surface close to streams, lakes, or marshes (Figure 10.5). The reason for this water table configuration is that water percolating down through the unsaturated zone tends to raise the water table, whereas seepage into streams tends to draw off ground water and to lower its level.

Because ground water moves extremely slowly, a difference in water table level is built up and maintained between high and low points on the water table. In periods of abnormally high precipitation, the water table rises under divide areas; in periods of water deficit, occasioned by drought, the water table falls (Figure 10.5).

The subsurface phase of the hydrologic cycle is completed when the ground water emerges in places where the water table intersects the ground surface. Such places

are the channels of streams and the floors of marshes and lakes. By slow seepage and spring flow the water emerges fast enough to balance the rate at which water enters the ground water table by percolation.

Figure 10.6 shows paths of flow of ground water. Flow takes paths curved concavely upward. Water entering the hillside midway between divide and stream flows rather directly. Close to the divide point on the water table, however, the flow lines go almost straight down to great depths, from which they recurve upward to points under the streams. Progress along these deep paths is incredibly slow; that near the surface is much faster. The most rapid flow is close to the place of discharge in the stream, where the arrows are shown to converge.

Use of Ground Water as a Resource

Rapid withdrawal of ground water has begun to make serious impacts on the environment in many places. The drilling of vast numbers of wells, from which water is forced out in great volumes by powerful pumps, has profoundly altered nature's balance of ground water recharge and discharge. Increased urban populations and industrial developments require larger water supplies, needs that cannot always be met from construction of new surface water reservoirs.

In agricultural lands of the dry climates, heavy dependence is placed on irrigation water from pumped wells, especially since many of the major river systems have already been fully developed for irrigation from surface supplies. Wells can be drilled within the limits of a given agricultural or industrial property and can provide

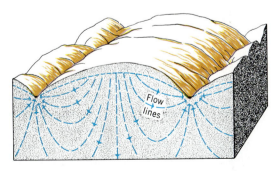

FIGURE 10.6 Paths of ground water movement under divides and valleys. (Drawn by A. N. Strahler.)

immediate supplies of water without any need to construct expensive canals or aqueducts.

Formerly, the small well needed to supply domestic and livestock needs of a home or farmstead was actually dug by hand as a large cylindrical hole, lined with masonry where required. By contrast, the modern well put down to supply irrigation and industrial water is drilled by powerful machinery that may bore a hole 40 cm (16 in.) or more in diameter to depths of 300 m (1000 ft) or more. Drilled wells are sealed off by metal casings that exclude impure near-surface water and prevent clogging of the tube by caving of the walls. Near the lower end of the hole, in the ground water zone, the casing is perforated to admit the water. The yields of single wells range from as low as a few liters per day in a domestic well to many millions of liters per day for large industrial or irrigation wells.

As water is pumped from a well, the level of water in the well drops. At the same time, the surrounding water table is lowered in the shape of a conical surface. Where many wells are in operation, their intersecting cones produce a general lowering of the water table.

Depletion often greatly exceeds the rate at which the ground water of the area is recharged by percolation from rain or from the beds of streams. In an arid region, much of the ground water for irrigation is from wells driven into thick sands and gravels. Recharge of these deposits depends on the seasonal flows of water from streams heading high in adjacent mountain ranges. The extraction of ground water by pumping can greatly exceed the recharge by stream flow. Deeper wells and more powerful pumps are then required. Overdrafts of water accumulate, and the result is exhaustion of a natural resource

not renewable except over long periods of time.

In humid areas, where a large annual water surplus exists, natural recharge is by general percolation over the entire ground area surrounding the well. Here the prospects of achieving a balance of recharge and withdrawal are highly favorable through the control of pumping. An important recycling measure is the return of waste waters or stream waters to the ground water table by means of recharge wells in which water flows downward.

Pollution of Ground Water

Disposal of solid wastes poses a major environmental problem in the United States because our advanced industrial economy provides an endless source of garbage and trash. Traditionally, these waste products were trucked to the town dump and burned there in continually smoldering fires that emitted foul smoke and gases. The partially consumed residual waste was then buried under earth.

In recent years, a major effort has been made to improve solid-waste disposal methods. One method is high-temperature incineration. Another is the *sanitary landfill* method in which waste is not allowed to burn. Instead, the waste is continually covered by protective overburden, usually sand or clay available on the landfill site. The waste is thus buried in the unsaturated zone. Here it is subject to reaction with percolating rainwater infiltrating the ground surface. This water picks up a wide variety of ions from the waste body and carries these down to the water table (Figure 10.7).

Once in the water table, the pollutants follow the flow paths of the ground water.

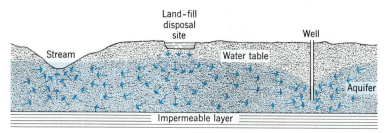

FIGURE 10.7 Polluted water, leached from a waste disposal site, moves toward supply well (right) and a stream (left). (From A. N. Strahler, *Planet Earth*, Harper & Row, New York. Copyright © 1972 by Arthur N. Strahler.)

As the arrows in Figure 10.7 indicate, the polluted water may flow toward a supply well, which is drawing in ground water from a large radius. Once the polluted water has reached the well, the water becomes unfit for human consumption. Polluted water may also move toward a nearby valley, causing pollution of the stream flowing there.

Forms of Overland Flow

We have now traced the subsurface movements of surplus water beneath the lands. We turn next to trace the surface flow paths of surplus water.

Runoff that flows down the slopes of the land in broadly distributed sheets is referred to as *overland flow.* We distinguish overland flow from *stream flow,* in which the water occupies a narrow channel confined by lateral banks. Overland flow can take several forms. It may be a continuous thin film, called sheet flow, where the soil or rock surface is smooth (Figure 10.8). Flow may take the form of a series of tiny rivulets connecting one water-filled hollow with another, where the ground is rough or pitted. On a grass-covered slope, overland flow is subdivided into countless tiny threads of water, passing around the stems. Even in a heavy and prolonged rain, you might not notice overland flow in progress on a sloping lawn. On heavily forested slopes, overland flow may pass entirely concealed beneath a thick mat of decaying leaves.

Stream Flow

Overland flow eventually contributes to a stream, which is a much deeper, more concentrated form of runoff. We define a *stream* as a long, narrow body of flowing water occupying a trenchlike depression, or channel, and moving to lower levels under the force of gravity.

The *channel* of a stream is a narrow trough, shaped by the forces of flowing water to be most effective in moving the quantities of water and sediment supplied to the stream (Figure 10.9). Channels may be so narrow that a person can jump across them, or, in the case of the Mississippi River, as wide as 1.5 km (1 mi).

The size of a stream channel can be stated in terms of the area of cross section, *A,* which is the area in square meters or square feet between the stream surface and bed, measured in a vertical slice across the stream (Figure 10.9). The rate of fall in altitude of the stream surface in the downstream direction is the *stream gradient.* As a stream flows under the influence of gravity, the water encounters resistance—a form of friction—with the channel walls. As a result, water close to the bed and banks moves slowly, and that in the deepest and most centrally located zone flows fastest. Figure 10.9 indicates by arrows the speed of flow at various points in the stream. The single line of maximum velocity is located in midstream, in the case where the channel is straight and symmetrical.

Our statement about velocity needs to be qualified. Actually, in all but the most

FIGURE 10.8 Overland flow taking the form of a thin sheet of water covers the nearly flat plain in the middle distance. This water is converging into stream flow in a narrow, steep-sided gully (left). The photograph was taken shortly after a summer thunderstorm had deluged the area. The locality, near Raton, New Mexico, shows steppe grassland vegetation. (Mark A. Melton.)

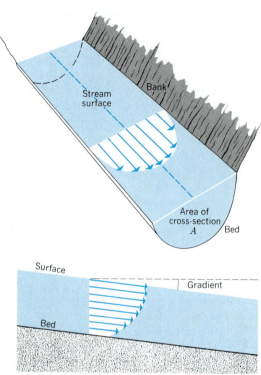

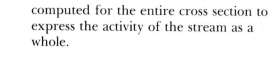

FIGURE 10.9 Stream flow within a channel is most rapid near the center.

sluggish streams, the water is affected by *turbulence*, a system of innumerable eddies that are continually forming and dissolving. A particular molecule of water, if we could keep track of it, would describe a highly irregular, corkscrew path as it is swept downstream. Motions include upward, downward, and sideward directions.

Turbulence in streams is extremely important because of the upward elements of flow that lift and support fine particles of sediment. The murky, turbid appearance of streams in flood is ample evidence of turbulence, without which sediment would remain near the bed. Only if we measure the water velocity at a certain fixed point for a long period of time, say several minutes, will the average motion at that point be downstream and in a line parallel with the surface and bed. Average values are shown by the arrows in Figure 10.9.

Because the velocity at a given point in a stream differs greatly according to whether it is being measured close to the banks and bed or out in the middle line, a single value, the *mean velocity*, is needed. Mean velocity is

computed for the entire cross section to express the activity of the stream as a whole.

Stream Discharge

A most important measure of stream flow is *discharge*, which is defined as the volume of water passing through a given cross section of the stream in a given unit of time. Commonly, discharge is stated in cubic meters per second (abbreviated to *cms*). (In English units discharge is stated in cubic feet per second, *cfs*.) Discharge, Q, may be obtained by taking the mean velocity, V, and multiplying it by cross-sectional area, A. This relationship is stated by the important equation $Q = AV$.

We realize that water will flow faster in a channel of steep gradient than in one of gentle gradient, because the force of gravity acts more strongly for the steeper gradient. As shown in Figure 10.10, velocity increases quickly where a stream passes from a wide pool of low gradient to a steep stretch of rapids. As velocity V increases, cross-sectional area A must decrease; otherwise their product, AV, would not be held constant. In the pool, where velocity is low, cross-sectional area is correspondingly increased.

An important activity of the U.S. Geological Survey is the measurement, or gauging, of stream discharge in the United

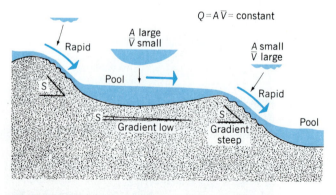

FIGURE 10.10 Schematic diagram of the relationships among cross-sectional area, mean velocity, and gradient. (From A. N. Strahler, *The Earth Sciences*, 2nd ed., Harper & Row, New York. Copyright © 1971 by Arthur N. Strahler.)

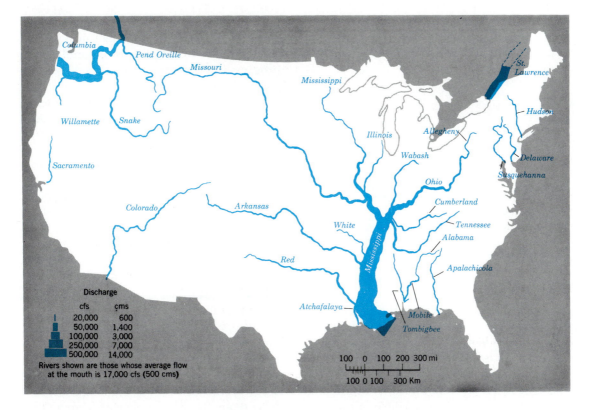

FIGURE 10.11 This schematic map shows the relative magnitude of United States rivers. Width of the color band is proportional to mean annual discharge. (U.S. Geological Survey.)

States. In cooperation with states and municipalities this organization maintains over 6000 gauging stations on principal streams and their tributaries. Information on daily discharge and flood discharges is essential for planning the distribution and development of surface waters as well as for design of flood-protection structures and for the prediction of floods as they progress down a river system.

Figure 10.11 is a map showing the relative discharge of major rivers of the United States. ("River" is a popular term applied to a large stream. The word "stream" is the scientific term designating channel flow of any magnitude of discharge.) The mighty Mississippi with its tributaries dwarfs all other North America rivers, although the MacKenzie, Columbia, and Yukon rivers, as well as the Great Lakes discharge through the St. Lawrence River, are also of major proportions. The Colorado River, a much smaller stream, crosses a vast semiarid and arid region in

which little tributary flow is added to the snowmelt source high in the Rocky Mountains.

From the standpoint of river flow and floods, mountain climates are of great importance in midlatitudes. The higher ranges serve as snow storage areas, keeping back the precipitation until early or midsummer, then releasing it slowly through melting. In this way a continuous river flow is maintained. As melting proceeds to successively higher levels, the meltwater is supplied to the drainage basin. Among the snow-fed rivers of the western United States are the Columbia, Snake, Missouri, Platte, Arkansas, and Colorado.

Drainage Systems

Seeking to escape to progressively lower levels and eventually to the sea, runoff becomes organized into a *drainage system*. The system consists of a branched network

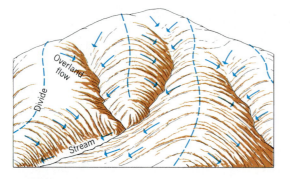

FIGURE 10.12 Overland flow from slopes in the headwater area of a drainage system supplies water to the smallest elements of the channel network.

watershed for overland flow. A drainage system is a converging mechanism funneling the weaker forms of runoff into progressively deeper and more intense paths of activity.

In a downstream direction, the gradient of the main channel of a drainage system becomes progressively gentler in gradient. The general rule is: the larger the cross section, the lower the gradient. Great rivers, such as the Mississippi and Amazon, have gradients so low that they can be described as "flat." For example, the water surface of the lower Mississippi River falls in elevation about 3 cm for each kilometer of downstream distance.

of stream channels, as well as the sloping ground surfaces that contribute overland flow to those channels (Figure 10.12). The entire system is bounded by a drainage divide, outlining a more-or-less pear-shaped *drainage basin*. The basin system is adjusted to dispose as efficiently as possible of the runoff and its contained load of mineral particles.

A typical stream network contributing to a single outlet is shown in Figure 10.13. Note that each fingertip tributary receives runoff from a small area of land surface surrounding the channel. This area may be regarded as the unit cell of the drainage system. The entire surface within the outer divide of the drainage basin constitutes the

Stream Flow and Precipitation

It seems obvious that the discharge of a stream will increase in response to a period of heavy rainfall or snowmelt. The response is delayed, of course, but the length of delay depends on a number of factors. The most important factor is the size of the drainage basin feeding the stream above the place where the gauging station is located.

The relationship between stream discharge and precipitation is best studied by means of a simple graph, called a *hydrograph*. Figure 10.14 is a hydrograph for a drainage basin about 800 km² (300 mi²) in area located in Ohio within the moist continental climate. The graph gives data

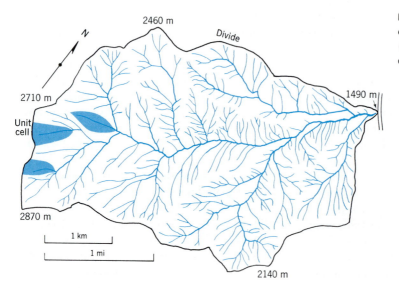

FIGURE 10.13 Channel network of a small drainage basin. (Data of U.S. Geological Survey and Mark A. Melton.)

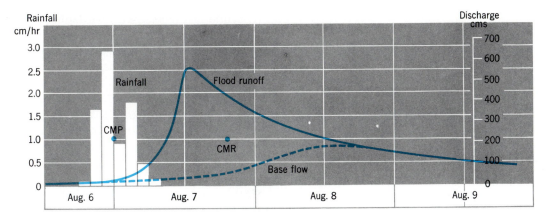

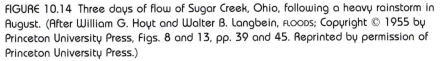

FIGURE 10.14 Three days of flow of Sugar Creek, Ohio, following a heavy rainstorm in August. (After William G. Hoyt and Walter B. Langbein, FLOODS; Copyright © 1955 by Princeton University Press, Figs. 8 and 13, pp. 39 and 45. Reprinted by permission of Princeton University Press.)

for a two-day summer storm. Rainfall is shown by a bar graph giving the number of centimeters of precipitation in each two-hour period. Also plotted on the graph (smooth line) is the discharge of Sugar Creek, the trunk stream of the drainage basin. The average total rainfall over the watershed of Sugar Creek was about 15 cm (6 in.); of this amount about half passed down the stream within three days' time. Some rainfall was held in the soil as soil water, some evaporated, and some infiltrated to the water table to be held in long-term storage in the ground water body.

Studying the rainfall and runoff graphs in Figure 10.14, we see that prior to the onset of the storm, Sugar Creek was carrying a small discharge. This flow, being supplied by the seepage of ground water into the channel, is termed *base flow*. After the heavy rainfall began, several hours elapsed before the stream gauge at the basin mouth began to show a rise in discharge. This interval, called the *lag time*, indicates that the branching system of channels was acting as a temporary reservoir. The channels were at first receiving inflow more rapidly than it could be passed down the channel system to the stream gauge.

Lag time is measured as the difference between center of mass of precipitation (CMP) and center of mass of runoff (CMR), as labeled in Figure 10.14. The peak flow of

Sugar Creek was reached almost 24 hours after the rain began; the lag time was about 18 hours. Note also that the rate of decline in discharge was much slower than the rate of rise.

In general, the larger a watershed, the longer is the lag time between peak rainfall and peak discharge, and the more gradual is the rate of decline of discharge after the peak has passed. Notice that the flow of Sugar Creek showed a slow but distinct rise in the amount of discharge contributed by base flow.

How Urbanization Affects Stream Flow

The growth of cities affects the flow of small streams in two ways. First, an increasing percentage of the surface is rendered impervious to infiltration by construction of roofs, driveways, walks, pavements, and parking lots. In a closely built-up residential area with small lot sizes the percentage of impervious surface may run as high as 80 percent.

An increase in proportion of impervious surface increases overland flow generally from the urbanized area. An important result is to increase the frequency and height of flood peaks during heavy storms. This effect applies to small watersheds lying largely within the urban area. There is also a reduction of recharge to the ground water body beneath, and this reduction, in turn,

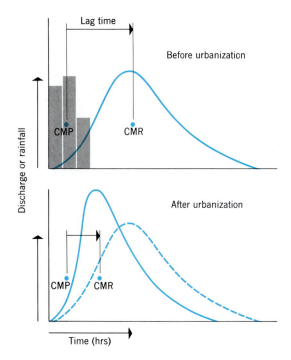

FIGURE 10.15 These schematic hydrographs show the effect of urbanization on lag time and peak discharge. Urbanization increases the peak flood flow and diminishes the lag time. Points CMP and CMR are centers of mass of rainfall and runoff, respectively, as in Figure 10.14. (After L. B. Leopold, U.S. Geological Survey.)

A second change caused by urbanization is brought about by the introduction of storm sewers that allow storm runoff from paved areas to be taken directly to stream channels for discharge. Runoff travel time to channels is shortened at the same time that the proportion of runoff is increased by expansion in impervious surfaces. The two changes together conspire to reduce the lag time, as shown by the schematic hydrographs in Figure 10.15.

Many rapidly expanding suburban communities are now finding that certain low-lying residential areas, formerly free of flooding, are being subjected to inundation by flooding of a nearby stream. The need for careful terrain study and land-use planning is obvious in such cases to protect the unwary homebuyer from locating in a flood-prone neighborhood.

The Annual Flow Cycle of a Large River

In regions of humid climates, where the water table is high and normally intersects the important stream channels, the hydrographs of larger streams will show clearly the effects of two sources of water: (1) base flow and (2) overland flow. Figure 10.16 is a hydrograph of the Chattahoochee River in Georgia, a fairly large river draining a watershed of 8700 km^2 (3350 mi^2), much of it in the humid southern Appalachian Mountains. The sharp, abrupt

decreases the base flow contribution to channels in the same area. Thus, the full range of stream discharges, from low stages in dry periods to flood stages, is made greater by urbanization.

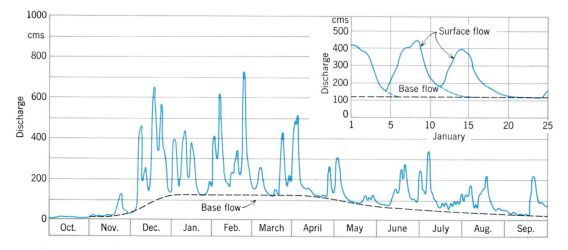

FIGURE 10.16 This hydrograph shows the fluctuating discharge of the Chattahoochee River, Georgia, throughout a typical year. The high peaks are caused by runoff from storms that produced heavy overland flow. (Data of U.S. Geological Survey in E. E. Foster, 1949, *Rainfall and Runoff*, Macmillan, New York.)

fluctuations in discharge are produced by overland flow following rain periods of one to three days' duration. These are each similar to the hydrograph of Figure 10.14, except that they are here shown much compressed by the time scale.

After each rain period the discharge falls off rapidly but, if another storm occurs within a few days, the discharge rises to another peak. The enlarged inset graph in Figure 10.16 shows details for the month of January. When a long period intervenes between storms, the discharge falls to a low value, the base flow, at which it levels off.

Throughout the year the base flow, which represents ground water inflow into the stream, undergoes a marked annual cycle. During the period of recharge (winter and early spring), water table levels are raised and the rate of inflow into streams is increased. For the Chattahoochee River, the rate of base flow during January, February, March, and April holds uniform at about 100 cms (4000 cfs). The base flow begins to decline in spring, as heavy evapo-transpiration losses reduce soil water and therefore cut off the recharge of ground water. The decline continues through the summer, reaching a low of about 30 cms (1000 cfs) by the end of October.

FIGURE 10.17 The city of Harrisburg, Pennsylvania, lies partly submerged beneath flood waters generated by Hurricane Agnes, June 1972. The Susquehanna River, seen in the distance, rose to nearly 5 m (16 ft) above flood stage, inundating the low river terrace on which the downtown portion of Harrisburg is built. (Department of the Army.)

River Floods

Everyone has seen enough news-media photos of river floods to have a good idea of the appearance of flood waters and the havoc wrought by their erosive power and by the silt and clay that they leave behind. Even so, it is not easy to define the term *flood*. Perhaps it is enough to say that a condition of flood exists when the discharge of a river cannot be accommodated within the margins of its normal channel, so that the water spreads over the adjoining ground on which crops or forests are able to flourish.

Most rivers of humid climates have a *floodplain*, a broad belt of low flat ground bordering the channel on one or both sides inundated by stream waters about once a year. This flood usually occurs in the season when abundant supplies of surface water combine with effects of a high water table to supply more runoff than can stay within

the channel. Such an annual inundation is considered a flood, even though its occurrence is expected and does not prevent the cultivation of crops after the flood has subsided. The seasonal inundation does not interfere with the growth of dense forests, which are widely distributed over low, marshy floodplains in all humid regions of the world. Still higher discharges of water, the rare and disastrous floods that may occur as infrequently as once in three to five decades, inundate ground lying well above the floodplain (Figure 10.17).

For practical purposes, the National Weather Service, which provides a flood-warning service, designates a particular river surface height at a given place as the *flood stage;* it is the critical level above which inundation of the floodplain may be expected to set in.

Flood Prediction

The National Weather Service operates a River and Flood Forecasting Service through 85 offices located at strategic points along major river systems of the United States. Each office issues river and flood forecasts to the communities within the associated district, which is laid out to cover one or more large watersheds. Flood warnings are publicized by every possible means. Close cooperation is maintained with various agencies to plan evacuation of threatened areas and the removal or protection of vulnerable property.

Graphs of flood stages tell the likelihood of occurrence of given stages of high water for each month of the year. Figure 10.18 shows expectancy graphs for two rivers. The meaning of the strange-looking bar symbols is explained in the key. The Mississippi River at Vicksburg illustrates a great river responding largely to spring floods so as to yield a simple annual cycle. All floods have occurred in the first six calendar months of the year; none in the second six months. The Sacramento River shows nicely the effect of the Mediterranean climate, with its winter wet season and long severe summer drought. Winter floods are caused by torrential rainstorms and snowmelt in the mountain watersheds of the Sierra Nevada and southern Cascades. By midsummer, river flow has shrunken to a very low stage.

Lakes and Ponds

Lakes are integral parts of drainage systems and participate in runoff of water in the hydrologic cycle. Lakes are of major environmental importance in many ways. They represent large bodies of fresh water in storage; they support ecosystems that provide food for humans. Today, the recreational value of lakes is assuming increasing importance.

Where lakes are not naturally present in the valley bottoms of drainage systems, we create lakes as needed by placing dams across the stream channels. Many regions that formerly had almost no natural lakes are now abundantly supplied. As you travel by airplane across such a region the glint of sunlight from hundreds of artificial lakes

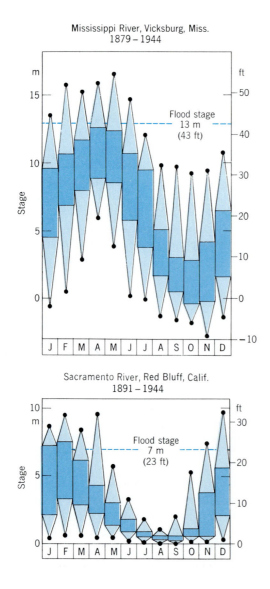

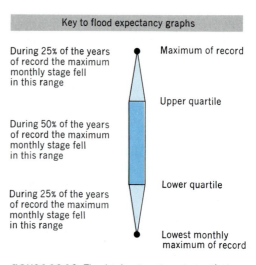

FIGURE 10.18 The highest water stage that occurred in each month is given in terms of percentages on these graphs. (After National Weather Service.)

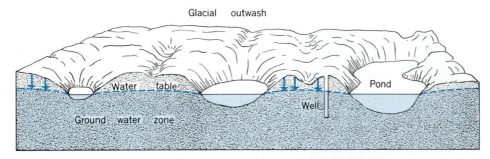

FIGURE 10.19 Freshwater ponds in sandy glacial deposits on Cape Cod, Massachusetts. (From *A Geologist's View of Cape Cod*, Doubleday & Co., New York. Copyright © 1966 by Arthur N. Strahler. Used by permission of Doubleday & Co.)

will catch your eye. Some are small ponds made to serve ranches and farms; others cover hundreds of square kilometers. Obviously, an abundance of such lakes represents a major environmental modification and has far-reaching consequences.

The term *lake* includes a very wide range of kinds of water bodies. Lakes have in common only the requirement that they have an upper water surface exposed to the atmosphere and no appreciable gradient with respect to a level surface of reference. Ponds (small, usually shallow water bodies), marshes, and swamps with standing water can be included.

Lake water may be fresh or saline, and we may have some difficulty in deciding whether a body of salt water adjacent to the open ocean is to be classed as a lake or an extension of the sea. A practical criterion rules that a coastal water body is not a lake if it is subject to influx of salt water from the ocean. Lake surfaces may, however, lie below sea level; an example is the Dead Sea with surface elevation of −396 m (−1300 ft). The largest of all lakes, the Caspian Sea, has a surface elevation of −25 m (−80 ft). Significantly, both of these large below-sea-level lakes are saline.

Basins occupied by lakes show a wide range of origins as well as a vast range in dimensions. Basins are created by geologic processes and it should not be surprising that there are lakes produced by every category of geologic process.

An important point about lakes in general is that they are for the most part short-lived features in terms of geologic time. Lakes disappear from the scene by one of two processes, or a combination of both. First, lakes that have stream channel outlets will be gradually drained as the outlet channels are eroded to lower levels. Where a strong bedrock threshold underlies the outlet, erosion will be slow but nevertheless certain. Second, lakes accumulate inorganic sediment carried by streams entering the lake and organic matter produced by plants within the lake.

Lakes also disappear by excessive evaporation accompanying climatic changes. Many former lakes of the southwestern United States flourished in moister periods of glacial advance during the Pleistocene Epoch, but today are greatly shrunken or have disappeared entirely under the present arid regime.

In moist climates, water level of lakes and ponds coincides closely with the water table in the surrounding area. Seepage of ground water, as well as direct runoff of precipitation, maintains these free water surfaces permanently throughout the year. Examples of such freshwater ponds are found widely distributed in North America and Europe, where plains of glacial sand and gravel contain natural pits and hollows left by the melting of stagnant ice masses (see Chapter 18). Figure 10.19 is a block diagram showing small freshwater ponds on Cape Cod. The surface elevation of these ponds coincides closely with the level of the surrounding water table.

Many former freshwater water-table ponds have become partially or entirely

FIGURE 10.20 A freshwater pond on Cape Cod, Massachusetts. Hygrophytic plants occupy a zone close to the water's edge. Pine forest in the background is on higher, well-drained ground. (Arthur N. Strahler.)

filled by the organic matter from growth and decay of water-loving plants (Figure 10.20). The ultimate result is a bog with a surface close to the water table.

Freshwater marshes and swamps, in which water stands at or close to the ground surface over broad areas, represent the appearance of the water table at the surface. Such areas of poor surface drainage have a variety of origins. For example, the broad, shallow freshwater swamps of the Atlantic and Gulf coastal plain represent regions only recently emerged from the sea. Other marshes are created by the shifting of river channels on floodplains.

Saline Lakes and Salt Flats

Lakes with no surface outlet are characteristic of arid regions. Here, on the

average year after year, the rate of water loss by evaporation balances the rate of stream inflow. If the rate of inflow should increase, the lake level will rise. At the same time the lake surface will increase in area, allowing a greater rate of evaporation. A new balance can then be achieved. Since dissolved solids are brought into the lake by streams—usually ones that head in distant highlands where a water surplus exists—and there is no surface outlet, the solids accumulate with resultant increase in salinity of the water. Salinity, or degree of "saltiness," refers to the abundance of certain common ions in the water. Eventually, salinity levels reach a point where salts are precipitated in the solid state.

Evaporation control is a subject of major importance in conserving the water supplies in a reservoir, particularly where the reservoir is situated in a region of arid climate. This situation occurs where an exotic river is dammed. (An *exotic river* is one that is sustained in its flow across an arid region through runoff derived from a distant region of water surplus.) The Colorado River in Arizona is a good example. A large reservoir, such as Lake Mead behind Hoover Dam, presents an enormous water surface exposed to intense evaporation.

Because the input of stream water is finite, a reservoir may be designed with such a large capacity that it will never completely fill in an arid climate. This situation is possible because there is a point at which annual evaporation equals the annual input.

Poorly drained, shallow basins (playas) accumulate highly soluble salts. These form sterile, white salt flats (see figure 15.32). On rare occasions these flats are covered by a shallow layer of water, brought by flooding streams heading in adjacent highlands. A number of desert salts are of economic value and have been profitably extracted. An example well known to most persons is borax (sodium borate), widely used as a water-softening agent. In shallow coastal estuaries in the desert climate, sea salt for human consumption is commercially harvested by allowing it to evaporate in shallow basins. One well-known salt source of this kind is the Rann of Kutch, a coastal lowland in the tropical desert of westernmost India, close to Pakistan. Here the evaporation of shallow water of the Arabian Sea has long provided a major source of salt for inhabitants of the interior.

Desert Irrigation and Salinization

Human interaction with the tropical desert environment is as old as civilization itself. Two of the earliest sites of civilization—Egypt and Mesopotamia—lie in the tropical deserts. The key to successful occupation of the deserts lies in the availability of large supplies of water from nondesert sources. This is a concept so familiar to all that it scarcely needs to be stated. For Egypt and Mesopotamia, the water sources of ancient times were exotic rivers deriving their flow from regions having a water surplus and flowing across the desert region because of geologic events and controls having nothing to do with climate.

A look at a global population map shows that population density is less than one person per square kilometer (two persons per square mile) over nearly all the area of the tropical deserts. Only where exotic streams cross the desert does the population density rise sharply. Valleys of the Nile, the Tigris and Euphrates, and the Indus are striking examples in the Old World. But in the coastal desert of Peru, we also find a substantial population long dependent on exotic streams fed from the Andes range and crossing the coastal desert to reach the Pacific Ocean.

Can we increase the production of food by expanding agriculture into the tropical deserts? Making the desert bloom is a romantic concept fostered on the American scene for generations by bureaucrats, politicians, and land developers. Were these promoters of vast irrigation schemes working in the long-term public interest? Only in recent years have the undesirable environmental impacts of desert irrigation come to the forefront. But we could have read the modern scenario in the history of the rise and fall of the Mesopotamian civilization.

Irrigation systems in arid lands divert the discharge of a large river, such as the Nile, Indus, Jordan, or Colorado, into a distributary system that allows the water to infiltrate the soil of areas under crop cultivation. Ultimately, such irrigation projects suffer from two undesirable side effects: salinization and waterlogging of the soil.

The irrigated area is subject to very heavy soil-water losses through evapotranspiration. Salts contained in the irrigation water remain in the soil and increase in concentration. This process is called *salinization*. (Areas of salinization show as white surfaces on the remote sensing imagery of Figure 2.22.) Ultimately, when salinity of the soil reaches the limit of tolerance of the plants, the land must be abandoned. Prevention or cure of salinization may be possible by flushing the soil salts downward to lower levels by the use of more water. This remedy requires greater water use than for crop growth alone.

Infiltration of large volumes of water causes a rise in the water table and may, in time, bring the zone of saturation close to the surface. This phenomenon is called *waterlogging*. Crops cannot grow in perpetually saturated soils. Furthermore, when the water table rises to the point at which upward movement under capillary action can bring water to the surface, evaporation is increased and salinization is intensified.

One of the largest of the modern irrigation projects affected adversely by salinization and waterlogging lies within the basin of the lower Indus River in Pakistan. Here the annual rate of rise of the water table averaged about 0.3 m (1 ft), and the annual increase in land area adversely affected was on the order of 20,000 hectares (50,000 acres).

The question of expanding irrigation agriculture in the southwestern United States was studied by the Committee on Arid Lands of the American Association for the Advancement of Science. In its 1972 report, this body recommended that additional large-scale importation of irrigation water to the American Southwest from distant sources should be made only where there are compelling reasons to do so. One such reason is to augment rapidly failing ground water supplies in districts already under irrigation. A second is to arrest the progress of salinization in areas already under irrigation. In short, the committee agreed that additional water should be imported only to prevent social and economic disruption in established irrigated areas. The lesson is that the search for new regions in which to expand agriculture should be directed to other, more favorable environments where the scales are not so heavily weighted by enormous evaporative water losses.

Pollution of Water Supplies

Streams, lakes, bogs, and marshes are specialized habitats of plants and animals; their ecosystems are particularly sensitive to changes induced by human activity in the water balance and in water chemistry. Not only does our industrial society make radical physical changes in water flow by construction of engineering works (dams, irrigation systems, canals, dredged channels), but we also pollute and contaminate our surface waters with a large variety of wastes. Some of these wastes are in the form of ions in solution. The subject of water quality is appropriate to discuss in this chapter. Introduction of mineral sediment into surface waters as a result of land disturbances is a suitable topic for Chapter 15.

Chemical pollution by direct disposal into streams and lakes of wastes generated in industrial plants is a phenomenon well known to the general public and can be seen firsthand in almost any industrial community in the United States. Direct outfall of sewage, whether raw (untreated) or partially treated, is another form of direct pollution of streams and lakes.

In urban and suburban areas pollutant matter entering streams and lakes includes deicing salt, lawn conditioners (lime and fertilizers), and sewage effluent. In agricultural regions, important sources of pollutants are fertilizers and the body wastes of livestock. Major sources of water pollution are associated with mining and processing of mineral deposits. In addition to chemical pollution, there is thermal pollution from discharge of heated water from nuclear generating plants. The possibility of contamination by radioactive substances released from nuclear processing plants also exists.

Among the common chemical pollutants of both surface water and ground water are sulfate, nitrate, phosphate, chloride, sodium, and calcium ions. (Ions are explained in Chapter 19.) Sulfate ions enter runoff both by fallout from polluted urban air and as sewage effluent. Important sources of nitrate ions are fertilizers and sewage effluent. Excessive concentrations of nitrate in water supplies are highly toxic and, at the same time, their removal is difficult and expensive. Phosphate ions are contributed in part by fertilizers and by detergents in sewage effluent. Phosphate and nitrate are plant nutrients and can lead to excessive growth of algae in streams and lakes. This process is *eutrophication*, often described as the aging of a lake. Chloride and sodium ions are contributed both by fallout from polluted air and by deicing salts used on highways. Instances are recorded in which water supply wells close to highways have become polluted from deicing salts.

A particular form of chemical pollution of surface water goes under the name of *acid mine drainage*. It is an important form of environmental degradation in parts of Appalachia where abandoned coal mines and strip mine workings are concentrated (Chapter 16). Ground water emerges from abandoned mines and as soil water percolating through strip mine waste banks. This water is charged with sulfuric acid and various salts of metals, particularly of iron. Acid of this origin in stream waters can have adverse effects on animal life. In sufficient concentrations it is lethal to certain species of fish and has at times caused massive fish kills.

Toxic metals, among them mercury, along with pesticides and a host of other industrial chemicals are introduced into streams and lakes in quantities that are locally damaging or lethal to plant and animal communities. In addition, sewage introduces live bacteria and viruses that are classed as biological pollutants; these pose a threat to health of humans and animals.

Thermal pollution is a term applied generally to the discharge of heat into the environment from combustion of fuels and from nuclear energy conversion into electric power. We have described thermal pollution of the atmosphere and its effects in Chapter 5. Thermal pollution of water is different in its environmental effects because it takes the form of heavy discharges of heated water locally into streams, estuaries, and lakes. The thermal environmental impact may thus be quite drastic in a small area.

Fresh Water as a Natural Resource

Fresh water is a basic natural resource essential to varied and intense agricultural and industrial activities. Runoff held in reservoirs behind dams provides water supplies for great urban centers, such as New York City and Los Angeles; when diverted from large rivers, it provides irrigation water for highly productive lowlands in arid lands, such as the Imperial Valley of California and the Nile Valley of Egypt. To these uses of runoff are added hydroelectric power, where the gradient of a river is steep, or routes of inland navigation, where the gradient is gentle.

Unlike ground water, which represents a large water storage body, fresh surface water in the liquid state is stored only in small quantities. (An exception is the Great Lakes system.) Referring back to Figure 10.1, note that the quantity of available ground water is about 20 times as large as that stored in freshwater lakes, whereas the water held in streams is only about one one-hundredth of that in lakes. Because of small natural storage capacities, surface water can be drawn only at a rate comparable with its annual renewal through precipitation. Dams are built to develop useful storage capacity for surplus runoff that would otherwise escape to the sea but, once the reservoir has been filled, water use must be scaled to match the natural supply rate averaged over the year. Development of surface water supplies brings on many environmental changes, both physical and biological, and these must be taken into account in planning for future water developments.

Death of a Civilization—
Its Lesson Unheeded

What causes a civilization to collapse after it has flourished for 2000 years? For the ancient hydraulic civilizations, those that relied on irrigation of crops in a desert land, one cause of collapse was self-inflicted, although not intentionally. No doubt you learned as a child about the Fertile Crescent, that lowland of antiquity spreading in a great arc across the Near East. The Fertile Crescent starts on the east at the head of the Persian Gulf in the lower plain of the Tigris and Euphrates rivers; it follows this broad valley to the northwest through what is now the nation of Iraq, but what was long known as Mesopotamia. Upon entering Syria, the Fertile Crescent turns west, then southward through the coastal region of Lebanon and Israel.

The Sumerian civilization evolved in the lower Tigris–Euphrates valley, where a village culture has been discovered dating to about the fifth millenium B.C. By 3000 B.C. the Sumerian civilization was well established. The agriculture upon which that civilization rested was dependent on a system of irrigation canals. Supported by this agricultural base, urban culture progressed to a high level and a cuneiform system of writing was invented. Skilled craftsmen produced pottery and—using gold, silver, and copper—jewelry and ornate weapons.

Trouble set in for the Sumerians about 2400 B.C. It came in the form of a deterioration of their croplands because of the accumulation of salt in the soil. This process, today called salinization, resulted from the increment, century after century, of salt dissolved in the irrigation water diverted from the two great rivers. Wheat, a staple crop, is quite sensitive to salt and its yield declines sharply as the salt concentration rises in the soil water. Barley, another staple crop of that time, is somewhat less sensitive to salt, but only up to a limit.

The first indications by means of which historians can infer the onset of salinization in the Sumerian agricultural lands is a shift in the proportions of wheat and barley in

their national agricultural output. It seems that around 3500 B.C. these two grains were grown in about equal amounts. By 2500 B.C., however, wheat accounted for only about one-sixth of the total grain production. Evidently the more salt-tolerant barley had, of necessity, replaced wheat over much of the area.

At this point in time, the Sumerian grain yield began to decline seriously. Records showed that in 2400 B.C., around the city of Girsu, fields were yielding an average of about 300 kg of grain per hectare, a high rate, even by modern standards. By 2100 B.C. the grain yield had declined to 180 kg per hectare, and by 1700 B.C., to a mere 100 kg. Cities so declined in population that they finally dwindled into mere villages. As Sumerian civilization withered in the south, Babylonia in the northern part of the valley rose to ascendency. Soon political control passed from Sumer to Babylon.

Historians chronicle an event that seems also to have played a part in the decline of the Sumerian civilization. A group of cities along the Euphrates River needed more irrigation water than they had. After a long and fruitless series of conflicts with other water-diverting cities farther upstream, the problem was solved by running a canal across the valley from the Tigris River. Now there was plenty of water, and it was used so liberally that a rise of the ground water table set in. As the water table came closer to the ground surface, capillary force drew the ground water to the surface, where it evaporated rapidly in the hot sun. The result: an even faster accumulation of salt. When the water table came up still farther, it saturated the roots of the barley plants, devastating the crops. Unknowingly, through their greed for water, these Sumerians had sealed their own fate.

The rising civilization of Babylon was also doomed to fall in turn. Along the Tigris River, east of the modern city of Baghdad, an extremely elaborate system of irrigation canals was evolved. This system was begun in a pre-Babylonian period, 3000 to 2400 B.C. After a long history of abandonments and reconstructions, it was superseded between A.D. 200 and 500 by a final irrigation system based upon a central

An artist's visualization of the Ziggurat, part of a great Sumerian temple at Ur, built about 2200 B.C. A solid mass of brick, rising about 15 m (50 ft) high, is all that remains today. (Joint Expedition of the British Museum and the University Museum, University of Pennsylvania.)

canal, the Nahrwan Canal. The irrigation system featured long, branching distributaries that proved to be traps for silt. Frequent cleaning out of the silt was required, and it piled up in great embankments and mounds beside the channels. Gradually, silt from the mounds was washed by rains into nearby fields to add layer upon layer to the land surface. According to one estimate, about one meter of silt accumulated over the fields in a 500-year period. This rise in level of the farmland by silting made its irrigation all the more difficult. Until the strong central government authority collapsed, things went along well enough. But thereafter, the entire system gradually broke down, and by the 12th century it was abandoned completely. Not long after, Mongol hordes invaded the valley, but by then there remained little of the once prosperous civilization that had held sway for over 4000

Salinization, waterlogging, and silt accumulation have affected nearly every major irrigation scheme practiced on

alluvial lands in a desert climate. One might almost say that failure was a built-in feature from the start. Today Pakistan struggles with salinization and waterlogging in a vast irrigation system in the Indus Valley. The lower Colorado River lands, including areas in Mexico, have suffered from salinization. Israel faces the threat of salinization of vital agricultural lands only recently placed in irrigation. Some new methods may help. For example, soaking the ground through tubes, while keeping the soil covered under a plastic sheet, can reduce the buildup of salt.

Small wonder, then, that new proposals to increase the extent of irrigated desert soils arouse little enthusiasm. Perhaps a record of nearly total failure, spread over the entire span of civilization, is beginning to get its message across.

Data Source: T. Jacobsen and R. M. Adams (1958), Salt and silt in ancient Mesopotamian agriculture, *Science*, vol. 128, pp. 1251–1258; Erik P. Eckholm (1975), Salting the earth, *Environment*, vol. 17, no. 7, pp. 9–15.

CHAPTER 11

EARTH MATERIALS

THE LITHOSPHERE   As viewed by geographers, the lithosphere is a stable platform for ecosystems as well as a nutrient reservoir essential to life processes.

MAJOR ROCK CLASSES   Igneous, sedimentary, and metamorphic rocks are related by a continuous recycling of their mineral ingredients.

SILICATE MINERALS   These are compounds rich in silicon and oxygen; they make up the bulk of the lithosphere.

FELSIC AND MAFIC ROCKS   We recognize light-colored silicate rocks of low density—the felsic group—as well as those of dark color and high density—the mafic group.

IGNEOUS ROCKS   They are formed from molten rock (magma) that solidifies both below and on the surface.

SEDIMENTARY ROCKS   They come in almost infinite variety—most are deposited in water on the floors of oceans, lakes, and rivers.

METAMORPHIC ROCKS   They result from the alteration of igneous and sedimentary rocks at great depth through applications of intense pressure or heat.

WITH this chapter we turn to the lithosphere, or solid mineral realm of our planet. Which materials and structures of the lithosphere are most important to humans and to the processes of the life layer? The inorganic mineral matter of the lithosphere is a vitally important nutrient reservoir for life processes, as we will see in later chapters on soils and nutrient cycling in the biosphere. Rocks exposed at the earth's surface are the primary source of these nutrient materials.

The lithosphere is also a platform for life on the lands. The platform consists basically of the continents. Continents and ocean basins are global features of first order of magnitude. What materials make up the continents? Why are they shaped and distributed in such an odd way over the globe? These are important questions in the background of physical geography. The surfaces of the continents are shaped into a remarkable variety of surface configurations, called *landforms*. Landforms strongly influence the distribution of ecosystems; landforms exert strong controls over human occupation of the lands.

Our plan in this series of eight chapters

(Chapters 11–18) is first to examine the lithosphere on a large scale to determine its composition and structure and to interpret the major features of the continents. This section deals first with *geology*, the science of the solid earth. Then we investigate landforms in terms of their origin and environmental qualities. In following this plan, we will start with sources of energy and matter arising from deep within the planet, but we will end up with surface features shaped by solar energy through activity of the atmosphere and hydrosphere. In this way we will be prepared to investigate the soil layer and its cover of vegetation.

Composition of the Earth's Crust

From the standpoint of the human environment, the really significant zone of the solid earth is the thin outermost layer— the earth's crust. This mineral skin, averaging about 17 km (10 mi) in thickness for the globe as a whole, contains the continents and ocean basins and is the source of soil and other sediment vital to life, of salts of the sea, of gases of the atmosphere, and of all free water of the oceans, atmosphere, and lands.

Figure 11.1 displays in order the eight most abundant elements of the earth's crust in terms of percentage by weight. Oxygen, the predominant element, accounts for about half the total weight. It occurs in combination with silicon, the second most abundant element.

Aluminum and iron are third and fourth on the list. These metals are of primary importance in our industrial civilization, and it is most fortunate that they are comparatively abundant elements. Four metallic elements follow: calcium, sodium, potassium, and magnesium. All four are on the same order of abundance (2 to 4 percent). Their importance in soil fertility will be stressed in Chapter 19.

If we were to extend the list, the ninth-place element would be titanium, followed in order by hydrogen, phosphorus, barium, and strontium. Phosphorus is one of the essential elements in plant growth.

Rocks and Minerals

The elements of the earth's crust are organized into compounds that we recognize as minerals. *Rock* is broadly defined as any aggregate of minerals in the solid state. Rock comes in a very wide range of compositions, physical characteristics, and ages. A given rock is usually composed of two or more minerals, and usually many minerals are present; however, a few rock varieties consist almost entirely of one mineral. Most rock of the earth's crust is extremely old in terms of human standards, the times of formation ranging back many millions of years. But rock is also being formed at this very hour as a volcano emits lava that solidifies on contact with the atmosphere.

A mineral is perhaps easier to define than a rock. A *mineral* is a naturally occurring, inorganic substance, usually possessing a definite chemical composition and characteristic atomic structure. A vast number of minerals exist, together with a great number of their combinations into rocks. We must generalize and simplify this discussion so as to refer to rocks and minerals in a way meaningful in terms of their environmental properties and value as natural resources.

Rocks of the earth's crust fall into three major classes. (1) *Igneous rocks* are solidified

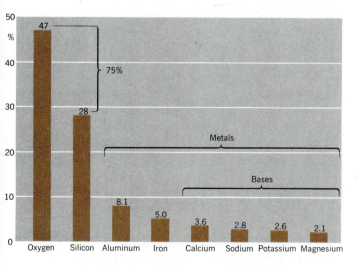

FIGURE 11.1 The average composition of the earth's crust is given here in terms of the percentage by weight of the eight most abundant elements.

from mineral matter in a high-temperature molten state. (2) *Sedimentary rocks* are layered accumulations of mineral particles derived in various ways from preexisting rocks. (3) *Metamorphic rocks* are igneous or sedimentary rocks that have been physically and chemically changed, usually by application of heat and pressure during mountain-making activities. Although we have listed these classes in a conventional sequence, it will become obvious in later pages that no one class has first place in terms of origin. Instead, they form a continuous circuit through which the crustal minerals have been recycled during many millions of years of geologic time. We begin with igneous rock for simplicity, not because this is the original class of rock.

The Silicate Minerals

The great bulk of igneous rock consists of silicate minerals, which are all compounds containing silicon atoms in combination with oxygen atoms in a close linkage. In the crystal structure of silicate minerals one atom of silicon is linked with four atoms of oxygen as the unit building block of the compound. Most of the silicate minerals also have one, two, or more of the metallic elements listed in Figure 11.1 (i.e., aluminum, iron, calcium, sodium, potassium, and magnesium). We are simplifying the list of silicate minerals to seven; these are listed in Figure 11.2.

Among the most common minerals of all the rock classes is *quartz;* its composition is

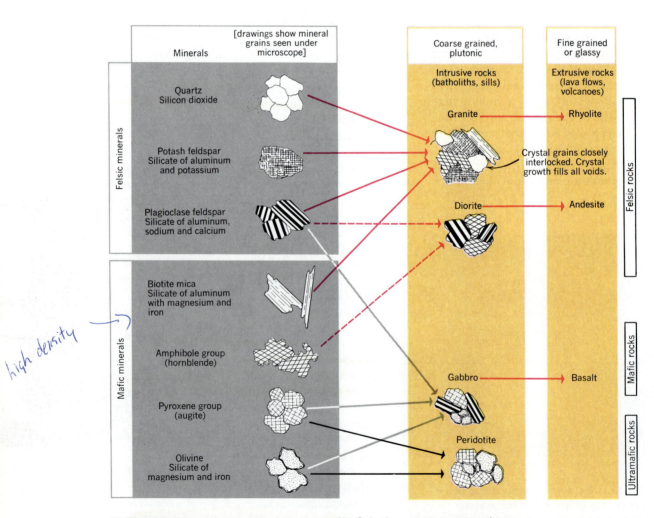

FIGURE 11.2 Silicate minerals and igneous rocks. Only the most important silicate mineral groups are listed, along with four common igneous rock types.

silicon dioxide. Two silicate–aluminum compounds called *feldspars* follow. One type, *potash feldspar*, contains potassium as the dominant metal besides aluminum. A second type, *plagioclase feldspar*, is rich in sodium, calcium, or both.

Quartz and the feldspars are light in color (white, pink, or grayish) and low in density compared with the other silicate minerals. (The concept of density of matter is explained in Figure 11.3. *Density* is the quantity of matter contained in a unit volume.) For this reason they form a silicate mineral group described as *felsic* ("fel" for feldspar; "si" for silicate).

The next three silicate minerals are actually mineral groups, with a number of mineral varieties in each. They are the *mica group* (example: biotite), *amphibole group* (example: hornblende), and *pyroxene group* (example: augite). All three are compounds of silicon, aluminum, magnesium, iron, and potassium or calcium. The seventh mineral, *olivine*, is a silicate of only magnesium and iron; it lacks aluminum. Altogether, these minerals are described as *mafic* ("ma" for magnesium; "f" from the chemical symbol for iron, Fe). The mafic minerals are dark in color (usually black) and are denser than the felsic minerals.

The Igneous Rocks

The silicate minerals occur in a high-temperature molten state as *magma*. From pockets a few kilometers below the earth's surface, magma makes its way upward through older solid rock and eventually solidifies as igneous rock. No single igneous rock is made up of all seven silicate minerals that we have listed but, instead, of three or four of those minerals as the major compounds.

Figure 11.2 lists four common igneous rocks and the major silicate minerals of which they are composed. These four serve as good examples, but there are other important mineral combinations. The first example is *granite*. The bulk of granite consists of quartz (27 percent), potash feldspar (40 percent), and plagioclase feldspar (15 percent). The remainder is mostly biotite and amphibole. Because most

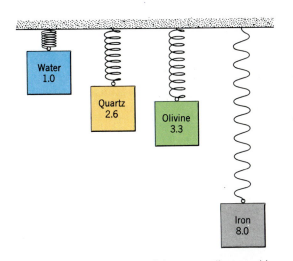

FIGURE 11.3 The concept of density is illustrated by several cubes of the same size, but of different substances, hung from a coil spring. The extent to which the spring is stretched is directly proportional to the weight of the cube and to its density in grams per cubic centimeter. Quartz and olivine are common minerals. (After A. N. Strahler, *Principles of Earth Science*, Harper & Row, New York. Copyright © 1976 by Arthur N. Strahler.)

FIGURE 11.4 A fresh exposure of massive gray granite high in the Sierra Nevada, California. The white bands are igneous dikes of felsic composition. The vertical boreholes were drilled to blast out the rock with explosives. (Arthur N. Strahler.)

of the volume of granite is of felsic minerals, we classify granite as a *felsic igneous rock*. Granite is a mixture of white, grayish or pinkish, and black grains, but the overall appearance is a light gray or pink color (Figure 11.4). The mineral grains are very tightly interlocked, and the rock is exceedingly strong.

Diorite, the second igneous rock on the list, lacks quartz. It consists largely of plagioclase feldspar (60 percent) and secondary amounts of amphibole and pyroxene. Diorite is a light-colored felsic rock, only slightly denser than granite.

The third igneous rock is *gabbro,* in which the major mineral is pyroxene (60 percent). A substantial amount of plagioclase feldspar (20 to 40 percent) is present and, in addition, there may be some olivine (0 to 20 percent). Gabbro is classed as a *mafic igneous rock;* it is dark in color and denser than the felsic rocks. The fourth igneous rock, *peridotite,* has a predominance of olivine (60 percent), and the rest is mostly pyroxene (40 percent). Peridotite is classed as an *ultramafic igneous rock,* denser even than the mafic types.

Our purpose in describing four common kinds of igneous rocks is to show that there is a range from felsic, through mafic, to ultramafic types, increasing in density from one end of the scale to the other. This arrangement is important because it is duplicated on a grand scale in the principal rock layers comprising the earth. We will mention this arrangement again in describing the earth's interior structure.

Intrusive and Extrusive Igneous Rocks

Magma that solidifies below the earth's surface and is surrounded by older, preexisting rock is called *intrusive igneous rock.* Where magma reaches the surface, it emerges as *lava,* which solidifies to form *extrusive igneous rock* (Figure 11.5).

Although both an intrusive rock and an extrusive rock can come from a single type of magma, their outward appearance is quite different. Intrusive igneous rocks cool very slowly and, as a result, develop large mineral crystals. A good example is the granite pictured in Figure 11.4. In an intrusive igneous rock the individual

FIGURE 11.5 This lava flow of basalt inundated a road near Kilauea Volcano in Hawaii National Park. The rough, blocky surface resulted from the continual solidification and fracturing of the lava surface as the flow moved over the ground. (John S. Shelton.)

mineral crystals can easily be distinguished with the unaided eye.

A body of intrusive igneous rock is called a *pluton.* Granite typically accumulates in enormous plutons, called *batholiths.* Figure 11.6 shows the relationship of a batholith to the overlying rock. As it made its way upward, the magma made room for its bulk by dissolving and incorporating the older rock above it. Batholiths are several kilometers in depth and may extend beneath an area of several thousand square kilometers. Figure 11.6 shows two other common forms of plutons. One is a *sill,* a platelike layer formed when magma forced its way between two rock layers, lifting the overlying rock to make room. A second is

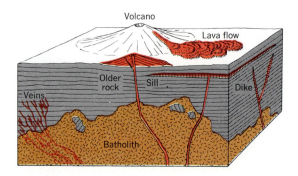

FIGURE 11.6 Forms of igneous rock bodies. (Drawn by A. N. Strahler.)

the *dike,* a wall-like body formed by the spreading apart of a rock fracture. Figure 11.7 shows a dike of mafic rock cutting across layers of sedimentary rock. The dike rock is fine textured because of rapid cooling. Dikes are commonly the conduits by means of which magma reaches the surface. Magma entering small fractures in the overlying rock solidifies in a branching network of thin veins.

Extrusive igneous rocks, emerging as lava, cool very quickly in contact with the air or water at the earth's surface. Quick cooling gives lava a very fine grained texture and sometimes even a glassy texture. Figure 11.8 shows two kinds of extrusive texture. One is a frothy, bubble-filled lava, called scoria (or sometimes pumice). The other is a natural volcanic glass. Most lava solidifies simply as a dense uniform rock of dull surface appearance in which the mineral grains are too small to be distinguished with the unaided eye.

Figure 11.2 names three extrusive rocks; each one is of the same mineral composition as a plutonic rock. *Rhyolite* is the name for lava of the same composition as granite; *andesite* is a lava of the composition of diorite. *Basalt* is lava of the composition of gabbro. Rhyolite and andesite are pale grayish or pink in color; basalt is black.

Lava flows, along with particles of solidified lava blown explosively from narrow vents, accumulate as large surface structures, called volcanoes. These are described in Chapter 13.

FIGURE 11.7 A dike of mafic igneous rock with nearly vertical parallel sides, intruded into flat-lying sedimentary rock layers. Arrows mark the contact between igneous rock and sedimentary rock. Spanish Peaks region, Colorado. (Arthur N. Strahler.)

available. Chemical change in a response to the changed environment is called *mineral alteration.*

Rock surfaces are also acted on by physical forces of disintegration that break up igneous rock into small fragments and separate the component minerals, grain from grain. Fragmentation is essential for the chemical reactions of mineral alteration, since it results in a great increase in mineral surface area exposed to chemically active solutions. The processes of physical disintegration of rocks are discussed in

Chemical Alteration of Igneous Rocks

As a transition to the second class of rocks, the sedimentary rocks, it is important that we investigate the ways in which the silicate minerals of igneous rocks are changed into a new set of minerals upon exposure at the earth's surface.

The surface environment of the lands is poorly suited to the preservation of intrusive rocks formed under conditions of high pressure and high temperature. Most silicate minerals do not last long, geologically speaking, in the low temperatures and pressures of atmospheric exposure, particularly since free oxygen, carbon dioxide, and water are abundantly

FIGURE 11.8 A frothy, gaseous lava solidifies into a light, porous scoria (left). Rapidly cooled lava may form a dark volcanic glass (right). (Arthur N. Strahler.)

Chapter 14. At this point, we are concerned with the chemical nature of mineral alteration as a process leading to production of sedimentary rock.

The presence of dissolved oxygen in water in contact with mineral surfaces leads to oxidation beneath the soil surface. *Oxidation* is the combination of oxygen with the metallic elements, such as calcium, magnesium, and iron, which are abundant in the silicate minerals. At the same time, carbon dioxide in solution forms carbonic acid, a weak acid in soil water. *Carbonic acid action* is capable of dissolving certain minerals, especially calcium carbonate. In addition, where decaying vegetation is present, soil water contains complex organic acids that are capable of reacting with mineral compounds. Certain common minerals, such as rock salt (sodium chloride), dissolve directly in water, but simple solution is not particularly effective for the silicate minerals.

Water itself combines with silicate mineral compounds in a reaction known as *hydrolysis.* This process is not merely a soaking or wetting of the mineral, but a true chemical change that produces a different compound and a different mineral. The reaction is not readily reversible under atmospheric conditions, so the products of hydrolysis are stable and long-lasting, as are the products of oxidation. In other words, these changes represent a permanent adjustment of mineral matter to a new surface environment of low pressures and temperatures.

Certain of the alteration products of silicate minerals are clay minerals. A *clay mineral* is one that has plastic properties when moist, because it consists of thin flakes of colloidal size, lubricated by layers of water molecules. (We explain the ion-holding property of these clay colloids in Chapter 19.)

Potash feldspar undergoes hydrolysis to become *kaolinite,* a white mineral with a greasy feel. Kaolinite becomes plastic when moistened. It is an important ceramic mineral used to make chinaware, porcelain, and tile.

Bauxite is an important alteration product of feldspars, occurring typically in warm climates of tropical and equatorial zones where rainfall is abundant year-round or in a rainy season. The dominant constituent of bauxite is oxide of aluminum in combination with water; it is an unusually stable compound. Unlike kaolinite, which is a true clay, bauxite forms massive rocklike layers.

Another important clay mineral is *illite,* which is formed as an alteration product of feldspar and mica. Illite is a silicate of aluminum and potassium with water and is an abundant mineral in sedimentary rocks. It occurs as minute, thin flakes and is carried long distances in streams. *Montmorillonite* is another common clay mineral (more correctly a group of minerals) derived from alteration of feldspar, mafic minerals, or volcanic ash.

A most important alteration product of the mafic minerals is *limonite,* consisting of oxide of iron with water. This stable form of iron oxide is found widely distributed in rocks and soils. It is closely associated with bauxite in low latitudes. Limonite supplies the typical reddish to chocolate-brown colors of soils and rocks. Some shallow accumulations of limonite were formerly mined as a source of iron.

Sediments and Sedimentary Rocks

The second great class of rocks is the sedimentary rocks. Their substance is derived both from preexisting rock of any origin and from newly formed organic matter. Igneous rock is the most important source of the inorganic mineral matter that makes up sedimentary rock. This explains why the sedimentary rocks are best studied after the igneous rocks.

In the process of mineral alteration, solid rock is softened and fragmented, yielding particles of many sizes. When transported in a fluid medium—air, water, or ice—these particles are known collectively as *sediment.* Used in its broadest sense, sediment includes both inorganic and organic matter. Dissolved mineral matter in the form of ions in solution is also included. (Ions are explained in Chapter 19.)

Streams carry sediment to lower levels and to locations where accumulation is possible. Wind and glacial ice also transport sediment, but not necessarily to lower

FIGURE 11.9 These flat-lying strata consist of banded shale layers that are easily eroded by running water. San Rafael Group, Upper Jurassic age, Paria River, southern Utah, west of Glen Canyon City. (A. N. Strahler.)

elevations or to places suitable for accumulation. Usually the most favorable sites of sediment accumulation are in shallow seas bordering the continent, but they may also be inland seas and large lakes. Thick accumulations of sediment may become deeply buried under newer sediments. Over long spans of time the sediments undergo physical or chemical changes; they become compacted and hardened, forming sedimentary rock.

There are three major classes of sediment: (1) *clastic sediment* consists of mineral particles derived by breakage from a parent rock source. Examples are the materials in a sand bar on a river bed or on a beach. (2) *Chemically precipitated sediment* consists of inorganic mineral compounds precipitated from a saltwater solution or as hard parts of organisms. One example is a layer of rock salt; another is the white lime rock of a coral reef. (3) *Organic sediment* consists of the tissues of plants and animals,

accumulated and preserved after the death of the organism. An example is a layer of peat in a bog. Sedimentary rocks include a wide range of varieties in terms of both physical and chemical properties. We can only touch on a few of the important kinds of sedimentary rocks.

Despite their great variety, the sedimentary rocks share some physical features in common. The sediment accumulates in nearly horizontal layers, called *strata* (or simply "beds") (Figure 11.9). Strata are separated by stratification planes or "bedding planes," which allow one layer to be easily removed from the next. Strata of widely different compositions can occur one above the next, so that the eroded strata of a great accumulation of sedimentary rocks show a diversity of bands and ledges. A fine example is seen in the upper walls of the Grand Canyon, in Arizona (Figure 11.10).

FIGURE 11.10 Cliffs and ledges of upper Paleozoic sedimentary rock formations in the North Rim of Grand Canyon, Arizona. The great sheer cliff of sandstone (center) is the Coconino Formation, an ancient dune deposit. Below it are deep red shales and thin sandstones of the Hermit and Supai formations. Above lie marine limestones of the Kaibab Formation, forming the canyon rim. (A. N. Strahler.)

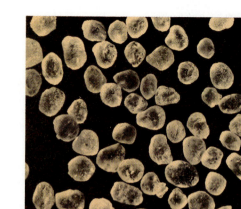

FIGURE 11.11 Rounded quartz grains from an ancient sandstone. The grains average about 1 mm (0.04 in.) in diameter. (Andrew McIntyre, Columbia University.)

The Clastic Sedimentary Rocks

Clastic sediments are derived from one or more of the rock groups—igneous, sedimentary, and metamorphic—and thus include a very wide range of parent minerals. One sediment source is from the silicate minerals and the alteration products of those minerals. Because quartz is hard and is immune to alteration, it is usually the most important single component of the clastic sediments (Figure 11.11). Second in abundance are fragments of unaltered fine-grained parent rocks. Feldspar and mica are also commonly present. Clay minerals, particularly kaolinite, illite, and montmorillonite, are major constituents of the very fine clastic sediments.

The natural range of particle sizes in clastic sediment determines the ease and distance of travel of particles in transport by water currents. Obviously, the finer the particles, the more easily they are held in suspension in the fluid; the coarser particles tend to settle to the bottom of the fluid layer. In this way a separation of grades, called *sorting*, occurs. Sorting determines the texture of the sediment deposit and of the sedimentary rock derived from that sediment. Colloidal clay particles do not settle out unless they are made to clot together into larger clumps. This clotting process, called *flocculation*, usually occurs when clays carried by rivers mix with the salt water of the ocean.

Compaction and cementation of layers of sediment leads to formation of sedimentary rocks. The following are important varieties. *Sandstone* is formed of the sand grades of sediment cemented into a solid rock by silica (silicon dioxide) or calcium carbonate (Figure 11.12). The sand grains are commonly of quartz, such as those shown in Figure 11.11. (Refer to Figure 19.3 for names and diameters of the various grades of sediment particles.)

A mixture of water with particles of silt and clay, along with some sand grains, is called *mud*. The sedimentary rock hardened from such a mixture is called mudstone. Compacted and hardened clay layers

FIGURE 11.12 Natural arches in massive sandstone can be closely examined by boat in a drowned side canyon of the Colorado River. Glen Canyon, Lake Powell, Utah. Navajo Formation, Jurassic age. (A. N. Strahler.)

become claystone. Sedimentary rocks of mud composition are commonly laminated in such a way that they break apart into small flakes and plates; they are described as being fissile. Rock with this structure and composition is called *shale;* it is the most abundant of the sedimentary rocks. Shale is formed largely of the clay minerals kaolinite, illite, and montmorillonite. The compaction of the mud to form shale involves a considerable loss of volume as water is driven out of the clay.

Chemically Precipitated Sedimentary Rocks

Under favorable conditions, mineral compounds are deposited from the salt solutions of seawater and of salty inland lakes in desert climates. One of the most common sedimentary rocks formed by chemical precipitation is *limestone,* composed largely of the mineral *calcite.* Calcite is *calcium carbonate* ($CaCO_3$). Limestone strata formed on the seafloor—marine limestones—have accumulated in thick layers in many ancient seaways in past geologic eras (Figure 11.13). A closely related rock is dolomite, composed of calcium–magnesium carbonate. Limestone and dolomite are grouped together as the *carbonate rocks.* They are dense rocks, sometimes white, sometimes pale gray, and occasionally black.

Many varieties of limestones are formed of the mineral hard parts of organisms, either of plants or of animals. For example, chalk, a pure white rock, is formed of countless skeletons of marine algae. Reef limestone is formed from the hard parts of corals.

Seawater also yields sedimentary layers of silica (silicon dioxide) in hard, noncrystalline form called *chert.* Chert is a variety of sedimentary rock, but also commonly occurs combined with limestone (cherty limestone). Another chemical precipitate is phosphate rock, a marine sediment rich in the mineral phosphorus. Its principal use is in phosphate fertilizers, which are essential to modern agriculture.

Where evaporation is sustained and intense, shallow water bodies acquire a very high level of salinity. Shallow bays and

FIGURE 11.13 Montezuma Castle, a cliff dwelling in the Verde Valley of Arizona, was built by Indians of pre-Columbian time. They took advantage of a natural succession of deep niches running along the bedding planes of limestone strata. By walling off the recesses with masonry, the inhabitants created a fortified village accessible only by crude ladders. (Donald L. Babenroth.)

estuaries in coastal deserts are one such saline body; another type is the playa lake of inland desert basins (see Figure 15.32). Sedimentary minerals and rocks deposited from such concentrated solutions are called *evaporites.* Ordinary rock salt, the mineral halite (sodium chloride), has accumulated in this way into important sedimentary rock units.

Hydrocarbon Compounds in Sedimentary Rocks

Hydrocarbon compounds (compounds of carbon, hydrogen, and oxygen) form a most important type of organic sediment. These substances occur both as solids (peat and coal) and as liquids and gases (petroleum and natural gas), but only coal qualifies physically as a rock.

Peat, a soft, fibrous substance of brown to black color, accumulates in a bog environment where the continual presence of water inhibits decay and oxidation of plant remains. One form of peat is of freshwater origin and represents the filling of shallow lakes. Thousands of such peat bogs are found in North America and Europe; they occur in depressions remaining after recession of the great ice sheets of the Pleistocene Epoch (Figure 9.31). This peat has been used for centuries as a low-grade fuel. Peat of a different sort is formed in the saltwater environment of tidal marshes (Chapter 17).

At various times and places in the geologic past, conditions were favorable for the large-scale accumulation of plant remains, accompanied by subsidence of the area and burial of the compacted organic matter under thick layers of inorganic sediments. *Coal* is the end result of this process. Coal seams are interbedded with shale, sandstone, and limestone strata (Figure 11.14). *Petroleum*, or *crude oil*, as the liquid form is often called, includes many hydrocarbon compounds. *Natural gas*, found in close association with accumulations of liquid petroleum, is a mixture of gases. The principal gas is methane (marsh gas). Petroleum and natural gas are not classed as minerals, but they originated as organic compounds in sediments and are classed as mineral fuels.

It is generally agreed that petroleum and natural gas are of organic origin, but the nature of the process is not fully understood. A favored explanation for petroleum is that the oil originated within microscopic floating marine organisms (plankton). Upon their death, these organisms sink to the ocean floor and what remains of them is incorporated into mud or clay. Hydrocarbon compounds within the

FIGURE 11.14 A coal seam (black layer at base of cliff) exposed in the wall of an open-pit mine, near Piedra, Colorado. (John S. Shelton.)

organisms—oils, fats—are thus added to the bottom sediment. Much later, the buried sediments experience compaction and heating, causing the parent hydrocarbon compounds to be reformed into new hydrocarbon compounds. These accumulations are known as crude oils. Today we find oil shales holding petroleum in a dispersed state, and these give support to the organic hypothesis. Crude oil in concentrations that can be brought to the surface is found only in a porous reservoir rock. This fact implies that the oil migrated from the source body to the reservoir rock.

The simplest arrangement of strata favorable to trapping petroleum and natural gas is an uparching of the type shown in Figure 16.25. Shale forms an impervious cap rock. A porous sandstone beneath the cap rock serves as a reservoir. Natural gas occupies the highest position, with the oil below it.

Everyone interested in energy resources has heard of *oil shale* and of the tremendous reserve of hydrocarbon fuel it holds. The fact is that this sedimentary rock in the Rocky Mountain region is not really shale at all, and the hydrocarbon it holds is not really petroleum. Strata of the Rocky Mountains called "oil shales" are composed of calcium carbonate and magnesium carbonate. The strata were formed millions of years ago as lake deposits of lime mud (marl) in an ancient lake. These soft, laminated deposits belong to the Green River Formation. The oil shale beds occur largely in northeastern Utah, northwestern Colorado, and southwestern Wyoming.

The hydrocarbon matter of the Green

River Formation occurs in a particular bed, the Mahogany Zone, about 20 m (70 ft) thick. It is a waxy substance, called *kerogen,* which adheres to the tiny grains of carbonate material. When the shale is crushed and heated to a temperature of 480°C (900°F), the kerogen is altered to petroleum and driven off as a liquid. The rock may be mined and processed in surface plants, or burned in underground mines, from which the oil is pumped to the surface.

Yet another form of occurrence of hydrocarbon fuels is *bitumen,* a variety of petroleum that behaves much as a solid, although it is actually a highly viscous liquid. Bitumen goes by other common names, such as tar, asphalt, or pitch. In some localities bitumen occupies pore spaces in layers of sand or porous sandstone. It remains immobile in the enclosing sand and will flow only when heated. Outcrops of *bituminous sand (oil sand)* exposed to the sun will show bleeding of the bitumen. Perhaps the best known of the great bituminous sand deposits are those occurring in Alberta, Canada. Where exposed along the banks of the Athabasca River, the oil sand is being extracted from surface mines. Extraction of oil from wells will require that the sand be heated by steam or other heat sources. Many difficult problems remain to be solved before this vast energy resource can make a major contribution to the global energy supply.

From the standpoint of energy resources for human use, the outstanding concept relating to the hydrocarbon compounds within the earth—*fossil fuels,* they are collectively called—is that they have required millions of years to accumulate, whereas they are being consumed at a prodigious rate by our industrial society. These fuels are nonrenewable resources. Once they are gone there will be no more, because the quantity produced in a thousand years by geologic processes is scarcely measurable in comparison to the quantity stored through geologic time.

Metamorphic Rocks

Any of the types of igneous or sedimentary rocks may be altered by the tremendous pressures and high temperatures that accompany mountain-building processes of the earth's crust. The result is a rock so changed in appearance and structure as to be classified as a metamorphic rock. Typically, metamorphism results in the development of new textures and structures within the rock. Mineral components of the parent rock are, in many cases, reconstituted into different mineral varieties. Recrystallization of the original minerals can also occur.

Shale, after being subjected to the unequal application of stress, is altered to *slate,* a fine-textured rock that splits neatly into thin cleavage plates so familiar as roofing shingles and as flagstones of patios and walks.

With application of increased heat and unequal stress, slate changes into *schist,* representing a more advanced stage of metamorphism. Schist has a structure called *foliation,* consisting of thin but rough and irregularly curved planes of parting in the rock. Schist is set apart from slate by the coarse texture of the mineral grains, the abundance of mica, and occasionally the presence of scattered large crystals of newly formed minerals, such as garnet.

At a still more advanced level of metamorphism, schist may be altered into *gneiss,* a rock characterized by alternate light and dark streaks or bands (Figure 11.15). The light bands are composed largely of felsic minerals (quartz and feldspar), and the dark bands of mafic minerals (biotite, amphibole, and pyroxene). Schist and gneiss can also form from igneous rock such as basalt or granite.

Two common metamorphic rocks are recognized on the basis of being composed largely of a single mineral. The metamorphic equivalent of pure quartz sandstone or chert is *quartzite,* formed by recrystallization of the quartz as a result of heat and pressure. *Marble* is formed by the recrystallization or growth in crystal size of the mineral calcite in limestone during the application of heat and pressure. Marble has a coarse texture on freshly broken surfaces. Bedding planes are obscured, and masses of mineral impurities may be drawn out into darker streaks or bands.

FIGURE 11.15 Intensely deformed structures in schist and gneiss of Precambrian age, revealed by weathering of this glacially polished rock surface near Breckenridge, Colorado, in the Rocky Mountains. (Copyright © 1988 by Mark A. Melton.)

The Cycle of Rock Transformation

We can now bring together the formative processes of rocks into a single unified concept of recycling of matter through geologic time. A schematic diagram, shown in Figure 11.16, distinguishes between a surface environment of low pressures and temperatures and a deep environment of high pressures and temperatures. The deep environment is the realm of the igneous and metamorphic rocks. The surface environment is one of rock alteration and sediment deposition.

Seen in its complete form, the total circuit of rock changes in response to environmental stress constitutes the *cycle of rock transformation.* Our diagram emphasizes that mineral matter is continually recycled through the three major rock classes. Igneous rocks of the common kinds we see around us today are by no means the "original" rocks of the earth's crust. Actually, there is no known record of the rocks that first formed the earth's crust; they were consumed and recycled long ago.

In the next chapter we will expand our scale of vision to examine the earth's crust more broadly. How do the three great classes of rocks fit into the global patterns of continents and ocean basins and of mountain chains and low plains? The geographer is interested in these distributional patterns. Armed with a basic knowledge of minerals and rocks, we can effectively turn to these larger configurations of earth materials.

FIGURE 11.16 Schematic diagram of the cycle of rock transformation.

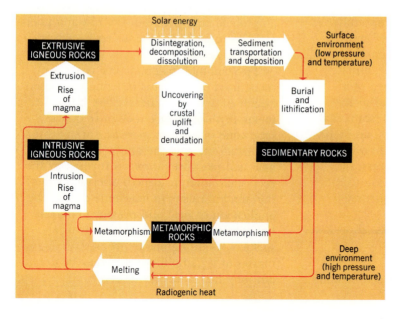

CHAPTER 12

THE LITHOSPHERE AND PLATE TECTONICS

LITHOSPHERE AND ASTHENOSPHERE   The lithosphere is a brittle outer layer of rock, bearing the continents; it overlies a soft, yielding rock layer, the asthenosphere.

THE CHANGING CRUST   Powerful internal forces shape the earth's outermost rock layer—the crust—and have been responsible for the growth of continents and ocean basins through eons of geologic time.

MOUNTAIN-BUILDING   Lofty mountain chains have been formed either by recent volcano-building or by tectonic activity, the bending and breaking of the crust.

CONTINENTAL SHIELDS   Large expanses of our continents are composed of very ancient rocks, deeply eroded and quite stable.

OCEAN BASINS   Much younger than the continental shields are the ocean basins; they have been formed of new mafic igneous rock, rising along mid-ocean rifts.

PLATE TECTONICS   A field of recent scientific discovery, plate tectonics investigates the rifting apart and colliding of enormous segments of lithosphere.

CONTINENTAL COLLISIONS   Masses of continental lithosphere can collide as the ocean basin between them narrows and disappears. Collision intensely deforms sedimentary strata and produces lofty alpine mountain chains.

ENVIRONMENTAL regions of the globe depend for their distribution on configurations of the earth's crust dictated by geologic processes. These processes are powered by energy sources deep within the earth. The internal earth forces have shaped the continents and ocean basins without conforming in the least to the orderly latitude zones of climate. We can think of latitude zones of temperature, winds, and precipitation as concentric color bands painted on a circular dinner plate as the potter's wheel spins.

Suppose that the dinner plate falls to the floor and is shattered and that then the fragments are put back into place and cemented together. The fracture patterns cut across the circular color zones in a discordant and random pattern.

This same unique combination of disorder on order characterizes the earth's solid surface. We find volcanoes erupting today in the cold desert of Antarctica as well as near the equator in Africa. An alpine mountain range has been pushed up in the cold subarctic zone of Alaska, where it trends east–west; but another lies astride the equator in South America and runs north–south. Both ranges lie in belts of intense crustal activity, where numerous strong earthquakes are generated. Although the geologic processes that create high mountains are quite insensitive to latitude and climate, climates do respond to mountain ranges through the orographic effect. In this way, the chance configurations of the earth's relief features bring diversity to the global climate.

In this chapter we will survey the major geologic features of our planet, starting with its deep interior as a layered structure. We then examine the outermost layer, or crust, and compare crustal forms of the continents with those of the ocean basins. Only within the past decade have geologists provided a unified theory to explain the differences between continents and ocean basins and to interpret the major forms of crustal unrest in a meaningful way. Fortunately, we are reviewing geologic processes just at the moment in history when a new revolution in geology has accomplished a scientific upheaval. The revolutionary findings can now furnish us with a complete scenario of earth history on a grand scale of both time and spatial dimensions.

The Earth's Interior

Figure 12.1 is a cutaway diagram of the earth to show its major parts. The earth is an almost spherical body approximately 6400 km (4000 mi) in radius. The center is occupied by the *core*, a spherical zone about 3500 km (2200 mi) in radius. Because of the sudden change in behavior of earthquake waves upon reaching this zone, it has been concluded that the outer core has the properties of a liquid, in abrupt contrast to a solid mass that surrounds it. However, the innermost part of the core is probably in the solid state.

Outside of the core lies the *mantle*, a layer about 2895 km (1800 mi) thick, composed of mineral matter in a solid state. Judging from the behavior of earthquake waves, the mantle is probably composed largely of the mineral olivine and resembles the ultramafic igneous rock peridotite.

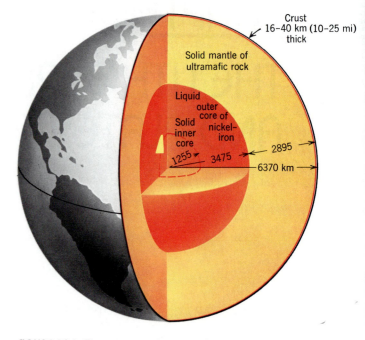

FIGURE 12.1 This cutaway diagram of the earth shows the inner core of liquid iron and the thick outer mantle of ultramafic rock. The crust is too thin to show to correct scale.

The Crust

The outermost and thinnest of the earth zones is the *crust*, a layer about 8 to 40 km (5 to 25 mi) thick. It is formed largely of igneous rock. The base of the crust, where it contacts the mantle, is sharply defined. This contact is established from the way in which earthquake waves change velocity abruptly at that level (Figure 12.2). The surface of demarcation between the crust and mantle is called the *Moho*, a simplification of the name of the seismologist who discovered it.

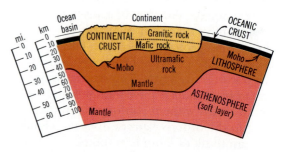

FIGURE 12.2 Idealized cross section of the earth's crust and mantle. The lithosphere includes the upper part of the mantle as well as the crust.

From a study of earthquake waves geologists conclude that the crust beneath the continents consists of two rock zones: a lower, continuous rock zone of mafic composition, and an upper zone of felsic rock. Because the felsic portion has a chemical composition about like that of granite, it is commonly described as being *granitic rock*. Much of the granitic rock is metamorphic rock. There is no sharply defined surface of demarcation between these zones. The crust of the ocean basins consists of basalt and gabbro.

The crust is much thicker beneath the continents than beneath the ocean floors, as Figure 12.2 shows. Whereas 40 km (25 mi) is a good average figure for crustal thickness beneath the continents, 5 km (3 mi) is an average figure for thickness of the basaltic crust beneath the deep ocean floors. The reasons for differences in both thickness and rock composition between continental and oceanic crust are explained later in terms of processes that have created the crust.

The Lithosphere

As broadly defined in Chapter 2, the lithosphere consists of the entire solid earth from outer surface to innermost core. In recent years, however, the word "lithosphere" has been restricted by geologists to mean an outer layer, or shell, of rigid brittle rock.

The lithosphere includes not only the crust, but also the upper part of the mantle (Figure 12.2). Rock beneath the brittle lithosphere is highly heated to a physical state that is semiplastic—much like white-hot iron that can be drawn and shaped with little difficulty. Using the same analogy, the lithosphere resembles cold cast iron that responds to strong unequal forces by breaking abruptly along sharp fractures. Some tens of kilometers deep in the earth, the brittle condition of the rock gives way gradually to the softened, semiplastic state; at still further depth, the strength of the rock material again increases. Thus we recognize a *soft layer* beneath the lithosphere; it is named the *asthenosphere*. (This word is derived from the Greek root *asthenēs*, meaning "weak.")

In terms of states of matter, the asthenosphere is not a liquid, for its temperature, about 1400°C (2600°F), is below the melting point. It behaves much like an ingot of white-hot iron that holds its shape when resting on a flat surface prior to being passed through the rollers that will form it into bars or sheets.

To show actual dimensions, the layers of the outer earth sphere, defined here according to their physical state, are drawn to true scale in Figure 12.3. The lithosphere ranges in thickness from 40 to 80 km (25 to 50 mi); the diagram uses an average depth figure of 60 km (37 mi). The lithosphere is thickest (80 km) under the continents and thinnest (40 km) under the ocean basins. The asthenosphere extends down to a depth of about 300 km (185 mi), but we stress the point that both upper and lower boundaries are gradational. The weakest portion of the asthenosphere lies at a depth of roughly 200 km (125 mi). Here the rock is close to its melting point.

An important concept derived from the facts we have stated is that the rigid, brittle lithosphere has the capability of moving bodily over the soft, plastic asthenosphere. Yielding of the asthenosphere is distributed through a thickness of many tens of kilometers.

If the lithosphere formed a single continuous shell over the entire earth, that shell would—in theory at least—be capable of rotating bodily over the entire inner mass of the earth. Instead, the lithosphere is broken into large fragments called *lithospheric plates*. A single large plate is typically of continental dimensions and capable of moving independently of the plates that surround it. Like great slabs of floating ice on the polar sea, lithospheric plates can be seen in one place to be separating from one another, whereas elsewhere they are colliding in crushing impacts that raise great welts, or one plate is diving down beneath the edge of its neighbor.

The Geologic Timetable

To place crustal rocks and structures in their positions in time, we need to refer to some major units in the scale of geologic

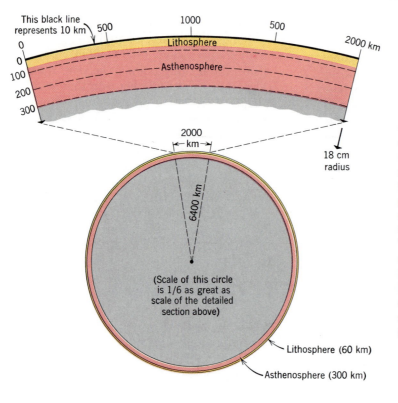

FIGURE 12.3 The lithosphere and asthenosphere drawn to true scale. The curvature of the upper diagram fits a circle 18 cm in radius. The black line at the top is scaled to represent a thickness of 10 km (6 mi); it will accommodate about 98 percent of the earth's surface features, from ocean floors to high mountains and plateaus. Only a few lofty mountains would project above the black line and only a few deep ocean trenches would project below the line. The complete circle below is drawn on a scale one-sixth as large as the upper diagram. Here we gain an appreciation of the extreme thinness of the mobile lithospheric plates that move over the asthenosphere.

time. Table 12.1 lists the major time divisions. All time older than 570 million years (m.y.) is *Precambrian time*. Three eras of time follow: Paleozoic, Mesozoic, and Cenozoic. These eras saw the evolution of life-forms in the oceans and on the lands.

The geologic eras are subdivided into periods; their names, ages, and durations are also given in Table 12.1. Throughout the entire course of geologic time, brief but intense episodes of crustal deformation occurred in which strata were crumpled and broken. A mountain-making episode of this kind is known as an *orogeny*. Names of a few important orogenies are given in Table 12.1.

The Cenozoic Era is particularly important in terms of the continental surfaces, because nearly all landscape features seen today were produced in the 65 million years since that era began. Because the Cenozoic Era is comparatively short in duration—scarcely more than the average duration of a single period in older eras—it is subdivided directly into lesser time units called epochs. Details of the Pleistocene and Holocene epochs are given in Chapter 18.

The human race, genus *Homo*, evolved during the late Pliocene Epoch and throughout the Pleistocene Epoch. As you can see, the period of human occupation of the earth's surface is an insignificant moment in the vast duration of geologic history.

Distribution of Continents and Ocean Basins

The first-order relief features of the earth are the continents and ocean basins. Using a globe, we can compute that about 29 percent of the globe is land and 71 percent oceans. If the seas were to drain away, however, it would become obvious that broad areas lying close to the continental shores are actually covered by shallow water, less than 150 m (500 ft) deep. From these relatively shallow continental shelves the ocean floor drops rapidly to depths of thousands of meters. In a way, the ocean basins are brimful of water. The oceans have even spread over the margins of ground that would otherwise be assigned to the continents. If the ocean level were to

TABLE 12.1 Table of geologic time

Era	Period	Epoch	Duration m.y.	Age m.y.	Orogenies
CENOZOIC	Quaternary	Holocene	(10,000 yr)		
		Pleistocene	2	2	Cascadian
		Pliocene	3	5	
	Tertiary	Miocene	19	24	
		Oligocene	13	37	
		Eocene	21	58	
		Paleocene	8	66	
MESOZOIC	Cretaceous		78	144	Laramide (Rocky Mountains)
	Jurassic		64	208	
	Triassic		37	245	Alleghanian or Hercynian
PALEOZOIC	Permian		41	286	
	Carboniferous		74	360	
	Devonian		48	408	Caledonian
	Silurian		30	438	
	Ordovician		67	505	
	Cambrian		65	570	

PRECAMBRIAN TIME   (Extends to oldest known rocks, about 3.6 billion years)

Age of earth as a planet: 4.6 to 4.7 billion years.

Age of universe: 17 to 18 billion years.

drop by 150 m (500 ft) the surface area of continents would increase to 35 percent; the ocean basins would decrease to 65 percent. We can use these figures as representative of the true relative proportions.

Scale of the Earth's Relief Features

Before turning to a description of the major subdivisions of the continents and ocean basins, we need to grasp the true scale of the earth's relief features in comparison with the earth as a sphere. Most relief globes and pictorial relief maps are greatly exaggerated in vertical scale.

For a true-scale profile around the earth we might draw a chalk-line circle 6 m (20 ft) in diameter, representing the earth's circumference. A chalk line 1 cm (⅜ in.) wide would include within its limits not only the highest point on the earth, Mount Everest (8840 m; 29,000 ft), but also the deepest known ocean trenches, somewhat deeper than 11,000 m (36,000 ft).

Figure 12.4 shows profiles correctly curved and scaled to fit a globe whose diameter is 6.4 m (21 ft). The surface profile is drawn to natural scale, without vertical exaggeration. Although the most imposing landscape features of Asia and North America are shown, they are only trivial irregularities on the great global circle.

Second-Order Relief Features of the Continents

The continents and ocean basins are first-order relief features. We can recognize subdivisions within each that are relief features of a second order of magnitude. We will deal first with the relief features of the continents, which are familiar to most persons from direct experiences of travel and from photographs. Relief features of the second order—mountain ranges, plains, and plateaus, for example—are collectively designated by geographers as landforms.

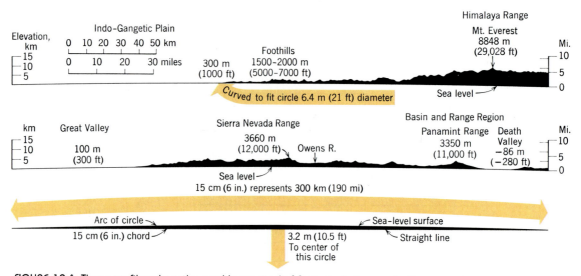

FIGURE 12.4 These profiles show the earth's great relief features in true scale. Sea-level curvature is fitted to a globe about 6 m (20 ft) in diameter.

Broadly viewed, the continental masses consist of two basic subdivisions: (1) active belts of mountain-making and (2) inactive regions of old rocks. The growth of mountain ranges occurs through one of two very different geologic processes. First is *volcanism*, the formation of massive accumulations of volcanic rock by extrusion of magma. Many lofty mountain ranges consist of chains of volcanoes built of extrusive igneous rocks. Second of the mountain-building processes is *tectonic activity*, the breaking and bending of the earth's crust under internal earth forces. Crustal masses that are raised by tectonic activity form mountains and plateaus; masses that are lowered form crustal depressions. In some instances, volcanism and tectonic activity have combined to produce a mountain range. Landforms produced by volcanic and tectonic activity are the subject of Chapter 13.

Alpine Chains

Active mountain-making belts are narrow zones; most lie along continental margins. These belts are sometimes referred to as *alpine chains*, because they are characterized by high, rugged mountains, such as the Alps of central Europe. These mountain belts were formed in the Cenozoic Era by volcanic or tectonic activity, or a combination of both; this activity has continued to the present day in many places.

The alpine chains are characterized by broadly curved patterns on the world map (Figure 12.5). Each curved section of an alpine chain is referred to as a *mountain arc*; the arcs are linked in sequence to form the two principal mountain belts. One is the *circum-Pacific belt;* it rings the Pacific Ocean basin. In North and South America, this belt is largely on the continents and includes the Andes and Cordilleran ranges. In the western part of the Pacific basin the mountain arcs lie well offshore from the continents and take the form of *island arcs,* running through the Aleutians, Kuriles, Japan, the Philippines, and many lesser islands. Volcanic activity is important in these arcs. Between the large islands, these arcs are represented by volcanoes rising above the sea as small, isolated islands.

The second chain of major mountain arcs forms the *Eurasian–Indonesian belt,* shown in Figure 12.5. It starts in the west at the Atlas Mountains of North Africa and continues through the European Alps, the Near East, and Iran to join the Himalayas. The belt then continues through Southeast Asia into Indonesia, where it joins the circum-Pacific belt in a T-junction. We will return later to the location of these active belts of mountain-making and explain them in terms of lithospheric plate motions.

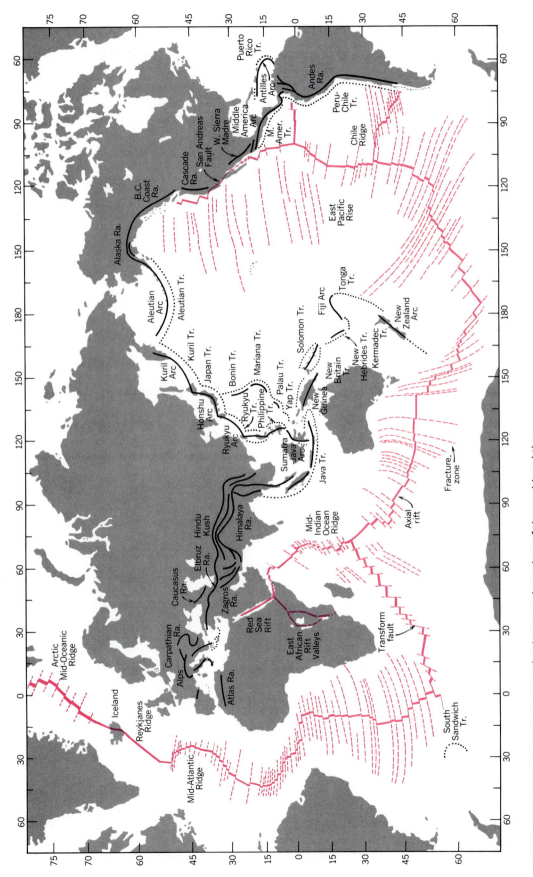

FIGURE 12.5 Principal mountain arcs, island arcs, and trenches of the world and the mid-oceanic ridge. (Mid-oceanic ridge map from A. N. Strahler, *Physical Geology.* Harper & Row, New York. Copyright © 1981 by Arthur N. Strahler.)

Puerto Rico Tr.

Antilles Arc

Andes Ra.

M. Amer. Tr.

Peru-Chile Tr.

W. Sierra Madre

Middle America Arc

Chile Ridge

San Andreas Fault

Cascade Ra.

East Pacific Rise

B.C. Coast Ra.

Alaska Ra.

Aleutian Arc

Aleutian Tr.

Tonga Tr.

Fiji Arc

Kuril Tr.

Solomon Tr.

New Zealand Arc

Kermadec Tr.

Kuril Arc

Japan Tr.

Bonin Tr.

Mariana Tr.

New Hebrides Tr.

New Britain Tr.

Honshu Arc

Ryukyu Tr.

Philippine Tr.

Palau Tr.

Yap Tr.

New Guinea

Ryukyu Arc

Sumatra Java Arc

Java Tr.

Fracture zone

Mid-Indian Ocean Ridge

Axial rift

Hindu Kush

Himalaya Ra.

Elbruz Ra.

Caucasus Ra.

Zagros Ra.

Carpathian Ra.

Red Sea Rift

East African Rift Valleys

Transform fault

Arctic Mid-Oceanic Ridge

Alps

Atlas Ra.

Iceland

Reykjanes Ridge

Mid-Atlantic Ridge

South Sandwich Tr.

259

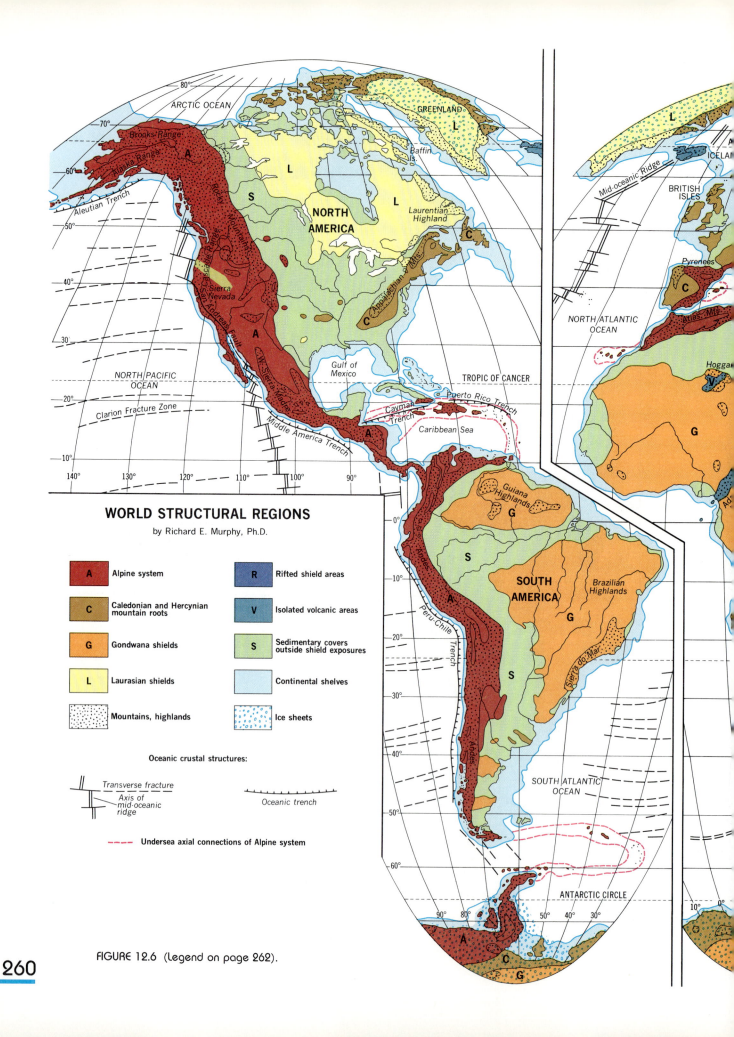

WORLD STRUCTURAL REGIONS

by Richard E. Murphy, Ph.D.

A Alpine system	**R** Rifted shield areas
C Caledonian and Hercynian mountain roots	**V** Isolated volcanic areas
G Gondwana shields	**S** Sedimentary covers outside shield exposures
L Laurasian shields	Continental shelves
Mountains, highlands	Ice sheets

Oceanic crustal structures:

Transverse fracture
Axis of mid-oceanic ridge

Oceanic trench

— — — Undersea axial connections of Alpine system

260   FIGURE 12.6 (legend on page 262).

Map labels: ARCTIC OCEAN, Brooks Range, Alaska Range, Aleutian Trench, Rocky Mountains, Cascade Ranges, Sierra Nevada, San Andreas Fault, W. Sierra Madre, NORTH PACIFIC OCEAN, Clarion Fracture Zone, Middle America Trench, NORTH AMERICA, GREENLAND, Baffin Is., Laurentian Highland, Appalachian Mts., Gulf of Mexico, TROPIC OF CANCER, Puerto Rico Trench, Cayman Trench, Caribbean Sea, Mid-oceanic Ridge, ICELAND, BRITISH ISLES, Pyrenees, NORTH ATLANTIC OCEAN, Atlas Mts., Hogga, Guiana Highlands, SOUTH AMERICA, Brazilian Highlands, Andes, Peru-Chile Trench, Sierra do Mar, SOUTH ATLANTIC OCEAN, ANTARCTIC CIRCLE

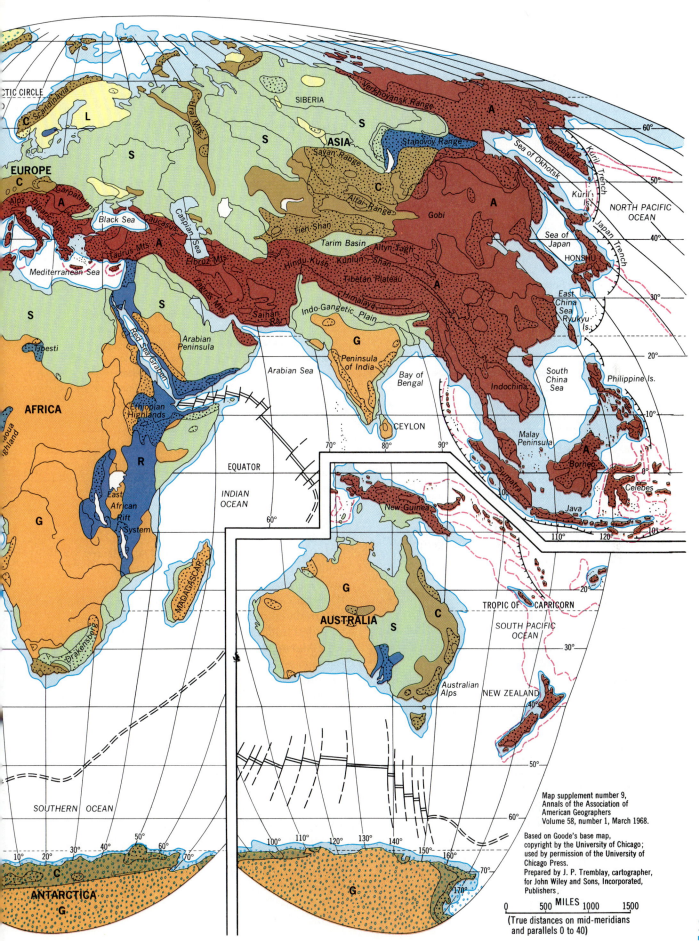

ARCTIC CIRCLE

Scandinavia

C

L

S

EUROPE

C

Alps

A

Carpathians

Dinaric Alps

Black Sea

Caucasus

Taurus Mts

A

Ural Mts

SIBERIA

S

ASIA

S

Sayan Range

C

Altai Range

Tien Shan

Stanovoy Range

Verkhoyansk Range

A

Kamchatka

Sea of Okhotsk

Kuril Trench

Kuril Is

S

Sea of Japan

NORTH PACIFIC OCEAN

Japan Trench

HONSHU

Mediterranean Sea

Caspian Sea

Elbruz Mts

Tarim Basin

Altyn Tagh

Hindu Kush

Kunlun Shan

Gobi

A

A

40°

30°

East China Sea

Ryukyu Is.

60°

50°

Zagros Mts

Arabian Peninsula

S

Saihan Ra.

Tibetan Plateau

Himalaya

Indo-Gangetic Plain

Peninsula of India

G

South China Sea

Indochina

20°

Philippine Is.

Red Sea Graben

Tibesti

S

Arabian Sea

Bay of Bengal

CEYLON

80°

90°

10°

Malay Peninsula

Borneo

A

AFRICA

Ethiopian Highlands

East African Rift System

R

EQUATOR

INDIAN OCEAN

70°

Sumatra

Java

Celebes

-10°

Ndaoua Highland

G

60°

New Guinea

100°

110°

120°

MADAGASCAR

Drakensberg

G

AUSTRALIA

G

S

C

Australian Alps

NEW ZEALAND

TROPIC OF CAPRICORN

SOUTH PACIFIC OCEAN

-20°

30°

40°

SOUTHERN OCEAN

10°

20°

30°

40°

50°

60°

70°

100°

110°

120°

130°

140°

150°

160°

170°

50°

60°

70°

ANTARCTICA

C

G

G

Map supplement number 9,
Annals of the Association of
American Geographers
Volume 58, number 1, March 1968.

Based on Goode's base map,
copyright by the University of Chicago;
used by permission of the University of
Chicago Press.
Prepared by J. P. Tremblay, cartographer,
for John Wiley and Sons, Incorporated,
Publishers.

0   500   MILES   1000   1500

(True distances on mid-meridians
and parallels 0 to 40)

FIGURE 12.6 World structural regions. (From R. E. Murphy, 1968, *Annals, Association of American Geographers*, Map Supplement No. 9. Based on Goode Base Map.)

A. Alpine system. World-girdling system of mountain chains formed since late Mesozoic time (Cretaceous Period, or younger). Faulted areas, plateaus, basins, and coastal plains enclosed by such ranges are included in the system.
C. Caledonian and Hercynian mountain roots, the remains of mountain chains and ranges formed during the Paleozoic and Mesozoic eras, prior to the Cretaceous Period. Faulted areas, plateaus, basins, and coastal plains enclosed by these mountain remnants are included.
G. Gondwana shields. Areas of stable, massive blocks of continental crust, lying south of the great east–west portion of the alpine system, where Precambrian rocks form either the entire surface rock or where Precambrian rocks form an encircling enclosure with no gap of more than

320 km (200 mi) between outcroppings of younger rock layers.
L. Laurasian shields. Areas of stable, massive blocks of continental crust lying north of the great east–west portion of the Alpine system. (Remainder of definition same as in G, above.)
R. Rifted shield areas. Block-faulted areas of shields forming grabens together with associated horsts and volcanic features. Rifting is a result of extension and thinning of continental lithospheric plates.
V. Isolated volcanic areas. Areas of volcanoes, active or extinct, with associated volcanic features, lying outside the Alpine or older mountain systems and outside the rifted shield areas. Volcanism is an expression of hot spots above mantle plumes.
S. Sedimentary covers. Areas of sedimentary layers that have not been subjected to orogeny and that lie outside either the crystalline rock enclosures of the shields or the enclosing mountains and hills of the Alpine or older orogenic systems. These areas of sedimentary rock form continuous covers over underlying structures.

Our world map of structural regions, Figure 12.6, recognizes alpine chains as belonging to the *Alpine system.* Because the Alpine system also includes some adjacent inactive regions produced by orogenies of the Mesozoic Era, it appears on the world map in broad belts rather than as the narrow, linear arcs of late Cenozoic activity suggested in Figure 12.5.

Continental Shields and Mountain Roots

Belts of recent and active mountain-making account for only a small portion of the continental crust. The remainder consists of comparatively inactive regions of much older rock. Within these stable regions we recognize two structural types of crust: shields and mountain roots. *Continental shields* are low-lying continental surfaces beneath which lie igneous and metamorphic rocks in a complex arrangement (see Figure 12.6). The rocks are very old, mostly of Precambrian age, and have had a very involved geologic history. For the most part, the shields are regions of low hills and low plateaus, although there are some

exceptions where large crustal blocks have been uplifted. Many thousands of meters of rock have been eroded from the shields during their exposure throughout a half-billion years.

Large areas of the continental shields are under a cover of younger sedimentary layers, ranging in age from Paleozoic through Cenozoic eras. These strata accumulated at times when the shields subsided and were inundated by shallow seas. Marine sediments were laid down on the ancient shield rocks in thicknesses ranging from hundreds to thousands of meters. These shield areas were then broadly arched and again became land surfaces. Erosion has since removed large sections of the sedimentary cover, but it remains intact over vast areas. We refer to such areas as *covered shields,* to distinguish them from the *exposed shields* in which the Precambrian rocks lie bare. An example of an exposed shield is the Canadian Shield of North America. Exposed shields are also extensive in Scandinavia, South America, Africa, peninsular India, and Australia (see Figure 12.6).

Remains of older mountain belts run within the shields in many places. These *mountain roots* are mostly formed of Paleozoic and early Mesozoic sedimentary rocks that have been intensely deformed and locally changed into metamorphic rocks.

One important system of mountain roots was formed in the Paleozoic Era, during an intense mountain-making episode—the Caledonian orogeny—that took place about 400 million years ago. These roots, called Caledonides, form belts across the northern British Isles and Scandinavia; they are also present in the Maritime Provinces of eastern Canada and in New England (see Figure 12.6). A second, but younger root system was formed during the Alleghanian (or Hercynian) orogeny, near the close of the Paleozoic Era, about 250 million years ago. In North America, this system is represented by the Appalachian Mountains. Thousands of meters of overlying rocks have been removed from these old tectonic belts, so that only basal structures remain. Roots appear as chains of long, narrow ridges, rarely rising over a thousand meters above sea level. (Landforms of mountain roots are described in Chapter 16.)

A Global System of Continental Structural Classes

The world map of structural regions, Figure 12.6, brings together the various second-order components of the continents within a complete global classification system. Six structural classes are identified

by letters. Descriptions of each class are given in the map legend. Class *A*, the alpine system, includes not only the narrow alpine chains of active volcanic and tectonic activity, but also adjoining belts of high mountains formed near the end of the Mesozoic Era and throughout the entire Cenozoic Era. Class *C*, named for the Caledonides in Europe, consists of mountain roots produced in the Paleozoic and Mesozoic eras. Exposed shields are represented by Classes *G* and *L;* covered shields by Class *S*. Two other classes, rifted shield areas (*R*) and isolated volcanic areas (*V*), are explained later in this chapter.

Second-Order Relief Features of the Ocean Basins

Crustal rock beneath the ocean floors consists almost entirely of basalt, covered over large areas by a comparatively thin accumulation of sediments. Age determinations of the basalt and its sediment cover show that the oceanic crust is quite young, geologically speaking. Much of that crust was formed during the Cenozoic Era and is less than 60 million years old. Over large areas, the rock age is of Mesozoic age, mostly from 65 to about 135 million years old. When we consider that the bulk of the continental crust is of Precambrian age—mostly over one billion years old—the young age of the oceanic crust is all the more remarkable. We will need to fit this fact into the general theory of global tectonic activity.

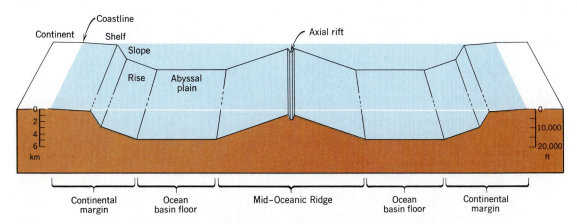

FIGURE 12.7 This schematic block diagram shows the ocean basins as symmetrical elements on a central axis. The model applies particularly well to the North and South Atlantic oceans.

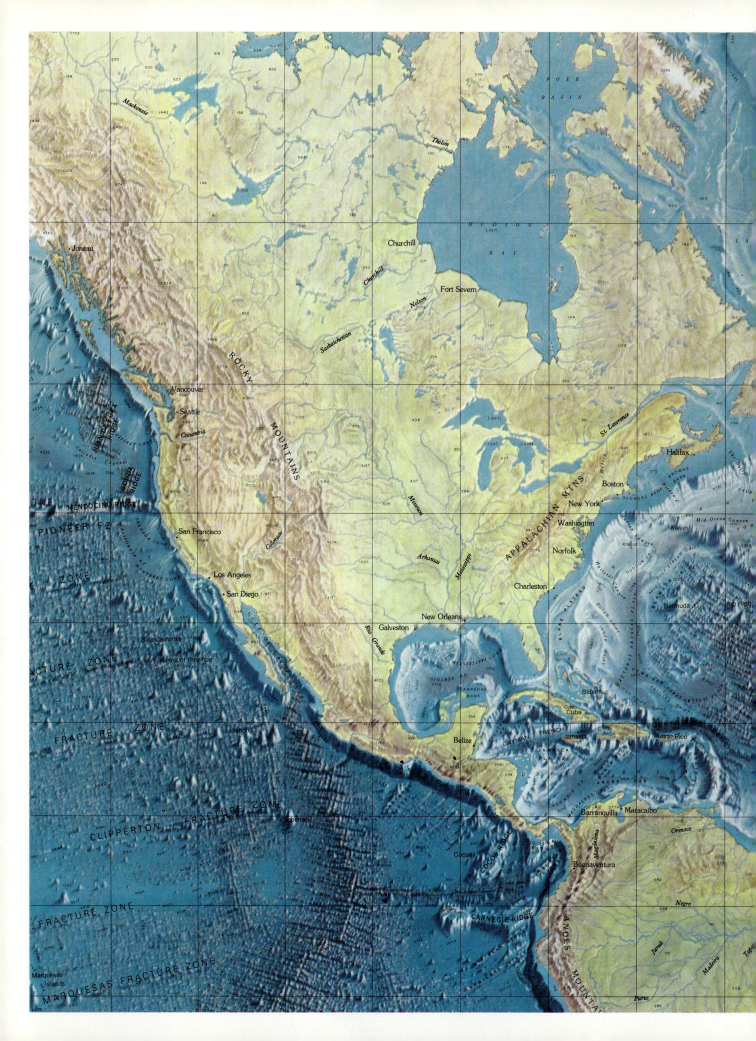

WORLD OCEAN FLOOR
by B. C. Heezen and Marie Tharp

FIGURE 12.8 A portion of Map of the World Ocean Floor by Bruce C. Heezen and Marie Tharp. Based on Mercator map projection. (Copyright © 1977 by Marie Tharp. Used by permission.)

In overall plan, the ocean basins are characterized by a central ridge structure that divides the basin about in half. This feature is shown by schematic diagram in Figure 12.7. The *mid-oceanic ridge* consists of submarine hills rising gradually to a mountainous central zone. Precisely in the center of the ridge, at its highest point, is an *axial rift,* which is a trenchlike feature. The form of this rift suggests that the crust is being pulled apart along the line of the rift axis. The mid-oceanic ridge and its principal branches can be traced through the ocean basins for a total distance of about 60,000 km (38,000 mi). Figure 12.5 shows the extent of the ridge. It divides the Atlantic Ocean basin from Iceland to the South Atlantic, where it turns east and enters the Indian Ocean. There, one branch penetrates Africa and the other continues east between Australia and Antarctica, then swings across the South Pacific. Nearing South America, it turns north and penetrates North America at the head of the Gulf of California.

The axial rift is broken in many places along its length by crustal fractures. Motion on these fracture lines (faults) has caused the rift to be sharply offset. Traces of the offsetting fractures extend far out on either side of the mid-oceanic ridge. As we will explain later, the axial rift and its offsetting fractures represent the boundary between adjacent lithospheric plates that are undergoing separation.

On either side of the mid-oceanic ridge are broad, deep plains and hill belts belonging to the *ocean basin floor.* Their average depth is about 5 km (17,000 ft). The flat surfaces are called *abyssal plains;* they are extremely smooth because they have been built up of fine sediment.

Many details of the ocean basins and their submarine landforms are shown in the accompanying color plate, Figure 12.8. The map of topography of the ocean floor is constructed from thousands of bottom profiles made by automatic depth-recording apparatus carried on all oceanographic research vessels. The apparatus uses reflected sound waves and is highly accurate. Because modern navigation methods, using orbiting satellites, are now extremely precise, the mapping of the ocean floor has become remarkably accurate.

The Passive Continental Margins

Nearing the continents the ocean floor begins to slope gradually upward, forming the *continental rise.* The floor then steepens greatly in the *continental slope.* At the top of this slope we arrive at the brink of the *continental shelf,* a gently sloping platform some 120 to 160 km (75 to 100 mi) wide along the eastern margin of North America. Water depth is about 150 m (500 ft) at the outer edge of the shelf.

The *continental margin,* shown in Figure 12.7 as the third element of the typical ocean basin, can be defined as the narrow zone in which oceanic lithosphere is in contact with continental lithosphere (see Figure 12.2). Thus, the continental margin is a feature shared by continent and ocean basin.

The symmetrical model illustrated in Figure 12.7 is fully exemplified in the North Atlantic and South Atlantic ocean basins. It also applies rather well to the Indian Ocean and Arctic Ocean basins. The margins of these symmetrical basins are described as *passive continental margins,* meaning that they have not been subjected to Cenozoic tectonic and volcanic activity.

The passive continental margins are underlain by great thicknesses of sedimentary strata derived from the continents. The strata range in age from the late Mesozoic (Jurassic, Cretaceous) through the Cenozoic. The shelf strata form a wedge-shaped deposit, thinning landward and feathering out over the continental shield (Figure 12.9). The sediments have been brought from the land by rivers and spread over the shallow seafloor by currents. A great deal of attention is now being paid to the continental-shelf wedge as a potential source of rich petroleum

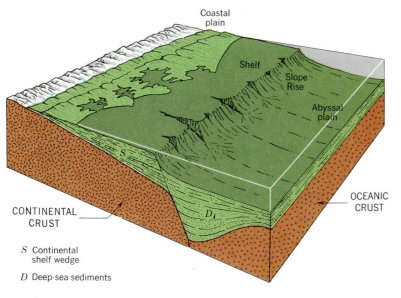

FIGURE 12.9 This block diagram shows an inner wedge of sediments beneath the continental shelf and an outer wedge of deep-sea sediments beneath the continental rise and abyssal plain.

accumulations, reached only from offshore drilling platforms. Below the continental rise and its adjacent abyssal plain is another thick sediment deposit; it is formed of deep-sea sediments carried down the continental slope by swift muddy currents, called *turbidity currents.*

Major rivers of the continents bring large amounts of sediment to the inner continental shelves, where large deltas are formed (Chapter 17). Seaward of the delta there is often present a narrow troughlike feature called a *submarine canyon,* representing the seaward extension of the river channel across the outer shelf. (Both the Hudson River and the Congo River have major submarine canyons of this kind.) Sediment carried down this canyon by turbidity currents descends the continental slope and accumulates on the continental rise in the form of a *submarine fan* (or *cone*). The fan of a large river may extend out over the deep ocean floor for a distance of many hundreds of kilometers (Figure 12.10). Two of the greatest known fans are those of the Indus and Ganges–Brahmaputra rivers. A similar, but smaller, fan lies off the Mississippi delta in deep water of the Gulf of Mexico (see Figure 12.8).

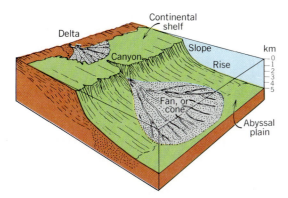

FIGURE 12.10 Block diagram of a deep-sea fan (cone) and its relationship to the continental shelf. (Redrawn from A. N. Strahler, *Physical Geology,* Harper & Row, New York. Copyright © 1981 by Arthur N. Strahler.)

The Active Continental Margins

The Pacific Ocean basin, although having a mid-oceanic ridge with ocean basin floors on either side, has quite different continental margins; they are characterized by mountain arcs or island arcs with offshore *oceanic trenches.* Geologists refer to these ocean basin limits as *active continental margins.*

The locations of the major trenches are shown in Figure 12.5. Trench floors reach depths of 7 km (23,000 ft) and even more (see Figure 12.8). Many lines of scientific evidence show that the oceanic crust is sharply downbent to form these trenches and that they mark the boundary between two lithospheric plates that are being brought together.

In the western Pacific Ocean basin are several subdivisions of the deep ocean floor known as *backarc basins.* A typical backarc basin is bounded on the continental side by either a deep trench or a passive continental margin and on the oceanward side by an island arc and its adjacent trench. One example is the Bering Abyssal Plain lying between the Alaskan–Siberian continental margin and the Aleutian volcanic island arc with its bordering Aleutian Trench. A second example is the deep ocean basin that lies between the Ryukyu–Philippine Trench and the Bonin–Mariana–Yap–Palau island arc system.

Plate Tectonics

Both crustal spreading along the axial rift of the mid-oceanic ridge and crustal downbending beneath oceanic trenches involve the entire thickness of lithospheric plates. The general theory of lithospheric plates with their relative motions and boundary interactions is *plate tectonics.* *Tectonics* is a noun meaning "the study of tectonic activity." Tectonic activity, in turn, refers to all forms of breaking and bending of the entire lithosphere, including the crust.

Figure 12.11 shows the major features of plate interactions. The vertical dimension of the block diagram (*A*) is greatly exaggerated, as are the landforms. A true-scale cross section (*B*) shows the correct relationships between crust and lithosphere, but surface relief features can scarcely be shown. A global diagram (*C*) with greatly exaggerated relief gives the correct impression of plates as being curved to fit the earth-sphere. The diagrams show two plates, Plate *X* and Plate *Y*, both made up of *oceanic lithosphere,* which is comparatively thin (about 50 km, 30 mi thick). Plate *Z*, made up of *continental lithosphere,* is much thicker (150 km, 95 mi). Because the continental lithosphere bears a thick crust, much of which is felsic (granitic) rock, it is comparatively buoyant. Oceanic lithosphere, on the other hand, is made up only of mafic and ultramafic rock; it is comparatively

dense and has a low-lying upper surface.

Plates *X* and *Y* are pulling apart along their common boundary, which lies along the axis of a mid-oceanic ridge. This plate activity tends to create a gaping crack in the crust, but magma continually rises from the mantle beneath. The magma appears as basaltic lava in the floor of the rift and quickly congeals. At greater depth under the rift, magma solidifies into gabbro, an intrusive rock of the same composition as basalt. Together, the basalt and gabbro continually form new oceanic crust.

At the right, the oceanic lithosphere of Plate *Y* is moving toward the thick mass of continental lithosphere that comprises Plate *Z*. Because the oceanic plate is comparatively thin and dense, in contrast to the thick, buoyant continental plate, the oceanic lithosphere bends down and plunges into the soft asthenosphere. The process of downplunging of one plate beneath another is called *subduction.*

The leading edge of the descending plate is cooler than the surrounding asthenosphere—enough cooler, in fact, that this descending slab of brittle rock is denser than the surrounding asthenosphere. Consequently, once subduction has begun, the slab "sinks under its own weight," so to speak. Gradually, however, the slab is heated by the surrounding asthenosphere and thus it eventually softens. The underportion, which is mantle rock in composition, simply reverts to asthenosphere as it softens. The thin upper crust, formed of less dense mineral matter, may actually melt and become magma. This magma tends to rise because it is less dense than the surrounding material. Figure 12.11 shows some magma pockets formed from the upper edge of the slab. They are pictured as rising like hot-air balloons through the overlying continental lithosphere. As they reach the earth's surface, quantities of this magma build volcanoes, which tend to form a volcano chain lying about parallel with the deep oceanic trench that marks the line of descent of the oceanic plate.

Viewing Plate *Y* as a unit in Figure 12.11, it appears that a single lithospheric plate is simultaneously undergoing *accretion* (growth by addition) and *consumption* (by softening and melting), so that the plate might

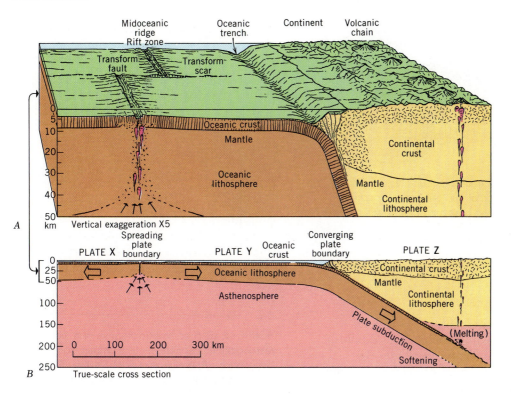

A

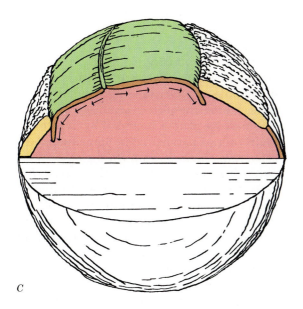

B

C

FIGURE 12.11 Schematic cross sections showing some of the important elements of plate tectonics. Diagram *A* is greatly exaggerated in vertical scale so as to emphasize surface and crustal features. Only the uppermost 30 km (20 mi) is shown. Diagram *B* is drawn to true scale and shows conditions to a depth of 250 km (155 mi). Here the actual relationships between lithospheric plates can be examined, but surface features can scarcely be shown. Diagram *C* is a pictorial rendition of plates on a spherical earth and is not to scale. (Redrawn from A. N. Strahler, *Physical Geology*, Harper & Row, New York. Copyright © 1981 by Arthur N. Strahler.)

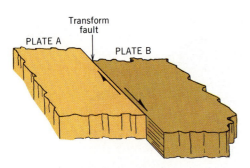

FIGURE 12.12 A transform fault involves the horizontal motion of two adjacent lithospheric plates, one sliding past the other. (Redrawn from A. N. Strahler, *Physical Geology*, Harper & Row, New York. Copyright © 1981 by Arthur N. Strahler.)

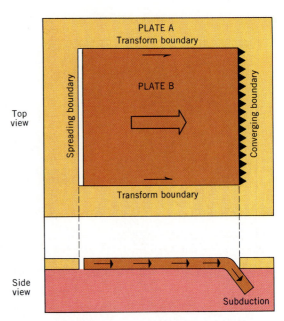

FIGURE 12.13 A schematic diagram of a single rectangular lithospheric plate with two transform boundaries. (Redrawn from A. N. Strahler, *Physical Geology*, Harper & Row, New York. Copyright © 1981 by Arthur N. Strahler.)

conceivably maintain its size, without necessarily having to expand or diminish. Actually, plate tectonics includes the possibility that a plate of oceanic lithosphere can either grow or diminish in extent. There are also tectonic models that allow for the creation of new plates of oceanic lithosphere where none existed and for plates to disappear completely. In this respect the theory is quite flexible.

We have yet to consider a third type of lithospheric plate boundary. Two lithospheric plates may be in contact along a common boundary on which one plate merely slides past the other with no motion that would cause the plates either to separate or to converge (Figure 12.12). The plane along which motion occurs is a nearly vertical fracture extending down through the entire lithosphere; it is called a *transform fault*. A *fault* is a plane or rock fracture along which there is motion of the rock mass on one side with respect to that on the other (see also Chapter 13).

In summary, there are three major kinds of active plate boundaries:

Spreading boundaries. New lithosphere is being formed by accretion.

Converging boundaries. Subduction is in progress; lithosphere is being consumed.

Transform boundaries. Plates are gliding past one another on a transform fault.

Let us put these three boundaries into a pattern to include an entire lithospheric

plate. As shown in Figure 12.13, we have visualized a moving rectangular plate set in the middle of a surrounding stationary plate, that is, the plate resembles a window. The moving plate is bounded by transform faults on two parallel sides. Spreading and converging boundaries form the other two parallel sides. Several familiar mechanical devices come to mind in visualizing this model. One is the sunroof in the top of an automobile, it is a window that opens by sliding backward along parallel sidetracks to disappear under the fixed roof area at the rear. Another familiar device is the old-fashioned rolltop desk. Boundaries can be curved as well as straight, and individual plates can pivot as they move. There are many geometric variations in the shapes and motions of individual plates.

The Global System of Lithospheric Plates

The global system of lithospheric plates consists of six great plates. These are listed in Table 12.2 and shown on a world map in

TABLE 12.2 The lithospheric plates

Great plates:

  Pacific

  American (North, South)

  Eurasian

    Persian subplate

  African

    Somalian subplate

  Austral-Indian

  Antarctic

Lesser plates:

  Nazca

  Cocos

  Philippine

  Caribbean

  Arabian

  Juan de Fuca

  Caroline

  Bismark

  Scotia

Figure 12.14. Several lesser plates are also recognized, ranging from intermediate in size to comparatively small. Several subplates are also recognized within the great plates. Two shown in Figure 12.14 are the Persian subplate and the Somalian subplate. Plate boundaries are shown by standard symbols, explained in the key accompanying the map. Keep in mind that the Mercator grid distorts the areas of the plates, making them appear greatly expanded in high latitudes.

Figure 12.15 is a schematic circular cross section of the lithosphere along a great circle in low latitudes. It shows several of the great plates and their boundaries.

The great Pacific plate occupies much of the Pacific Ocean basin and consists almost entirely of oceanic lithosphere. Its relative motion is northwesterly, so that it has a subduction boundary along most of the western and northern edge. The eastern and southern edge is mostly a spreading boundary. A sliver of continental lithosphere is included and makes up the coastal portion of California and all of Baja California. The California portion of the boundary is an active transform fault (the San Andreas Fault).

The American plate includes most of the continental lithosphere of North and South America as well as the entire oceanic lithosphere lying west of the mid-oceanic ridge that divides the Atlantic Ocean basin down the middle. For the most part, the western edge of the American plate is a subduction boundary; the eastern edge is a spreading boundary. (Some classifications recognize separate North American and South American plates.) The Eurasian plate is mostly continental lithosphere, but is fringed on the west and north by a belt of oceanic lithosphere.

The African plate can be visualized as having a central core of continental lithosphere nearly surrounded by oceanic lithosphere. The Austral-Indian plate takes the form of an elongate rectangle. It is mostly oceanic lithosphere but contains two cores of continental lithosphere—Australia and peninsular India. The Antarctic plate has an elliptical shape and is almost completely enclosed by a spreading plate boundary. The continent of Antarctica forms a central core of continental lithosphere completely surrounded by oceanic lithosphere.

Of the remaining six plates, the Nazca and Cocos plates of the eastern Pacific are rather simple fragments of oceanic lithosphere bounded by the Pacific mid-oceanic spreading boundary on the west and by a subduction boundary on the east. The Philippine plate is noteworthy as having subduction boundaries on both east and west edges. Two small but distinct lesser plates—Caroline and Bismark—lie to the southeast of the Philippine plate, but these can be included within the Pacific plate. The Arabian plate resembles the "sunroof" model shown in Figure 12.13; it has two transform fault boundaries and its relative motion is northeasterly. The Caribbean plate also has important transform fault boundaries. The tiny Juan de Fuca plate is steadily diminishing in size and will eventually disappear by subduction beneath the American plate.

Subduction Tectonics and Volcanic Arcs

Converging plate boundaries, with subduction in progress, are zones of intense

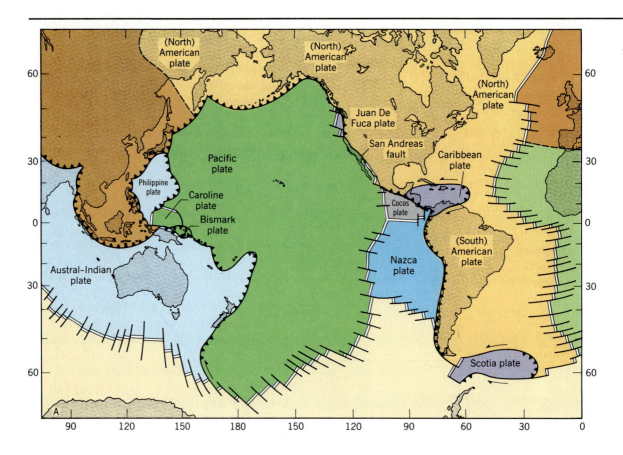

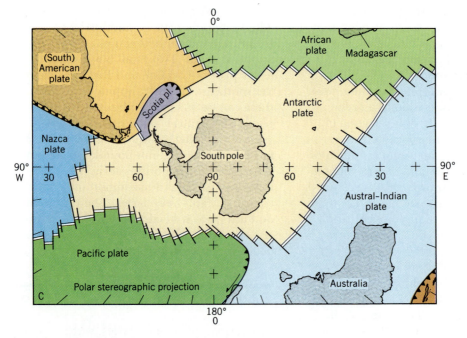

FIGURE 12.14 World map of lithospheric plates.

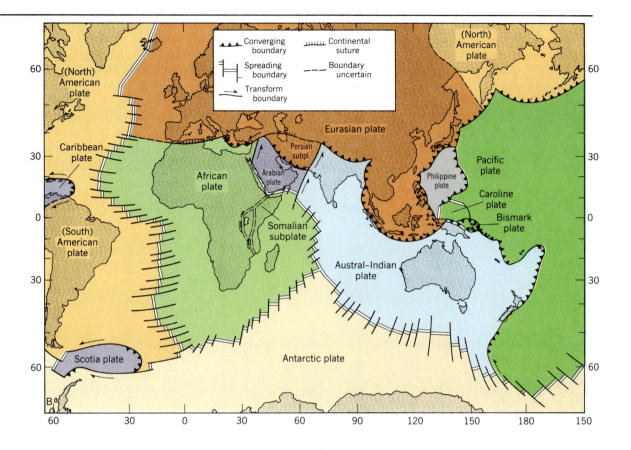

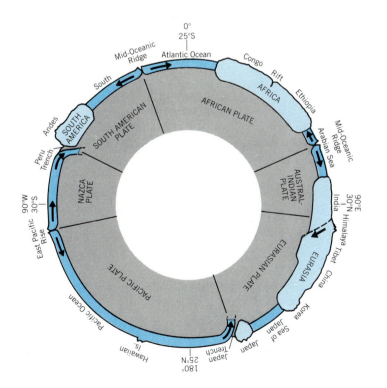

FIGURE 12.15 Schematic circular cross section of the major plates on a great circle tilted about 30 degrees with respect to the equator. (A. N. Strahler.)

tectonic and volcanic activity. The narrow zone of a continent that lies above a plate undergoing subduction is therefore an active continental margin. Figure 12.16 shows some details of the geologic processes that are associated with plate subduction. Two diagrams are used: (A) exaggerated to show crustal and surface details, and (B) drawn to true scale to show the lithospheric plates.

The trench axis represents the line of meeting of sediment coming from two sources. Carried along on the moving oceanic plate is deep-ocean sediment—fine clay and ooze—that has settled to the ocean floor. From the continent comes terrestrial sediment in the form of sand and mud brought by streams to the shore and then swept into deep water by currents. In the bottom of the trench both types of sediment

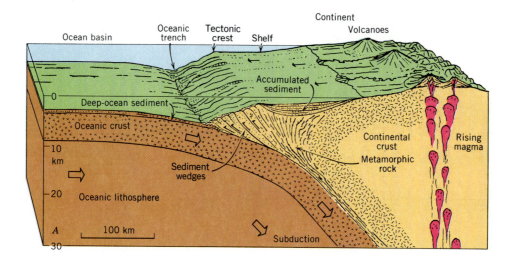

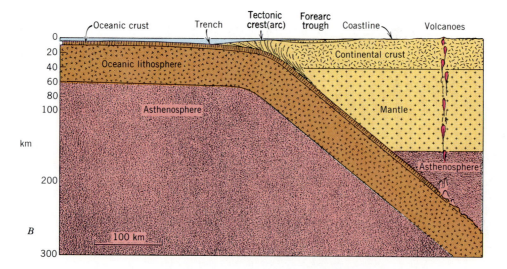

FIGURE 12.16 Some typical features of an active subduction zone. Diagram A uses a great vertical exaggeration to show surface and crustal details. Sediments scraped off the moving plate form tilted wedges that accumulate in a rising tectonic mass. Between the tectonic crest and the mainland is a shallow trough in which sediment brought from the land is accumulating. Metamorphic rock is forming above the descending plate. Magma rising from the top of the descending plate reaches the surface to build a chain of volcanoes. Diagram B is a true-scale cross section showing the entire thickness of the lithospheric plates. (Redrawn from A. N. Strahler, *Physical Geology*, Harper & Row, New York. Copyright © 1981 by Arthur N. Strahler.)

are intensely deformed and are carried downward with the moving plate. The deformed sediment is then shaped into wedges that ride, one over the other, on steep fault planes. The wedges accumulate into an *accretionary prism* in which metamorphism takes place. In this way, the continental margin is built outward and new continental crust of metamorphic rock is formed. The accretionary prism is of relatively low density and tends to rise, forming a *tectonic crest*. The tectonic crest is shown to be submerged, but in some cases it forms an island chain paralleling the coast; that is, a *tectonic arc*. Between the tectonic crest and the mainland is a shallow trough, the *forearc trough*. This trough traps a great deal of terrestrial sediment, which accumulates in a basinlike structure. The bottom of the forearc trough continually subsides under the load of the added sediment. In some cases, the seafloor of the trough is flat and shallow, forming a type of continental shelf. Sediment carried across the shelf moves down the steep outer slope of the accretionary prism in tonguelike flows of turbidity currents.

The lower diagram of Figure 12.16 shows the descending lithospheric plate entering the asthenosphere. Intense heating of the upper surface of the plate melts the oceanic crust, forming basaltic magma. As this magma rises, some of the denser mafic minerals remain at the base of the crust and the rising magma takes on the composition of andesite. The andesite magma reaches the surface to form volcanoes of andesite lava, such as those we see in the Andes of South America.

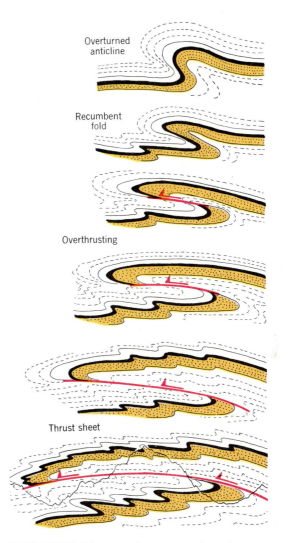

FIGURE 12.17 Schematic diagrams to show the development of a recumbent fold and nappe, broken by a low-angle overthrust fault to produce a thrust sheet in alpine structure. (Based on diagrams by A. Heim, 1922, *Geologie der Schweiz*, vol. II-1, Tauschnitz, Leipzig.)

Orogeny

Generally speaking, high-standing mountain masses (other than volcanic mountains) are elevated by one of two basic tectonic processes: compression and extension. Compressional tectonic activity—"squeezing together" or "crushing"—acts at converging plate boundaries; extensional tectonic activity—"pulling apart"—occurs where oceanic plates are separating or where a continental plate is undergoing breakup into fragments. We look first at the compressional tectonic processes.

The alpine mountain chains typically consist of intensely deformed strata of marine origin. The strata are tightly compressed into wavelike structures, called *folds*. Typically, the folds become overturned and take on recumbent attitude, as shown in Figure 12.17. Accompanying the folding is a form of faulting in which slices of rock move over the underlying rock on fault surfaces of low inclination; these are *overthrust faults*.

Individual rock slices, called *thrust sheets*, are carried many tens of kilometers over

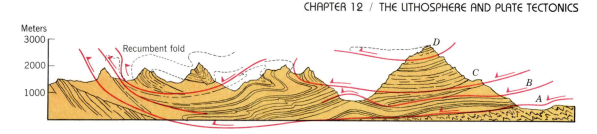

FIGURE 12.18 A cross section through a portion of the Helvetian Alps, Switzerland, shows thrust slices. Horizontal and vertical scales are the same. (Simplified from A. Heim, 1922, *Geologie der Schweiz*, vol. II-1, Tauschnitz, Leipzig.)

the underlying rock (Figure 12.17). In the European Alps, thrust sheets of this kind were named *nappes* (from the French word meaning "cover sheet" or "tablecloth"). Nappes may be thrust one over the other to form a great pile (Figure 12.18). The entire deformed rock mass produced by such compressional mountain-making is called an *orogen;* the event that produced it is an orogeny.

Thrust sheets are known to have moved nearly horizontally for distances of several tens of kilometers. Geologists postulate that at the time the sheets were in motion, the thrust planes actually sloped downward in the direction of motion of the sheet. You might visualize the thrust sheets as sliding "downhill" under the force of gravity. The phenomenon is known as *gravity gliding*. The presence of ground water under great pressure beneath the thrust sheet is thought to have reduced the friction so much that the enormous rock layer could glide easily for a long distance.

Orogens and Collisions

As you can easily visualize in a situation where two lithospheric plates are converging along a subduction boundary, any high-standing mass of continental lithosphere projecting above the level of the oceanic lithosphere will come closer and closer to the continent beneath which that plate is descending. Ultimately, the two masses of continental lithosphere must collide, because the impacting mass is too thick and too buoyant to pass down beneath the continental plate it is impacting. The result is orogeny in which various kinds of crustal rocks are crumpled into folds and

sliced into nappes. Quite aptly, this process has been called "telescoping."

Two types of orogens are recognized. One type results from the impact of a relatively small mass of high-standing continental lithosphere against a full-sized mass of continental lithosphere. Following this type of collision, the small mass is firmly welded to the continent and will become a permanent part of the continental shield. The second type of orogen results from the collision of two very large bodies of continental lithosphere (full-sized plates), permanently uniting them and terminating further tectonic activity along that collision zone.

To prepare for the first type of orogeny, a simple mechanical model will be useful. Imagine we are in a custom bakery that makes cakes on special order. As each cake is completed, it is placed on a continually moving conveyer belt that transports it to the packaging department. Here, the cake slides off onto a short table surface at the level of the belt. The table will accommodate only a few cakes, so that if the attendant leaves the station for too long, disaster soon sets in. As another cake arrives, it slams into those already on the table, crushing them severely and compacting them into a single mass.

The conveyer belt in our analog represents oceanic lithosphere disappearing into a subduction zone. Each cake represents one of several possible crustal features that can project above the otherwise uniform abyssal surface. Small projecting objects, such as seamounts, usually cause no problem—they may be knocked off at the base by impact with the overlying plate to become incorporated into the accretionary prism. The problem is with

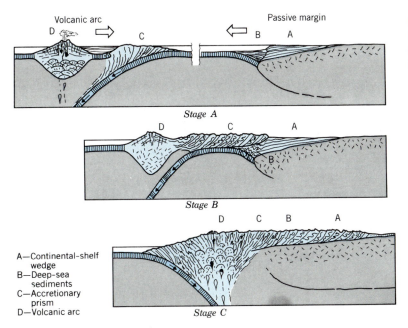

A—Continental-shelf
    wedge
B—Deep-sea
    sediments
C—Accretionary
    prism
D—Volcanic arc

FIGURE 12.19 Schematic cross sections of an arc–continent collision. Not shown to true scale. (Drawn by A. N. Strahler.)

much larger crustal protrusions that are too massive and too firmly rooted to be subducted.

What kinds of traveling crustal masses do we have in mind as capable of colliding with a large continental mass and adhering to it? One is the volcanic island arc; for example, the Kuril arc or the Lesser Antilles arc. An entirely different class of objects consists of small fragments or bits of continental crust called *microcontinents*. Some of the older island arcs with a long history of accretion qualify as microcontinents—Honshu, the Philippines, or Hispaniola, for example. In other cases, a fragment of continent has been pulled away from the mainland on a widening rift that develops into a backarc basin. These islands of continental crust are thus surrounded by oceanic crust. Over millions of years, the impacting object can travel thousands of kilometers.

The impact of an island arc is called an *arc–continent collision*. It is illustrated by a set of diagrams in Figure 12.19. The setting, shown as Stage *A*, is one of a passive continental margin (right) closing with a volcanic arc (left). In this case, the subduction is occurring along the volcanic arc, thus narrowing the expanse of oceanic lithosphere separating the two features. An accretionary prism is growing larger as the volcanic arc increases in size and depth. In Stage *B*, the intervening ocean is completely closed and the accretionary prism is forced to ride up the continental margin, gliding on a basal thrust fault. As shown in Stage *C*, impact of the volcanic arc causes the overthrusting to "telescope" the strata of the passive margin, extending the low-angle thrust slice far over the continental shield. Strata of the thrust slice as well as rocks of the accretionary prism and the volcanic arc itself are severely deformed and converted into metamorphic rock.

Notice also that in Stage *C*, a new subduction boundary has come into existence on the oceanward side of the former volcanic arc. Magmas rising from the downplunging plate penetrate the metamorphic zone of the orogen. These give rise to bodies of felsic magma that rise even higher and become granitic batholiths. Magma also reaches the surface to produce outpourings of felsic lava. The process of intrusion by magma has greatly thickened the crust beneath the orogen, making it a permanent addition to the continent.

Examples of orogens resulting from arc–continent collisions can be found on both eastern and western sides of North America. The relationships shown in Figure 12.19 are patterned in a general way after Paleozoic arc–continent collisions that affected the Appalachian and Ouachita mountain belts. Another example is the Cordilleran Orogeny that (as the name indicates) built the Cordilleran Ranges of western North America. In the northern Rocky Mountains of Montana, Alberta, and British Columbia the overthrust zone is remarkably displayed in high, glaciated mountains that hold Waterton–Glacier International Peace Park, Banff Park, and Jasper Park.

Accreted Terranes of Western North America

In recent years, geologists who have been involved in mapping the bedrock features of western North America from Mexico to Alaska have come to recognize that it consists of a mosaic of crustal patches called *terranes*. A particular terrane is quite distinct geologically from those that surround it, in that it has its own special rock type or suite of component rock types. If we think of this assemblage of terranes as a mosaic of tiles, we see that whereas most of the tiles are unique, others are duplicated in widely separated places. Figure 12.20 is a map of these terranes. There are at least 50 of them and each is separated from its neighbors by a fault contact. In many cases, the fault is similar to a transform fault and trends more or less parallel with the continental margin.

Each terrane has a name. For example, the terrane called "Wrangellia," shown on the map by special color, occurs in five separate patches. It is a terrane consisting of basaltic island-arc volcanics and sedimentary rocks of types that include both cherts of deep-sea origin and limestones of shallow-depth origin. Using paleomagnetic methods, geologists have been able to establish the original global location of certain of the terranes. Much to their initial surprise, Wrangellia was found to have come from a location at 10 degrees latitude and as far distant as a spot equivalent to where New Guinea is now located— traveling since then a distance of nearly 10,000 km! It occupied that location in Triassic time when some three kilometers' thickness of basaltic island-arc lavas accumulated.

Each of the many terranes is now regarded as a former microcontinent— using that term in a general sense.

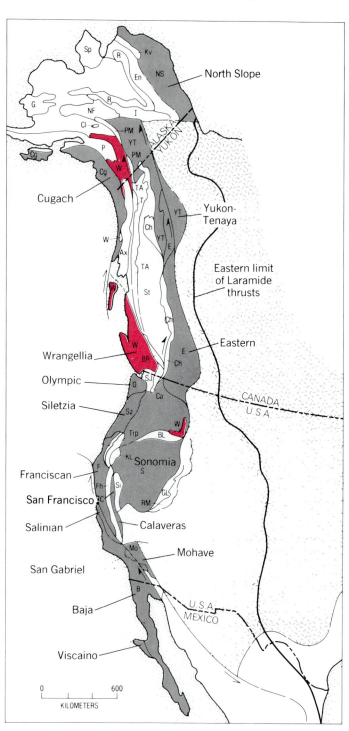

FIGURE 12.20 Map of microcontinent terranes of western North America. Fragments of Wrangellia are shown in solid color. Names of several of the terranes are given. (Based on data of U.S. Geological Survey; M. Beck, A. Cox, and D. L. Jones, 1980, *Geology*, vol. 8, p. 455; and others.)

Obviously, a terrane can only travel by being carried along with the lithospheric plate in which it is embedded. These bits of continental lithosphere were embedded in oceanic lithosphere, which moved toward North America, eventually bringing each microcontinent to a location close to a subduction boundary at the active western margin of North America. Here a collision took place and the microcontinent was welded to the continent. Many such collisions took place, eventually forming the mosaic of terranes.

Once the microcontinents were added to the continent, they were subjected to being sheared into two or more small fragments by faults trending about parallel with the continental margin. (The San Andreas Fault is an example of this type of fault.) Dragged along within the fault slices, terrane fragments became separated and distributed up and down the entire continental marginal zone. This is what seems to have happened in the case of Wrangellia. The accretion of microcontinents to form a mosaic pattern in the continental crust may have also been one of the modes by which the Precambrian shield was constructed.

Whether the arrival and impact of a microcontinent was capable of generating a full-blown orogeny is questionable. For a small microcontinent, a few tens of kilometers across, the tectonic effect would perhaps have been localized. In the case of an impacting volcanic island arc thousands of kilometers long, a major orogeny would have ensued that strongly affected thousands of kilometers along the continental margin. A phenomenon of this magnitude seems quite distinct from the repeated impacts of many microcontinents. Evidently, such collisions come in a wide range of magnitudes and produce a wide range of effects.

Continent—Continent Collisions

A second type of orogen results from the collision of two full-sized masses of continental lithosphere, both being of the dimensions of a great plate or a subplate. This kind of orogen is the result of a *continent—continent collision*. Collisions of this type have occurred in the Cenozoic Era along a great tectonic line that marks the southern boundary of the Eurasian plate. The line begins with the Atlas Mountains of

North Africa, runs through the European Alps, and extends across the Aegean Sea region into western Turkey. Beyond a major gap in Turkey, the line takes up again in the Zagros Mountains of Iran. Jumping another gap in southeastern Iran and Pakistan, the collision line sets in again in the great Himalayan Range.

Each segment of this collision zone represents the collision of a different north-moving plate against the single and relatively immobile Eurasian plate. A European segment containing the Alps was formed by collision of the African plate with the Eurasian plate in the Mediterranean region. A Persian segment resulted from the collision of the Arabian plate with the Eurasian plate. A Himalayan segment represents the collision of the Indian continental portion of the Austral-Indian plate with the Eurasian plate.

Figure 12.21 is a series of cross sections in which the tectonic events of a typical continent—continent collision are reconstructed. Diagram *A* shows a passive margin at the left and an active subduction margin at the right. As the ocean between the converging continents is eliminated, a succession of overthrust faults cuts through the oceanic crust (diagram *B*). The thrust slices ride up, one over the other, telescoping the oceanic crust and the sediments above it. As the slices become more and more tightly squeezed, they are forced upward. The upper part of each thrust sheet assumes a horizontal attitude to form a nappe, which then glides forward under gravity on a low downgrade. The final nappes in the series consist in part of the remains of slices of oceanic crust. A mass of metamorphic rock is formed between the joined continental plates, welding them together. This new rock mass is called a *continental suture;* it is a distinctive type of orogen.

The Himalayan collision segment is still highly active, with the Indian plate being underthrust deeply beneath the mountain range. Immediately to the north lies the Tibetan Plateau, with an extremely thick continental crust, perhaps resulting from compressional shortening and thickening.

Continent—continent collisions have occurred since the late Precambrian. Many ancient sutures have been identified in the

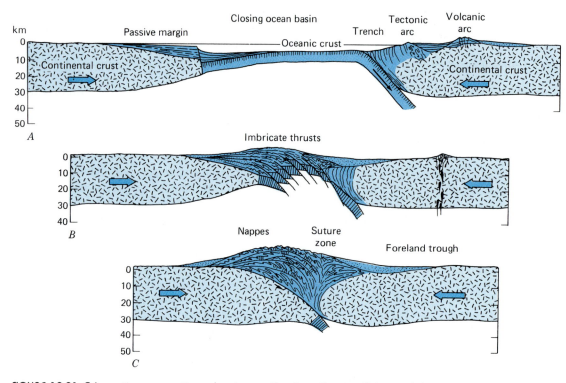

FIGURE 12.21 Schematic cross sections showing continent-continent collision and the formation of a suture zone with nappes. (Redrawn from A. N. Strahler, *Physical Geology*, Harper & Row, New York. Copyright © 1981 by Arthur N. Strahler.)

continental shields. The Ural Mountains are one such suture, trending north–south as the arbitrary dividing line between Europe and Asia. It was formed near the close of the Paleozoic Era.

Continental Rupture and New Ocean Basins

We have already noted that the continental margins bordering the Atlantic Ocean basin on both its eastern and western sides are very different from the active margin of a subduction zone. The Atlantic margins have no important tectonic activity at present; they are passive continental margins. Even so, they represent the contact between continental lithosphere and oceanic lithosphere, with continental crust meeting oceanic crust, as shown in Figure 12.9. To understand how passive margins are formed, we must go back into tectonic history that involved the rifting apart of a

single continental lithospheric plate. This process is called *continental rupture.*

Figure 12.22 shows by three schematic block diagrams how continental rupture takes place and leads to the development of passive continental margins. At first the crust is both lifted and stretched apart as the lithospheric plate is arched upward. Then a long narrow valley, called a *rift valley*, appears (Block A). The widening crack in its center is continually filled in with magma rising from the mantle below. The magma solidifies to form new crust in the floor of the rift valley. Crustal blocks slip down along a succession of steep faults, creating a mountainous landscape. As separation continues, a narrow ocean appears; down its center runs a spreading plate boundary (Block B). Plate accretion takes place to produce new oceanic crust and lithosphere. We find in the Red Sea today an example of a narrow ocean formed by continental rupture. Its straight, steep coasts are features we would expect of

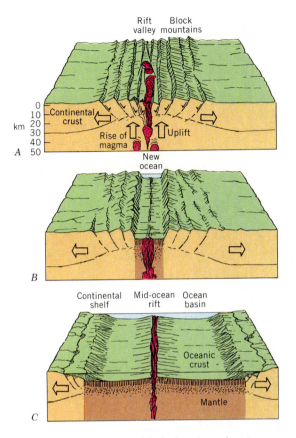

FIGURE 12.22 Schematic block diagrams showing stages in continental rupture and the opening up of a new ocean basin. The vertical scale is greatly exaggerated to emphasize surface features. (A) The crust is uplifted and stretched apart, causing it to break into blocks that become tilted on faults. (B) A narrow ocean is formed, floored by new oceanic crust. (C) The ocean basin widens, while the passive continental margins subside and receive sediments from the continents. (Redrawn from A. N. Strahler, *Physical Geology*, Harper & Row, New York. Copyright © 1981 by Arthur N. Strahler.)

These *transform scars* take the form of narrow ridges or scarps (clifflike features) and may extend for hundreds of kilometers across the ocean floors. Before the true nature of these scarps was understood, they were named *fracture zones*. That name still persists and can be seen on maps of the ocean floor (Figure 12.8). The scars are not, for the most part, associated with active faults, although some undergo minor vertical fault movements at a later time.

Notice in Block C of Figure 12.22 that the overall appearance of the ocean basin and its continental margins resembles the schematic diagram in Figure 12.7. The passive margins are accumulating terrestrial sediment in the form of a continental shelf that rests on the continental crust, while the oceanic crust is accumulating a wedge of deep-sea sediments. The continental margin gradually subsides as these sediment wedges thicken, until the total sediment thickness reaches several kilometers (Figure 12.9). A wide, shallow continental shelf is typical of the passive continental margins. Large deltas built by rivers contribute a great deal of the shelf sediment. Turbidity currents carry the sediment down the steep continental slope and spread it out on the continental rise, producing deep-sea fans (Figure 12.10).

Continental Drift— The Breakup of Pangaea

Although modern plate tectonics became an acceptable scientific theory within only the past two decades, the concept of breakup of an early supercontinent into fragments that drifted apart is many decades old. Almost as soon as good navigational charts became available to show the continental outlines, persons of learning became intrigued with the close correspondence in outline between the eastern coastline of South America and the western coastline of Africa. In 1668, a Frenchman interpreted the matching coastlines as proof that the two continents became separated during the biblical flood. In 1858 Antonio Snider-Pelligrini produced a map to show the American continents nested closely against Africa and Europe. He went beyond the purely geometrical fitting to suggest that the reconstructed

such a history (Figure 12.23). The widening of the ocean basin can continue until a large ocean has formed and the continents are widely separated (Figure 12.22, Block C).

During the process of opening of an ocean basin, the spreading boundary develops a series of offsets, one of which is shown in the upper left-hand part of Diagram A of Figure 12.11. The offset ends of the axial rift are connected by an active transform fault. As spreading continues, a scarlike feature is formed on the ocean floor as an extension of the transform fault.

FIGURE 12.23 Astronauts aboard the Gemini XII space vehicle took this south-looking photo of the Red Sea, which separates the Arabian Peninsula (left) from Africa (right). The narrow sea is about 200 km (125 mi) wide. Its straight, parallel coastlines suggest its origin—as a widening belt of new ocean between separating lithospheric plates. At the lower left we see the triangular Sinai Peninsula, bounded by two narrower fault depressions—the Gulf of Suez (bottom of photo) and the Gulf of Aqaba (left). (NASA)

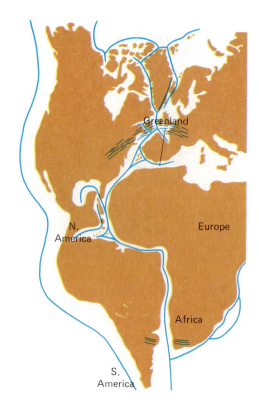

single continent explains the close similarity of fossil plant types in coal-bearing rocks in both Europe and North America.

Moving ahead to the early 20th century, we come to the ideas of two Americans, Frank B. Taylor and Howard B. Baker, whose published articles presented evidence favoring the hypothesis that the New World and Old World continents had drifted apart. Nevertheless, credit for a full-scale hypothesis of breakup of a single supercontinent and the drifting apart of

FIGURE 12.24 Alfred Wegener's 1915 map fitting together the continents that today border the Atlantic Ocean basin. The sets of dashed lines show the fit of Paleozoic tectonic structures between Europe and North America and between southernmost Africa and South America. (From A. Wegener, 1915, De Entstehung der Kontinente und Ozeane, F. Vieweg, Braunschweig.)

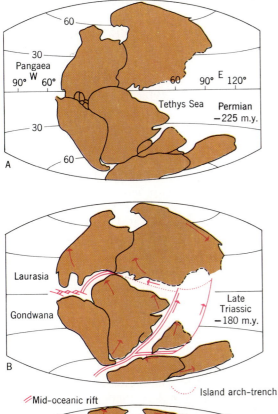

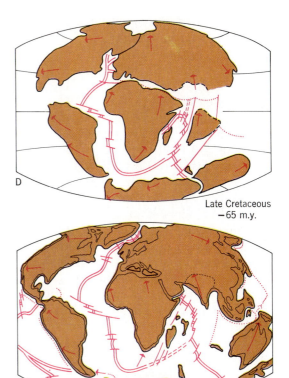

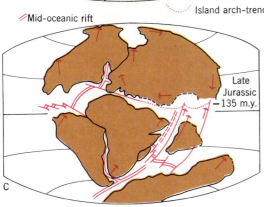

FIGURE 12.25 The breakup of Pangaea is shown in five stages. Inferred motion of lithospheric plates is indicated by arrows. (Redrawn and simplified from maps by R. S. Dietz and J. C. Holden (1970). *Jour. Geophysical Research*, vol. 75, pp. 4943–4951, figures 2 to 6. Copyrighted by the American Geophysical Union.)

individual continents belongs to a German scientist, Alfred Wegener, a meteorologist and geophysicist who became interested in the various lines of geologic evidence that the continents had once been united. He first presented his ideas in 1912 and his major work on the subject appeared in 1922 (Figure 12.24). A storm of controversy followed, and many American geologists denounced the hypothesis.

Wegener had reconstructed a supercontinent named *Pangaea,* which existed intact about 300 million years ago in a period of geologic time called the Carboniferous Period (see Figure 12.24). Wegener visualized the Americas as fitted closely against Africa and Europe, whereas the continents of Antarctica and Australia, together with the subcontinent of peninsular India and the island of Madagascar, were grouped closely around the southern tip of Africa. Starting about 200 million years ago, continental rifting began as the Americas pulled away from the rest of Pangaea, leaving a great rift that became the Atlantic Ocean. Later, the other fragments rifted apart, pulling away from Africa and from each other and causing the opening up of the ancestral Indian Ocean, as shown in Figure 12.25, *B* through *E*.

Several lines of hard geologic evidence favored the existence of Pangaea. The evidence for a single supercontinent seemed quite convincing to many geologists during the 1920s and 1930s, but the separation of the continents—a process then known as *continental drift*—was strongly opposed on physical grounds. Wegener had proposed that the continental layer of less dense rock had moved like a great floating raft through a "sea" of denser oceanic crustal rock. Geologists could show by use of principles of physics that this mechanism was physically impossible, because rigid crustal rock could not behave in such a fashion.

Wegener's scenario of continental drift took on new meaning in the 1960s and 1970s, when plate tectonics emerged as a leading theory. The modern interpretation is, of course, that continental drift involves entire lithospheric plates, much thicker than merely the outer crust of either the continents or the ocean basins. Plate motions over a soft, plastic asthenosphere have allowed the continents to be carried along according to the general timetable postulated by Wegener. Some changes have been made in Wegener's timetable of events. There have also been numerous improvements in the fitting together of the original pieces of the supercontinent Pangaea.

The Tectonic System

The system of lithospheric plates in motion represents an enormous material-flow system powered by an internal energy-flow system. The scheme of the cycling of mineral matter is fairly well understood in a general way, although many details remain speculative.

Figure 12.26 is a schematic diagram (not to scale) showing some of the major features of the material-flow system. Diagram *A* shows how a plate of oceanic lithosphere undergoing subduction transfers matter to the margin of the continental lithosphere by volcanic and tectonic processes. Magma formed by melting of the upper surface of the plate penetrates the continental lithosphere and is added to the continental crust in the form of igneous plutons and extrusive masses (volcanoes). Offscraping of the upper surface of the subducting plate contributes to the growth of accretionary prisms, which become permanent additions to the continental crust as metamorphic rock.

Most of the downgoing plate is softened by heating and is reabsorbed into the asthenosphere. Slow currents deep within the asthenosphere, moving in a direction generally opposite to the plate motion, return the enriched mantle rock to the spreading plate boundaries.

Diagram *B* of Figure 12.26 shows that under certain conditions, the dragging action of the downgoing plate tears loose blocks and slabs of the adjacent continental lithosphere. This material is carried down into the asthenosphere. Thus by *tectonic erosion* much felsic rock of the continental crust can enter the mantle and be recycled.

The energy system that causes plate motions is generally agreed to have its source in the phenomenon of radioactivity. Radioactive elements in the crust and upper mantle constantly give off heat. This is a process of transformation of matter into energy. As the temperature of mantle rock rises, the rock expands. As in the case of the atmosphere, upward motion of less dense material takes place by convection. It is thought that mantle rock is rising steadily beneath the spreading plate boundaries. How this rise of heated rock causes plates to move is not well understood, but one hypothesis states that as the lithospheric plate is lifted to a higher elevation above the rising mantle it tends to move horizontally away from the spreading axis under the influence of gravity. At the opposite edge of the plate, subduction occurs because the oceanic plate is denser than the asthenosphere through which it is sinking. Motion of the plate exerts a drag on the underlying asthenosphere, setting in motion flow currents in the upper mantle. Thus slow convection currents probably exist in the asthenosphere beneath the moving plates, but their pathways and depths of operation are not well understood.

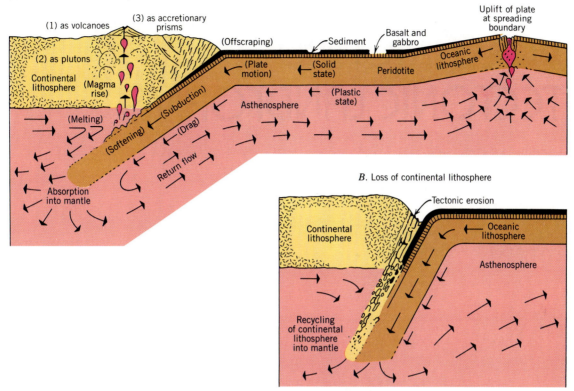

A. Gains to continental lithosphere:

B. Loss of continental lithosphere

FIGURE 12.26 Schematic cross sections to show how the plate tectonic system results in gain or loss of crustal material to the continental lithosphere. The diagram is not drawn to scale.

CHAPTER

VOLCANIC AND TECTONIC LANDFORMS

13

INITIAL LANDFORMS They are landforms resulting directly from volcanic and tectonic activity—volcanoes and fault blocks, for example.

COMPOSITE VOLCANOES Tall, steep-sided composite volcanoes are built of felsic lava and ash. They are highly explosive and can emit searing avalanches of glowing ash.

BASALTIC SHIELD VOLCANOES Volcanoes built of basalt take the form of large, broad domes with gentle side slopes.

TECTONIC LANDFORMS Crustal compression and shortening produce foreland folds; crustal pulling apart and rifting produce rift valleys in ancient continental crust.

LANDFORMS OF FAULTING Crustal dislocation along normal faults produces fault scarps and block mountains.

EARTHQUAKES A major environmental hazard, great earthquakes are generated largely close to active subduction zones and along major transform faults.

I N this chapter we continue to investigate geologic processes, but with emphasis on the detailed crustal features created by those processes, in contrast to the global perspective of enormous lithospheric plates. Volcanic and tectonic processes create a wide variety of both crustal rock masses and landforms.

Landforms are the surface configurations of the land, for example, mountain peaks, cliffs, canyons, and plains. *Geomorphology* is the scientific study of landforms, including their history and processes of origin. Here, we examine landforms produced directly by volcanic and tectonic processes.

In chapters to follow, landforms shaped by processes acting through the medium of the atmosphere and hydrosphere will be the objects of study. These activities of land sculpture can be described collectively as a process of *denudation,* the lowering of the continental surfaces by removal and transportation of mineral matter through the action of running water, waves and currents, glacial ice, and wind.

Initial and Sequential Landforms

The configuration of continental surfaces reflects the balance of power, so to speak, between internal earth forces, acting through volcanic and tectonic processes, and external forces, acting through the agents of denudation. Seen in this perspective, landforms in general fall into two basic categories.

Landforms produced directly by volcanic and tectonic activity are *initial landforms* (Figure 13.1). Initial landforms include volcanoes and lava flows, as well as downdropped rift valleys and elevated fault-block mountains in zones of recent crustal deformation. The energy for lifting molten rock and rigid crustal masses to produce the initial landforms has an internal heat source. As we explained in Chapter 12, this heat is believed to be produced largely by natural radioactivity in rock of the earth's crust and mantle; it is the fundamental energy source for the motions of lithospheric plates.

Landforms shaped by processes and agents of denudation belong to the class of *sequential landforms*, meaning that they follow in sequence after the initial

landforms are created and a crustal mass—a landmass—has been raised to an elevated position. As shown in Figure 13.1, a single uplifted crustal block (an initial landform) is set on by agents of denudation and carved up into a large number of sequential landforms.

Any landscape is really nothing more than the existing stage in a great contest. As lithospheric plates collide or pull apart, the internal earth forces spasmodically elevate parts of the crust to create initial landforms. The external agents persistently keep wearing these masses down and carving them into vast numbers of smaller sequential landforms.

All stages of this struggle can be seen in various parts of the world. High alpine mountains and volcanic chains exist where the internal earth forces have recently dominated. Rolling low plains of the continental interiors reflect the temporary victory of agents of denudation. All intermediate stages can be found. Because the internal earth forces act repeatedly, new initial landforms keep coming into existence as old ones are subdued.

Volcanic Activity

We have identified volcanism as one of the forms of mountain-building. The extrusion of magma builds landforms, and these collectively can accumulate both as volcanoes and as thick lava flows to make imposing mountain ranges. Most of these volcanic chains are within the circum-Pacific belt. Here, subduction of the Pacific, Nazca, Cocos, and Juan de Fuca plates is active. The Cascade Mountains of northern California, Oregon, and Washington represent one such chain. The Aleutian Range of Alaska is another. Important segments of the Andes Mountains in South America and the island of Java in Indonesia consist of volcanoes.

Volcanoes are conical or dome-shaped structures built by the emission of lava and its contained gases from a constricted vent in the earth's surface (Figure 13.2). The magma rises in a narrow, pipelike conduit from a magma reservoir lying beneath. Upon reaching the surface, igneous material may pour out in tonguelike lava

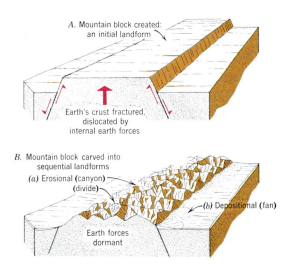

A. Mountain block created: an initial landform

Earth's crust fractured, dislocated by internal earth forces

B. Mountain block carved into sequential landforms
(a) Erosional (canyon)
(divide)
(b) Depositional (fan)

Earth forces dormant

FIGURE 13.1 Initial and sequential landforms. (Drawn by A. N. Strahler.)

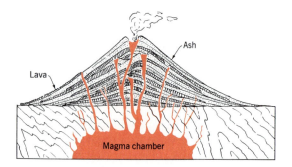

FIGURE 13.2 Idealized cross section of a composite volcanic cone with feeders from magma chamber beneath. (Redrawn from A. N. Strahler, *Planet Earth*, Harper & Row, New York. Copyright © 1972 by Arthur N. Strahler.)

flows or may be ejected under pressure of confined gases as solid fragments. Ejected solid fragments ranging in size from gravel and sand down to fine silt size are collectively called *tephra*. Form and dimensions of a volcano are quite varied, depending on the type of lava and the presence or absence of tephra. The nature of volcanic eruption, whether explosive or quiet, depends on the type of magma.

The important point is that the felsic lavas (rhyolite and andesite) have a high degree of viscosity (property of tackiness, resisting flowage) and hold large amounts of gas under pressure. As a result, these lavas produce explosive eruptions. In contrast, mafic lava (basalt) is highly fluid (low viscosity) and holds little gas, with the result that the eruptions are usually quiet and the lava can travel long distances to spread out in thin layers.

Composite Volcanoes

Tall, steep-sided volcanic cones are produced by felsic lavas. These cones usually steepen toward the summit, where a bowl-shaped depression, the *crater*, is located. In these volcanic eruptions tephra falls on the area surrounding the crater and contributes to the structure of the cone (Figure 13.3). The interlayering of ash layers (tephra) and lava streams produces a *composite volcano*. Volcanic bombs are also included in the tephra. These solidified masses of lava range up to the size of large boulders and fall close to the crater. Very fine volcanic dust rises high into the troposphere and stratosphere, traveling hundreds or thousands of kilometers before settling to the earth's surface (Figure 13.4).

Another important form of emission from the explosive type of volcanoes is a cloud of incandescent gases and fine ash. This intensely hot cloud, sometimes called a "glowing avalanche," travels rapidly down the flank of the volcanic cone, searing everything in its path. On the island of Martinique, in 1902, a glowing cloud issued without warning from Mount Pelée; it swept down on St. Pierre, destroying the city and killing all but two of its 30,000 inhabitants.

Most lofty conical volcanoes, well known for their scenic beauty, are of the composite type. Examples are Mount Hood and Mount St. Helens in the Cascade Range, Fujiyama in Japan, Mount Mayon in the Philippines (Figure 13.3), and Mount Shishaldin in the Aleutian Islands. Most of the world's active composite

FIGURE 13.3 Mount Mayon, in southeastern Luzon, the Philippines, is often considered the world's most nearly perfect composite volcanic cone. An active volcano, its summit rises to an altitude of nearly 2400 m (8000 ft). Volcanic ash, which forms a fresh layer with each eruption, is rapidly furrowed by water erosion and later acquires soil and a forest cover. (Consular General of the Philippines.)

FIGURE 13.4 Mount St. Helens, a composite volcano of the Cascade Range in southwestern Washington, erupted without warning on the morning of May 18, 1980, emitting a great column of condensed steam, heated gases, and ash from the summit crater. Within a few minutes the plume had risen to a height of 20 km (12 mi) and its contents were being carried eastward by stratospheric winds. The eruption was initiated by explosive demolition of the northern portion of the cone, concealed from this viewpoint. (J. G. Rosenbaum, U.S. Geological Survey.)

volcanoes lie above subduction zones. In Chapter 12 we explained the rise of andesitic magmas beneath volcanic arcs of active continental margins and island arcs. One good example is the volcanic arc of Sumatra and Java, lying over the subduction zone between the Australian plate and the Eurasian plate; another is the Aleutian volcanic arc, located above the subduction zone between the Pacific plate and the North American plate.

One of the most catastrophic of natural phenomena is a volcanic explosion so violent that it destroys the entire central portion of the volcano. There remains only a great central depression named a *caldera*. Although some of the upper part of the volcano is blown outward in fragments, most of it subsides into the ground beneath the volcano. Vast quantities of ash and dust are emitted and fill the atmosphere for many hundreds of square kilometers.

Krakatoa, a volcanic island in Indonesia, exploded in 1883, leaving a huge caldera. It is estimated that 75 km^3 (18 mi^3) of rock disappeared during the explosion. Great seismic sea waves generated by the explosion killed many thousands of persons living in low coastal areas of Sumatra and Java.

A classic example of a caldera produced in prehistoric times is Crater Lake, Oregon (Figure 13.5). Mount Mazama, the former volcano, is estimated to have risen 1200 m

FIGURE 13.5 Crater Lake, Oregon, is surrounded by the high, steep wall of a great caldera. Wizard Island (center) is an almost perfect basaltic cinder cone with basalt lava flows; it was built on the floor of the caldera after the major explosive activity had ceased. (Arthur N. Strahler.)

(4000 ft) higher than the present caldera rim. The event occurred about 6600 years ago.

Flood Basalts and Shield Volcanoes

Geologists postulate that at various points beneath the lithosphere there occur *mantle plumes*, which are isolated columns of heated rock rising slowly within the asthenosphere. Directly above a mantle plume, crustal basalt can be heated to the point of melting and produce a magma pocket. The site of magma is called a *hot spot*. Magma of basaltic composition makes its way through the overlying lithosphere to emerge at the surface as lava.

Where a mantle plume lies beneath a continental lithospheric plate, the hot spot may generate enormous volumes of basaltic lava that accumulate layer upon layer. The basalt may ultimately attain a thickness of thousands of meters and may cover thousands of square kilometers. These accumulations are called *flood basalts*. An important example is found in the Columbia Plateau region of southeastern Washington, northeastern Oregon, and westernmost Idaho; basalts of Cenozoic age cover an area of about 130,000 km^2 (50,000 mi^2), about the same area as the state of New York. Individual basalt flows, exposed

in the walls of river gorges, are expressed as cliffs, in which vertical joint columns are conspicuous (Figure 13.6).

Hot spots also form above mantle plumes in the oceanic lithosphere. The emerging basalt builds a class of volcanoes known as *shield volcanoes*. These are constructed on the deep ocean floor, far from the plate boundaries, and may be built high enough to rise above sea level as volcanic islands. As a lithospheric plate drifts slowly over a mantle plume beneath, a succession of shield volcanoes is formed. Thus a chain of volcanic islands comes into existence. Several chains of volcanic islands exist in the Pacific Ocean basin. Best known of the island chains is the Hawaiian group.

A few basaltic volcanoes also occur along the mid-oceanic ridge, where seafloor spreading is in progress. Perhaps the outstanding example is Iceland, in the North Atlantic Ocean. Iceland is constructed entirely of basalt flows superimposed on other basaltic rocks in the form of dikes and sills that entered the spreading rift at deeper levels. Mount Hekla, an active volcano on Iceland, is a shield volcano somewhat similar to those of Hawaii. Farther south along the Mid-Atlantic Ridge are other islands consisting of basaltic volcanoes: the Azores, Ascension Island, and Tristan da Cunha.

Shield volcanoes of the Hawaiian Islands

FIGURE 13.6 Basalt lava flows exposed in cliffs bordering the Columbia River in Washington. Columnar jointing of the massive basalt is conspicuous in the arid climate. (Arthur N. Strahler.)

FIGURE 13.7 Basaltic shield volcanoes of Hawaii. At lower left, the Halemaumau fire pit formed in the floor of the central depression of Kilauea volcano. On the distant skyline is the snow-capped summit of Mauna Kea volcano, its elevation over 4000 m (13,000 ft). (Werner Stoy/Camera Hawaii.)

FIGURE 13.8 A fire fountain of molten basalt lava in the floor of an active fire pit east of the summit of Kilauea volcano. The fountain rises to a height of about 75 m (250 ft). A thin layer of recently congealed lava covers the magma pool. (John S. Shelton.)

deep (Figure 13.7). These large depressions are a type of caldera produced by subsidence accompanying the removal of molten lava from beneath. Molten basalt is actually seen in the floors of deep pit craters that occur on the floor of the central depression or elsewhere over the surface of the lava dome (Figure 13.8). Most lava flows issue from fissures (long, gaping cracks) on the sides of the volcano.

Cinder Cones

Associated with flood basalts, shield volcanoes, and scattered occurrences of basalt lava flows is a small volcano known as a *cinder cone* (Figures 13.9 and 13.10). It forms where frothy basalt magma is ejected under high pressure from a narrow vent. The rain of tephra accumulates around the vent to form a circular hill with a central crater. Cinder cones rarely grow to heights of more than a few hundred meters. An exceptionally fine example of a cinder cone is Wizard Island, built upon the floor of Crater Lake, long after the caldera was formed (Figure 13.5).

are characterized by gently rising, smooth slopes that flatten near the top, producing a broad-topped volcano (Figure 13.7). Domes on the island of Hawaii rise to summit elevations 4000 m (13,000 ft) above sea level. Including the basal portion lying below sea level, they are more than twice that high. In width they range from 16 to 80 km (10 to 50 mi) at sea level and up to 160 km (100 mi) wide at the submerged base. The basalt lava of the Hawaiian volcanoes is highly fluid and travels far down the gentle slopes.

Lava domes have a wide, steep-sided central depression that may be 3 km (2 mi) or more wide and several hundred meters

The surfaces of cinder cones and basaltic lava flows remain barren and sterile for long periods after their formation. Recent basalt lava surfaces are extremely rough and difficult to traverse (Figure 13.9); the Spaniards who encountered such terrain in the southwestern United States named it "malpais" (bad ground).

FIGURE 13.9 A young cinder cone surrounded by rough-surfaced basalt lava flows. Lava Beds National Monument, northern California. (Alan H. Strahler.)

FIGURE 13.10 A cinder cone with its lava flows has dammed a valley, making a lake. Farther downvalley, another lava flow has made a second dam. (Drawn by W. M. Davis.)

Volcanic Eruption as an Environmental Hazard

The eruptions of volcanoes and lava flows are environmental hazards of the severest sort, often taking a heavy toll of plant and animal life and devastating human habitations. What natural phenomenon can compare with the Mount Pelée disaster in which thousands of lives were snuffed out in seconds? Perhaps only an earthquake or storm surge of a tropical cyclone is equally disastrous.

Wholesale loss of life and destruction of towns and cities are frequent in the history of peoples who live near active volcanoes. Loss occurs principally from sweeping clouds of incandescent gases that descend the volcano slopes like great avalanches, from lava flows whose relentless advance engulfs whole cities, from the descent of showers of ash, cinders, and bombs, and from violent earthquakes associated with volcanic activity. For habitations along low-lying coasts there is the additional peril of great seismic sea waves, generated elsewhere by explosive destruction of volcanoes.

In 1985, an explosive eruption of Ruiz Volcano in the Colombian Andes caused the rapid melting of ice and snow in the summit area. Mixing with volcanic ash, the water formed a variety of mudflow known as a *lahar*. Rushing downslope at speeds up to 145 km (90 mi) per hour, the lahar became channeled into a valley on the lower slopes, where it engulfed a town and killed more than 20,000 persons.

Despite their potential for destructive activity, volcanoes are a valuable natural resource in terms of recreation and tourism. Few landscapes can rival in beauty the mountainous landscapes of volcanic origin. National parks have been made of Mount Rainier, Mount Lassen, and Crater Lake in the Cascade Range, a mountain mass largely of volcanic construction. Hawaii Volcanoes National Park recognizes the natural beauty of Mauna Loa and Kilauea; their displays of molten lava are a living textbook of igneous processes.

Geothermal Energy Sources

Geothermal energy is energy in the form of sensible heat that originates within the earth's crust and makes its way to the surface by conduction. Heat may be conducted upward through solid rock or carried up by circulating ground water that is heated at depth and makes its return to the surface. Concentrated geothermal heat sources are usually associated with igneous activity, but there also exist deep zones of heated rock and ground water that are not directly related to igneous activity. We will examine briefly some forms of intensified internal heat that are potentially useful for industrial purposes.

From observations made in deep mines and boreholes, we know that the temperature of rock increases steadily with depth. Although the rate of increase falls off quite rapidly with increasing depth, temperatures attain very high values in the upper mantle, where rock is close to its melting point. As noted in Chapter 12, heat within the earth's crust and mantle is thought to be produced largely by radioactive decay. Slow as this internal heat production is, it scarcely diminishes with time and the basic energy resource can be regarded as limitless.

It might seem simple enough to obtain all our energy needs by drilling deep holes at any desired location into the crust and letting the hot rock turn injected fresh water into steam, which we could use to generate electricity as our primary energy resource. Unfortunately, at the depths usually required to furnish the needed heat intensity, crustal rock tends to close any cavity or opening by rupture and slow flowage; this phenomenon would either prevent the holes from being drilled, or would close them in short order. Generally, then we must look for *geothermal localities*, where special conditions have caused hot rock and hot ground water to lie within striking distance of conventional drilling methods.

At widely separated geothermal localities over the globe, ground water reaches the surface in *hot springs* at temperatures not far below the boiling point of water, which is 100°C (212°F) at sea-level atmospheric pressure (Figure 13.11). At some of these same places, jetlike emissions of steam and hot water occur at intervals from small vents; these are *geysers* (Figure 13.11). The water that emerges from hot springs and geysers is largely ground water that has been heated in contact with hot rock and forced to the surface. In other words, this water is recycled surface water. Little, if any, is water that was originally held in rising bodies of magma.

Natural hot water and steam localities were the first type of geothermal energy source to be developed and at present account for nearly all production of electrical power. Wells are drilled to tap the hot water, which flashes into steam under reduced atmospheric pressure as it reaches the surface. The steam is separated from

the water in large towers and fed into generating turbines to produce electricity (Figure 13.12). The hot water is usually released into surface stream flow, where it may create a pollution problem. The larger steam fields have sufficient energy to generate at least 15 megawatts of electric power, and a few can generate 200 megawatts or more.

Much greater energy sources than those we have just described lie in deeper zones of hot ground water, but these must be tapped by deep drilling. One region of deep, hot ground water, currently under development, is beneath the Imperial Valley of southern California. An area of 500 km^2 is involved, and extends over the border into Mexico in the Mexicali Valley. This region is tectonically active and has been interpreted as a zone of crustal spreading in which the lithospheric plate is being fractured. Rising basalt magma, which is

FIGURE 13.11 Geothermal activity in Yellowstone National Park, Wyoming. (left) Mammoth Hot Springs. Terraces of siliceous sinter hold steaming pools of hot water. (right) Old Faithful Geyser. (Arthur N. Strahler.)

FIGURE 13.12 An electricity generating plant at The Geysers, California. Steam pipes in the foreground lead to the plant. After use in generating turbines, the steam is condensed in large cylindrical towers. (Pacific Gas & Electric Company.)

found elsewhere in active spreading zones such as Iceland in the North Atlantic, may be responsible for the geothermal condition, but this interpretation is speculative. Test wells show that a large reservoir of extremely hot ground water, 260° to 370°C (500° to 700°F), is present here. This water readily flashes into steam when penetrated by a drill hole. Steam pressure forces both steam and hot water to the surface, much like the action of a coffee percolator. At East Mesa in the Imperial Valley of California this resource has already been developed, and produced 9 megawatts of electricity in 1986. The prospect of a large power development beneath Imperial Valley looks very good, and the salinity of the hot water is quite low. If fully developed, the Imperial Valley geothermal field could probably produce as much as 20 megawatts of electricity. If so, the needs of southern California could be fully met by this source alone. Heat remaining in the water after it has been used to generate electricity could be used to distill the waste water and produce a substantial yield of irrigation water, a valuable commodity in this desert agricultural region.

In certain areas, the intrusion of magma has been sufficiently recent that solid igneous rock of a batholith is still very hot in a depth range of perhaps 2 to 5 km (1 to 3 mi). At this depth, the rock is strongly compressed and contains little, if any, ground water. Rock in this zone may be as

hot as 300°C (600°F) and could supply an enormous quantity of heat energy. The planned development of this resource includes drilling into the hot zone and then shattering the surrounding rock by hydrofracture—a method using water under pressure that is widely used in petroleum development. Surface water would be pumped down one well into the fracture zone and heated water pumped up another well.

Experimental holes in northern New Mexico were drilled in the middle 1970s, and the dry granite rock at the bottom was fractured by water injected under pressure. This successful experiment was followed by flow tests in 1986, leading to increased optimism that deeper zones of hot rock can be developed. Promising locations have been found in Montana and Idaho. The potential for electrical power generation from deep hot-rock areas is believed to be many times greater than that for hot-water areas.

Landforms of Tectonic Activity

Our introduction to global plate tectonics in Chapter 12 brought out the essential distinction between two basically different expressions of tectonic activity. Along converging lithospheric plate boundaries, tectonic activity is basically that of compression, illustrated schematically in

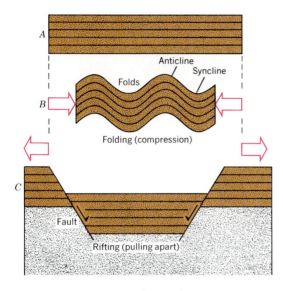

FIGURE 13.13 Two basic forms of tectonic activity that produce initial landforms. Crustal compression results in folding, and pulling apart (rifting) results in faulting.

Foreland Fold Belts

When continental collision begins to take place, wedges of strata of a passive continental margin come under strong forces of compression. The strata, which were originally more or less flat-lying, experience *folding*, as shown in Figure 13.13. The wavelike undulations imposed on the strata consist of alternating archlike upfolds, called *anticlines*, and troughlike downfolds, called *synclines*. Thus the initial landform associated with an anticline is a broadly rounded mountain ridge; the landform corresponding to a syncline is an elongate, open valley.

Two well-known examples of open folds of comparatively young geologic age have long attracted the interest of geographers. One of these is the Jura Mountains of France and Switzerland. Figure 13.14 is a block diagram of a small portion of that fold belt. The strata are mostly limestone layers of Jurassic age and were capable of being deformed by bending with little brittle fracturing. Folding occurred in late Cenozoic (Miocene) time. Notice that each mountain crest is associated with the axis of an anticline, whereas each valley lies over the axis of a syncline. Some of the anticlinal arches have been partially removed by erosion processes. The rock structure can be seen clearly in the walls of the winding gorge of a major river that crosses the area. The Jura folds lie just to the north of the main collision orogen of the Alps. In this respect they are called *foreland folds*.

Our second example of a belt of foreland folds produced during continental collision is the Zagros Mountains of southwestern

Figure 13.13. In subduction zones, sedimentary layers of the ocean floor are subject to compression within a trench as the descending plate forces them against the fixed plate. In continental collision, compression is of the severest kind.

In zones of rifting of continental plates, explained in Chapter 12, the brittle continental crust is pulled apart and yields by faulting. In the simple model shown in Figure 13.13, rifting is expressed in a pair of opposite-facing faults. The crustal block between them moves down to form a depressed area.

FIGURE 13.14 Block diagram of anticlinal ridges (A) and synclinal valleys (S). Jura Mountains, France and Switzerland. (Drawn by Erwin Raisz.)

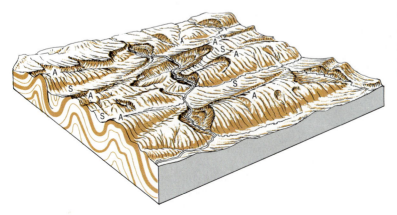

Iran. Seen from an orbiting space vehicle, the Zagros folds resemble an army of caterpillars crawling in parallel tracks (Figure 13.15). They were formed during late Cenozoic time when the Arabian plate smashed into the Eurasian plate, closing an ocean basin that formerly separated the two continental masses. Flat-lying strata on the passive margin of the Arabian plate were crumpled into open folds over a belt more than 200 km (125 mi) wide.

Faults and Fault Landforms

A fault in the brittle rocks of the earth's crust is a result of sudden yielding under unequal stresses. Faulting is accompanied by a displacement along the plane of breakage, or *fault plane*. Faults are often of great horizontal extent, so the surface trace, or *fault line*, can sometimes be followed along the ground for many kilometers. Most major faults extend down into the crust for at least several kilometers.

Faulting occurs in sudden slippage movements that generate earthquakes. A particular fault movement may result in a slippage of as little as a centimeter or as much as 15 m (50 ft). Successive movements may occur many years or decades apart, even several centuries apart, but accumulate into total displacements of hundreds or thousands of meters. In some places clearly recognizable sedimentary rock layers are offset on opposite sides of a fault, and the amount of displacement can be accurately measured.

One common type of fault associated with crustal rifting is the *normal fault*. The plane of slippage, or fault plane, is steep or nearly vertical. One side is raised or upthrown relative to the other, which is downthrown. A normal fault results in a steep, straight, clifflike feature called a *fault scarp* (Figure 13.16). Fault scarps range in

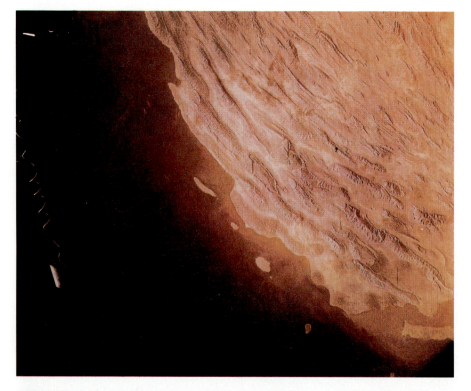

FIGURE 13.15 The Zagros Mountains of Iran, photographed by astronauts of the Gemini XII orbiting space vehicle in 1966. The view is toward the northwest and shows a portion of the Persian Gulf. The elongate ridges are anticlines, partly eroded by streams; they were elevated by folding which began late in the Cenozoic Era. (NASA S66—6383).

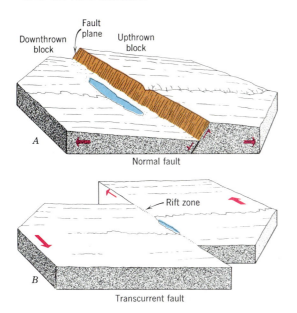

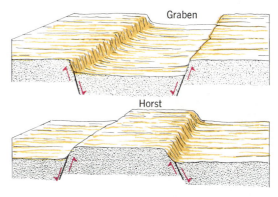

FIGURE 13.18 Graben and horst. (A. N. Strahler.)

FIGURE 13.16 (A) Normal fault. (B) Transcurrent fault. (Drawn by A. N. Strahler.)

height from a few meters to a few thousand meters (Figure 13.17). Their length is measurable in kilometers; often they attain lengths of 300 km (200 mi).

Normal faults rarely are isolated features. More often they occur in multiple arrangements, commonly as a parallel series of faults. This gives rise to a grain or pattern of rock structure and topography. A narrow block dropped down between two normal faults is a *graben* (Figure 13.18). A narrow block elevated between two normal faults is a *horst*. Grabens make conspicuous topographic trenches, with straight, parallel walls. Horsts make blocklike plateaus or mountains, often with a flat top, but steep, straight sides.

In rifted zones of the continents, regions where normal faulting is on a grand scale, mountain masses called *block mountains* are produced. These faulted mountain blocks can be described as tilted and lifted (Figure 13.19). A tilted block has one steep face, the fault scarp, and one gently sloping side. A lifted block, which is a type of horst, is bounded by steep slopes on both sides.

Recall that lithospheric plates slide past one another along major transform faults, and that these features comprise one type of lithospheric boundary. Long before the principles of plate tectonics became known, geologists referred to such faults as

FIGURE 13.17 This fault scarp was formed during the Hebgen Lake, Montana, earthquake of August 1959. In a single instant, a displacement of 6 m (19 ft) took place on a normal fault. (J. R. Stacy, U.S. Geological Survey.)

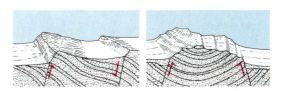

FIGURE 13.19 Fault block mountains may be of tilted type (left) or lifted type (right). (Drawn by W. M. Davis.)

FIGURE 13.20 The San Andreas Fault runs straight as an arrow across the Carrizo Plain in Kern County, California. You are looking in a northwesterly direction; the land on the left side of the fault line is moving away from you. Can you spot evidence to prove this statement? (John S. Shelton.)

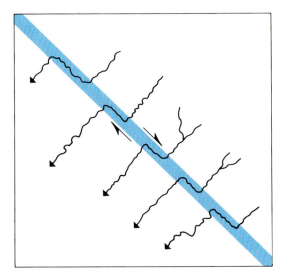

FIGURE 13.21 Schematic map of streams offset by long-continued movement along the San Andreas Fault. (From A. N. Strahler, *Principles of Earth Science*, Harper & Row, New York. Copyright © 1976 by Arthur N. Strahler.)

transcurrent faults (Figure 13.16*B*). In a transcurrent fault the movement is predominantly in a horizontal direction. No scarp, or a very low one at most, results. Instead, only a thin fault line is traceable across the surface. In some places a narrow trench, or rift, marks the fault.

Best known of the active transcurrent faults is the great San Andreas Fault, which can be followed for a distance of about 1000 km (600 mi) from the Gulf of California to well north of the San Francisco area. The San Andreas Fault has been interpreted as a transform fault, and forms the active boundary between the Pacific plate and the North American plate (see Figure 12.14). The Pacific plate is moving toward the northwest, which means that a great portion of the state of California and all of Lower (Baja) California

is moving bodily northwest with respect to the North American mainland.

Throughout many kilometers of its length, the San Andreas Fault appears as a straight, narrow scar, which in places is a trenchlike feature and elsewhere a low scarp (Figure 13.20). In some places, a stream valley makes an abrupt jog when crossing the fault line, showing that many meters of movement have occurred in fairly recent time (Figure 13.21).

Environmental Aspects of Faults and Block Mountains

Faults are of environmental and economic significance in several ways. Fault planes are usually zones along which the rock has been pulverized, or at least considerably fractured. This breakage has the effect of permitting ore-forming solutions to rise along fault planes. Many important ore deposits lie in fault planes or in rocks that faults have broken across.

Another related phenomenon is the easy rise of ground water along fault planes. Springs, both cold and hot, are commonly situated along fault lines. They occur along the base of a fault-block mountain. Examples are Arrowhead Springs along the base of the San Bernardino Range and Palm Springs along the foot of the San Jacinto Mountains, both in southern California.

Petroleum, too, finds its way along fault planes where the rocks have been crushed or it becomes trapped in porous beds that have been faulted against impervious shale beds. Some of the most intensive searches for oil center about areas of faulted sedimentary strata because of the great production that has come from reservoirs of this type.

Fault scarps can form imposing topographic barriers across which it is difficult to build roads and railroads. The great Hurricane Ledge of southern Utah is a feature of this type, in places a steep wall 800 m (2500 ft) high. Grabens may be so large as to form broad lowlands. An illustration is the Rhine graben of West Germany. Here a belt of rich agricultural land 30 km (20 mi) wide and 240 km (150 mi) long lies between the Vosges and Black Forest ranges, both of which are block

mountains faulted up in contrast with the downdropped Rhine graben block.

The Rift Valley System of East Africa

The rifting of continental lithosphere that is the very first stage in splitting apart of a continent to form a new ocean basin is beautifully illustrated by the East African rift valley system. It has attracted the attention of geologists since the early 1900s. They gave the name *rift valley* to what is basically a graben, but with a more complex history that includes the building of volcanoes in the graben floor. Figure 13.22 is a sketch map of the East African rift

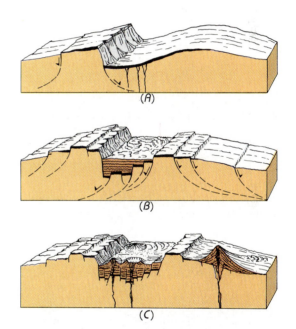

FIGURE 13.23 Development of a typical rift valley in East Africa. The diagrams are schematic and combine elements found in several localities. Width of the area shown is about 150 km. (*A*) Late Miocene and early Pliocene. Normal faulting has produced a tilted fault block on the left. Crust at right is deformed into a broad monocline with a cap of lava. (*B*) Late Pliocene. Renewed normal faulting has broken the valley floor into narrow blocks and raised the eastern side. The rift valley is now a graben structure. Lava flows have filled the valley floor. (*C*) Pleistocene and Holocene. After another episode of minor faulting, extrusive activity has built volcanoes in the rift valley and on the flank of the uplift. (Drawn by A. N. Strahler. Based on data of B. H. Baker, in *East Africa Rift System*, 1965, UNESCO Seminar, University College, Nairobi, p. 82.)

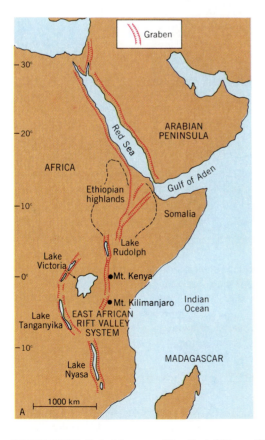

FIGURE 13.22 A sketch map of the East African rift valley system and the Red Sea to the north.

FIGURE 13.24 A portion of Gregory's Rift in southern Kenya displayed in false-color remote-sensing imagery. The rift floor, about 40 km (25 mi) wide, is bounded by a series of fault steps leading up to forested highlands on either side. The red color, indicating green foliage, becomes deeper with increasing elevation. Fault scarps are most sharply defined on the eastern side of the valley. In the valley floor lies Lake Naivasha (black); south of it are two volcanoes with prominent circular craters: Longonot (small) and Susuwa (large). (NASA ERTS 2188-07055, 29 July 1975. Reproduced by permission of EOSAT.)

valley system, which is a full 3000 km (1900 mi) long and extends from the Red Sea southward to the Zambezi River. The system consists of a number of grabenlike troughs, each a separate rift valley ranging in width from 30 to 60 km. As geologists had noted in earlier field surveys of this system, the rift valleys are like keystone

FIGURE 13.25 The rift valley wall in Ethiopia. Multiple fault scarps give the landscape a stepped appearance. (George Gerster/Photo Researchers, Inc.)

blocks of a masonry arch that have slipped down between neighboring blocks because the arch has spread apart somewhat (Figure 13.23). Thus, the floors of the rift valleys are above the elevation of most of the African continental surface, even though some of the valley floors are occupied by long, deep lakes and by major rivers (Figure 13.24). The side of the rift valleys may consist of multiple fault steps (Figure 13.25).

The rift-valley system consists of a number of domelike swells in the crust, the highest of which forms the Ethiopian Highlands on the north. Basalt lavas have risen from fissures in the floors of the rift valleys and from the flanks of the domes. Sediments, derived from the high plateaus that form the flanks of troughs, make thick fills in the floors of the valleys. Lake Victoria is flanked by two rift valleys, which join south of the lake. A single rift valley extends southward from this junction.

Two great composite volcanoes have been built close to the rift valley east of Lake Victoria. One is Mount Kilimanjaro, whose summit rises to over 6300 m (19,300 ft). The other, Mount Kenya, is only a little lower; it lies right on the equator. Geologists think the acidic magma that built these two great volcanoes came from thinned crust, domed upward by highly heated mantle rock that moved upward as the continental lithosphere was pulled apart.

Earthquakes

Everyone has read many news accounts about disastrous earthquakes and has seen pictures of their destructive effects (Figure 13.26). Californians know about severe earthquakes from firsthand experience, but many other areas in North America have also experienced earthquakes, and a few of these have been severe. An *earthquake* is a

FIGURE 13.26 Earthquake devastation in the downtown area of Managua, Nicaragua, December 1972. The number of persons killed in this disaster has been estimated as between 4000 and 6000, and property damage between $400 and $600 million. (Don Goode/Black Star.)

motion of the ground surface, ranging from a faint tremor to a wild motion capable of shaking buildings apart.

The earthquake is a form of energy of wave motion transmitted through the surface layer of the earth in widening circles from a point of sudden energy release—the focus. Like ripples produced when a pebble is thrown into a quiet pond, these *seismic waves* travel outward in all directions, gradually losing energy.

Earthquakes are produced by sudden movements along faults; commonly these are normal faults or transcurrent faults. The devastating San Franciso earthquake of 1906 resulted from slippage along the San Andreas Fault, which also passes about 60 km (40 mi) inland of the Los Angeles metropolitan area. Associated with the San Andreas Fault are several important transcurrent parallel and branching faults, all capable of generating severe earthquakes.

We will not go into the details of mechanics of faults and how they produce earthquakes. It must be enough to say that rock on both sides of the fault is slowly bent over many years as tectonic forces are applied in the movement of large crustal masses. Energy accumulates in the bent rock, just as it does in a bent crossbow. When a critical point is reached, the strain is relieved by slippage on the fault and a large quantity of energy is instantaneously released in the form of seismic waves. Slow bending of the rock takes place over many

decades. Its release then causes offsetting of features that formerly crossed the fault in straight lines, for example, a roadway or fence (Figure 13.27). Faults of this type can also show a slow, steady displacement known as fault creep, which tends to reduce the accumulation of stored energy.

The Richter scale of earthquake magnitudes was devised in 1935 by the distinguished seismologist Charles F. Richter to indicate the quantity of energy released by a single earthquake. Scale numbers range from 0 to over 9.0, but there is no

FIGURE 13.27 During the San Francisco earthquake of 1906, this fence was offset 2.4 m (8 ft) by lateral movement along the San Andreas Fault near Woodville, California. (G. K. Gilbert, U.S. Geological Survey.)

upper limit except for nature's own limit of energy release. A value of 9.5 is the largest observed to date—the Chilean earthquake of 1960. The great San Francisco earthquake of 1906 is now rated as magnitude 7.9.

Earthquakes and Plate Tectonics

Seismic activity, the repeated occurrence of earthquakes, shows a close geographic relationship to lithospheric plate boundaries (Figure 13.28). The greatest intensity of seismic activity is found along converging plate boundaries where oceanic plates are undergoing subduction. Strong pressures build up at the downslanting contact of the two plates, and these are relieved by sudden fault slippages that generate earthquakes of large magnitude. This mechanism explains the great earthquakes experienced in Japan, Alaska, Chile, and other narrow zones close to trenches and volcanic arcs of the Pacific Ocean basin.

Transform boundaries that cut through the continental lithosphere are sites of intense seismic activity, with moderate to strong earthquakes. The most familiar example is the San Andreas Fault, which forms the transform boundary between the American plate and the Pacific plate in California.

Spreading boundaries are a third class of narrow zones of seismic activity related to lithospheric plates. Most of these boundaries are identified with the mid-oceanic ridge and its branches. Earthquakes are generated both along the ridge axis and on the transform faults that connect offset ends of the ridge, but they are mostly small earthquakes set off at shallow depths.

Earthquakes also occur at scattered locations over the continental plates, far from active plate boundaries. In many cases, no active fault is visible and the geologic cause of the earthquake is obscure. Some large earthquakes of southern Asia are probably related to the continental suture between the Eurasian plate and the Arabian and Austral-Indian plates.

Earthquake as an Environmental Hazard

One of the great earthquakes of recent times was the Good Friday earthquake of

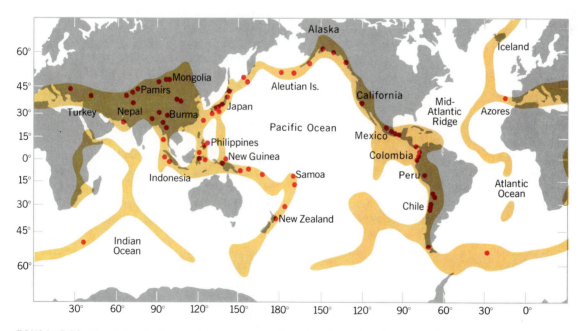

FIGURE 13.28 World distribution of shallow-focus earthquakes. Dots show locations of major earthquakes, measuring 7.9 or over on the Richter scale. The colored zones are principal areas of abundant earthquakes. (Generalized from data of C. F. Richter, 1958, *Elementary Seismology*, W. H. Freeman, San Francisco. From A. N. Strahler, *Planet Earth*, Harper & Row, New York. Copyright © 1972 by Arthur N. Strahler.)

FIGURE 13.29 Ground shaking during the Good Friday earthquake of March 27, 1964, at Anchorage, Alaska, set off large-scale slumping of unconsolidated sediments beneath these homes. (Gene Daniels/Black Star.)

March 27, 1964, with the surface point of origin located about 120 km (75 mi) from Anchorage, Alaska. Its magnitude was 9.2 on the Richter scale, approaching the maximum known. Of particular interest in connection with earthquakes as environmental hazards are the secondary effects. At Anchorage most of the damage was from secondary effects, because small buildings of modern wooden frame construction may experience little damage when built on solid rock. Damage was largely from earth movements in weak clays underlying one part of the city (Figure 13.29). These clays developed liquid properties on being shaken and allowed great segments of ground to subside and to pull apart in a succession of steps, thus tilting and rending houses. Other secondary effects were from the rise of water level and landward movement of large waves, which destroyed shipping and low-lying structures.

The great Anchorage earthquake illustrates yet another major environmental hazard often associated with a major earthquake centered on a subduction plate boundary. It is the *seismic sea wave,* or *tsunami* as it is known to the Japanese. A train of these water waves is often generated in the ocean at a point near the earthquake source by a sudden movement of the seafloor. The waves travel over the ocean in ever-widening circles, but they are not perceptible at sea in deep water. When a wave arrives at a distant coastline, the effect is to cause a rise of water level. Wind-driven waves, superimposed on the heightened water level, allow the surf to

attack places inland that are normally above the reach of waves. For example, the particularly destructive seismic sea wave of 1933 in the Pacific Ocean caused waves to attack ground as high as 9 m (30 ft) above normal tide level, causing widespread destruction and many deaths by drowning in low-lying coastal areas. It is thought that coastal flooding that occurred in Japan in 1703, with an estimated life loss of 100,000 persons, may have been caused by seismic sea waves.

Predicting Earthquakes along the San Andreas Fault

Eight decades have passed since the great San Francisco earthquake of 1906 was generated by movement on the San Andreas Fault. Since that event, this sector of the fault has been locked; that is, devoid of sudden slips (Figure 13.30). The two lithospheric plates that meet along the fault have in the meantime been moving steadily with respect to one another and a large amount of unrelieved strain has already accumulated in the crustal rock on either side of the fault. While the time of occurrence of another major earthquake cannot now be predicted within a time window of even one or two decades, it is inevitable. As each decade passes, the probability of that event becomes greater.

The last major slip on the San Andreas Fault in the section closest to Los Angeles County occurred in 1857; it was the great Fort Tejon earthquake with a Richter magnitude now estimated to have been 8.3. Studies of the past history of the San Andreas Fault indicate that an earthquake like that of 1857 has a recurrence interval from 100 to 230 years, with an average of 140 years. Since 1979, geophysicists monitoring this locked sector of the fault have detected a number of physical changes that suggest increased possibility for a major earthquake.

Another locked sector of the San Andreas Fault system lies in the region of the San Gorgonio Pass and extends southeastward into the Imperial Valley. Here, the last great earthquake occurred some 250 to 300 years ago, so an earthquake of magnitude 8.0 or larger is considered even more imminent than for

FIGURE 13.30 A sketch map of the San Andreas Fault showing the locked sections alternating with active sections of frequent small earthquakes and slow creeping motion. (Based on data of the U.S. Geological survey.)

the Fort Tejon locked section. A recent estimate places at about 50% the likelihood that a very large earthquake will occur within the next 30 years somewhere along the southern California portion of the San Andreas Fault.

For residents of the Los Angeles area, an additional serious threat lies in the large number of active faults close at hand. Movements on these local faults have produced more than 40 damaging earthquakes since 1800, including the Long Beach earthquake of the 1920s and the San Fernando earthquake of 1971. The latter measured 6.6 on the Richter scale and produced severe structural damage near the earthquake center (epicenter); the total damage was set at $500 million. In 1987, an earthquake of magnitude 6.1 struck the vicinity of Pasadena and Whittier, located within about 20 km (12 mi) of downtown Los Angeles. Known as the Whittier Narrows earthquake, it was generated along

a local fault system that had not previously shown significant seismic activity. The brief but intense primary shock and aftershocks that followed damaged beyond repair many older structures built of unreinforced brick masonry. Although a slip along the San Andreas Fault, some 50 km (30 mi) to the north of this densely populated area, will release an enormously larger quantity of energy, its destructive effects in downtown Los Angeles will be somewhat moderated by the greater travel distance. Actually, the ground-shaking violence of a local earthquake of intensity 6.6 may be just as great as that from one of magnitude 8.3 on the distant San Andreas Fault, but we should not overlook the much longer duration of the San Andreas shaking and the enormously greater inhabited area it will reach.

The Earth's Crust in Review

In three chapters we have made a survey of the composition, structure, and geologic activity of the earth's crust. One underlying concept is that of a rock cycle continuously in operation over some three billion years or more of geologic time. The recycling of crustal mineral matter has taken place through the mechanism of plate tectonics. The continents have gradually grown in extent through accumulation of felsic rock produced in subduction zones, and in this way, the first-order relief features of the globe have come into existence.

Active tectonic and volcanic belts are dominant environmental features of the continents. Cutting across the belts of prevailing winds, high mountain chains induce orographic precipitation and, at the same time, create rainshadow deserts. Landforms produced by volcanic activity are important in these new mountain chains. Landforms produced by faulting are important in areas where continental rifting is occurring. These activities also pose environmental hazards for those human populations living nearby.

Another great class of landforms remains to be investigated—those formed by agents of erosion through processes of interaction of the land surface with the hydrosphere and atmosphere. These processes are the subject of the next five chapters.

CHAPTER

WEATHERING AND MASS WASTING

14

DENUDATION   Weathering and mass wasting work with the fluid agents to carry out denudation of the continental surfaces.

REGOLITH   The wasting of hillslopes involves transformation of bedrock to soft regolith. This transformation process requires both physical breakup of rock and chemical change of minerals.

FROST ACTION   The evidence of intense frost action is striking on arctic land surfaces; in contrast, chemical decay of rocks is dominant in warm, moist climates.

LIMESTONE CAVERNS   Besides serving as a natural resource, ground water excavates limestone cavern systems of scenic value.

MASS WASTING   Under the influence of gravity, downhill movements of regolith and rock range from extremely slow creep of soil to catastrophic flows and slides of huge masses of rock.

INDUCED MASS WASTING   Human activity, particularly urban expansion into unstable mountainsides, aggravates disastrous earthflows and debris floods.

SCARIFICATION   Deep scarring of the land during the extraction of ores and coal is a severe form of environmental impact; it has many undesirable side effects.

NOW that we have completed a study of the earth's crust, its mineral composition, and its tectonic and volcanic landforms, we can focus on the shallow life layer itself. At this interface, the externally acting, solar-driven energy systems of the atmosphere and oceans mesh with the internally driven geologic system that has created and raised the continental masses. The geologic processes have caused varied rock types to become exposed to the surface environment. Our study of the interaction of these two great planetary systems began in Chapter 11 with a study of the chemical alteration of rock and the production of sediment. This is a process essential to the rock transformation cycle and to growth of the continental crust. We will now look at the processes that shape the surface of the lands.

Weathering and Mass Wasting

Weathering is the general term applied to the combined action of all processes causing rock to be disintegrated physically and decomposed chemically because of exposure at or near the earth's surface. The products of rock weathering tend to accumulate in a soft surface layer, called *regolith* (Figure 14.1). The regolith grades downward into solid, unaltered rock, known simply as *bedrock*. Regolith, in turn, provides the source for sediment consisting of detached mineral particles transported and deposited in a fluid medium, which may be water, air, or glacial ice. Both regolith and sediment comprise parent materials for the formation of the true soil, a surface layer capable of supporting the growth of plants. (See Chapter 19 and Figure 19.1).

In Chapter 11, processes of chemical weathering (as mineral alteration) were explained to provide an understanding of the production of various kinds of sediments. Weathering also plays a major role in denudation. Disintegration and decomposition of various kinds of hard bedrock greatly facilitate the erosion of the land surface by running water. Besides this function, weathering leads to a number of distinctive landforms, which we will describe in this chapter.

In addition to the chemical and physical changes in mineral matter that result from weathering, there is a continued agitation in the soil and regolith because of changes in temperature and water content. These daily and seasonal rhythms of change continue endlessly.

The spontaneous downhill movement of soil, regolith, and rock under the influence of gravity (but without the dynamic action of moving fluids) is included under the general term *mass wasting*. Important here is the role of gravity as a pervasive environmental factor. All processes of the life layer take place in the earth's gravity field, and all particles of matter tend to respond to gravity. Movement to lower levels takes place when the internal strength of a mass of soil or rock declines to a critical point below which the force of gravity cannot be resisted. This failure of strength under the ever-present force of gravity takes many forms and scales, and we will see that human activity causes or aggravates several forms of mass wasting.

The Wasting of Slopes

As used in geomorphology, the term *slope* designates some small element or area of the land surface that is inclined from the horizontal. Thus, we speak of "mountain slopes," "hillslopes," or "valley-side slopes" with reference to the inclined ground surfaces extending from divides and summits down to valley bottoms.

Slopes guide the flow of surface water under the influence of gravity. Slopes fit together to form drainage systems in which surface-water flow converges into stream channels. These, in turn, conduct the water and rock waste to the oceans to complete the hydrologic cycle. Natural processes have

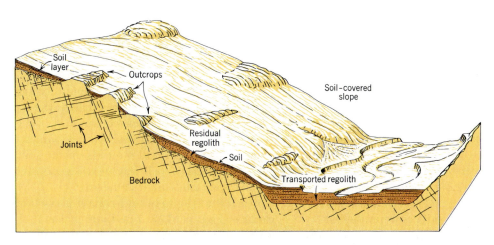

FIGURE 14.1 Soil, regolith, and outcrops on a hillslope. Alluvium, a form of transported regolith, lies in the floor of an adjacent stream valley. (Drawn by A. N. Strahler.)

so completely provided the earth's land surfaces with slopes that perfectly horizontal or vertical surfaces are extremely rare.

Figure 14.1 shows a typical hillslope forming one wall of the valley of a small stream. Soil and regolith mantle the bedrock except in a few places where the bedrock is particularly hard and projects in the form of *outcrops*. *Residual regolith* is derived from the rock beneath and moves very slowly down the slope toward the stream. Beneath the valley bottom are layers of transported regolith—alluvium—transported and deposited by the stream. This sediment had its source in regolith prepared on hillslopes many kilometers or tens of kilometers upstream. All terrestrial accumulations of sediment, whether deposited by streams, waves and currents, wind, or glacial ice, can be designated *transported regolith* in contrast to residual regolith.

Soil or regolith, or both, may be missing. In some places everything is stripped off down to the bedrock, which then appears at the surface as an outcrop. In some places, following cultivation or forest fires, the soil is partly eroded away. Severe erosion exposes the regolith. The thickness of soil and regolith is quite variable. Although the soil is rarely more than one or two meters thick, residual regolith on decayed and fragmented rock may extend down 5 to 100 m (15 to 300 ft), or more. Formation of regolith is greatly aided by the presence of innumerable bedrock cracks, called *joints* (Figure 14.1). Water can move easily through joints to promote rock decay.

Physical Weathering

Physical weathering produces fine particles from massive rock by the action of forces strong enough to fracture the rock. One of the most important physical weathering processes in cold climates is *frost action*, the repeated growth and melting of ice crystals in the pore spaces of soil and in rock fractures. Frost action is believed to be capable of rupturing even extremely hard rocks. Frost action produces a number of conspicuous effects and forms in all climates with cold winters. Features caused by ground-ice accumulation and frost action

FIGURE 14.2 Frost-shattered blocks of quartzite on the summit of the Snowy Range, Wyoming, elevation 3700 m (12,000 ft). (Arthur N. Strahler.)

are particularly conspicuous in the tundra climate or arctic coastal fringes and islands, and above timberline in high mountains.

Frost shattering of hard rocks exposed above timberline leads to surface accumulations of large angular fragments, including huge boulders (Figure 14.2). Frost action on cliffs of bare rocks in high mountains detaches rock fragments that fall to the cliff base. Where production of fragments is rapid, they accumulate to form *talus slopes*. Most cliffs are notched by narrow ravines that funnel the fragments into separate tracks, and so produce conelike talus bodies arranged side by side along the cliff (Figure 14.3). Fresh talus slopes are unstable, so that the disturbance created by walking across the slope or dropping a large rock fragment from the cliff above will easily set off a sliding of the surface layer of particles.

In fine-textured soils and sediments, composed largely of silt and clay, freezing of soil water takes place in horizontal layers or lenses. As these ice layers thicken, the overlying soil layer is heaved upward. Prolonged soil heaving can produce minor irregularities and small mounds on the soil surface. Where a rock fragment lies on the surface, perpendicular ice needles grow beneath the fragment and raise it above the surface (Figure 14.4). The same process acting on a rock fragment below the soil surface will eventually bring the fragment to the surface.

FIGURE 14.3 Talus cones of angular boulders built against the steep headwall of a glacial cirque. Snowy Range, Wyoming. (Arthur N. Strahler.)

A related process acting in coarse-textured regolith of the barren tundra causes the coarsest fragments—pebbles and cobbles—to move horizontally and to become sorted out from the finer particles. This type of sorting produces ringlike arrangements of coarse fragments. Linked with adjacent rings, the gross pattern becomes netlike to form a system of *stone rings* (Figure 14.5).

In silty alluvium, such as that formed on river floodplains and delta plains in the arctic environment, ice has accumulated in vertical wedge-forms in deep cracks in the sediment. These ice wedges are interconnected into a system of polygons, called *ice-wedge polygons* (Figure 14.6). Subsurface ice of the tundra is a permanent feature, called permafrost (permanently frozen ground). Only a shallow surface layer experiences summer melting. (The environmental aspects of permafrost were described in Chapter 9.) Ice wedges are thought to originate as shrinkage cracks formed during extreme winter cold. During the spring melt, water enters the cracks and becomes frozen.

Closely related to the growth of ice crystals is the weathering process of rock disintegration by growth of salt crystals. This process operates extensively in dry climates and is responsible for many of the niches, shallow caves, rock arches, and pits seen in sandstone formations. During long drought periods, ground water is drawn to the surface of the rock by capillary force.

FIGURE 14.5 Sorted circles of gravel form a system of netlike stone rings on this nearly flat land surface where water drainage is poor. The circles in the foreground are 3 to 4 m across; the gravel ridges are 20 to 30 cm high. Broggerhalvoya, western Spitsbergen, latitude 78°N. (Bernard Hallet, Quaternary Research Center, University of Washington, Seattle.)

FIGURE 14.4 Frost heaving above timberline. Needle ice, attached to the underside of a rock fragment, lifted the mass as the crystals lengthened. Below is the cavity in which the ice formed. (Mark A. Melton.)

FIGURE 14.6 Patterned ground near Barrow Point, Alaska. Ice-wedge polygons rise higher than the intervening ground, which contains small lakes. (William R. Farrand.)

As evaporation of the water takes place in the porous outer zone of the sandstone, tiny crystals of salts are left behind. The growth force of these crystals is capable of producing grain-by-grain breakup of the sandstone, which crumbles into a sand and is swept away by wind and rain. Especially susceptible are zones of rock lying close to the base of a cliff, because there the ground water seeps outward to reach the rock surface (Figure 14.7). In the southwestern United States, many of the deep niches or cavelike recesses formed in this way were occupied by Indians. Their cliff dwellings

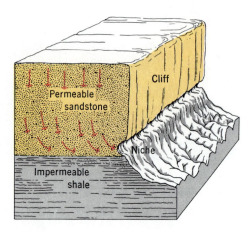

FIGURE 14.7 In dry climates there is a slow seepage of water from the cliff base. Here niches develop through rock weathering. (Drawn by A. N. Strahler.)

gave protection from the elements and safety from armed attack (Figure 14.8).

Salt crystallization also acts adversely on masonry buildings and highways. Brick and concrete in contact with moist soil are highly susceptible to grain-by-grain disintegration from this cause. The salt crystals can be seen as a soft, white, fibrous layer on basement floors and walls. Man has added to these destructive effects by spreading deicing salts on streets and highways. Sodium chloride (rock salt), widely used for this purpose, is particularly destructive to concrete pavements and walks, curbstones, and other exposed masonry structures.

Most rock-forming minerals expand when heated and contract when cooled. Where rock surfaces are exposed daily to the intense heating of the sun alternating with nightly cooling, the resulting expansion and contraction exerts powerful disruptive forces on the rock. Although firsthand evidence is lacking, it seems likely that temperature changes cause breakup of rock already weakened by other agents of weathering.

The wedging of plant roots also deserves consideration as a mechanism causing joint blocks to be separated. You may have seen a tree whose lower trunk and roots are firmly wedged between two great joint blocks of massive rock (Figure 14.9). Whether the tree has actually been able to spread the blocks farther apart or has merely occupied the available space is open to question. However, it is certain that pressure exerted by growth of tiny rootlets in joint fractures causes the loosening of countless small rock scales and grains.

A widespread process of rock disruption related to physical weathering results from *unloading*, the relief of confining pressure of overlying rock. Unloading occurs as rock is brought nearer to the earth's surface through the erosional removal of overlying rock. Rock formed at great depth beneath the earth's surface (particularly igneous and metamorphic rock) is in a slightly compressed state because of the confining pressure of overlying rock. On being brought to the surface, the rock expands slightly in volume. Expansion causes thick shells of rock to break free from the parent mass below. The new surfaces of fracture

FIGURE 14.8 The White House Ruin, a former Indian habitation, occupies a great niche in sandstone in the lower wall of Canyon de Chelly, Arizona. (Mark A. Melton.)

are a form of jointing called *sheeting structure*. The rock sheets show best in massive rocks such as granite and marble (Figure 14.10).

Sheeting structure is well developed in granite quarries, where it greatly facilitates the removal of rock. Individual sheets sometimes break free and arch upward with explosive violence. A similar phenomenon in deep mines and tunnels is the explosive breaking away of ceilings. Known as "popping rock," this spontaneous rock disintegration is a hazard to miners.

Where sheeting structure has formed over the top of a single large body of massive rock, an *exfoliation dome* is produced

(Figure 14.10). Domes are among the largest of the landforms caused primarily by weathering. In Yosemite Valley, California, where domes are spectacularly displayed, the individual rock shells may be as thick as 15 m (50 ft).

Forms Produced by Chemical Weathering

We investigated *chemical weathering* processes in Chapter 11 under the heading of

FIGURE 14.9 Jointing in sandstone resembles pavement blocks at Artists View, Catskill Mountains, New York. A tree has grown up between two joint blocks. (Arthur N. Strahler.)

FIGURE 14.10 Thick sheets of granite have fallen away from this cliff in Yosemite National Park. An exfoliation dome forms a summit at the upper right. (Arthur N. Strahler.)

chemical alteration. Recall that the dominant processes of chemical change affecting silicate minerals are oxidation, carbonic acid action, and hydrolysis. Feldspars and the mafic minerals are very susceptible to chemical decay (Figure 14.11). On the other hand, quartz is a highly stable mineral, almost immune to decay.

Decomposition by hydrolysis and oxidation changes strong rock into very weak regolith. This change allows erosion to operate with great effectiveness, wherever the regolith is exposed. Weakness of the regolith also makes it susceptible to natural forms of mass wasting.

In warm, humid climates of the equatorial, tropical, and subtropical zones hydrolysis and oxidation often result in the decay of igneous and metamorphic rocks to depths as much as 100 m (300 ft). To the civil engineer, deeply weathered rock is of major importance in constructing highways, dams, or other heavy structures. Although the regolith is soft and can be removed by power shovels with little blasting, there is serious danger of failure of foundations under heavy loads. This regolith also has undesirable plastic properties because of a high content of clay minerals.

The hydrolysis of exposed granite surfaces is accompanied by the grain-by-grain breakup of the rock. This process creates many interesting boulder and pinnacle forms by rounding of angular joint blocks (Figure 14.12). These forms are particularly conspicuous in arid regions. There is ample moisture in most deserts for

FIGURE 14.12 Large joint blocks of granite are gradually being rounded into huge boulders through grain-by-grain disintegration in a desert environment. Joshua Tree National Monument, California. (Arthur N. Strahler.)

hydrolysis to act, given sufficient time. The products of grain-by-grain breakup form a coarse desert gravel, which consists largely of quartz and partially decomposed feldspars.

Atmospheric carbon dioxide is dissolved in all surface waters of the lands, including rainwater, soil water, and stream water. The solution of carbon dioxide in water produces a weak acid, called carbonic acid. Carbonate sedimentary rocks (limestone, marble) are particularly susceptible to the acid action. Mineral calcium carbonate is dissolved, yielding calcium ions and bicarbonate ions, both of which are carried away in solution in water of streams.

Carbonic acid reaction with limestone produces many interesting surface forms, mostly of small dimensions. Outcrops of limestone typically show cupping, rilling, grooving, and fluting in intricate designs (Figure 14.13). In a few places the scale of deep grooves and high wall-like rock fins reaches proportions that prevent passage of people and animals.

Limestone Caverns

Most of you are familiar with the names of famous caverns, such as Mammoth Cave or Carlsbad Caverns (Figure 14.14). Millions of Americans have visited these famous tourist

FIGURE 14.11 Mafic igneous rock undergoing decay in place to produce a thick regolith. White dikes of felsic rock are little affected. Sangre de Cristo Mountains, New Mexico. (Arthur N. Strahler.)

FIGURE 14.13 This desert outcrop of massive white limestone shows rills, cups, and sharp ridges. Although the climate is dry, occasional rainshowers drench the rock surface, allowing carbonic acid to do its work. Charleson Range, Nevada. (John S. Shelton.)

FIGURE 14.14 The main entrance to Carlsbad Caverns, New Mexico. Horizontal limestone strata form a supporting lintel over the gaping cavity. (Arthur N. Strahler.)

attractions. *Limestone caverns* are interconnected subterranean cavities in bedrock formed by the action of circulating ground water on limestone. Figure 14.15 suggests how caverns may develop. In the upper diagram the action of carbonic acid is shown to be particularly concentrated in the saturated zone just below the water table. Products of solution are carried along in the ground water flow paths to emerge in streams.

In a later stage, shown in the lower diagram, the stream has deepened its valley and the water table has been correspondingly lowered to a new position. The cavern system previously excavated is now in the unsaturated zone. Deposition of carbonate matter, known as *travertine*, now begins to take place on exposed rock surfaces in the caverns. Encrustations of travertine take many beautiful forms—stalactites, stalagmites, columns, drip curtains, and terraces (Figure 14.16).

The environmental importance of caves is felt in several ways. Throughout the early development of the human species, caves were an important habitation. Now we find the skeletal remains of these people, together with their implements and cave drawings, preserved through the centuries

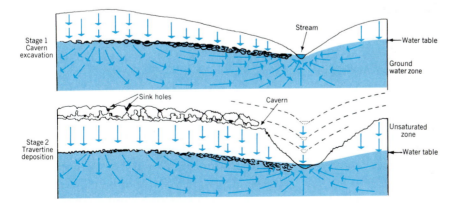

FIGURE 14.15 Cavern development in the ground water zone, followed by travertine deposition in the unsaturated zone. (From A. N. Strahler, *The Earth Sciences*, 2nd ed., Harper & Row, New York. Copyright © 1971 by Arthur N. Strahler.)

FIGURE 14.16 Travertine deposits in Carlsbad Caverns include stalactites (slender rods hanging from the ceiling) and sturdy columns. (Mark A. Melton.)

FIGURE 14.17 A shallow sinkhole on the Kaibab Plateau, Arizona. Clay impedes the drainage of water, allowing a pond to form. A needleleaf forest of spruce is seen in the distance. (Arthur N. Strahler.)

in caves in many parts of the world. Today, caverns are being used as storage facilities, living quarters, and factories.

Caverns have provided some valuable deposits of guano, the excrement of birds or bats, which is rich in nitrates. Guano deposits have been used in the manufacture of fertilizers and explosives. Bat guano was taken from Mammoth Cave for making gunpowder during the war of 1812.

Karst Landscapes

Where limestone solution is very active, we find a landscape with many unique landforms. This is especially true along the Dalmatian coastal area of Yugoslavia, where the landscape is called *karst*. The term may be applied to the topography of any limestone area where sinkholes are numerous and small surface streams are nonexistent. A *sinkhole* is a surface depression in limestone of a cavernous region (Figure 14.17). Some sinkholes are filled with soil washed from nearby hillsides. Others are steep-sided, deep holes.

Development of a karst landscape is shown in Figure 14.18. In an early stage,

funnel-like sinkholes are numerous. Later, the caverns collapse, leaving open, flat-floored valleys. Some important regions of karst or karstlike topography are the Mammoth Cave region of Kentucky, the Yucatan Peninsula, and parts of Cuba and Puerto Rico.

Mass Wasting

Everywhere on the earth's surface, gravity pulls continually downward on all materials. Bedrock is usually so strong and well supported that it remains fixed in place but, where a mountain slope becomes too steep, bedrock masses break free and fall or slide to new positions of rest. In cases where huge masses of bedrock are involved, the result can be catastrophic in loss to life and property in towns and villages in the path of the slide. Such slides are a major form of environmental hazard in mountainous regions. Because soil, regolith, and many forms of sediment are held together poorly, they are much more susceptible to the force of gravity than is bedrock. Abundant evidence shows that on most slopes at least a small amount of downhill movement is

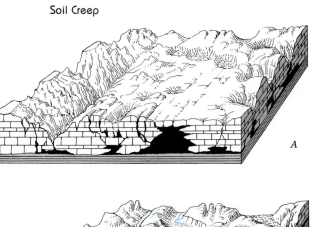

FIGURE 14.18 Features of a karst landscape. (A) Rainfall enters the cavern system through sinkholes in the limestone. (B) Extensive collapse of caverns reveals surface streams flowing on shale beds beneath the limestone. The flat-floored valleys can be cultivated. (Drawn by E. Raisz.)

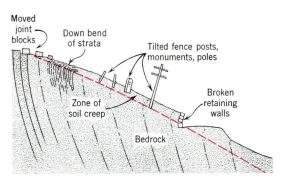

FIGURE 14.19 The slow, downhill creep of soil and regolith shows up in many ways on a hillside. (After C. F. S. Sharpe.)

going on constantly. Although much of this motion is imperceptible, the regolith sometimes slides or flows rapidly.

Taken altogether, the various kinds of downhill movements occurring under the pull of gravity, collectively called mass wasting, constitute an important process in lowering the continental surfaces. Humans have added to the natural forms of mass wasting by moving enormous volumes of rock and soil on construction sites of dams, canals, highways, and buildings.

Soil Creep

On almost any steep, soil-covered slope, you can find evidence of extremely slow downhill movement of soil and regolith, a process called *soil creep*. Figure 14.19 shows some of the evidence that the process is going on. Joint blocks of distinctive rock types are found moved far downslope from the outcrop. In some layered rocks such as shales or slates, edges of the strata seem to bend in the downhill direction (Figure

14.20). This is not true plastic bending, but is the result of slight movement on many small joint cracks. Fence posts and telephone poles lean downslope and even shift measurably out of line. Retaining walls of road cuts buckle and break under pressure of soil creep.

What causes soil creep? Heating and cooling of the soil, growth of frost needles, alternate drying and wetting of the soil, trampling and burrowing by animals, and shaking by earthquakes all produce some disturbance of the soil and regolith. Because gravity exerts a downhill pull on every such rearrangement, the particles are urged slowly downslope.

FIGURE 14.20 Slow downhill creep of regolith on this mountainside near Downieville, California, has caused vertical rock layers to seem to "bend over." (Copyright © 1988 by Mark A. Melton.)

Earthflow

In regions of humid climate, a mass of water-saturated soil, regolith, or weak shale may move down a steep slope during a period of a few hours in the form of an *earthflow*. Figure 14.21 is a sketch of an earthflow showing how the material slumps away from the top, leaving a steplike terrace bounded by a curved, wall-like scarp. The saturated material flows sluggishly to form a bulging toe.

Shallow earthflows, affecting only the soil and regolith, are common on sod-covered and forested slopes that have been saturated by heavy rains. An earthflow may affect a few square meters, or it may cover an area of several hectares (Figure 14.22). If the bedrock of a mountainous region is rich in clay (shale or deeply weathered volcanic rocks), earthflow sometimes involves millions of tons of bedrock moving by plastic flowage like a great mass of thick mud.

Earthflows are a common cause of blockage of highways and railroad lines, usually during periods of heavy rains. Usually the rate of flowage is slow, so that the flows are not often a threat to life. Property damage to buildings, pavements, and utility lines is often severe where construction has taken place on unstable soil slopes.

A special variety of earthflow found in the arctic tundra is *solifluction* (from Latin

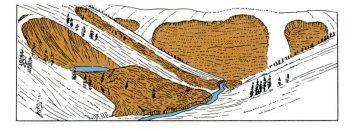

FIGURE 14.22 Earthflows in a mountainous region leave great scars. (Drawn by W. M. Davis.)

words meaning "soil" and "flow"). In early summer, when thawing has penetrated the upper few decimeters, the soil is fully saturated. Soil water cannot escape downward because the underlying frozen mass will not allow water to drain through it. Flowing almost imperceptibly, this saturated soil forms terraces and lobes that give the mountain slope a stepped appearance (Figure 14.23).

Mudflow

One of the most spectacular forms of mass wasting and one that is potentially a serious environmental hazard is the *mudflow*. This is a mud stream of fluid consistency that pours down canyons in mountainous regions (Figure 14.24). In deserts, where vegetation does not protect the mountain soils, local convectional storms produce rain much faster than it can be absorbed by the soil. As the water runs down the slopes, it forms a thin mud, which flows down to the canyon floors. Following stream courses, the mud continues to flow until it becomes so thick it must stop. Great boulders are carried along buoyed up in the mud. Roads, bridges, and houses in the canyon floor are engulfed and destroyed. Where the mudflow emerges from the canyon and spreads across a piedmont plain, there may be severe property damage and even loss of life (Figure 14.25).

Mudflows also occur on the slopes of erupting volcanoes. Freshly fallen volcanic ash and dust are turned into mud by heavy rains and flow down the slopes of the volcano. Herculaneum, a city at the base of Mount Vesuvius, was destroyed by a mudflow during the eruption of A.D. 79. At

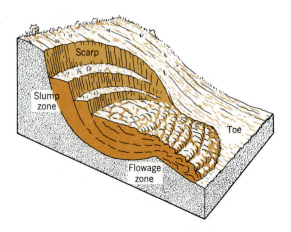

FIGURE 14.21 An earthflow with slump features well developed in the upper part. Flowage has produced a bulging toe. (Drawn by A. N. Strahler.)

FIGURE 14.23 A solifluction lobe on Baffin Island in the arctic tundra. While bearing intact is cover of plants and soil, a bulging mass of water-saturated regolith has slowly moved downslope, overriding the ground surface below it. A backpack marks the base of the advancing lobe. (M. Church, University of British Columbia.)

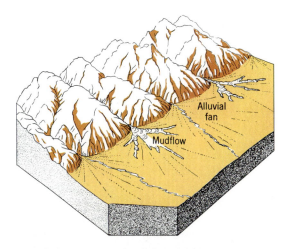

FIGURE 14.24 Thin, streamlike mudflows issue occasionally from canyon mouths in arid regions. The mud spreads out on the piedmont slopes in long, narrow tongues. (Drawn by A. N. Strahler.)

the same time, the neighboring city of Pompeii was buried under volcanic ash.

Mudflows show varying degrees of consistency, from a mixture about like the concrete that emerges from a mixing truck to thinner consistencies that are little different than those in turbid stream floods. The watery type of mudflow is called a *debris flood* in the western United States, and particularly in southern California, where it occurs commonly and with disastrous effects.

Landslide

Landslide is the rapid sliding of large masses of bedrock. Wherever mountain slopes are steep there is a possibility of large, disastrous landslides (Figure 14.26). In Switzerland, Norway, or the Canadian Rockies, for example, villages built on the floors of steep-sided valleys have been destroyed and their inhabitants killed by the sliding of millions of cubic meters of rock set loose without any warning.

A great landslide disaster occurred in 1959 in Montana when a severe earthquake (Hebgen Lake earthquake) caused an entire mountainside to slide into the Madison

River gorge, killing 27 persons. The Madison Slide, as it is known, formed a debris dam over 60 m (200 ft) high and produced a new lake. The volume of this slide was about 27 million m^3 (35 million yd^3). The lake has now been made permanent by construction of a protected spillway. Severe earthquakes in mountainous regions are a major immediate cause of landslides and earthflows.

Aside from occasional great catastrophes, landslides have rather limited environmental influence because of their sporadic occurrence in thinly populated mountainous regions. Small slides can, however, repeatedly block or break an important mountain highway or railway line.

FIGURE 14.25 This mudflow, carrying numerous large boulders, issued from a steep mountain canyon in the Wasatch Mountains, Utah. (Orlo E. Childs.)

FIGURE 14.26 A great landslide descended from the high mountain summit at the upper left. The rock became pulverized into loose debris, which came to rest at the mountain base. Beaver Lake, British Columbia. (Mark A. Melton.)

Induced Mass Wasting

Human activities induce mass wasting in forms ranging from mudflow and earthflow to landslide. These activities include (1) piling up of waste soil and rock into unstable accumulations that fail spontaneously, and (2) removal of support by undermining natural masses of soil, regolith, and bedrock.

At Aberfan, Wales, a major disaster occurred when a hill 180 m (600 ft) high, built of rock waste (culm) from a nearby coal mine, spontaneously began to move and quickly developed into a mudflow of thick consistency. The waste pile had been constructed on a steep hillslope and on a spring line as well, making a potentially unstable configuration. The debris tongue overwhelmed part of the town below, destroying a school and taking over 150 lives (Figure 14.27). Phenomena of this type are often called "mudslides" in the news media.

In Los Angeles County, California, real estate development has been carried out on very steep hillsides and mountainsides by the process of bulldozing roads and homesites out of the deep regolith. The excavated regolith is pushed out into adjacent embankments where its instability poses a threat to slopes and stream channels below. When saturated by heavy winter rains, these embankments can give way, producing earthflows, mudflows, and debris floods that travel far down the canyon floors and spread out on the piedmont surfaces, burying streets and yards in bouldery mud. Many debris floods of this area are also produced by heavy rains falling on undisturbed mountain slopes denuded by fire of vegetative cover in the preceding dry summer. Some of these fires are set by humans. Disturbance of slopes by construction practices is simply an added source of debris and serves to enhance what is already an important environmental hazard.

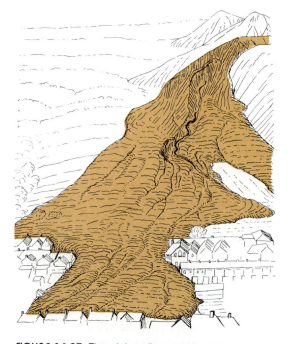

FIGURE 14.27 This debris flow at Aberfan, Wales, buried a school and killed many children. (From A. N. Strahler, *Planet Earth*, Harper & Row, New York. Copyright © 1972 by Arthur N. Strahler.)

FIGURE 14.28 The Ray Mine, a great open-pit copper mine near Teapot Mountain, Arizona. Because the copper minerals are disseminated through an enormous body of igneous rock, large volumes of rock must be excavated. (Mark A. Melton.)

Scarification of the Land

Industrial societies now possess enormous machine power and explosives capable of moving great masses of regolith and bedrock from one place to another. One such activity is the extraction of mineral resources. Another is the reorganization of terrain into suitable configurations for highway grades, airfields, building foundations, dams, canals, and various other large structures. Both activities involve removal of earth materials, which destroys entirely the preexisting ecosystems and habitats of plants and animals. The same activities include building up of new land on adjacent surfaces using those same earth materials. This process destroys ecosystems and habitats by burial. What distinguishes artificial forms of mass wasting from the natural forms is that machinery is used to raise earth materials against the force of gravity. Explosives used in blasting can produce disruptive forces many times more powerful than the natural forces of physical weathering.

Scarification is a general term for excavations and other land disturbances produced for purposes of extracting mineral resources; it includes accumulation of waste matter as spoil or tailings. Among the forms of scarification are open-pit mines, strip mines, quarries for structural materials, borrow pits along highway grades, sand and gravel pits, clay pits, phosphate pits, scars from hydraulic mining, and stream gravel deposits reworked by dredging. Open-pit mining of low-grade copper ores is illustrated in Figure 14.28.

Scarification is on the increase. Demands for coal to meet energy requirements are on the rise. There are also increased demands for industrial minerals used in manufacturing and construction. At the same time, as the richer and more readily available mineral deposits are consumed, industry turns to poorer grades of ores and to less easily accessible coal deposits. As a result, the rate of scarification is further increased.

Strip Mining and Mass Wasting

Where coal seams lie close to the surface or actually outcrop along hillsides, the *strip mining* method is used. Here, earthmoving equipment removes the covering strata (overburden) to bare the coal, which is lifted out by power shovels. There are two kinds of strip mining, each adapted to the given relationship between ground surface and coal seam.

Area strip mining is used in regions of flat land surface under which the coal seam lies horizontally (Figure 14.29A). After the first trench is made and the coal removed, a parallel trench is made, the overburden of which is piled as a spoil ridge into the first trench. In this way the entire seam is gradually uncovered, and a series of parallel spoil ridges remains.

The contour strip mining method is used where a coal seam outcrops along a steep hillside (Figure 14.29B). The coal is uncovered as far back into the hillside as possible, and the overburden is dumped on the downhill side. There results a bench bounded on one side by a freshly cut rock wall and on the other by a ridge of loose spoil with a steep outer slope leading down into the valley bottom. The benches form sinuous patterns following the plan of the outcrop (Figure 14.30). Strip mining is carried to depths as great as 30 m (100 ft) below the surface.

FIGURE 14.30 Contour strip mining near Lynch, Kentucky. A highway makes use of the winding bench at the base of the high rock wall. (Billy Davis/Black Star.)

The spoil bank produced by contour strip mining is unstable and is a constant threat to the lower slope and valley bottom below. When saturated by heavy rains and melting snows, the spoil generates earthflows and mudflows; these descend on houses, roads, and forest. The spoil also supplies sediment that clogs stream channels far down the valleys.

Weathering and Mass Wasting in Review

In this chapter we have compared natural processes of wasting of the continental surfaces with changes of a similar nature

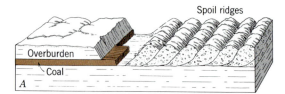

FIGURE 14.29 Two kinds of strip mining: (A) Area strip mining; (B) contour strip mining. (Drawn by A. N. Strahler.)

induced by human activity. The processes of weathering and soil creep are for the most part slow-acting and produce effects that are visible only when accumulated over centuries. Natural mass-wasting processes also include catastrophic events. Indeed, some large landslides dwarf anything that humans have accomplished in equal time by earth-moving with machines and explosives. But machines work constantly with enormous quantities of energy, and the cumulative environmental damage and destruction continues to mount. Only now are the conflicts of interest beginning to emerge and to be squarely faced. Only now are we designing and implementing environmental protection measures necessary to control the spread of scarification and its many harmful side effects.

CHAPTER 15

LANDFORMS MADE BY RUNNING WATER

FLUVIAL PROCESSES   Running water creates both erosional and depositional landforms.

SOIL EROSION   The geologic norm of soil erosion can give way to devastating accelerated soil erosion where human activity impacts the soil surface.

WORK OF STREAMS   Erosion, transportation, and deposition by streams produce a host of fluvial landforms as the process of gradation transforms the stream profile through time.

DENUDATION   Landmasses evolve through a sequence of denudational stages, closing with a land surface of low elevation and faint slope—a peneplain.

AGGRADATION AND DEGRADATION   Changing conditions can cause streams to deposit alluvium or to remove alluvium, altering the valley landscape in the process.

FLOODPLAINS   Landforms of alluvial rivers with meandering channels are responses to seasonal flooding, a hazard to humans who occupy the floodplain.

DESERT LANDFORMS   Landscapes of mountainous deserts consist of unique assemblages of landforms reflecting the arid climate.

GEOMORPHOLOGY deals largely with the action of *fluid agents* that erode, transport, and deposit mineral and organic matter. There are four fluid agents: (1) running water in surface and underground flow systems; (2) waves, acting with currents in oceans and lakes; (3) glacial ice, moving sluggishly in great masses; and (4) wind, blowing over the ground.

Of the four agents, three are forms of water. Consequently, the science of hydrology is inseparably interwoven with geomorphology. One might be not far wrong in saying simply that the hydrologist is preoccupied with "where water goes," and the geomorphologist with "what water does." Hydrology concerns itself with the hydrologic cycle in an attempt to calculate the water balance and to measure rates of flow of water in all parts of that cycle (Chapter 10). Geomorphology concerns itself with geologic work that the water in motion performs on the land.

Denudation, defined in Chapter 13, is a useful term for the total action of all processes by which the exposed rocks of the continents are worn away and the resulting sediments are transported to the sea by the fluid agents. Denudation is an overall lowering of the land surface. Denudation tends toward reducing the continents to nearly featureless sea-level surfaces and, ultimately, through wave action, to submarine surfaces. If it had not been repeatedly counteracted by tectonic activity throughout geologic time, denudation would have eliminated all terrestrial environments.

An important point that emerges as we look back through geologic time is that terrestrial life environments have been in constant change, even as plants and animals have undergone their evolutionary development. The varied denudation processes have produced, maintained, and changed a wide variety of landforms, which have been the habitats for evolving life-forms. In turn, the life-forms have become adapted to those habitats and have diversified to a degree that matches the diversity of the landforms themselves.

Geomorphic and hydrologic systems operating in the life layer have long been subjected to quite radical modification by the works of humans. Agriculture has for centuries altered the surface properties of areas of subcontinental size. Agriculture has modified the action of running water and the water balance, to say nothing of radically changing the character of the soil. Urbanization is an even more radical alteration by seriously upsetting hydrologic processes. Engineering and mining activities, such as strip mining and the construction of highways, dams, and canals, not only upset hydrologic systems but can completely destroy or submerge entire assemblages of landforms. Of the four fluid agents of landform sculpture, only glaciers of ice have so far successfully resisted changes of activity imposed by humans.

Invariably, attempts by humans to control the action of running water, waves, and currents produce unpredicted and undesirable side effects, some of which are physical and others ecological. An important reason to study geomorphology is to predict the consequences of such changes and to plan wisely for the future.

Fluvial Processes and Landforms

Landforms shaped by running water are conveniently described as *fluvial landforms* to distinguish them from landforms made by the other fluid agents—glacial ice, wind, and waves. Fluvial landforms are shaped by the *fluvial processes* of overland flow and stream flow. Weathering and the slower forms of mass wasting, such as soil creep, operate hand in hand with overland flow, and cannot be separated from the fluvial processes.

Fluvial landforms and fluvial processes dominate the continental land surfaces the world over. Throughout geologic history, glacial ice has been present only in comparatively small global areas located in the polar zones and in high mountains. Landforms made by wind action occupy only trivially small parts of the continental surfaces, whereas landforms made by waves and currents are restricted to a very narrow contact zone between oceans and continents. In terms of area, the fluvial landforms are dominant in the environment of terrestrial life and are the major source areas of human food resources through the practice of agriculture. Almost all lands in crop cultivation and almost all grazing lands have been shaped by fluvial processes.

Fluvial processes perform the geological activities of erosion, transportation, and deposition. Consequently, there are two major groups of fluvial landforms: erosional landforms and depositional landforms (Figure 15.1). Valleys are formed where rock is eroded away by fluvial agents.

FIGURE 15.1 Erosional and depositional landforms. (Drawn by A. N. Strahler.)

Between the valleys are ridges, hills, or mountain summits representing unconsumed parts of the crustal block. All such sequential landforms shaped by progressive removal of the bedrock mass are *erosional landforms*.

Fragments of soil, regolith, and bedrock that are removed from the parent mass are transported by the fluid agent and deposited elsewhere to make an entirely different set of surface features, the *depositional landforms*. Figure 15.1 illustrates the two groups of landforms. The ravine, canyon, peak, spur, and col are erosional landforms; the fan, built of rock fragments below the mouth of the ravine, is a depositional landform. The floodplain, built of material transported by a stream, is also a depositional landform.

Slope Erosion

Fluvial action starts on the uplands as *soil erosion*. Overland flow, by exerting a dragging force over the soil surface, picks up particles of mineral matter ranging in size from fine colloidal clay to coarse sand or gravel, depending on the speed of the flow and the degree to which the particles are bound by plant rootlets or held down by a mat of leaves. Added to this solid matter is dissolved mineral matter in the form of ions produced by acid reactions or direct solution.

This slow removal of soil is part of the natural geological process of denudation; it is both inevitable and universal. Under stable natural conditions, the erosion rate in a humid climate is slow enough that a soil with distinct horizons is formed and maintained, enabling plant communities to maintain themselves in a stable equilibrium. Soil scientists refer to this state of activity as the *geologic norm*.

By contrast, the rate of soil erosion may be enormously speeded up by human activities or by rare natural events to result in a state of *accelerated erosion* that removes the soil much faster than it can be formed. This condition comes about most commonly from a forced change in the plant cover and in the physical state of the ground surface and uppermost soil horizons. Destruction of vegetation by clearing of land for cultivation or by forest fires sets the stage for a series of drastic changes. No foliage remains to intercept rain, and protection afforded by a ground cover of fallen leaves and stems is removed. Consequently, the rain falls directly on the mineral soil.

Direct force of falling drops on bare soil causes a geyserlike splashing in which soil particles are lifted and then dropped into new positions, a process termed *splash erosion* (Figure 15.2). It is estimated that a torrential rainstorm has the ability to disturb as much as 225 metric tons of soil per hectare (100 tons per acre). On a sloping ground surface, splash erosion shifts the soil slowly downhill. A more important effect is to cause the soil surface to become much less able to absorb water because the natural soil openings become sealed by particles shifted by raindrop splash.

FIGURE 15.2 A large raindrop (above) lands on a wet soil surface, producing a miniature crater (below). Grains of clay and silt are thrown into the air and the soil surface is disturbed. (Official U.S. Navy photograph.)

Reduced infiltration permits a much greater proportion of overland flow to occur from a given amount of rain. Increased overland flow intensifies the rate of soil removal.

Another effect of destruction of vegetation is to reduce greatly the resistance of the ground surface to the force of erosion under overland flow. On a slope covered by grass sod, even a deep layer of overland flow causes little soil erosion because the energy of the moving water is dissipated in friction with the grass stems, which are tough and elastic. On a heavily forested slope, countless check dams made by leaves, twigs, roots, and fallen tree trunks take up the force of overland flow. Without much vegetative cover the eroding force is applied directly to the bare soil surface and easily dislodges the grains and sweeps them downslope.

We can get a good appreciation of the contrast between normal and accelerated erosion rates by comparing the quantity of sediment derived from cultivated surfaces with that derived from naturally forested or reforested surfaces. The comparison is made within a single region in which climate, soil, and topography are fairly uniform. *Sediment yield* is a technical term for the quantity of sediment removed by overland flow from a unit area of ground surface in a given unit of time. Yearly sediment yield is stated in metric tons per hectare, or tons per acre.

Figure 15.3 gives data of annual average sediment yield and runoff by overland flow from several types of upland surface in northern Mississippi. Notice that both surface runoff and sediment yield decrease greatly with increased effectiveness of the protective vegetative cover. Sediment yield from cultivated land undergoing accelerated erosion is over ten times greater than that from pasture and about one thousand times greater than that from pine plantation land. The reforested land has a sediment-yield rate representing the geologic norm of soil erosion for this region; it is about the same as for mature pine and hardwood forests that have not experienced cultivation.

The distinction between normal and accelerated slope erosion applies to regions in which the water balance shows an annual surplus. Under a midlatitude semiarid climate with summer drought the natural plant cover consists of short-grass prairie (steppe). Although it is sparse and provides rather poor ground cover of plant litter, the grass cover is strong enough that the geologic norm of erosion can be sustained. In these semiarid environments, however, the natural equilibrium is highly sensitive to upset. Depletion of plant cover by fires or the grazing of herds of domesticated animals can easily set off rapid erosion. These sensitive, marginal environments require cautious use, because they lack the potential to recover rapidly from accelerated erosion once it has begun.

Erosion at a very high rate by overland flow is actually a natural geological process in certain favorable localities in semiarid and arid lands; it takes the form of *badlands*. One well-known area of badlands is the Big Badlands of South Dakota, along the White River. Badlands are underlain by clay formations, which are easily eroded by overland flow. Erosion rates are too fast to

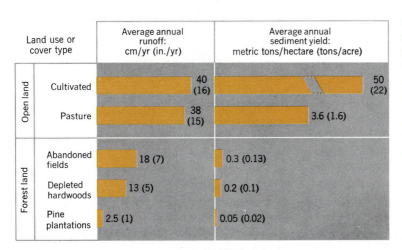

Land use or cover type		Average annual runoff: cm/yr (in./yr)	Average annual sediment yield: metric tons/hectare (tons/acre)
Open land	Cultivated	40 (16)	50 (22)
	Pasture	38 (15)	3.6 (1.6)
Forest land	Abandoned fields	18 (7)	0.3 (0.13)
	Depleted hardwoods	13 (5)	0.2 (0.1)
	Pine plantations	2.5 (1)	0.05 (0.02)

FIGURE 15.3 The bar graph shows that both runoff and sediment yield are much greater for open land than for land covered by shrubs and forest. Cultivated land has an enormous sediment yield, as compared with any of the other types. (Data of S. J. Ursic, 1965, Department of Agriculture.)

FIGURE 15.4 Badlands at Zabriskie Point, Death Valley National Monument, California. (Alan H. Strahler.)

permit plants to take hold, and no soil can develop. A maze of small stream channels is developed and ground slopes are very steep (Figure 15.4). Badlands such as these are self-sustaining and have been in existence at one place or another on continents throughout much of geologic time.

Forms of Accelerated Soil Erosion

Humid regions of substantial water surplus, for which the natural plant cover is forest or prairie grasslands, experience accelerated soil erosion when humans expend enough energy to remove the plant cover and keep the land barren by annual cultivation. With fossil fuels to power machines of plant and soil destruction, human activity has easily overwhelmed the restorative forces of nature over vast expanses of continental surfaces. We now consider the consequences of these activities.

When a plot of ground is first cleared of forest and plowed for cultivation, little erosion will occur until the action of rain splash has broken down the soil aggregates and sealed the larger openings. Overland flow then begins to remove the soil in rather uniform thin layers, a process termed *sheet erosion*. Because of seasonal cultivation, the effects of sheet erosion are often little noticed until the upper horizons of the soil are removed or greatly thinned. When soil particles reach the base of the slope, where the surface slope is sharply

reduced to meet the valley bottom, they come to rest and accumulate in a thickening layer termed *colluvium*. This deposit, too, has a sheetlike distribution and may be little noticed, except where it can be seen that fence posts or tree trunks are being slowly buried.

Material that continues to be carried by overland flow to reach a stream in the valley axis is then carried farther downvalley and may accumulate as alluvium in layers on the valley floor. *Alluvium* is a word applied generally to any stream-laid deposit. Deposition of alluvium results in burial of fertile soil under infertile, sandy layers. The alluvium chokes the channels of small streams and causes the water to flood broadly over the valley bottoms.

Where slopes are steep, runoff from torrential rains produces a more intense activity, *rill erosion*, in which innumerable, closely spaced channels are scored into the soil and regolith (Figure 15.5). If these rills are not destroyed by soil tillage, they may soon begin to integrate into still larger channels, called *gullies*. Gullies are steep-walled, canyonlike trenches whose upper ends grow progressively upslope (Figure 15.6). Ultimately, a rugged, barren topography, like the badland forms of the dry climates, results from accelerated soil erosion that is allowed to proceed unchecked.

The natural soil with its well-developed horizons is a nonrenewable natural resource. The rate of soil formation is

FIGURE 15.5 Shoestring rills on a steep bank of mine wastes illustrate the susceptibility of artificial accumulations of disturbed soil and regolith to accelerated erosion. Sediment from this source forms unwanted alluvial deposits in the nearest stream bed. Oatman, Arizona. (Arthur N. Strahler.)

extremely slow in comparison with the rate of its destruction once accelerated erosion has begun and is allowed to go unchecked. Soil erosion as a potentially disastrous form of environmental degradation was brought to public attention decades ago in the United States.

Curative measures developed by the Soil Conservation Service have proved partially effective in halting accelerated soil erosion and permitting the return to slow erosion rates approaching the geologic norm. These measures include construction of terraces to eliminate steep slopes, permanent restoration of overly steep slope belts to dense vegetative cover, and the healing of gullies by placing check dams in the gully floors.

Renewed hope for conserving agricultural soils in the corn-producing areas of the Appalachian region and midwestern plains has arisen with the introduction of a new farming technique. Known as no-till farming, the planting of corn dispenses with deep tilling and overturning of the soil. Instead, the seed is planted in a narrow, knifelike cut, while the stubble of the former crop remains intact to protect the soil from erosion. In coming decades, losses of soil by erosion on bare fields may be greatly reduced by use of this alternate method.

Geologic Work of Streams

The geologic work of streams consists of three closely interrelated activities: erosion, transportation, and deposition. *Stream erosion* is the progressive removal of mineral material from the floor and sides of the

FIGURE 15.6 Deep branching gullies have carved up an overgrazed pasture near Shawnee, Oklahoma. Contour terracing and check dams have halted the headward growth of the gullies. (Mark A. Melton.)

Sheep, Goats, and Fire—Degradation of the Mediterranean Landscape

Lands rimming the Mediterranean Sea draw pilgrims by the thousands to view the wonders of antiquity—or what is left of them. Unwittingly, they are paying homage to those Romans, Carthaginians, Greeks, and Moors who for century after century denuded and degraded the Mediterranean hillsides. The ruined landscape is now a fitting shambles in which to display their ruined temples and baths.

It has not been easy for geographers and historians to reconstruct the Mediterranean landscape of 4000 B.C., a time before the spread of farming cultures westward from the Fertile Crescent to Iberia. The summer-dry Mediterranean climate lends itself to a native vegetation of woodland—an open forest—and grassland. You can see this kind of landscape today in the coastal hills of southern California, where parklike woodlands of evergreen oak with grasses clothe dark brown residual soils. You can also see hillsides and mountainsides clothed in woody shrubs—the chaparral. Forest of oak, often quite dense, favor the north-facing hill slopes that retain moisture; chaparral favors the south-facing hill slopes that become bone-dry under months of the summer sun's direct rays. A surprisingly complete cover of topsoil mantles the lower foothills. Increasingly, irrigated groves of avocados and citrus are now replacing the native chaparral and woodland on these hill slopes.

In contrast to southern California, the hill slopes you will see around you in Italy, Greece, Sicily, Turkey, Lebanon, and Israel seem extraordinarily rocky. Ledges of bedrock and a litter of boulders seem to be everywhere, often marked off into sterile plots by disrupted stone walls. Woody shrubs grow in places; elsewhere, gnarled but productive olive trees sprinkle the hillsides. In a few sheltered spots, groves of Aleppo pines rise sharply above the other plants. In many areas, vineyards cover more favorable hillsides. Moving nimbly over the rocky terrain, goats browse off the miserable patches of shrub. In contrast,

many of the valley bottoms have thick accumulations of fine-grained soil, and here one finds the agricultural land of the Mediterranean region. Under irrigation these bottom lands are richly productive of cereal crops—wheat, corn, oats, and barley—and citrus fruits.

The impression one receives, by and large, is that whatever soil the Mediterranean hill slopes may once have had is now mostly deposited in the valley bottoms. Those who study the region through history seem generally agreed that extensive deforestation, accompanied by severe soil erosion, has been the history of this landscape from very early times of human occupancy. They point to a number of factors contributing to this prolonged environmental degradation.

The need for wood as fuel seems to have been a potent force in the deforestation of Mediterranean lands. The Romans, wherever they formed colonies throughout their empire, needed wood to heat their baths. Some scholars have attributed severe deforestation directly to this need for large amounts of firewood. (Today we affluent Americans follow a not-so-different course of action as we consume natural gas and fuel oil in prodigious quantities to heat our swimming pools, spas, and hot tubs.) Also, the Romans mined metals extensively in the Mediterranean lands; their ore smelters needed large supplies of charcoal since coal was not then in use.

As to the grasses and woody shrubs of the Mediterranean lands, sheep and goats seem to have been the principal machines of destruction, aided and abetted by their herders. The story of the sheep is a particularly interesting one in Spain, where sheepraising for wool assumed major importance in the economy from the 13th to the 18th centuries. Organized sheep owners of Spain forced the opening of vast areas of Spain to grazing by herds of Merino sheep. Their activity is said to have been a major factor in destroying the forest and reducing the shrub cover by repeated burning to improve pasture conditions. A similarly destructive role seems to have been played elsewhere by herds of goats and

Sheep grazing a steep slope in the southern foothills of the Pyrenees Mountains of Spain. Close cropping by these animals in the dry summer season, shown here, sets the stage for devastating soil erosion during the rainy winter. (Dennis Stock/Magnum.)

their managers. Annual burning of the shrub cover was carried out to encourage the growth of succulent new vegetation.

Whatever the factors and forces may have been, and in what order and intensity they acted, prolonged destruction of the plant cover by human agencies led to severe soil erosion. Although soil erosion was also a major form of environmental degradation in other parts of the civilized world—perhaps most strikingly in China—the Mediterranean region stands as a salient example.

Those settlers from western Europe who colonized the eastern fringe of the North American continent did not receive the benefit of any lesson from their neighbors in the Mediterranean lands, for that lesson had not yet been interpreted by students of history and geography. So, in a highly vulnerable environment, the American colonists put on a full-scale, repeat performance of the Mediterranean scenario, compressing the degradation of the forested land of the Piedmont upland into scarcely more than two centuries of intensive cultivation. Both in the United States and in the Mediterranean lands, the past few decades have seen major programs of reforestation and erosion control. Much of that is too late, for soil is a nonrenewable natural resource in terms of the life span of a civilization.

Data Source: Ralph Zon (1920), Forests and human progress, *Geological Review*, vol. 10, pp. 139–166; Sheldon Judson (1963), Erosion and deposition of Italian stream valleys during historic time, *Science*, vol. 140, pp. 898–899; Charles F. Bennett, Jr. (1975), *Man and Earth Ecosystems*, John Wiley & Sons, New York, Chapter 4.

channel, whether bedrock or regolith. *Stream transportation* consists of movement of the eroded particles by dragging along the bed, by suspension in the body of the stream, or in solution. *Stream deposition* is the accumulation of transported particles on the stream bed and floodplain, or on the floor of a standing body of water into which the stream empties. Obviously, erosion cannot occur without some transportation taking place, and the transported particles must eventually come to rest. Erosion, transportation, and deposition are simply three phases of a single activity.

Stream Erosion

Streams erode in various ways, depending on the nature of the channel materials and the tools with which the current is armed. The force of the flowing water alone, exerting impact and a dragging action on the bed and banks, can erode poorly consolidated alluvial materials, such as gravel, sand, silt, and clay. This erosion process, called *hydraulic action*, is capable of excavating enormous quantities of unconsolidated materials in a short time. The undermining of the banks causes large masses of alluvium to slump into the river, where the particles are quickly separated and become a part of the stream's load. This process of bank caving is an important source of sediment during high river stages and floods (Figure 15.7).

Where rock particles carried by the swift current strike against bedrock channel walls, chips of rock are detached. The rolling of cobbles and boulders over the stream bed will further crush and grind smaller grains to produce an assortment of grain sizes. This process of mechanical wear constitutes *abrasion;* it is the principal means of erosion in bedrock too strong to be affected by simple hydraulic action (Figure 15.8).

Finally, the chemical processes of rock weathering—acid reactions and solution—are effective in removal of rock from the stream channel and may be referred to as *corrosion.* Effects of corrosion are conspicuous in limestone, which develops cupped and fluted surfaces.

Stream Transportation

The solid matter carried by a stream is the *stream load;* it is carried in three forms. Dissolved matter is transported invisibly in the form of chemical ions. All streams carry some dissolved salts resulting from mineral alteration. Clay and silt are carried in *suspension;* that is, they are held up in the water by the upward elements of flow in turbulent eddies in the stream. This

FIGURE 15.7 Rapid bank caving undermined this cabin on Big Thomson Creek, Colorado, during a disastrous summer flood in 1976. Alluvium was flushed out from the channel floor in the foreground leaving only scoured bedrock and a litter of large boulders. (David C. Shelton, Colorado Geological Survey.)

FIGURE 15.8 These potholes in lava bedrock attest to the abrasion that takes place on the bed of a swift mountain stream. McCloud River, California. (Alan H. Strahler.)

FIGURE 15.9 In this vertical air view of the gorge of the Salt River, Arizona, the turbid storm runoff of a small tributary contrasts sharply with the clear water of the main stream. (Mark A. Melton.)

fraction of the transported matter is the *suspended load* (Figure 15.9). Sand, gravel, and cobbles move as *bed load* close to the channel floor by rolling or sliding and an occasional low leap (Figure 15.10).

The load carried by a stream varies enormously in the total quantity present and the size of the fragments, depending on the discharge and stage of the river. In flood, when velocities are high, the water is turbid with suspended load.

Suspended Loads of Large Rivers

A large river such as the Mississippi carries most of its load—about 90 percent—in suspension. Most of the suspended load comes from its great western tributary, the Missouri River, which is fed from semiarid lands, including the Dakota Badlands.

The Yellow River (Huanghe) of China heads the world list in annual suspended sediment load, and its watershed sediment yield is one of the highest known for a large river basin. The explanation lies in a high soil-erosion rate on intensively cultivated upland surfaces of wind-deposited silt. (See Chapter 17 and Figure 17.50.) Much of the

FIGURE 15.10 Carrying a heavy charge of suspended load, this turbulent mountain stream is beginning to cut into a bouldery layer carried and deposited as bed load in a higher stage of flow. Rincon Creek, Santa Barbara and Ventura counties, California. (Arthur N. Strahler.)

upper watershed is in a semiarid climate with dry winters; vegetation is sparse, and the runoff from heavy summer rains sweeps up a large amount of sediment.

Most investigators generally agree that land cultivation has greatly increased the sediment load of rivers of Asia, Europe, and North America. The increase because of human activities is thought to be greater by a factor of two and one-half than the geologic norm for the entire world land area. For the more strongly affected river basins, the factor may be 10 or more times larger than the geologic norm. This form of environmental impact has attracted little attention because it has accumulated over many centuries of agricultural development.

Capacity of a Stream to Transport Load

The maximum solid load of debris that can be carried by a stream at a given discharge is a measure of the *stream capacity*. Load is stated as the weight of material moved through the stream cross section in a given unit of time, commonly in units of metric tons per day. Total solid load includes both the bed load and the suspended load.

Where a stream is flowing in a channel of hard bedrock, it may not be able to pick up enough alluvial material to supply its full capacity for bed load. When a flood occurs, the channel cannot be quickly deepened in response. Such conditions exist in streams occupying deep gorges and having steep gradients. Conditions are different where thick layers of silt, sand, and gravel underlie the channel. As the speed of flow increases, the stream easily picks up and sets in motion all the alluvial material that it is capable of moving. In other words, the

increasing capacity of the stream for load is easily satisfied.

Capacity for load increases sharply with an increase in the stream's velocity, because the swifter the current is the more intense is the turbulence and the stronger is the dragging force against the bed. Capacity to move bed load goes up about as the third to fourth power of the velocity. This means that when a stream's velocity is doubled in flood, its ability to transport bed load is increased from eight to sixteen times. It is no wonder, then, that most of the conspicuous changes in the channel of a stream occur in flood stage, with few important changes occurring in low stages.

Stream Gradation

A trunk stream, fully developed within its drainage basin, has undergone a long period of adjustment of its channel so that it can discharge not only the surplus water produced by the basin but also the solid load that the tributary channels supply. The transport of bed load requires a gradient. A stream channel system adjusts its gradient to achieve an average balanced state of operation, year in and year out and from decade to decade. In this equilibrium condition the stream is referred to as a *graded stream*.

It will be helpful in developing the concept of a graded stream system to investigate the changes that will take place along a stretch of stream that is initially poorly adjusted to the transport of its load. Such an ungraded channel is illustrated in Figure 15.11. The starting profile is imagined to be brought about by crustal uplift in a series of fault steps, bringing to

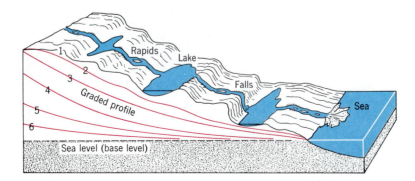

FIGURE 15.11 Schematic diagram of gradation of a stream. Originally the channel consists of a succession of lakes, falls, and rapids. (Drawn by A. N. Strahler. From A. N. Strahler, *Planet Earth*, Harper & Row, New York. Copyright © 1972 by Arthur N. Strahler.)

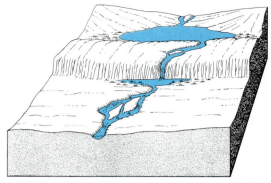

A. Stream established on a land surface dominated by landforms of recent tectonic activity.

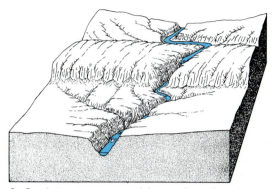

B. Gradation in progress; lakes and marshes drained; deepening gorge; tributary valleys extending.

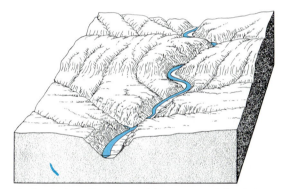

C. Graded profile attained; beginning of floodplain development; valley widening in progress.

D. Floodplain widened to accommodate meanders; floodplains extended up tributary valleys.

FIGURE 15.12 Evolution of a stream and its valley. (Drawn by E. Raisz.)

view a surface that was formerly beneath the ocean and exposing it to fluvial processes for the first time. Overland flow collects in shallow depressions, which fill and overflow from higher to lower levels. In this way a throughgoing channel originates and begins to conduct runoff to the sea.

Figure 15.12 illustrates the gradation process in a series of block diagrams. Over *waterfalls* and *rapids,* which are simply steep-gradient portions of the channel, flow velocity is greatly increased and abrasion of bedrock is most intense (Block *A*). As a result, the falls are cut back and the rapids trenched, and the ponded reaches are filled by sediment and are lowered in level as the outlets are cut down. In time the lakes disappear and the falls are transformed into

rapids. Erosion of rapids reduces the gradient to one more closely approximating the average gradient of the entire stream (Block *B*). At the same time, branches of the stream system are being extended into the landmass, carving out a drainage basin and transforming the original landscape into a fluvial landform system.

In the early stages of gradation and extension, the capacity of the stream exceeds the load supplied to it, so that little or no alluvium accumulates in the channels. Abrasion continues to deepen the channels, with the result that they come to occupy steep-walled *gorges* or *canyons* (Figure 15.13). Weathering and mass wasting of these rock walls contribute an increasing supply of rock debris to the channels. Debris shed

FIGURE 15.13 The Grand Canyon of the Yellowstone River has been carved in volcanic rock (rhyolite) of the Yellowstone Plateau. Through the chemical action of rising hot-water (hydrothermal) solutions, the rock has received its pastel shades of color. Yellowstone National Park, Wyoming. (Arthur N. Strahler.)

from land surfaces contributing overland flow to the newly developed branches is also on the increase.

We can anticipate a gradual decrease in the stream's capacity for bed load, resulting from the gradual reduction in the channel gradient. This decrease will be converging on the increasing load with which it is being supplied. There will come a time at which the supply of load exactly matches the stream's capacity to transport it. At this point the stream has achieved the graded condition and possesses a *graded profile* and descends smoothly and uniformly in the downstream direction (Figure 15.11).

It is important to understand that the balance between load and a stream's capacity exists only as an average condition over periods of many years. We are aware that streams scour their channels in flood and deposit load when in falling stage.

Perhaps in terms of conditions of the moment, a stream is rarely in equilibrium; but, over long periods of time, the graded stream maintains its level. After attaining this state of balance, the stream continues to cut sidewise into its banks on the outside of bends. This *lateral cutting* does not appreciably alter the gradient and therefore does not materially affect the equilibrium.

The first indication that a stream has attained a graded condition is the beginning of floodplain development. On the outside of a bend the channel shifts laterally into a curve of larger radius and thus undercuts the valley wall. On the inside of the bend alluvium accumulates in the form of a sand bar. Widening of the bar deposit produces a crescentic piece of low ground, which is the first stage in floodplain development. This stage is illustrated in Figure 15.12, Block *C*. As lateral cutting continues, the floodplain strips are widened, and the channel develops sweeping bends, called *alluvial meanders* (Block *D*). The floodplain is then widened into a continuous belt of flat land between steep valley walls.

Floodplain development reduces the frequency with which channel scour attacks and undermines the adjacent valley wall. Weathering, mass wasting, and overland flow then act to reduce the steepness of the valley-side slopes (Figure 15.14). As a result, in a humid climate, the gorgelike aspect of the valley gradually disappears and eventually gives way to an open valley with soil-covered slopes protected by a dense plant cover.

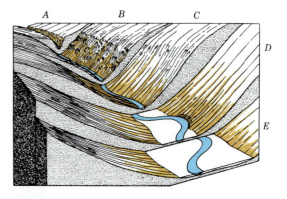

FIGURE 15.14 Following stream gradation the valley walls become gentler in slope and the bedrock is covered by soil and weathered rock. (Drawn by W. M. Davis.)

FIGURE 15.15 Victoria Falls is located on the Zambesi River, which forms the border between Zambia and Zimbabwe in southern Africa. The water falls 128 m (420 ft) to the bottom of a deep, narrow gorge excavated along a fault line. (Harm J. deBlij.)

Environmental Significance of Gorges and Waterfalls

Deep gorges and canyons of major rivers exert environmental controls in a variety of ways. There is little or no room for roads or railroads between the stream and the valley sides, so that roadbeds must be cut or blasted at great expense and hazard from the sheer rock walls. Maintenance is expensive because of flooding or undercutting by the stream and the sliding and falling of rock, which can damage or bury the roadbed. But a gorge may afford the only feasible passage through a mountain range. Consider, also, that a deep canyon is a barrier to movement across it and may require construction of expensive bridges.

Although small waterfalls are common features of alpine mountains carved by glacial erosion (Chapter 18), large waterfalls on major rivers are comparatively rare the world over. Faulting and dislocation of large crustal blocks has caused spectacular waterfalls on several African rivers (Figure 15.15). New river channels resulting from flow diversions caused by ice sheets of the Pleistocene Epoch provide one class of falls and rapids of large discharge. Certainly the preeminent example is Niagara Falls. Overflow of Lake Erie into Lake Ontario happened to be situated over a gently inclined layer of limestone, beneath which lies easily eroded shale (Figure 15.16). The fall is maintained by continual undermining of the limestone by erosion in the plunge-pool at the base of the fall (Figure 15.17). The height of the falls is now 52 m (170 ft) and its discharge is about 17,000 cms (200,000 cfs). The drop of Niagara Falls is utilized for the production of hydroelectric power by the Niagara Power Project, in which water is withdrawn upstream from the falls and carried in tunnels to generating plants located about 6 km (4 mi) downstream from the falls.

Most large rivers of steep gradient do not possess falls, and it is necessary to build dams to create artifically the vertical drop

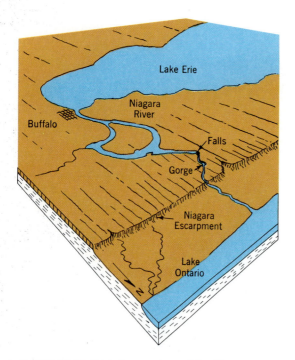

FIGURE 15.16 A bird's-eye view of the Niagara River with its falls and gorge carved in strata of the Niagara Escarpment. View is toward the southwest from a point over Lake Ontario. (Redrawn from a sketch by G. K. Gilbert, 1896. From A. N. Strahler, 1971, *The Earth Sciences,* 2nd ed., Harper & Row, New York.)

necessary for turbine operation. An example is the Hoover Dam, behind which lies Lake Mead, occupying the canyon of the Colorado River.

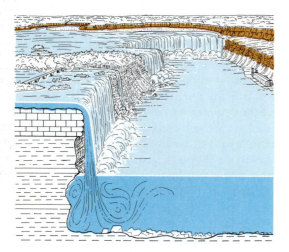

FIGURE 15.17 Niagara Falls is formed where the river passes over the eroded edge of a massive limestone layer. Continual undermining of weak shales at the base keeps the fall steep. (Drawn by E. Raisz.)

We should not lose sight of the esthetic and recreational values of gorges, rapids, and waterfalls of major rivers. The Grand Canyon of the Colorado River, probably more than any single product of fluvial processes, highlights the scenic value of a great river gorge. Against the advantages in obtaining hydroelectric power and fresh water for urban supplies and irrigation by construction of large dams, we must weigh the permanent loss of some large segments of our finest natural scenery, along with the destruction of ecosystems adapted to the river environment. It is small wonder, then, that new dam projects are meeting with stiff opposition from concerned citizen groups who fear that the harmful environmental impacts of such structures far outweigh their future benefits.

Goal of Fluvial Denudation

With the concept of a graded stream in mind, this is a good place to expand our thinking to take in some of the broader aspects of fluvial denudation. Consider a landscape made up of many drainage basins and their branching stream networks. The region is rugged, with steep mountainsides and high, narrow crests (Figure 15.18, Block *A*). Rock debris is being transported out of each drainage basin by the graded trunk streams at the same average rate as the debris is being contributed from the land surfaces within the basin. Obviously, this equilibrium cannot be sustained, since the export of debris must lower the land surface generally. This being the case, the average altitude of the land surface is steadily declining, and this decline must be accompanied by a reduction in the average gradients of all streams (Block *B*). In essence, the fluvial system is consuming its own mass. The mass of rock lying above sea level and available for removal by fluvial action is called the *landmass*.

To describe the ensuing events briefly, as time passes, the streams and valley-side slopes of the drainage basins must undergo gradual change to lower gradients. In theory, the ultimate goal of the denudation process is a reduction of the landmass to a featureless plain at sea level. In this process, sea level projected beneath the landmass

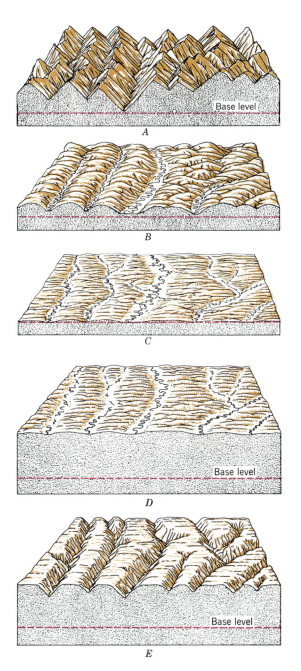

FIGURE 15.18 Reduction of a landmass by fluvial erosion. (*A*) In early stages, relief is great, slopes are steep, and the rate of erosion is rapid. (*B*) In an advanced stage, relief is greatly reduced, slopes are gentle, and rate of erosion is slow. Soils are thick over the broadly rounded hill summits. (*C*) After many millions of years of fluvial denudation, a penelain is formed. Slopes are very gentle and the landscape is an undulating plain. Floodplains are broad, and the stream gradients are extremely low. All of the land surface lies close to base level. (*D*) The peneplain is uplifted. (*E*) Streams trench a new system of deep valleys in the phase of landmass rejuvenation. (Drawn by A. N. Strahler.)

represents the lower limiting level, or *base level*, of the fluvial denudation (labeled on Figures 15.11 and 15.18). But, because the rate of denudation becomes progressively slower, the land surface approaches the base level surface of zero elevation at a slower and slower pace. Under this program, the ultimate goal can never be reached. Instead, after the passage of some millions of years, the land surface is reduced to an undulating surface of low elevation, called a *peneplain* (Block *C*). This term was coined by an American geographer, W. M. Davis, by combining the words "penultimate" and "plain."

Production of a peneplain requires a high degree of crustal and sea-level stability for a period of many millions of years. One region that has been cited as a possible example of a contemporary peneplain is the Amazon–Orinoco basin of South America. This region is a stable continental shield of ancient rock. Peneplains that have been uplifted after their development and are now high-standing land surfaces are numerous within the continental shields. These regions are characterized by an upland surface of uniform elevation. The upland is trenched by stream valleys graded with respect to a lower base level.

Figure 15.18, Block *D*, shows the peneplain of Block *C* uplifted to an elevation of several hundred meters. The base level is now far below the land surface and a large thickness of new landmass has been created. Soon, however, streams begin to trench the landmass and to carve deep, steep-walled valleys, shown in Block *E*. This process is called *landmass rejuvenation*. Landscapes in various stages of rejuvenation are recognized throughout the continental shield area of North America. With the passage of many millions of years, the landscape returns to the rugged stage shown in Block *A* and may ultimately be reduced to a second peneplain.

Aggradation and Alluvial Terraces

A graded stream, delicately adjusted to its supply of water and rock waste from upstream sources, is highly sensitive to changes in those inputs. Changes in climate and in surface characteristics of the

FIGURE 15.19 This braided river channel in the Yukon Territory carries meltwater from a glacier terminus, visible in the distance. Notice that an alluvial fan (left) has been built out into the valley floor by a tributary stream. (John S. Shelton.)

watershed bring changes in discharge and load at downstream points, and these changes in turn require channel readjustments.

Consider first the effect of an increase in bed load beyond the capacity of the stream. At the point on a channel where the excess load is introduced, the coarse sediment accumulates on the stream bed in the form of bars of sand, gravel, and pebbles. These deposits raise the elevation of the stream bed, a process called *aggradation*. As more bed materials accumulate, the stream channel gradient is increased; the increased flow velocity enables bed materials to be dragged downstream and spread over the channel floor at progressively more distant downstream reaches.

Aggradation typically changes the channel cross section from one of narrow and deep form to wide, shallow cross section. Because bars are continually being formed, the flow is divided into multiple threads, and these rejoin and subdivide repeatedly to give a typical *braided stream* (Figure 15.19). The coarse channel deposits spread across the former floodplain and bury fine-textured alluvium with coarse material. We have already seen that alluvium accumulates as a result of accelerated soil erosion.

How is aggradation induced in a stream system by natural processes? Figure 15.19 illustrates one natural cause of aggradation that has been of major importance in stream systems of North America and Eurasia during the Pleistocene Epoch.

Advance of a valley glacier has resulted in the input of a large quantity of coarse rock debris at the head of the valley. Valley

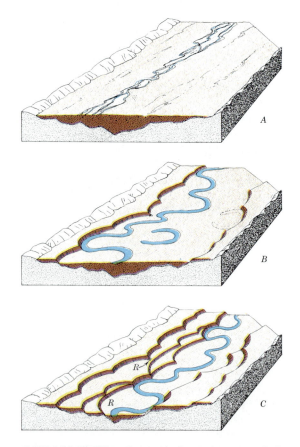

FIGURE 15.20 Alluvial terraces form when a graded stream slowly cuts away the alluvial fill in its valley. The letter "R" marks points where bedrock lies at the surface, protecting the terrace above it from being undercut. (Drawn by A. N. Strahler.)

FIGURE 15.21 Alluvial terraces of the Rakaia River gorge on the South Island of New Zealand. The flat terrace surface in the foreground is used as pasture for sheep. Two higher terrace levels can be seen at the left. (F. Kenneth Hare.)

aggradation was widespread in a broad zone marginal to the great ice sheets of the Pleistocene Epoch; the accumulated alluvium filled most valleys to depths of several tens of meters. Figure 15.20, Block A, shows a valley filled in this manner by an aggrading stream. The case could represent any one of a large number of valleys in New England or the Middle West.

Suppose, next, that the source of bed load is cut off or greatly diminished. In the case illustrated in Figure 15.20, the ice sheets have disappeared from the headwater areas and, with them, the supplies of coarse rock debris. Reforestation of the landscape has restored a protective cover to valley-side and hill slopes of the region, thus holding back coarse mineral particles from entering overland flow. Now the streams have copious water discharges but little bed load. In other words, they are operating below capacity. The result is channel scour and deepening. The channel form becomes deeper and narrower.

Gradually, the stream profile level is lowered, a process of *degradation*. Because the stream is very close to being in the graded condition at all times, its dominant activity is lateral (sidewise) cutting by growth of meander bends, as shown in Figure 15.20, Block B. The valley alluvium is gradually excavated and carried downstream, but it cannot all be removed because the channel encounters hard bedrock in many places. Consequently, as shown in Block C, steplike surfaces remain on both sides of the valley. The treads of these steps are *alluvial terraces*.

Alluvial terraces have always attracted human occupation because of the advantages over both the valley-bottom floodplain, which is subject to annual flooding, and the hill slopes beyond, which may be too steep and rocky to cultivate. Besides, terraces were easily tilled and made prime agricultural land (Figure 15.21). Towns were easily laid out on the flat ground of a terrace, and roads and railroads were easily run along the terrace surfaces parallel with the river.

Induced Aggradation

Channel aggradation is a common form of environmental degradation brought on by human activities. Accelerated soil erosion following cultivation, lumbering, and forest fires is the most widespread source of sediment for valley aggradation. Aggradation of channels has also been a serious form of environmental degradation in coal-mining regions, along with water pollution (acid mine drainage) mentioned in Chapter 10. Throughout the Appalachian coal fields, channel aggradation is widespread because of the huge supplies of coarse sediment from mine wastes. Strip mining has enormously increased the aggradation of valley bottoms because of the vast expanses of crushed rock available to entrainment by runoff.

Urbanization and highway construction are also major sources of excessive sediment that causes channel aggradation. Large earth-moving projects are involved in creating highway grades and preparing sites for industrial plants and housing developments. Although these surfaces are eventually stabilized, they are vulnerable to erosion for periods ranging from months to years. The regrading involved in these projects often diverts overland flow into different flow paths, further upsetting the activity of streams in the area.

Mining, urbanization, and highway construction not only cause drastic increases in bed load, which cause channel aggradation close to the source, but also increase the suspended load of the same streams. Suspended load travels downstream and is eventually deposited as silt and clay in lakes, reservoirs, and estuaries far from the source areas. This sediment is particularly damaging to the bottom environments of aquatic life. The accumulation of sediment reduces the capacity of reservoirs and limits the useful life of a large reservoir to perhaps a century or so. Excess sediment also results in rapid filling of tidal estuaries, thus requiring increased dredging of channels.

Alluvial Rivers and Their Floodplains

An *alluvial river* is one that flows on a thick accumulation of alluvial deposits constructed by the river itself in earlier stages of its activity. A characteristic of an alluvial river is that it experiences overbank floods with a frequency ranging between annual and biennial occurrence during the season of large water surplus over the watershed. Overbank flooding of an alluvial river normally inundates part or all of a floodplain that is bounded on either side by rising slopes, called *bluffs*.

Typical landforms of an alluvial river and its floodplain are illustrated in Figure 15.22. Dominating the floodplain is the meandering river channel itself, and also

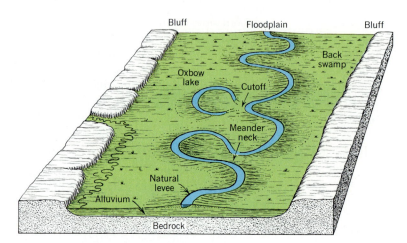

FIGURE 15.22 Floodplain landforms of an alluvial river. (Drawn by A. N. Strahler.)

FIGURE 15.23 An oxbow lake of the Snake River, near Roberts, Idaho. The Snake River can be seen at the upper left. At both ends, the lake is blocked by hygrophytic (swamp) forest. The pattern of nested arcs within the former meander bend are sandbars and intervening low troughs (swales), showing the stages in growth of the meander. (John S. Shelton.)

abandoned reaches of former channels. Meanders develop narrow necks, which are cut through, thus shortening the river course and leaving a meander loop abandoned. This event is called a *cutoff*. It is quickly followed by deposition of silt and sand across the ends of the abandoned channel, producing an *oxbow lake* (Figure 15.23). The oxbow lake is gradually filled in with fine sediment brought in during high floods and with organic matter produced by aquatic plants. Eventually the oxbows are converted into swamps, but their identity is retained indefinitely (Figure 15.24).

During periods of overbank flooding, when the entire floodplain is inundated, water spreads from the main channel over adjacent floodplain deposits (Figure 15.25). As the current rapidly slackens, sand and silt are deposited in a zone adjacent to the channel. The result is an accumulation known as a *natural levee*. Because deposition is heavier closest to the channel and decreases away from the channel, the levee surface slopes away from the channel (Figure 15.22). Figure 15.24 shows broad cultivated levees of the lower Mississippi River. Between the levees and the bluffs is lower ground, called the *backswamp*.

Overbank flooding results not only in the deposition of a thin layer of silt on the floodplain, but also brings an infusion of dissolved mineral substances that enter the soil. As a result of the resupply of nutrients,

floodplain soils retain their remarkable fertility in regions of soil-water surplus from which these nutrients are normally leached away.

Alluvial Rivers and Civilization

Alluvial rivers attracted human habitations long before the dawn of recorded history. Early civilizations arose in the period 4000 to 2000 B.C. in alluvial valleys of the Nile River in Egypt, the Tigris and Euphrates Rivers of Mesopotamia, the Indus River of what is now Pakistan, and the Yellow River of China. Fertile and easily cultivated alluvial soils, situated close to rivers of reliable flow and from which irrigation water was easily lifted or diverted, led to intense utilization of these fluvial zones. With respect to human culture, the alluvial river and its floodplain are sometimes referred to as the riverine environment. Today, about half of the world's population lives in southern and southeastern Asia; the bulk of these persons are small farmers cultivating alluvial soils of seven great river floodplains.

To the agricultural advantages of the alluvial zone is added the value of the river itself as an artery of transportation. Navigability of large alluvial rivers led to growth of towns and cities, many situated at the outsides of meander bends, where deep water lies close to the bank, or at points on

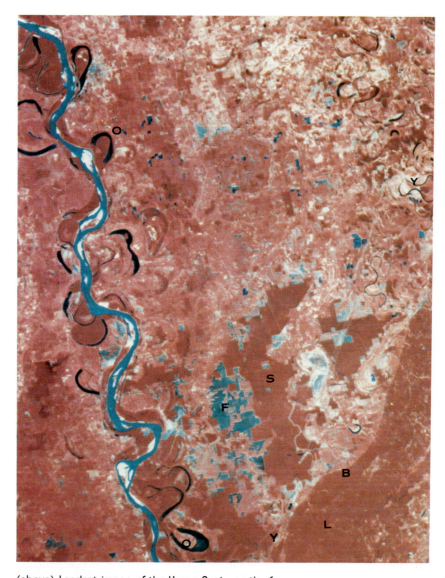

(above) An abandoned section of the Mississippi River near Mound Bayou, Louisiana. Soil color differences in the cultivated area reveal old bars and swales. (Arthur N. Strahler.)

FIGURE 15.24 The Mississippi River floodplain.

(above) Landsat image of the Yazoo Basin, north of Vicksburg, Mississippi, taken in late August 1973. Blue represents water surfaces, such as the Mississippi River (left) and patches of flood water (F) remaining from the great flood of March–April 1973. Oxbow lakes (O) appear as dark blue crescents. Forested land forms solid red patches on backswamp areas (S) of the floodplain. The eastern bluffs (B) of the floodplain show as a sharp line, with forested upland underlain by loess (L) at the right. The Yazoo River (Y) follows the eastern edge of the floodplain. Sand bars in the Mississippi River channel show as pure white. Other white areas are cultivated fields with a large proportion of bare soil. Highways and railroad embankments form thin white lines. (NASA.)

(below) Cultivated natural levees (right) and forested backswamp (background) of the Mississippi river, just south of New Orleans, Louisiana. An artificial levee runs close to the river channel (Orlo E. Childs.)

FIGURE 15.25 This river floodplain in Bangladesh is largely under water during a 1973 flood. Villages occupy the higher ground of the natural levees bordering the river channel. (A. Moldvay/Photo Researchers.)

the floodplain bluffs where the river is close by (e.g., Memphis, Tennessee, and Vicksburg, Mississippi).

Flood Abatement Measures

In the face of repeated disastrous floods, vast sums of money have been spent on a wide variety of measures to reduce flood hazards on the floodplains of alluvial rivers. The economic, social, and political aspects of flood abatement are beyond the scope of physical geography; we will review only the engineering principles applied to the problem. Two basic forms of regulation are: first, to detain and delay runoff by various means on the ground surfaces and in smaller tributaries of the watershed; and second, to modify the lower reaches of the river where floodplain inundation is expected.

The first form of regulation aims at treatment of watershed slopes, usually by reforestation or planting of other vegetative cover to increase the amount of infiltration and reduce the rate of overland flow. This type of treatment, together with construction of many small flood-storage dams in the valley bottoms, can greatly reduce the flood crests in smaller streams of the watershed.

The second type of flood control is designed to protect the floodplain areas directly. The building of *artificial levees,* parallel with the river channel on both sides, can function to contain the overbank flow and prevent inundation of the adjacent floodplain. Artificial levees are broad embankments built of earth and must be high enough to contain the greatest floods; otherwise they will be breached rapidly by great gaps at the points where water spills over. Under the control of the Mississippi River Commission, which began in 1879, a vast system of levees was built along the Mississippi River in the expectation of containing all floods (Figure 15.26). Levees have been continuously improved and now total more than 4000 km (2500 mi) in length and in places are as high as 10 m (30 ft).

Because of natural and artificial levees, the river channel is slowly built up to a higher level than the floodplain, making these "bottom lands," as they are called, subject to repeated inundations and consequent heavy loss of life and property. Floodplains of many of the world's great rivers present serious problems of this type.

FIGURE 15.26 Taken in 1903 on the lower Mississipi River, this photo shows an artificial levee built close to the river channel. A major flood is in progress and the water level of the river has risen nearly to the crest of the levee. In the distance, river steamers have brought work crews to raise the levee crest with planks and sandbags. If a breach occurs, all the lower land on the left will soon be under one to two meters of water. (A breach did occur soon after the photo was taken.) (Photographer not known.)

In China, in 1887, the Huanghe river inundated an area of 130,000 km^2 (50,000 mi^2), causing the direct death of a million people and the indirect death of a still greater number through ensuing famine.

Another measure taken to abate floods on the Mississippi River is the artificial cutoff of meander bends. The river course is thus shortened and the velocity increased, with the effect of reducing the flood crests. Certain parts of the floodplain are set aside as temporary basins into which the river is to be diverted according to plan to reduce the flood crest. In the delta region, floodways are designed to conduct flow from river channels to the ocean by alternate routes.

Fluvial Processes in an Arid Climate

The general appearance of desert regions is strikingly different from that of humid regions, reflecting differences in both vegetation and landforms. Rain falls in dry climates as well as in moist, and most landforms of desert regions are formed by running water. A particular locality in a dry desert may experience heavy rain only once in several years but, when it does fall, stream channels carry water and perform important work as agents of erosion, transportation, and deposition. Although running water is a rather rare phenomenon in dry deserts, it works with more spectacular effectiveness on the fewer occasions when it does act. This is explained by the meagerness of vegetation in dry deserts. The few small plants that survive offer little or no protection to soil or bedrock. Without a thick vegetative cover to protect the ground and hold back the swift downslope flow of water, large quantities of coarse rock debris are swept into the streams. A dry channel is transformed in a few minutes into a raging flood of muddy water heavily charged with rock fragments (Figure 15.27).

An important contrast between regions of arid and humid climates lies in the way in which the water enters and leaves a stream channel. In a humid region with a high water table sloping toward the stream channels, ground water moves steadily

FIGURE 15.27 A flash flood has filled this desert channel with raging, turbid waters. A distant thunderstorm produced the runoff. Coconino Plateau, Arizona. (Arthur N. Strahler.)

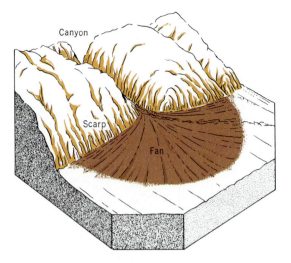

FIGURE 15.28 A simple alluvial fan. (Drawn by A. N. Strahler.)

toward the channels, into which it seeps, producing permanent (perennial) streams.

In arid regions, where streams flow across plains of gravel and sand, water is lost from the channels by seepage to a water table that lies below the level of the channel floor. Loss of discharge by seepage and evaporation strongly affects streams in alluvium-filled valleys of arid regions. Aggradation occurs and braided channels are conspicuous. Streams of desert regions are often short and terminate in alluvial deposits or on shallow, dry lake floors.

Alluvial Fans

One very common landform built by braided, aggrading streams is the *alluvial fan,* a low cone of alluvial sands and gravels resembling in outline an open Japanese fan (Figure 15.28). The apex, or central point of the fan, lies at the mouth of a canyon or ravine. The fan is built out on an adjacent plain. Alluvial fans are of many sizes; some desert fans are many kilometers across (Figure 15.29).

Fans are built by streams carrying heavy loads of coarse rock waste from a mountain or upland region. The braided channel shifts constantly, but its position is firmly fixed at the canyon mouth. The lower part of the channel, below the apex, sweeps back and forth. This activity accounts for the semicircular fan form and the downward slope in all radial directions from the apex.

Large, complex alluvial fans also include mudflows (Figure 15.30). Mud layers are interbedded with sand and gravel layers. Water infiltrates the fan at its head, making its way to lower levels along sand layers that serve as aquifers. The mudflow layers serve as barriers to ground water movement, or aquicludes. Ground water trapped beneath an aquiclude is under pressure from water higher in the fan apex. When a well is drilled into the lower slopes of the fan, water rises spontaneously as artesian flow. (See Chapter 16 and Figure 16.18.)

Alluvial fans are the dominant class of ground water reservoirs in the southwestern United States. Sustained heavy pumping of these reserves for irrigation has lowered the water table severely in many fan areas. Rate of recharge is extremely slow in comparison. Efforts are made to increase this recharge by means of water-spreading structures and infiltrating basins on the fan surfaces.

A serious side effect of excessive ground water withdrawal is that of subsidence of the ground surface. Important examples of subsidence are found in alluvial valleys filled to great depth with alluvial fan and lake-deposited sediments. For example, in one locality in the San Joaquin Valley of California, the water table has been drawn down over 30 m (100 ft). The resulting ground subsidence has amounted to about 3 m (10 ft) over a 35-year period.

FIGURE 15.29 Great alluvial fans extend out upon the floor of Death Valley. The canyons from which they originate have carved deeply into a great uplifted fault block. (Mark A. Melton.)

The Landscape of Mountainous Deserts

Where tectonic activity in the form of block faulting has been active in an area of continental desert, the assemblage of fluvial landforms is particularly diverse. The basin-and-range region of the western United States is such an area; it includes large parts of Nevada and Utah, southeastern California, southern Arizona and New Mexico, and adjacent parts of Mexico. Between uplifted and tilted blocks are down-dropped tectonic basins.

Fluvial denudation has carved up the mountain blocks into a rugged landscape of deep canyons and high divides, as shown in Block A of Figure 15.31. Rock waste furnished by these steep mountain slopes is carried from the mouths of canyons to form numerous large alluvial fans. The fan deposits form a continuous apron extending far out into the basins.

In the centers of desert basins lie the saline lakes and dry lake basins referred to in Chapter 10. Accumulation of fine sediment and precipitated salts produces an extremely flat land surface, referred to in the southwestern United States and in Mexico as a *playa*. Salt flats are found where an evaporite layer forms the surface. In some playas shallow water forms a salt lake.

Figure 15.32 is an air photograph of a mountainous desert landscape in the Death Valley region of eastern California. Three environmental zones can be seen: (1) rugged mountain masses dissected into canyons with steep rocky walls; (2) a piedmont zone of alluvial fans; and (3) the playa occupying the central part of the basin.

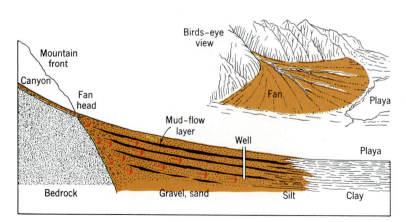

FIGURE 15.30 An idealized cross section of an alluvial fan showing mudflow layers (aquicludes) interbedded with sand layers (aquifers). (From A. N. Strahler, *Planet Earth*, Harper & Row, New York. Copyright © 1972 by Arthur N. Strahler.)

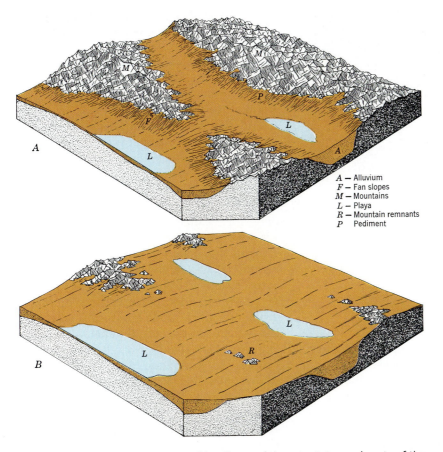

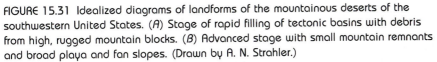

A — Alluvium
F — Fan slopes
M — Mountains
L — Playa
R — Mountain remnants
P Pediment

FIGURE 15.31 Idealized diagrams of landforms of the mountainous deserts of the southwestern United States. (A) Stage of rapid filling of tectonic basins with debris from high, rugged mountain blocks. (B) Advanced stage with small mountain remnants and broad playa and fan slopes. (Drawn by A. N. Strahler.)

FIGURE 15.32 Racetrack Playa, a flat, white plain, is surrounded by alluvial fans and rugged mountains. This desert valley lies in the northern part of the Panamint Range, not far west of Death Valley, California. In the distance rises the steep eastern face of the Inyo Mountains, a great fault block. (John S. Shelton.)

FIGURE 15.33 Bedrock is exposed in places on this pediment, carved upon ancient schist. Fragments of schist form a desert pavement. Cabeza Prieta Range, Arizona. (Mark A. Melton.)

surrounded by areas of alluvial fan and playa surfaces.

Fluvial Processes in Review

Running water as a fluid agent of denudation dominates the sculpturing of the continents. We have examined fluvial systems in which overland flow becomes organized into channel systems that converge the flow into larger and larger streams. As streams erode, transport, and deposit rock material, they shape a wide variety of secondary landforms. Gorges and canyons dominate the earlier stages in carving up of a landmass. These transient forms finally give way to broad floodplains on which meandering rivers move on low gradients. More subtle landforms, such as the natural levees, now dominate the fluvial landscape. Overbank flooding and widespread inundation bring major environmental impact to these flat alluvial lands, where they are intensively occupied by humans. Efforts to control flooding by engineering works often fail, despite enormous inputs of money. Doubts are now raised as to the wisdom of continuing to protect lands naturally subject to river inundations.

It should be obvious that fluvial processes in such a region are limited to local transport of rock particles from a mountain range to the nearest adjacent basin, which receives all the sediment and must be gradually filling as the mountains are diminishing in elevation. Because there is no outflow to the sea, the concept of a base level of denudation has no meaning, and a peneplain does not represent the penultimate stage of denudation, as it does in the humid environment. Each arid basin becomes a closed system so far as mass transport is involved.

As the desert mountain masses are lowered in height and reduced in extent, the fan slopes encroach farther upon the mountain fringes. Close to the receding mountain front there develops a sloping rock floor, called a *pediment* (Figure 15.33). As the remaining mountains shrink further in size, the pediment surfaces expand to form wide rock platforms thinly veneered with alluvium. This advanced stage is shown in Block *B* of Figure 15.31. The desert land surface produced in an advanced stage of fluvial activity is an undulating plain, called a *pediplain*. It consists of areas of pediment

The impact of humans on the action of running water is severe in agricultural and urban areas. Soil erosion is easily induced, but controlled only with great difficulty. Channels are damaged by aggradation from upstream sources of sediment. Reservoirs impounded by great dams are being steadily filled with sediment, and we have good reason to question the desirability of building more of these expensive works.

In this chapter we have also compared the action of running water in deserts with that in humid climates. Although the basic principles are the same everywhere, aridity emphasizes the transport and accumulation of bed load with only short distances separating the supply region from the accumulation region.

CHAPTER 16

LANDFORMS AND ROCK STRUCTURE

LANDFORMS AND ROCK STRUCTURE The nature of the bedrock and its internal structure strongly influence the appearance of landforms in many areas.

STRUCTURAL CONTROL Fluvial processes rapidly lower a surface underlain by weak rock varieties, whereas resistant rock masses remain high to form ridges and plateaus.

HORIZONTAL STRATA Sedimentary strata lying in a horizontal attitude form lines of cliffs rimming benches and table-topped plateaus.

COASTAL PLAINS Sedimentary strata with alternating layers of weak and resistant rock form cuestas and lowlands on coastal plains.

FOLDS AND DOMES Belts of folded strata give rise to a ridge-and-valley landscape, whereas strata uplifted into dome structures produce a circular pattern of ridges and valleys.

FAULT BLOCKS Areas of ancient block faulting show long, straight scarps, which delimit fault trenches and uplifted fault blocks.

METAMORPHIC AND IGNEOUS MASSES Landscapes of metamorphic rock show a distinctive grain of ridges and valleys, whereas large masses of granite lack a grain.

FLUVIAL denudation acts on any landmass exposed to the atmosphere, but those landmasses differ greatly from place to place in rock composition and rock structure. As you might expect, landforms of fluvial denudation can take on a wide variety of shapes and patterns, expressing the complexity of the crustal rock from which they are carved.

Recall from the previous chapter that the mass of rock lying above base level constitutes the landmass. In theory, denudation cannot reduce the land surface below the base level, which is the inland extension of sea level. But the earth's crust can be lifted higher above sea level by means of tectonic movements. Crustal rise can increase the available landmass. Through repeated crustal uplifts and repeated episodes of denudation, deep-seated rocks and structures appear at the surface. In this way the root structures of mountain belts come to be exposed, along with batholiths and other plutons.

In Chapter 13 we classified all landforms as being either initial or sequential in origin. As a landmass is first brought into existence by tectonic activity or volcanic activity, its surface configuration is made up of initial landforms. Initial landforms produced directly by volcanic activity, folding, and faulting were described in Chapter 13. Denudation soon converts the initial landforms into sequential landforms. In this chapter our interest is in the erosional class of sequential landforms, controlled in shape, size, and arrangement by the underlying rock structure.

Denudation acts more rapidly on the weaker rock types, lowering them to produce valley floors and leaving the stronger rocks to stand out in bold relief as ridges and uplands. This is a subject we will pursue further in the present chapter. Rock structure controls the placement of streams and the shapes and heights of the intervening divides. A distinctive assemblage of landforms and stream patterns is developed for each of the major types of crustal structures. The habitats of plants and animals are varied by these landforms. Humans occupy these habitats and exploit them for agricultural lands, communication lines, and urban development. In all these activities a distinctive set of constraints and opportunities is imposed by each of the types of structurally controlled landforms.

Structural Groups of Landmasses

Landmasses fall into three major groups, and each of these in turn contains two or more distinctive types. As major groups, we can recognize the following.

Group A. Undisturbed sedimentary strata. These are thick covers of sedimentary rocks overlying ancient shield rocks of continental lithosphere. The strata are most commonly of marine origin deposited on the floors of continental shelves of passive continental margins and shallow inland seas; they have been brought above sea level by crustal rise and are now in the process of undergoing fluvial denudation. Because no significant amount of bending or faulting has affected these strata, their attitude is nearly horizontal over very large areas.

Group B. Disturbed structures of tectonic activity that has long since ceased. These landmasses show strongly the effects of bending and breaking of the crust by mountain-building processes. Sedimentary strata showing folding and faulting are included in this group. Metamorphic rocks produced in root zones of ancient mountain belts are another type.

Group C. Eroded igneous masses. Exposed large plutons—bodies of intruded igneous rock—are one important type of igneous mass. A quite different type includes extinct composite and shield volcanoes undergoing deep erosion.

Figure 16.1 illustrates eight landmass types within the three major groups. Within Group A, two important types are coastal plains and horizontal strata. Group B includes deeply eroded folds, domes, fault blocks, and metamorphic belts. Group C includes exposed plutons and eroded volcanoes.

A brief overview of each landmass type will give insight into the meaning of diverse landforms of the continents and the ways in which these landforms affect the environment and provide a variety of mineral resources. Each of our national parks and wilderness areas owes its distinctive scenery to the influence of a particular landmass type. These protected natural areas are a national resource on which no dollar value can be placed; their diversity depends largely on the kinds of rocks and structures out of which their landforms have been carved by fluvial denudation.

Rock Structure as a Landform Control

As denudation takes place, landscape features develop in close conformity with patterns of bedrock composition and structure. Figure 16.2 shows five types of sedimentary rock, together with a mass of much older igneous rock on which the sediments were deposited in horizontal layers. The diagram shows the usual landform habit of each rock type and whether it forms valleys or mountains. The cross section shows conventional rock symbols used by geologists. These rock

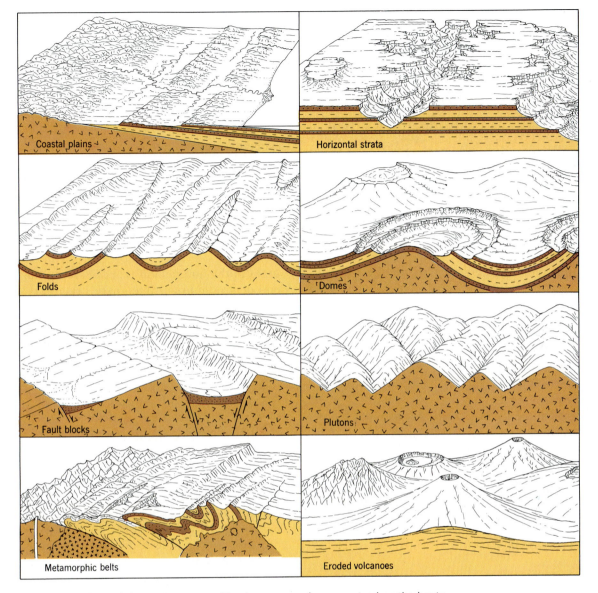

FIGURE 16.1 Several distinctive types of landmasses can be recognized on the basis of rock composition and structure. (Drawn by A. N. Strahler.)

strata have been strongly tilted and deeply eroded.

Shale is a weak rock and is reduced to the lowest valley floors of the region. Limestone, easily subject to carbonic acid action, is also a valley-former in humid climates. On the other hand, limestone is highly resistant and usually stands high in arid climates. Sandstone and conglomerate are typically resistant rocks; they form ridges and uplands. As a group, the igneous rocks are resistant to denudation; they typically form uplands rising above sedimentary strata.

The metamorphic rocks are, as a group, more resistant to denudation than their sedimentary parent types. However, as shown in Figure 16.28, there are conspicuous differences among the types of metamorphic rocks.

Natural layers and planes of weakness are characteristic of the structure of each type of rock. We need a system of geometry to enable us to measure and describe the attitude of these natural planes and to indicate them on maps. Examples of such planes are the bedding layers of sedimentary strata, the sides of a dike, and

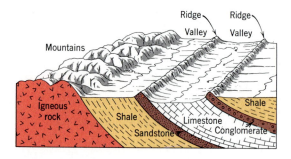

FIGURE 16.2 Landforms evolve through the slow erosional removal of weaker rock, leaving the more resistant rock standing as ridges or mountains. (Drawn by A. N. Strahler.)

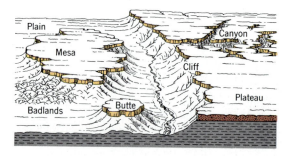

FIGURE 16.4 In arid climates, distinctive erosional landforms develop in horizontal strata. (Drawn by A. N. Strahler.)

the joints in granite. Rarely are these planes truly horizontal or vertical.

The acute angle formed between a natural rock plane and an imaginary horizontal plane is termed the *dip*. The amount of dip is stated in degrees ranging from 0° for a horizontal plane to 90° for a vertical plane. Figure 16.3 shows the dip angle for an outcropping layer of sandstone, against which a horizontal water surface rests. The compass direction of the line of intersection between the inclined rock plane and an imaginary horizontal plane is the *strike*. In Figure 16.3 the strike is north.

Horizontal Strata

Extensive areas of the ancient continental shields are covered by thick sequences of horizontal sedimentary strata. At various

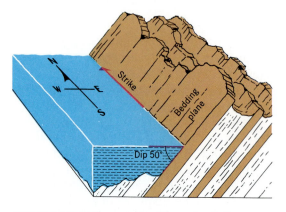

FIGURE 16.3 Strike and dip. (Drawn by A. N. Strahler.)

times in the 600 million years following the end of Precambrian time, these strata were deposited in shallow inland seas. Following crustal uplift, with little disturbance other than minor warping or faulting, these areas became continental surfaces undergoing denudation.

In arid climates, where vegetation is sparse and the action of overland flow especially effective, sharply defined landforms develop on horizontal sedimentary strata (Figure 16.4). The normal sequence of landforms is a sheer rock wall, called a *cliff*. At the base of the cliff is an inclined slope. This slope flattens out to make a bench, terminated at the outer edge by the cliff of the next lower set of forms. In the walls of the great canyons of the Colorado Plateau region, these forms are wonderfully displayed (Figure 16.5).

In these arid regions erosion strips away successive rock layers leaving behind *plateaus* capped by hard rock layers. Cliffs retreat as near-perpendicular surfaces because the weak clay or shale formations exposed at the cliff base are rapidly washed away by storm runoff and channel erosion. When undermined, the rock in the upper cliff face repeatedly breaks away along vertical fractures.

Cliff retreat produces *mesas*, table-topped plateaus bordered on all sides by cliffs (Figure 16.4). Mesas represent the remnants of a formerly extensive layer of resistant rock. As a mesa is reduced in area by retreat of the rimming cliffs, it maintains its flat top. Before its complete consumption the final landform is a small steep-sided hill known as a *butte* (Figure 16.6).

FIGURE 16.5 The Grand Canyon of the Colorado River, Arizona, seen from the South Rim. (Arthur N. Strahler.)

Regions of horizontal strata have branching stream networks formed into a *dendritic drainage pattern* (Figure 16.7). The smaller streams in this pattern take a great variety of directions.

Resources of Horizontal Strata

Horizontal strata have only those minerals and rocks of economic value that are

FIGURE 16.6 A butte of sandstone not far from Ship Rock, northwestern New Mexico. Weak shales form the base of the butte. (Arthur N. Strahler.)

associated with sedimentary rocks. Building stone, such as the Bedford limestone in Indiana or the Berea sandstone in Ohio, is a valuable resource. Limestone may be quarried for use in manufacture of portland cement or as flux in iron smelting. Some important deposits of lead, zinc, and iron ores occur in sedimentary rocks. For example, the lead and zinc mines of the Tristate district (Missouri, Kansas, Oklahoma) are in horizontal limestones. Uranium ores are important in strata of the Colorado Plateau.

Perhaps the greatest mineral resources occurring in sedimentary strata are coal and petroleum. In Chapter 11 we described the occurrence of these fossil fuels. Recall from Chapter 14 that strip mining of coal is an extreme form of environmental degradation, with various undesirable side effects besides the massive scarification of the landscape. As the map of North American coal fields shows (Figure 16.8), large reserves of coal lie close to the surface in areas of horizontal strata in the Great Plains region west of the Mississippi, in the northern High Plains still further west, in sedimentary basins of the Rocky Mountains, and in the Colorado Plateau near the

FIGURE 16.7 In this Gemini VII space photo, canyons carved into horizontal limestone strata show a dendritic drainage pattern. The area is part of the Hadramawt Plateau, near the southern coast of the Arabian peninsula. At the upper left is a broad dry valley floored with sandy alluvium. Width of the area is about 160 km (400 mi). (NASA No. S65–64010.)

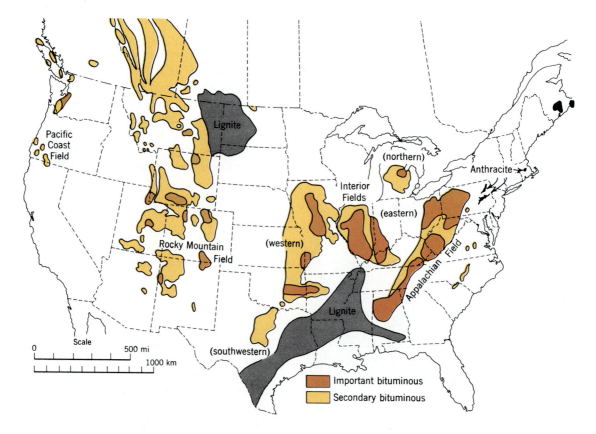

FIGURE 16.8 Coal fields of the United States and southern Canada.

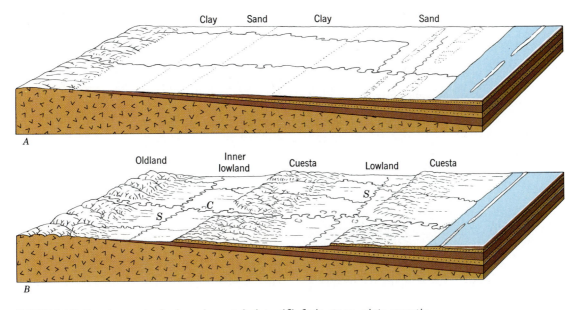

Clay   Sand   Clay   Sand

A

Oldland   Inner lowland   Cuesta   Lowland   Cuesta

B

FIGURE 16.9 Development of a broad coastal plain. (*A*) Early stage: plain recently emerged. (*B*) Advanced stage: cuestas and lowlands developed. *S* = subsequent stream; *C* = consequent stream. (Drawn by A. N. Strahler.)

common corner of the states of Utah, Arizona, Colorado, and New Mexico. The great demands for this coal in decades to come make it likely that large areas will be strip mined, and that power generating plants of large capacity will be built close to the supplies of coal. Environmentalists are greatly concerned with the land scarification and air pollution that such developments can bring.

Coastal Plains

Coastal plains are found along passive continental margins largely free of tectonic activity. Block *A* of Figure 16.9 shows a coastal zone that has recently emerged from beneath the sea. Formerly this zone was a shallow continental shelf accumulating successive layers of sediment brought from the land and distributed by currents. On the newly formed land surface, streams flow directly seaward, down the gentle slope. A stream of this origin is a *consequent stream,* defined as any stream whose course is controlled by the initial slope of a land surface. Consequent streams occur on many landforms, such as volcanoes, fault blocks, or beds of drained lakes.

In an advanced stage of coastal-plain development a new series of streams and topographic features has developed (Block *B*). Where more easily eroded strata (usually clay or shale) are exposed, denudation is rapid, making *lowlands*. Between them rise broad belts of hills called *cuestas*. Cuestas are commonly underlain by sand, sandstone, limestone, or chalk. The lowland lying between the area of older rock and the first cuesta is called the *inner lowland*.

Streams that develop along the trend of the lowlands, parallel with the shoreline, are of a class known as *subsequent streams*. They take their position along any belt or zone of weak rock and therefore follow closely the pattern of rock exposure. Subsequent streams occur in many regions, and we will mention them again in the discussion of folds, domes, and faults. The drainage lines on a fully dissected coastal plain combine to form a *trellis drainage pattern*. In this pattern the subsequent streams trend at about right angles to the consequent streams.

The coastal plain of the United States is a major geographical region, ranging in width from 160 to 500 km (100 to 300 mi) and extending for 3000 km (2000 mi) along the Atlantic and Gulf coasts. The coastal plain starts at Long Island, New York, as a partly

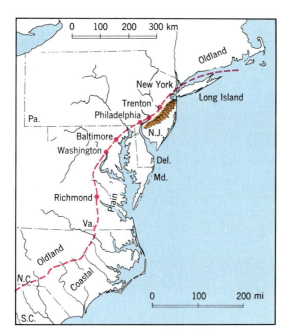

FIGURE 16.10 The coastal plain of the Atlantic seaboard states shows little cuesta development except in New Jersey. The inner limit of the coastal plain is marked by a series of Fall Line cities. (After A. K. Lobeck.)

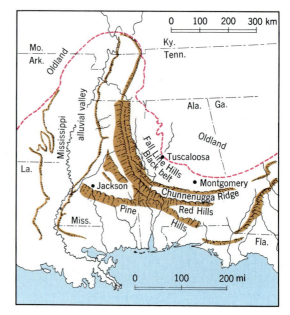

FIGURE 16.11 The Alabama–Mississippi coastal plain is belted by a series of sandy cuestas and shale lowlands. (After A. K. Lobeck.)

submerged cuesta, and widens rapidly southward to include much of New Jersey, Delaware, Maryland, and Virginia (Figure 16.10). Throughout this portion the coastal plain has but one cuesta. The inner lowland is a continuous broad valley developed on weak clay strata.

In Alabama and Mississippi the coastal plain is fully dissected. Cuestas and lowlands run in belts roughly parallel with the coast (Figure 16.11). The cuestas are underlain by sandy formations and support pine forests. Limestone forms lowlands, such as the Black Belt in Alabama. This belt is named for its dark, fertile soils.

Environmental and Resource Aspects of Coastal Plains

Broad coastal plains, such as those of the eastern United States, show intensive agricultural development because of the soil fertility and easy cultivation of broad lowlands. Cuestas provide valuable pine forests in the southern United States.

Transportation tends to follow the lowlands and to connect the larger cities

located there. For example, important highways and railroads connect New York City with Trenton, Philadelphia, Baltimore, Washington, and Richmond, all of which are situated in the inner lowland (Figure 16.10). The inner limit of coastal plain strata is known as the Fall Line, because here the major rivers have developed falls and rapids on the resistant oldland rocks. Cities grew at the Fall Line, where the rivers become navigable tidal estuaries. Thus geology has played a part in the location of the nuclei of Megalopolis.

The seaward dip of sedimentary strata in a coastal plain provides a structure favorable to the development of artesian water wells (Figure 16.12). Water penetrates

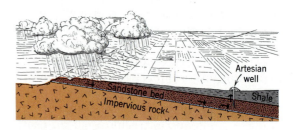

FIGURE 16.12 An artesian well requires a dipping sandstone layer. (Drawn by E. Raisz.)

deeply into sandy cuesta strata, overlain by shale or clay. (See Figure 16.18 for a geologic explanation.) When a well is drilled into the sand formation considerably seaward of its surface exposure, water under pressure reaches the surface. Artesian water in large quantities is available in many parts of the Atlantic and Gulf coastal plains, although it is no longer sufficient to supply the demands of densely populated and industrialized communities.

Of the economic mineral resources of broad coastal plains the most important are petroleum and natural gas. The Gulf Coastal Plain is a major oil-producing region. Here the strata beneath the coastal plain and offshore continental shelf are extremely thick. Lignite, a low grade of coal, occurs beneath a large area of Alabama, Mississippi, Louisiana, and Texas (see Figure 16.8).

Other mineral deposits of economic importance in coastal plains include: sulfur, occurring in the coastal plain of Louisiana and Texas; phosphate beds, found in Florida; and clays, used in manufacture of pottery, tile, and brick in New Jersey and the Carolinas.

Sedimentary Domes

A distinctive type of tectonic structure is the *sedimentary dome,* a circular or oval structure in which strata have been raised into a domed shape (Figure 16.13). Sedimentary domes occur in various places within the covered shield areas of the continents. Igneous intrusions at great depth may have been responsible for certain of these uplifts.

Erosion features of a sedimentary dome are illustrated in the two block diagrams of Figure 16.14. Strata are first removed from the summit region of the dome, exposing older strata beneath. Eroded edges of steeply dipping strata form sharp-crested sawtooth ridges called *hogbacks* (Figure 16.15). When the last of the strata have been removed, the ancient shield rock is exposed in the central core of the dome, which then develops a mountainous terrain.

The stream network on a deeply eroded dome shows dominant subsequent streams forming a circular system, called an *annular drainage pattern* (Figure 16.16). The shorter

FIGURE 16.13 A small, nearly circular sedimentary dome near Sundance, Wyoming. (John S. Shelton.)

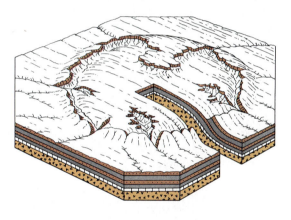

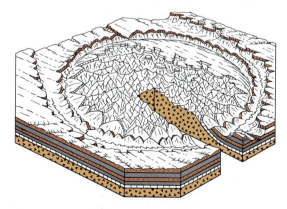

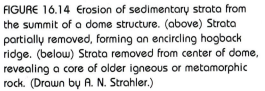

FIGURE 16.14 Erosion of sedimentary strata from the summit of a dome structure. (above) Strata partially removed, forming an encircling hogback ridge. (below) Strata removed from center of dome, revealing a core of older igneous or metamorphic rock. (Drawn by A. N. Strahler.)

tributaries make a radial arrangement. The total pattern resembles a trellis pattern bent into a circular form.

The Black Hills Dome

A classic example of a large and rather complex sedimentary dome is the Black Hills dome of western South Dakota and eastern Wyoming (Figure 16.17). Valleys that encircle this dome are ideal locations for railroads and highways, so it is natural that towns and cities should have grown in these valleys. One valley in particular, the Red Valley, is continuously developed around the entire dome and has been called the Race Track because of its shape. It is underlain by a weak shale, which is easily washed away. The cities of Rapid City, Spearfish, and Sturgis are located in the Red Valley. On the outer side of the Red Valley is a high, sharp hogback of Dakota sandstone, known simply as Hogback Ridge. It rises some 150 m (500 ft) above the level of the Red Valley. Farther out toward the margins of the dome the strata are less steeply inclined and form a series of cuestas. Artesian water is obtained from wells drilled in the surrounding plain.

The eastern central part of the Black Hills consists of a mountainous core of intrusive and metamorphic rocks. These mountains are richly forested, whereas the intervening valleys are beautiful open parks. Thus the region is attractive as a summer resort area. Harney Park, elevation 2207 m (7242 ft), is the highest peak of the core. In the northern part of the central core, in the vicinity of Lead and Deadwood, there are valuable ore deposits. At Lead is the fabulous Homestake Mine, one of the world's richest gold-producing mines.

The western central part of the Black Hills consists of a limestone plateau deeply carved by streams. The original dome had a flattened summit. The limestone plateau represents one of the last remaining sedimentary rock layers to be stripped from the core of the dome.

Artesian Ground Water in Domes

Where sedimentary strata are inclined, the ground water flow system may be quite

FIGURE 16.15 Hogbacks of sandstone lie along the eastern base of the Colorado Front Range, which is a large domed structure. The view is southward toward the city of Boulder. The rugged forested terrain in the distance is developed on igneous and metamorphic rock exposed in the core of the uplift. (John S. Shelton.)

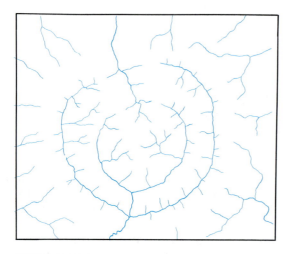

FIGURE 16.16 The drainage pattern on an eroded dome combines annular and radial elements.

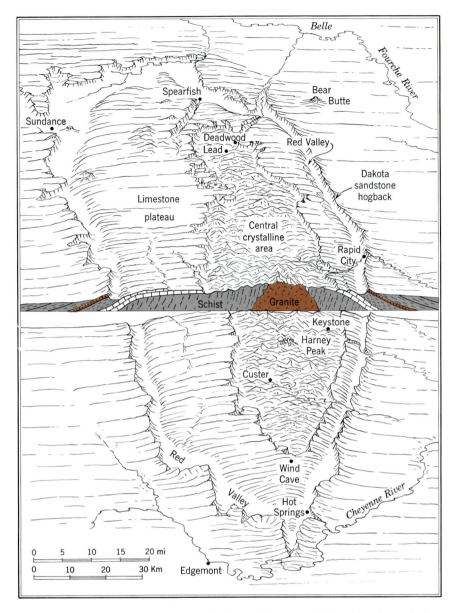

FIGURE 16.17 The Black Hills consist of a broad, flat-topped dome deeply eroded to expose a core of igneous and metamorphic rocks. (Drawn by A. N. Strahler.)

different from the simple flow pattern shown in Figure 10.6. On the flanks of a dome structure a favorable situation may exist for artesian ground water flow, illustrated in Figure 16.18. (The vertical scale is exaggerated to show the principle.) The eroded edge of a sandstone layer appears at the surface on high ground. In this case, the sandstone is porous, providing a large ground water storage reservoir through which ground water can move easily. A water-transmitting body is termed an *aquifer*. In contrast, the overlying shale layer is dense and lacks open pores; it impedes the flow of ground water and is called an *aquiclude*.

Water from precipitation enters the sandstone formation, then moves to great depths through the inclined aquifer to occupy a position below the adjacent valley floor. Here, the water is under confining pressure and cannot escape because it is trapped beneath the shale aquiclude. When a well is drilled in the valley, ground water

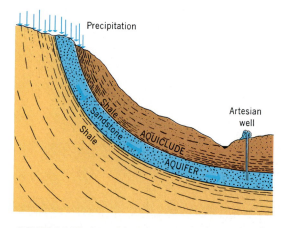

FIGURE 16.18 Ground water in a sandstone aquifer is held under pressure beneath a capping aquiclude of shale. Water rises above the ground surface from an artesian well.

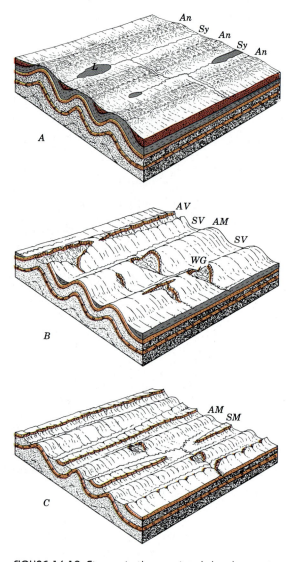

FIGURE 16.19 Stages in the erosional development of folded strata. (A) Initial folds of tectonic activity are shown here. While folding is still in progress, erosion cuts down the anticlines, and alluvium fills the synclines, keeping relief low. An = anticline; Sy = syncline; L = lake. (B) Long after folding has ceased, erosion exposes a highly resistant layer of sandstone or quartzite. AV = anticlinal valley; SV = synclinal valley; WG = watergap. (C) Continued erosion partly removes the resistant formation but reveals another below it. AM = anticlinal mountain; SM = synclinal mountain. (Drawn by A. N. Strahler.)

rises spontaneously and may issue copiously from the well. A well of this type is called an *artesian well*. The name comes from a French word, *artésien*, pertaining to the province of Artois in northern France, where the first artesian wells were drilled as early as 1750.

The Dakota sandstone of the flanks of the Black Hills and beneath large areas of the Great Plains region has been a major source of artesian ground water in past decades. Numerous irrigation wells were drilled into the Dakota formation early in this century, and artesian water was withdrawn at a rate greatly exceeding the rate of discharge. Consequently, few of these wells produce surface flow today and most must be pumped to furnish water.

Fold Belts

According to the model of mountain making developed in Chapter 12; strata of the continental margins are deformed into folds along narrow belts during arc–continent collision (see Figure 12.19). These are foreland folds, illustrated in Figures 13.14 and 13.15. Long after folding has ceased, denudation exposes the mountain roots, bringing the folded strata into relief. Alternating hard and soft rock layers give the crust a strong grain, like the grain in a wooden plank. Denudation removes the

weaker bands of the grain, allowing the harder bands to stand out in relief, much as surfwood is etched by sand and waves.

As we explained in Chapter 13, a troughlike downbend of strata is a syncline; the archlike upbend next to it is an anticline

FIGURE 16.20 Like an old wooden plank deeply etched by drifting sand to reveal the grain, the surface of south-central Pennsylvania shows zigzag ridges formed by bands of hard quartzite. Strata were crumpled into folds during a collision of continents that took place over 200 million years ago to produce the Appalachians. In this image, obtained by a Landsat orbiting satellite, the ridges may appear as trenches; if so, try inverting the page to see the relief as it actually is. The area shown is about 160 km (100 mi) in width. (NASA.)

(Figure 13.13). Obviously, in a belt of folded strata, synclines alternate with anticlines, like a succession of wave troughs and wave crests on the ocean surface.

A series of three block diagrams in Figure 16.19 shows some of the distinctive landforms resulting from fluvial denudation of a belt of folded strata. Erosion of these simple, open folds produces a *ridge-and-valley landscape*, as weaker formations of shale and limestone are eroded away, leaving hard sandstone strata to stand in bold relief as long, narrow ridges (Figure 16.20). On eroded folds, the stream network is distorted into a trellis drainage pattern (Figure 16.21). The principal elements of this pattern are long, parallel subsequent streams occupying the narrow valleys of weak rock.

The folds illustrated in Figure 16.19 are continuous and even-crested; they produce ridges that are approximately parallel in trend and continue for great distances. In some fold regions, however, the folds are not continuous and level-crested. Instead, the fold crests rise or descend from place to place. A descending fold crest is said to plunge. Plunging folds give rise to a zigzag line of ridges (Figure 16.22).

At this point you should refer back to

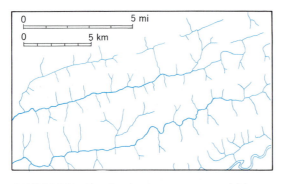

FIGURE 16.21 A trellis drainage pattern on folds.

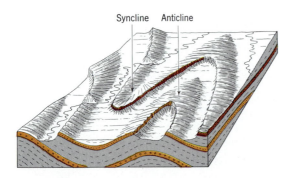

FIGURE 16.22 Folds with crests that plunge downward give rise to zigzag ridges following erosion. (Drawn by E. Raisz.)

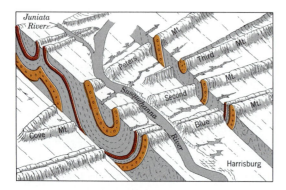

FIGURE 16.23 A great synclinal fold, involving three resistant quartzite formations and thick intervening shales, has been eroded to form bold ridges through which the Susquehanna River has cut a series of watergaps. Find this same area in Figure 16.20. (Based on a drawing by A. K. Lobeck.)

Figure 2.28, a radar image of a portion of the area shown in Figure 16.20; it lies about 4 cm (1.5 in.) from the top and 2.5 cm (1.0 in.) from the left. Using the block diagram of Figure 6.22 as a guide, identify three conspicuous plunging anticlines and three plunging synclines on the radar image.

Environmental and Resource Aspects of Fold Belts

Some of the environmental and resource aspects of dissected fold regions are illustrated by the Appalachians of southcentral and eastern Pennsylvania (Figures 16.20 and 16.23). The ridges, of resistant sandstones and conglomerates, rise boldly to heights of 600 m (2000 ft) above broad lowlands underlain by weak shales and limestones. Major highways run in the valleys, crossing from one valley to another through the watergaps of streams that have cut through the ridges. Important cities are situated near the watergaps of major streams. An example is Harrisburg, located where the Susquehanna River issues from a series of watergaps. Where no watergaps are conveniently located, roads must climb in long, steep grades over the ridge crests. The ridges are heavily forested; the valleys are rich agricultural belts.

In various fold regions of the world (e.g., in the Pennsylvania Appalachians), an important resource is anthracite, or hard coal (Figure 16.24). This coal occurs in strata that have been tightly folded. Pressure has converted the coal from bituminous into anthracite. Because of extensive erosion, all coal has been removed except that which lies in the central parts of synclines. The coal seams dip steeply; workings penetrate deeply to reach the coal that lies in the bottoms of the synclines. Seams near the surface are worked by strip mining.

Broad, gentle anticlinal folds may form important traps for the accumulation of petroleum. The principle is shown in Figure 16.25. Oil migrates in sandstone beds to the anticlinal crest, where it is trapped by an impervious cap rock of shale. Many of the oil and gas pools of western Pennsylvania, where petroleum production first succeeded, are on low anticlines.

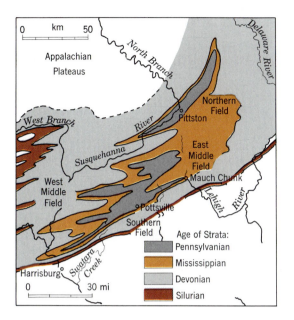

FIGURE 16.24 Anthracite coal basins of central Pennsylvania correspond with areas of Pennsylvanian strata downfolded into long synclinal troughs.

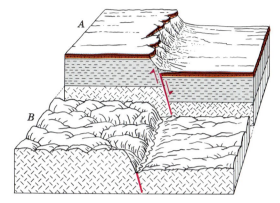

FIGURE 16.26 (A) Fault scarp. (B) Fault-line scarp. (From A. N. Strahler, *The Earth Sciences*, 2nd ed., Harper & Row, New York. Copyright © 1971 by Arthur N. Strahler.)

Erosion Forms on Fault Structures

In Chapter 13 we found that active normal faulting produces a sharp surface break called a fault scarp. Repeated faulting may produce a great rock cliff hundreds of meters high (see Figure 13.16). Erosion quickly modifies a fault scarp but, because the fault plane extends hundreds of meters down in bedrock, its effects on erosional landforms persist for long spans of geologic time. Figure 16.26 shows both the original fault scarp (above) and its later landform expression (below). Even though the cover of sedimentary strata has been completely removed, exposing the ancient shield rock, the fault continues to produce a landform. Because the fault plane is a zone of weak rock, crushed during faulting, it is occupied by a subsequent stream. The scarp persisting along the upthrown side is called a *fault-line scarp*. Fault-line scarps are numerous in parts of the exposed and covered continental shields.

Figure 16.27 shows some erosional features of a large, tilted fault block. The freshly uplifted block has a steep mountain face, but this is rapidly dissected into deep canyons. The upper mountain face reclines in angle as it is removed, and rock debris accumulates in the form of alluvial fans adjacent to the fault block. Vestiges of the fault plane are preserved in a line of triangular facets, the snubbed ends of ridges between canyon mouths.

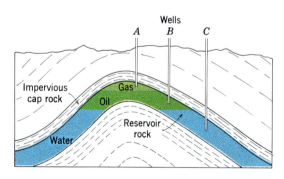

FIGURE 16.25 Idealized cross section of an oil pool on a dome structure in sedimentary strata. Well A will draw gas; well B will draw oil; and well C will draw water. The cap rock is shale; the reservoir rock is sandstone. (From A. N. Strahler, *Planet Earth*, Harper & Row, New York. Copyright © 1972 by Arthur N. Strahler.)

Metamorphic Belts

Where strata have been tightly folded and altered into metamorphic rocks during continental collision, denudation develops a landscape with a strong grain of ridges and

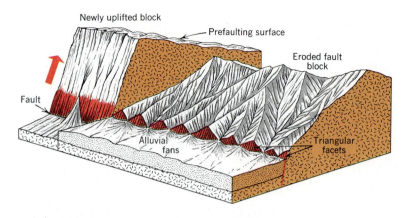

FIGURE 16.27 Erosion of a tilted fault block produces a rugged mountain range. A line of triangular facets at the mountain base marks the position of the original fault plane. (Drawn by A. N. Strahler.)

valleys. These features lack the sharpness and parallelism of ridges and valleys in belts of open folds, but each belt clearly reflects the greater or lesser resistance to denudation offered by each type of metamorphic rock (Figure 16.28). Ridges and valleys reflect different rates of denudation of parallel belts of metamorphic rocks, such as schist, slate, quartzite, and marble. Marble forms valleys; slate and schist make hill belts; quartzite stands out boldly and may produce conspicuous narrow hogbacks.

Much of New England, particularly the Taconic and Green mountains, illustrates the landforms of a metamorphic belt. The larger valleys trend north and south and are underlain by marble. These are flanked by ridges of gneiss, schist, slate, or quartzite.

Vast areas of the exposed continental shields show complex folding of ancient metamorphic rocks of Precambrian age (Figure 16.29). These structures are interpreted as the remains of numerous continental collisions that occurred during early growth stages of the major shield areas.

Exposed Batholiths

Batholiths, those huge plutons of intrusive igneous rock, are eventually uncovered by erosion and gain surface expression. Batholiths of granitic composition are a major ingredient in the mosaic of ancient rocks comprising the continental shields. A good example is the Idaho batholith, a granite mass exposed over an area of about 40,000 km^2 (16,000 mi^2)—a region almost as large as New Hampshire and Vermont combined (Figure 16.30). Another example is the Sierra Nevada batholith of California; it makes up most of the high central part of that great mountain block. The dendritic drainage pattern is often well developed on an eroded batholith (Figure 16.31).

Small bodies of granite, representing domelike projections of batholiths that lie below, are often found surrounded by ancient metamorphic rocks, into which the granite was intruded (Figure 16.32). A good example is Stone Mountain, which rises above the Piedmont Upland of Georgia (Figure 16.33).

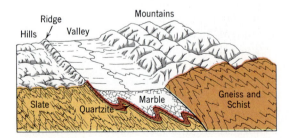

FIGURE 16.28 Metamorphic rocks form long, narrow parallel belts of valleys and mountains. (Drawn by A. N. Strahler.)

FIGURE 16.29 The Labrador fold belt in Quebec and Newfoundland, displayed by Landsat remote-sensing imagery. This region was reduced to a peneplain (perhaps repeatedly), which has since been etched by fluvial and glacial erosion to reveal complex folds of metamorphic rocks derived from Precambrian sedimentary strata. Lakes, occupying ice-scoured rock basins, appear in black. (NASA 1483–15013, October 4, 1973.)

FIGURE 16.30 The Idaho batholith, exposed in the Sawtooth Mountains of south-central Idaho, in the northern Rocky Mountains. (John S. Shelton.)

FIGURE 16.31 This dendritic drainage pattern is developed on the dissected Idaho batholith shown in Figure 16.30.

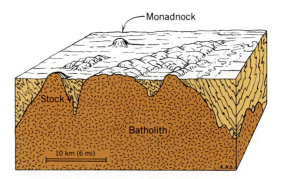

FIGURE 16.32 Batholiths appear at the land surface only after long-continued erosion has removed thousands of meters of overlying rocks. Small projections of the granite intrusion (known as stocks) appear first and are surrounded by older rock. (Drawn by A. N. Strahler.)

The name *monadnock* has been given to an isolated mountain or hill (such as Stone Mountain), that rises conspicuously above a peneplain. A monadnock develops because the rock within the monadnock is much more resistant to denudation processes than the bedrock of the surrounding region (Figure 16.32). The name, which was first used nearly a century ago by William M. Davis, is taken from Mount Monadnock in southern New Hampshire.

FIGURE 16.33 Stone Mountain, Georgia, is a striking erosion remnant, or monadnock, about 2.4 km (1.5 mi) long and rising 193 m (650 ft) above the surrounding Piedmont peneplain surface. The rock is granite, almost entirely free of joints, and has been rounded into a smooth dome by weathering processes. (Landis Aerial Photo.)

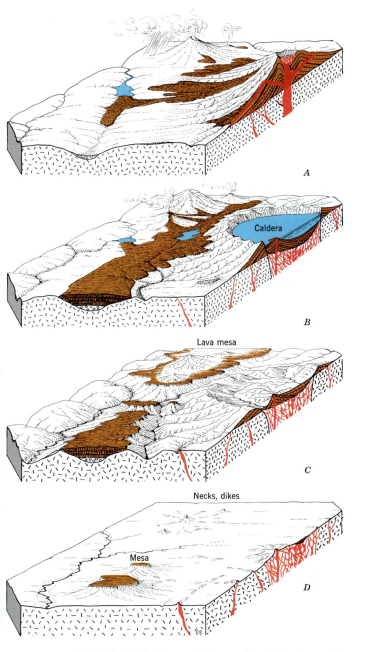

FIGURE 16.34 Stages in the erosional development of composite volcanoes and lava flows. (Drawn by E. Raisz.)

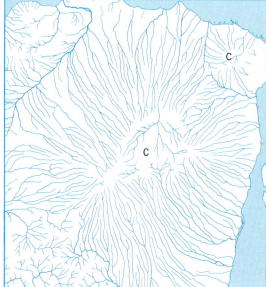

FIGURE 16.35 Radial drainage patterns of volcanoes in the East Indies. The letter C shows the location of a crater.

Erosional Landforms of Volcanoes and Lava Flows

Figure 16.34 shows successive stages in the erosion of composite volcanoes, lava flows, and a caldera. Shown in Block A are active volcanoes in the process of building. These are initial landforms. Lava flows issuing from the volcanoes have spread down into a stream valley, following the downward grade of the valley and forming a lake behind the lava dam.

In Block B some changes have taken place. The most conspicuous change is the destruction of the largest volcano to produce a caldera. A lake occupies the caldera, and a small cone has been built inside. One of the other volcanoes, formed earlier, has become extinct. It has been dissected by streams and has lost the smooth conical form. Smaller, neighboring volcanoes are still active, and the contrast in form is marked.

The system of streams on a dissected volcano cone is a *radial drainage pattern*. Because these streams take their position on a slope of an initial land surface, they are of the consequent variety. It is often possible to recognize volcanoes from a drainage map alone (Figure 16.35) because of the perfection of the radial pattern. A good example of a partly dissected volcano is Mount Shasta in California (Figure 16.36).

In Block C of Figure 16.34 all volcanoes are extinct and have been deeply eroded. The caldera lake has been drained and the rim worn to a low, circular ridge. The lava

FIGURE 16.36 Mount Shasta in the Cascade Range, northern California, is a dissected composite volcano. It has been carved into by streams and small alpine glaciers. Part way up the right-hand side is a more recent subsidiary volcanic cone, called Shastina, with a sharp crater rim. (John S. Shelton.)

flows have been able to resist erosion far better than the rock of the surrounding area and have come to stand as mesas high above the general level of the region.

Block *D* of Figure 16.34 shows an advanced stage of erosion of composite volcanoes. There remains now only a small sharp peak, called a *volcanic neck;* it is the solidified lava in the pipe of the volcano. Radiating from this are wall-like dikes, formed of magma, which filled radial fractures around the base of the volcano. Perhaps the finest illustration of a volcanic neck with radial dikes is Shiprock

New Mexico (Figure 16.37).

Shield volcanoes show erosion features quite different from those of the composite volcanoes. Figure 16.38 shows stages of erosion of Hawaiian shield volcanoes, beginning in the upper diagram with the active volcano and its central depression. These are initial landforms. Radial consequent streams cut deep canyons into the flanks of the extinct shield volcano, and these canyons are opened out into deep, steep-walled amphitheaters. Eventually, as the third diagram shows, the original surface of the shield volcano has been

FIGURE 16.37 Ship Rock, New Mexico, is a volcanic neck enclosed by a weak shale formation. The peak rises 520 m (1700 ft) above the surrounding plain. At the right, wall-like dikes extend far out from the central peak. (John S. Shelton.)

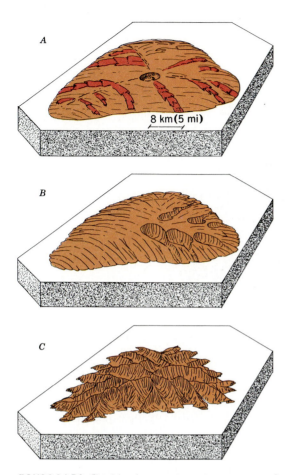

FIGURE 16.38 Shield volcanoes in various stages of erosion make up the Hawaiian Islands. (*A*) Newly formed dome with central depression. (*B*) Early stage of erosion with deeply eroded valley heads. (*C*) Advanced erosion stage with steep slopes and mountainous relief. (Drawn by A. N. Strahler.)

8 km (5 mi)

entirely obliterated, and a rugged mountain mass made up of sharp-crested divides and deep canyons remains (Figure 16.39).

Landforms and Rock Structure in Review

We have now added a new dimension to the realm of landforms produced by fluvial denudation. This dimension is the variety and complexity introduced by differences in rock composition and crustal structure. Because the various kinds of rocks offer different degrees of resistance to the forces of weathering, mass wasting, and running water, they exert a controlling influence on the shapes and sizes of landforms. We have seen, too, that streams carving up a landmass are controlled to a high degree by the structure of the rock on which they act. In this way distinctive drainage patterns evolve.

Diversity of landforms caused by variations in rock structure and composition have a profound influence on the environment. Each landform type is a distinctive habitat for plants; each landform type adds an element of uniqueness to the quality of the soil. Plant formations respond to these place-to-place differences in slope, drainage, and soil texture. Agriculture follows these same patterns by exploiting the most favorable habitats and rejecting those that are inhospitable. In this way the patterns of cultural features have come to reflect the patterns of rock structure.

FIGURE 16.39 Waimaea Canyon, over 760 m (2500 ft) deep, has been eroded into the flank of an extinct shield volcano on Kauai, Hawaii. Basaltic lava flows are exposed. (Arthur N. Strahler.)

CHAPTER

LANDFORMS MADE BY WAVES AND WIND

COASTAL LANDFORMS   Coastal scenery derives its character from the action of breaking waves and their powerful water surges.

MARINE CLIFFS   Most striking of coastal landforms, rock cliffs change little with the passage of centuries, whereas sand beaches come and go with the passage of seasons.

BEACHES   Transport of sand along a shoreline builds many landforms, such as beaches, sandspits, and sandbars, but at the same time can deprive a stretch of coast of its beach sand and allow the coast to retreat.

TIDAL CURRENTS   Along with the action of waves, tidal currents shape the shoreline and its depositional landforms.

WIND ACTION   Landforms shaped by the action of wind are most conspicuous in regions of dry climate and along coasts.

SAND DUNES   They show a great variety of shapes, ranging from wavelike ridges of bare sand to irregular mounds bearing a sparse plant cover.

LOESS   Layers of silt, representing centuries of accumulation of dust particles carried in turbulent winds, cover large areas of the midlatitude zone.

THE continental shoreline, where the salt water of the oceans contacts fresh water and the solid mineral base of the continents, is a complex environmental zone of great importance. Humans have occupied the shore zone for a number of reasons. First, there are food resources of shellfish, finfish, and waterfowl to be had in the shallow waters and estuaries. Second, the shoreline is a base from which ships embark to seek marine food resources farther from land and to transport people and goods between the continents; in time of war the shoreline is a critical barrier to be defended from invading forces arriving by sea. Third, the coastal zone is a recreational facility, with its sea breezes, bathing beaches, and opportunities for surfing, skin diving, sport fishing, and boating.

Along with its opportunities, the shore zone imposes restraints and hazards on humans and their structures. Some coasts are rocky and cliffed; they provide little or no shelter in the form of harbors. Along other coasts the enormous energy of storm waves can cut back the shore and undermine buildings and roads. High water levels in time of storm can cause inundation of low-lying areas and bring the force of breaking waves to bear on ground several meters above the normal levels reached by seawater. Storm surges and seismic sea waves, which cause such inundations, have already been described in earlier chapters. For centuries, humanity has been at war with the sea, building fortresslike walls to keep out the sea, and even forcing the sea to give up coastal land so as to produce more crops for food and forage.

Great segments of our coastlines face environmental degradation and destruction as urbanization of the coastal zone demands more land and expanded port facilities. The pressures of a growing population seeking its holiday pleasures along the shore threaten to destroy the very benefits that the shore offers. Management of the environment of the continental shorelines requires a knowledge of the natural forms and processes of this sensitive zone; the purpose of this chapter is to provide that basic knowledge.

Throughout this chapter we will use the term *shoreline* to mean the shifting line of contact between water and land. The broader term *coastline*, or simply *coast*, refers to a zone in which coastal processes operate or have a strong influence. The coastline includes the shallow water zone in which waves perform their work, as well as beaches and cliffs shaped by waves, and coastal dunes.

Wave Erosion

Waves travel across the deep ocean with little loss of energy. When waves reach shallow water, the drag of the bottom slows and steepens the wave until it leaps forward and collapses as a *breaker* (Figure 17.1). Many tons of water surge forward, riding up the beach slope. Where a cliff lies within reach of the moving water, it is impacted with enormous force. Rock fragments of all sizes, from sand to cobbles, are carried by the surging water and thrust against bedrock of the cliff. The impact breaks away new rock fragments and the cliff is undercut at the base.

Where weak, incoherent materials— various kinds of regolith, such as alluvium— make up the coastline the force of the moving water alone easily cuts into the coastline. Here, erosion is rapid, and the shoreline may recede rapidly. An example is the Atlantic shoreline of Cape Cod, Massachusetts. Here, in glacial sand and clay, a steep bank (a *marine scarp*) is retreating steadily under attack of storm waves (Figure 17.2).

In contrast, the retreat of a *marine cliff* formed of hard bedrock is exceedingly slow, when judged in terms of a human life span. Some details of such a cliff are shown in Figure 17.3. A deep basal indentation, the wave-cut notch, marks the line of most intense wave erosion. The waves find points of weakness in the bedrock and penetrate deeply to form crevices and sea caves. More resistant rock masses project seaward and are cut through to make picturesque arches (Figure 17.4). After an arch collapses the remaining rock column forms a stack, but this is ultimately leveled.

As the sea cliff retreats landward, continued wave abrasion forms an *abrasion platform*. This sloping rock floor continues to be eroded and widened by abrasion beneath the breakers. If a beach is present, it is little more than a thin layer of gravel and cobblestones.

Sea cliffs are features of spectacular beauty as well as habitats for many forms of life, including sea mammals and shorebirds. Only in recent years has the need to preserve these cliffed coasts in their natural

FIGURE 17.1 A breaking wave. (Drawn by W. M. Davis.)

FIGURE 17.2 This wave-eroded scarp of glacial sand and clay at Highland Light, Cape Cod, is about 40 m (130 ft) high. It recedes at a rate of about 1 m (3 ft) per year. The lighthouse will eventually be undermined, unless moved back to safer ground. (Drawn by A. N. Strahler.)

state been fully appreciated. Although the strength of these rocky features resists alteration, the cliff line is vulnerable to heavy use for summer homes, motels, and restaurants. Intensive use not only destroys the pristine scenery, but it also adds pollution.

Beaches

Where sand is in abundant supply, it accumulates as a thick, wedge-shaped deposit, or *beach*. Beaches absorb the energy of breaking waves. During short periods of storm, the beach is cut back, but the sand is restored during long periods when waves are weak. In this way a beach may retain a stable configuration, or equilibrium state, over many years' time.

Beaches are shaped by alternate landward and seaward currents of water generated by breaking waves (Figure 17.5). After a breaker has collapsed, a foamy, turbulent sheet of water rides up the beach slope. This *swash* is a powerful surge causing a landward movement of sand and gravel on the beach. When the force of the swash has been spent against the slope of the beach, a return flow, or *backwash*, pours down the beach (Figure 17.6). Sands and gravels are swept seaward by the backwash. Surf bathers are familiar with the backwash as a strong seaward current felt in the breaker zone. This "undercurrent" or "undertow," as it is popularly called, can be strong enough to sweep unwary bathers off their feet and carry them seaward beneath the next oncoming breaker.

FIGURE 17.3 Landforms of sea cliffs. A = arch; S = stack; C = cave; N = notch; P = abrasion platform. (Drawn by E. Raisz.)

FIGURE 17.4 Rock arches have been carved from this wave-cut cliff of horizontal sandstone strata on the Lake Superior shore. (Orlo E. Childs.)

Littoral Drift

The unceasing shifting of beach materials with swash and backwash of breaking waves also results in a sidewise movement known as *beach drift* (Figure 17.6). Wave fronts usually approach the shore obliquely rather than directly head on. The swash rides obliquely up the beach, and the sand is moved obliquely up the slope. After the water has spent its energy, the backwash flows down the slope of the beach in the most direct downhill direction. The particles are dragged directly seaward and come to rest at a position to one side of the starting place. On a particular day, wave fronts approach consistently from the same direction, so that this movement is repeated many times. Individual rock particles travel long distances along the shore. Multiplied many thousands of times to include the numberless particles of the beach, beach drift becomes a leading form of sediment transport.

When waves approach a shoreline under the influence of strong winds, the water level is slightly raised near shore by a slow shoreward drift of water. The excess water pushed shoreward must escape. A longshore current is set up parallel to shore in a direction away from the wind (Figure 17.7). When wave and wind conditions are favorable, this current is capable of moving sand along the sea bottom in a direction

FIGURE 17.5 A curving pocket beach on the island of Kauai, Hawaii. The foaming water of a collapsed breaker is approaching the beach, where it will form the landward swash. Receding before it is the backwash, leaving a thin water film on the sand surface. (Arthur N. Strahler.)

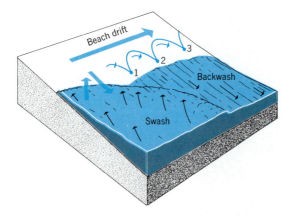

FIGURE 17.6 Oblique approach of waves allows the swash and backwash to move sand grains along the shoreline in a series of arched paths. (Drawn by A. N. Strahler.)

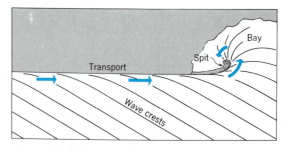

FIGURE 17.8 Along a straight coast, littoral drift carries sand in one direction to reach the mouth of a bay, where it is formed into a sandspit. (From A. N. Strahler, *The Earth Sciences*, 2nd ed., Harper & Row, New York. Copyright © 1971 by Arthur N. Strahler.)

parallel to the shore. The process is called *longshore drift.*

Both beach drift and longshore drift move particles in the same direction for a given set of onshore winds, and they supplement each other's influence. The total process is called *littoral drift.*

Littoral drift operates to shape shorelines in two quite different situations. Where the shoreline is straight or broadly curved for many kilometers at a stretch, littoral drift moves the sand along the beach in one direction for a given set of prevailing winds. This situation is shown in Figure 17.8. Where a bay exists, the sand is carried out into open water as a long finger, or *sandspit* (Figure 17.9). As the sandspit grows, it forms a barrier, called a *bar,* across the mouth of the bay (see Figure 17.20).

A second situation is shown in Figure 17.10. Here the coastline consists of prominent headlands, projecting seaward, and deep bays. Erosional energy of the waves is concentrated on the exposed headlands. Here wave-cut cliffs develop (Figure 17.11). Sediment from the eroding cliffs is carried by littoral drift along the sides of the bay, converging on the head of the bay. The result is a crescent-shaped beach, often called a *pocket beach* (Figures 17.5 and 17.11).

Littoral Drift and Shore Protection

When sand is arriving at a particular section of the beach more rapidly than it can be

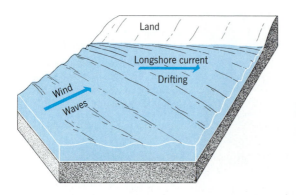

FIGURE 17.7 Longshore current drifting. (Drawn by A. N. Strahler.)

FIGURE 17.9 A white sandspit, growing in a direction toward the observer, leaves only a narrow inlet (foreground) for tidal currents. Sandy shoals lie in the bay at the right. Martha's Vineyard, Massachusetts. (Donald W. Lovejoy.)

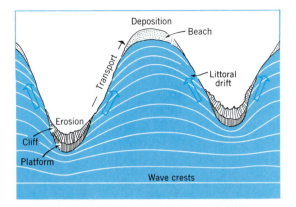

FIGURE 17.10 On an embayed coast, sediment is carried from eroding headlands to the bayheads, where pocket beaches can accumulate. (From A. N. Strahler, *The Earth Sciences*, 2nd ed. Harper & Row, New York. Copyright © 1971 by Arthur N. Strahler.)

FIGURE 17.12 Concrete seawalls of different styles protect pieces of shorefront property along this stretch of Lake Michigan shoreline of northern Indiana. (Ned L. Reglein.)

carried away, the beach is widened and built shoreward. This change is called *progradation.* When sand is leaving a section of beach more rapidly than it is being brought in, the beach is narrowed and the shoreline moves landward, a change called *retrogradation.*

Along stretches of shoreline affected by retrogradation, the beach may be seriously depleted or even entirely destroyed. When this occurs, cutting back of the coast can be rapid in weak materials, destroying valuable shore property. Protective engineering

structures, such as sea walls, designed for direct resistance to frontal wave attack, are prone to failure and are extremely expensive as well (Figure 17.12). In some circumstances, a successful alternative strategy is to install structures that will cause progradation, and so build a broad protective beach. The principle here is that the excess energy of storm waves will be dissipated in reworking the beach deposits. Cutting back of the beach in a single storm will be restored by beach-building between storms.

Progradation requires that sediment moving as littoral drift be trapped by the placement of baffles across the path of transport. To accomplish this, groins are installed at close intervals along the beach. A *groin* is simply a wall or embankment built at right angles to the shoreline; it may be constructed of large rock masses, of concrete, or of wooden pilings. Figure 17.13 shows the shoreline changes induced by groins. Sand accumulates on the updrift side of the groin, developing a curved shoreline. On the downdrift side of the groin, the beach will be depleted because of the cutting off of the normal supply of drift sand. The result may be harmful retrogradation and cutting back of the coast. For this reason groins must be closely spaced so that the trapping effect of one groin will extend to the next (Figure 17.14). Ideally, when the groins have trapped the maximum quantity of sediment, beach drift will be restored to its original rate for the shoreline as a whole.

FIGURE 17.11 A series of cliffed headlands on the Big Sur coast of northern California. A pocket beach lies in the foreground. Above the cliff is a marine terrace, partly buried in alluvium. (Alan H. Strahler.)

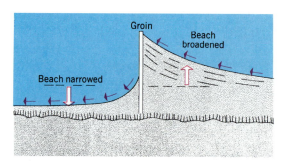

FIGURE 17.13 Construction of a groin causes marked changes in configuration of the sand beach. (From A. N. Strahler, *Planet Earth*, Harper & Row, New York. Copyright © 1972 by Arthur N. Strahler.)

In some instances the source of beach sand is from the mouth of a river. Construction of dams far upstream on the river may drastically reduce the sediment load and therefore also cut off the source of sand for littoral drift. Retrogradation may then occur on a long stretch of shoreline. The Mediterranean shoreline of the Nile Delta has suffered retrogradation because of reduction in sediment supply following dam construction far up the Nile.

Tidal Currents

Most marine coastlines are influenced by the *ocean tide,* a rhythmic rise and fall of sea level under the influence of changing attractive forces of moon and sun on the rotating earth. Where tides are great, the effects of changing water level and the currents set in motion are of major importance in shaping coastal landforms.

The tidal rise and fall of water level is graphically represented by the *tide curve.* We can make half-hourly observations of the position of water level against a measuring stick attached to a pier or sea wall. We then plot the changes of water level and draw the tide curve. Figure 17.15 is a tide curve for Boston Harbor covering a day's time. The water reached its maximum height, or high water, at 3.7 m (12 ft) on the tide staff, then fell to its minimum height, or low water, occurring about 6¼ hours later. A second high water occurred about 12½ hours after the previous high water, completing a single tidal cycle. In this example the range of tide, or difference between heights of successive high and low waters, is 2.7 m (9 ft).

The rising tide sets in motion, in bays and estuaries, currents of water known as *tidal currents.* The relationships between tidal currents and the tide curve are shown in Figure 17.16. When the tide begins to fall, an ebb current sets in. This flow ceases about the time when the tide is at its lowest point. As the tide begins to rise, a landward current, the flood current, begins to flow.

Tidal Power

Tidal power makes use of the tidal rise and fall of ocean level, an unending source of energy. To harness the power of the tides, a bay must be located along a coast subject to a large range of tide. An ideal location is to be found on the Atlantic coast of Maine, New Brunswick, and Nova Scotia, around the Bay of Fundy. Here the tide range is greater than 3 m (10 ft). The rugged coast near the boundary of Maine and Nova Scotia is deeply indented by narrow bays, with constricted places where the bays empty into the ocean. These constrictions

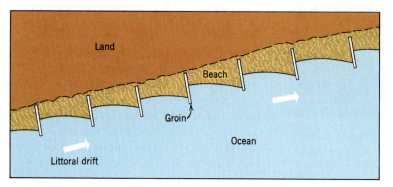

FIGURE 17.14 A system of multiple groins can successfully maintain a broad beach.

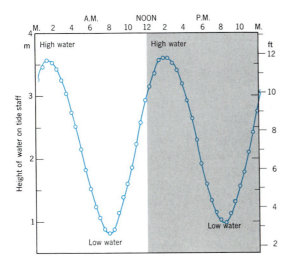

FIGURE 17.15 Height of water at Boston Harbor measured every half hour.

generating capacity of the plant, which has 24 conduits, is 240 megawatts when the turbines are running at full speed. The average output for the year is only 60 megawatts, which is about the output of one small nuclear power station. A second tidal power plant is operating in the Soviet Union in an inlet of the White Sea. In 1977 a Canadian study board reviewed the potential for development of tidal power in the Bay of Fundy and concluded that a major project is economically feasible. A tidal power conversion plant of 3800-megawatt capacity has been planned for possible operation by 1990.

The potential for tidal power development on a global scale is not great. For the world the annual tidal energy potentially available by exploitation of all suitable coastal sites is only about one percent of the total energy potentially available through hydropower development. For the United States, the developable tidal energy resources are about two-thirds as great as the hydropower resources.

can be closed off by dams and the tidal flow passed through conduits connecting the bay and the ocean. Turbines placed in the conduits can be turned by the swift tidal current.

Plans for a tidal power plant at Passamaquoddy Bay, at the entrance to the Bay of Fundy, were developed and modified over a long span of years. Work on a small part of the total project was begun in 1935 and suspended in 1937. Today, only two important tidal power plants are in operation. One is the La Rance plant on the Brittany coast of France. It was opened in 1966, and the maximum

Tidal Current Deposits

Ebb and flood currents generated by tides perform several important functions along a shoreline. First, the currents that flow in and out of bays through narrow inlets are very swift and can scour the inlet strongly to keep it open despite the tendency of shore drifting processes to close the inlet with sand.

Second, tidal currents carry large amounts of fine silt and clay in suspension, derived from streams that enter the bays or from bottom muds agitated by storm wave action. This fine sediment settles to the floors of the bays and estuaries where it accumulates in layers and gradually fills the bays. Much organic matter is present in this sediment.

In time, tidal sediments fill the bays and produce mud flats, which are barren expanses of silt and clay exposed at low tide but covered at high tide. Next, a growth of salt-tolerant plants takes hold on the mud flat. The plant stems entrap more sediment and the flat is built up to approximately the level of high tide, becoming a *salt marsh* (Figure 17.17). A thick layer of peat is

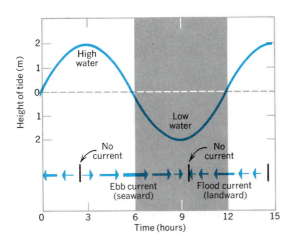

FIGURE 17.16 The ebb current flows seaward as the tide level falls; the flood current flows landward as the tide level rises.

FIGURE 17.17 Salt marsh of *Spartina* grass forms a yellowish-green border zone around this small tidal inlet on Cape Cod, Massachusetts. At the right, the salt-marsh peat has been removed to make a boat landing. (Arthur N. Strahler.)

FIGURE 17.18 Salt marsh with sinuous tidal channels (right) lies behind a protective barrier beach (left) on outer Cape Cod. A broad tidal inlet allows ebb and flood tide currents to drain and flood the marsh with each tidal cycle. A tidal delta can be seen on the landward side of the inlet. Sand dunes (foredunes) cover the barrier beach at the lower left. (John S. Shelton.)

eventually formed at the surface. Tidal currents maintain their flow through the salt marsh by means of a highly complex network of sinuous tidal streams (Figure 17.18).

Salt marsh is of interest to geographers because it is land that can be drained and made agriculturally productive. The salt marsh is first cut off from the sea by construction of an embankment of earth (a dike), in which gates are installed to allow the freshwater drainage of the land to exit during ebb flow. Gradually, the salt water is excluded and soil water of the diked land becomes fresh. Such diked lands are intensively developed in Holland (polders) and southeast England (fenlands).

Over many decades the surface of reclaimed salt marsh subsides because of compaction of the underlying peat layers and may come to lie well below mean sea level. The threat of flooding by salt water, when storm waves break the dikes, hangs constantly over the inhabitants of such low areas. The reclamation of salt marsh by dike construction and drainage was practiced by New World settlers in New England and Nova Scotia. In the industrial era, large expanses of salt marsh have been destroyed by landfill, a practice only recently inhibited by new legislation.

Common Kinds of Coastlines

There are many different kinds of coastlines. Each kind is unique because of the distinctive landmass against which the ocean water has come to rest. One group of coastlines derives its qualities from *submergence,* the partial drowning of a coast by a rise of sea level or a sinking of the crust. Another group derives its qualities from *emergence,* the exposure of submarine landforms by a falling of sea level or a rising of the crust. Another group of coastlines results when new land is built out into the ocean by volcanoes and lava flows, by the growth of river deltas, or by the growth of coral reefs.

A few important types of coastlines are illustrated in Figure 17.19. The *ria coast (A)* is a deeply embayed coast resulting from

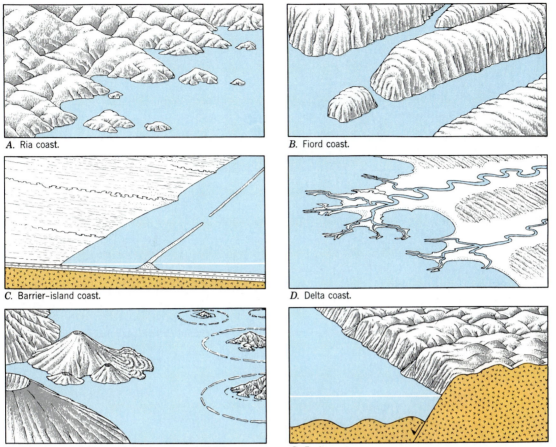

A. Ria coast.

B. Fiord coast.

C. Barrier-island coast.

D. Delta coast.

E. Volcano coast (left).   *F.* Coral-reef coast (right).

G. Fault coast.

FIGURE 17.19 Seven common kinds of coastlines are illustrated here. These examples have been selected to illustrate a wide range in coastal features. (Drawn by A. N. Strahler.)

submergence of a landmass dissected by streams. This coast has many offshore islands. A *fiord coast (B)* is deeply indented by steep-walled fiords, which are submerged glacial troughs (Chapter 18). The *barrier-island coast (C)* is associated with a recently emerged coastal plain. The offshore slope is very gentle, and a barrier island of sand is usually thrown up by wave action at some distance offshore. Large rivers build elaborate deltas, producing *delta coasts (D)*. The *volcano coast (E)* is formed by eruption of volcanoes and lava flows, partly constructed below water level. Reef-building corals create new land and make a *coral-reef coast (F)*. Down-faulting of the coastal margin of a continent can allow the shoreline to come to rest against a fault scarp, producing a *fault coast (G)*.

Development of a Ria Coast

The ria coast is formed when a rise of sea level or a crustal sinking (or both) brings the shoreline to rest against the sides of valleys previously carved by streams. This event is illustrated in Figure 17.20, Frame *A.* Soon wave attack forms cliffs on the exposed seaward sides of the islands and headlands (Frame *B*). Sediment produced by wave action then begins to accumulate in the form of beaches along the cliffed headlands and at the heads of bays. This sediment is carried by littoral drift and is built into sandspits across the bay mouths and as connecting links between islands and mainland (Frame *C*). Finally, all outlying islands are planed off by wave action and a nearly straight shoreline develops in which

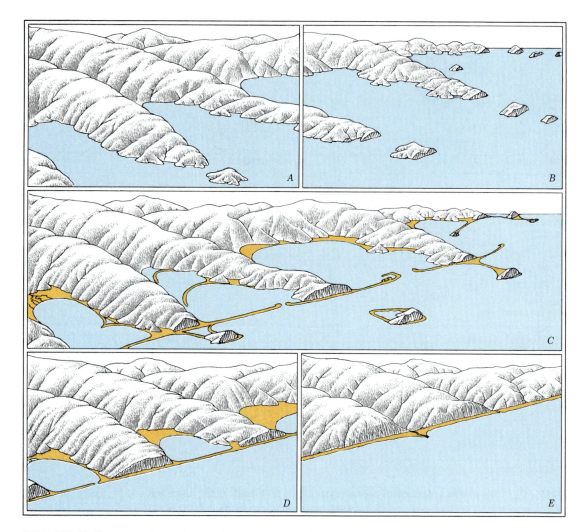

FIGURE 17.20 Stages in the evolution of a ria coastline. (Drawn by A. N. Strahler.)

the sea cliffs are fully connected by baymouth bars (Frame *D*). Now the bays are sealed off from the open ocean, although narrow tidal inlets may persist, kept open by tidal currents. Frame *E* shows a much later stage in which the coastline has receded beyond the inner limits of the original bays.

The influence of ria coastlines on human activity has been strong down through the ages. The deep embayments of the ria shoreline make splendid natural harbors. One of the finest of these is San Francisco Bay (Figure 17.21). Much of the ria coastline of Scandinavia, France, and the British Isles is provided with excellent harbor facilities. These peoples have a strong tradition of fishing, shipbuilding,

ocean commerce, and marine activity generally. Mountainous relief of ria and fiord coasts made agriculture difficult or impossible, inducing many people to turn to the sea for a livelihood. New England and the Maritime Provinces of Canada have a ria coastline with abundant good harbors. Here, too, we observe the same early development of fishing, whaling, ocean commerce, and shipbuilding seen in the British Isles and Scandinavian countries.

Barrier-Island Coasts

In contrast to ria and fiord coasts, with their bold relief and deeply embayed outlines, we find low-lying coasts from which the land

FIGURE 17.21 Golden Gate is the narrow, steep-walled entrance to San Francisco Bay, a branching arm of the sea that receives river flow from the Sierra Nevada far inland. This photo was taken in 1949, when the Golden Gate Bridge was twelve years old. (Ned Reglein.)

slopes gently beneath the sea. The coastal plain of the Atlantic and Gulf coasts of the United States presents a particularly fine example of such a gently sloping surface. As we explained in Chapter 16, this coastal plain is a belt of relatively young sedimentary strata, formerly accumulated beneath the sea as deposits on the continental shelf. Emergence as a result of repeated crustal uplifts has characterized this coastal plain during the latter part of the Cenozoic Era and into recent time.

Along much of the Atlantic Gulf Coast, there exists a *barrier island*, which is a low ridge of sand built by waves and further increased in height by the growth of sand dunes (Figure 17.22). Behind the barrier island lies a *lagoon*, which is a broad expanse of shallow water, often several kilometers wide, and in places largely filled with tidal deposits.

A characteristic feature of barrier islands is the presence of gaps, known as *tidal inlets*. Through these gaps strong currents flow alternately landward and seaward as the tide rises and falls (see Figure 17.16). In heavy storms, the barrier may be breached by new inlets. Tidal currents will subsequently tend to keep a new inlet open, but it may be closed by shore drifting of sand.

Figure 17.23 shows details of the barrier-island coast of North Carolina. Here, the barrier island is formed into sharply pointed headlands, or capes, whereas the inner shoreline is deeply embayed.

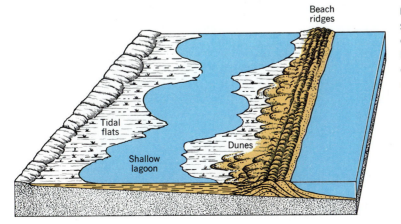

Beach ridges

Tidal flats

Dunes

Shallow lagoon

FIGURE 17.22 A barrier island is separated from the mainland by a wide lagoon. Sediments fill the lagoon, while dune ridges advance over the tidal flats. (Drawn by A. N. Strahler.)

(right) Seen from Apollo 9 spacecraft, the white barrier beach of sand— the Outer Banks— projects sharply seaward in two points: Cape Hatteras (H) and Cape Lookout (L). Pamlico Sound (P) lies between the barrier and the mainland shore, which is deeply embayed as a result of postglacial submergence. Extensive areas of salt marsh (S) make up the low coastal zone. Through two inlets (I) between Hatteras and Lookout sediment is being carried seaward in plumes of lighter color. The Gulf Stream boundary (G) is sharply defined. The area shown is about 160 km (100 mi) across (NASA AS9–20–3128.)

(below) A close look at the barrier beach on Cape Hatteras. A dune ridge, well protected by grass cover, lies to the left of the beach. The view is north; the Atlantic Ocean is at the right. (Mark A. Melton.)

(left) Near Cape Hatteras this stretch of barrier beach has been swept over by storm swash, topping the dune barrier and carrying sand in sheets toward the bay at the left. (Mark A. Melton.)

FIGURE 17.23 The barrier-island coast of North Carolina.

FIGURE 17.24 This Landsat image shows some details of the Texas barrier-island coast. Near the top of the frame is Corpus Christi Bay (circular shoreline), with the city of that name built on the southwestern shore. A ship channel can be traced diagonally across the lagoon to a pass (Aransas Pass) in the barrier island. Farther south, the barrier island (Padre Island) has an inner dune belt (white). Other dunes on the mainland appear as small white spots. Irrigated farmland near Corpus Christi appears as a pattern of dark squares. Ranchland lies to the south. (NASA Landsat 1092–16314, October 1972.)

FIGURE 17.25 The Gulf Coast of Texas is dominated by its offshore barrier island.

Perhaps the finest example of a barrier-island coast is the Gulf Coast of Texas (Figures 17.24 and 17.25). Here, the barrier island is unbroken for more than 160 km (100 mi) at a single stretch and natural passes are few. The lagoon is 8 to 16 km (5 to 10 mi) wide. As with most barrier-island coasts, deep natural harbors are lacking. Extensive channel dredging is required. Galveston is built on the barrier island adjacent to an inlet connecting Galveston Bay with the sea. Most other Texas ports are, however, located on the mainland shore. Corpus Christi, Rockport, Texas City, Port Lavaca, and other ports are located along the shores of river embayments. Oceangoing ships must enter and leave the lagoon through passes in the barrier island. These require sea walls and jetties to confine the tidal current and cause it to scour a deep channel.

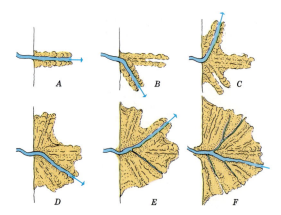

FIGURE 17.26 Stages in the formation of a simple delta. (After G. K. Gilbert.)

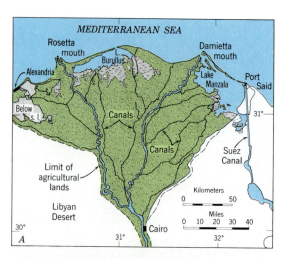

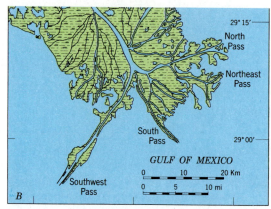

FIGURE 17.27 The Nile delta (*A*) and the Mississippi delta (*B*).

Delta Coasts

The deposit of clay, silt, and sand made by a stream where it flows into a body of standing water is known as a *delta* (Figure 17.26). Deposition is caused by rapid reduction in velocity of the current as it pushes out into the standing water. Typically, the river channel divides and subdivides into lesser channels called *distributaries*. The coarse particles settle out first; the fine clays continue out farthest and eventually come to rest in fairly deep water. Contact of fresh with salt water causes the finest clays to clot into larger aggregates, which settle to the sea floor.

Deltas show a wide variety of outlines, both because of varying configurations of the original shoreline and because of shoreline modifications by littoral drift of sediment at the distributary mouths. We will limit our description to two well-known deltas: the Nile River delta and the Mississippi River delta.

The Nile delta has the basic triangular shape of the Greek letter *delta* (Figure 17.27*A*). We are told that Herodotus, a Greek historian who lived nearly 2500 years ago, had visited Egypt and that he first applied the word "delta" to that fertile green plain surrounded by desert (Figure 17.28). The Nile delta has two major distributaries; these terminate in the Damietta Mouth and the Rosetta Mouth. From these mouths, which project as cusps

(tooth-forms) into the Mediterranean Sea, sediment has been carried along the shore by littoral drift to form curved barrier beaches. The total outline of the delta is thus quite like that of an alluvial fan.

The Mississippi delta is of the bird-foot type, with long projecting fingers growing far out into the Gulf of Mexico at the ends of the distributaries (Figure 17.27*B*). A satellite image of the delta, Figure 17.29, shows the great quantity of suspended sediment—clay and fine silt—discharged by the river into the Gulf, about one million metric tons per day.

Deltas of large rivers have been of importance from earliest historical times because their extensive flat fertile lands support dense agricultural populations.

FIGURE 17.28 In this photograph, taken by astronauts from an orbiting Gemini satellite, the Nile delta appears in true color as a dark blue-green triangle bounded by pale yellowish-brown desert areas. Two major distributaries (N) of the Nile River end in prominent cusps: the Damietta mouth (D) and the Rosetta mouth (R). Alexandria (A) is just within view in the foreground. Sand carried by littoral drift from the river mouths has accumulated in a broad barrier beach (B), separated from the delta plain by an open lagoon (L). The Suez Canal (C) connects the Mediterranean Sea (left) with the Gulf of Suez (right). Beyond lies the Sinai Peninsula. (NASA S—65—34776.)

Many coastal cities, linking ocean and river traffic, are situated on or near deltas. Examples are Alexandria on the Nile, Calcutta on the Ganges–Brahmaputra, Saigon on the Mekong, Bangkok on the Chao Phraya, Rangoon on the Irrawaddy, Karachi on the Indus, Amsterdam and Rotterdam on the Rhine, Shanghai on the Changjiang, Marseilles on the Rhone, and New Orleans on the Mississippi.

Delta growth is often rapid, for example, 60 m (200 ft) per year for the Po River delta. Some cities and towns that were at river mouths several hundred years ago are today several kilometers inland. An important engineering problem is to keep an open channel for oceangoing vessels that have to enter the delta distributaries to reach port. The mouths of the Mississippi River delta distributaries, known as passes, have been extended by the construction of jetties. The narrowed stream is forced to move faster, and thus scours a deep channel.

FIGURE 17.29 The Mississippi delta recorded from Landsat orbiting satellite and presented in false-color imagery. The natural levees of the bird-foot delta appear as lacelike filaments in a great pool of turbid river water. To the left (west and north) of the active modern delta are the remains of older deltas. The Chandeleur Islands, a system of arcuate barrier islands, have been built of material from the subsided remains of these older deltas. New Orleans can be seen at the upper left, occupying the natural levees of the Mississippi River and the southern shore of Lake Pontchartrain. (NASA Landsat 1177–16023, January 1973.)

Coral-Reef Coasts

Coral-reef coasts are unique in that the addition of new land is made by organisms: corals and algae. Growing together, these organisms secrete rocklike deposits of mineral carbonate, called *coral reefs*. As coral colonies die, new ones are built on them, accumulating as limestone. Coral fragments are torn free by wave attack, and the

FIGURE 17.30 The Island of Moorea and its fringing coral reef, Society Islands, South Pacific Ocean. The island is a deeply dissected volcano with a history of submergence. (Jack Fields/ Photo Researchers.)

pulverized fragments accumulate as sand beaches.

Coral-reef coasts occur in warm, tropical and equatorial waters between the limits of lat. 30° N and 25° S. Water temperatures above 20°C (68°F) are necessary for dense reef coral growth. Reef corals live near the water surface. The seawater must be free of suspended sediment and well aerated for vigorous coral growth. For this reason corals thrive in positions exposed to wave attack from the open sea. Because muddy water prevents coral growth, reefs are missing opposite the mouths of muddy streams. Coral reefs are remarkably flat on top. They are exposed at low tide and covered at high tide.

Three general types of coral reefs may be recognized: (1) fringing reefs, (2) barrier reefs, and (3) atolls. *Fringing reefs* are built as platforms attached to shore (Figure 17.30). They are widest in front of headlands where wave attack is strongest, and the corals receive clean water with abundant food supply. *Barrier reefs* lie out from shore and are separated from the mainland by a lagoon (Figure 17.31). Narrow gaps occur at intervals in barrier reefs. Through these openings excess water from breaking waves is returned from the lagoon to the open sea.

Atolls are more or less circular coral reefs enclosing a lagoon, but without any land inside (Figure 17.32). On large atolls, parts of the reef have been built up by wave action and wind to form low island chains connected by the reef. Most atolls have been built on a foundation of volcanic rock.

These foundations are thought to have been basaltic volcanoes built on the deep ocean floor. After being planed off by wave erosion, the extinct volcanoes slowly subsided. At the same time the fringing coral reef continued to build upward (Figure 17.33).

The environmental aspects of atoll islands are unique in some respects. First, there is no rock other than coral limestone, which is composed of calcium carbonate. This means that many food plants cannot be cultivated without the aid of fertilizers or some outside source of mineral nutrients from a larger island composed of volcanic or other igneous rock.

The palm tree is native to atoll islands because it thrives on brackish water, and the seed, or palm nut, is distributed widely by floating from one island to another. Native inhabitants cultivated the coconut palm to provide food, clothing, fibers, and building materials. Fresh water is scarce on small

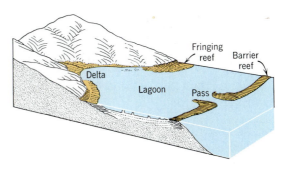

FIGURE 17.31 A barrier reef is separated from the mainland by a shallow lagoon. (Drawn by W. M. Davis.)

FIGURE 17.32 Atolls of the Banda Sea near the Celebes Islands in Indonesia. The location is about lat. 5° S. The atolls appear in this Landsat image as pale blue loops, partly obscured by cumuliform clouds. Islands, reddish in color because of the rain forest cover (green vegetation appers red on this false-color image), are almost completely obscured under cloud cover, but their fringing reefs can be discerned at a few points. The area shown is about 120 km (75 mi) across. (NASA ERTS E—1414— 01221, September 1973.)

atoll islands. Rainfall must be caught in open vessels or catchment basins and carefully conserved. Fish and other marine animals are an important part of the human diet on atoll islands. Calm waters of the

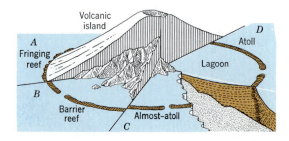

FIGURE 17.33 The subsidence theory of barrier-reef and atoll development is shown in four stages, beginning with a fringing reef attached to a volcanic island and ending with a circular reef. (Drawn by W. M. Davis.)

lagoon make a good place for fishing and for beaching canoes.

Coral islands of the western Pacific are in continual danger of devastation by tropical cyclones (typhoons), whose breaking waves wash over the low-lying ground, sweeping away palm trees and houses and drowning the inhabitants. There is no high ground for refuge. Great seismic sea waves of unpredictable occurrence also indunate atoll islands.

Raised Shorelines and Marine Terraces

The active life of a shoreline is sometimes cut short by a sudden rise of the coast. When this event occurs, a *raised shoreline* is formed. The marine cliff and abrasion platform are abruptly raised above the level

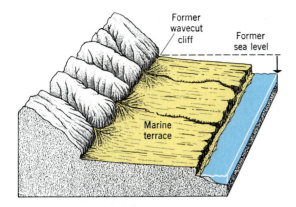

FIGURE 17.34 A raised shoreline becomes a cliff parallel with the newer, lower shoreline. The former abrasion platform is now a marine terrace. (Drawn by A. N. Strahler.)

of wave action. The former abrasion platform has now become a *marine terrace* (Figure 17.34). Fluvial denudation begins to destroy the terrace, and it may also undergo partial burial under alluvial fan deposits.

Marine terraces are important to human habitation of mountainous coasts because they offer strips of flat ground extending for tens of kilometers parallel with the shoreline. Highways and railroads are built on these terraces, and they are excellent sites for coastal towns and cities. Agriculture makes use of the flat terrace surfaces where the soil is good.

Raised shorelines are common along the continental and island coasts of the Pacific Ocean, because here tectonic processes are active along the mountain and island arcs.

Repeated uplifts result in a series of raised shorelines in a steplike arrangement. Fine examples of these multiple marine terraces are seen on the western slope of San Clemente Island, off the California coast (Figure 17.35).

Coastal Landforms in Review

Variety is the key word in describing the landforms produced by waves, currents, and organisms along the world's shorelines. We have dealt with only a few representative examples of the kinds of coastlines to be found on our planet. Each kind of coastal zone offers a different habitat for life-forms, and each offers a different situation to confront us in the expanding development of coastal resources.

In past centuries the coastal zone was used principally as a source of food or as a place from which to embark on the ocean in search of food or to trade with other lands and peoples. Today the situation is changing. The coastal zone is now in great demand as a recreation zone, and its economic value lies mostly in the worth of the waterfront land as real estate. Marinas spring up at every available sheltered spot. Here, recreational boating and sport fishing dwarf the shrinking commerical fishing activity. Luxury condominiums rise on filled land where the tidal flat and salt marsh once supported a complex marine ecosystem. Industry, too, presses for its

FIGURE 17.35 Marine terraces on the western slope of San Clemente Island, off the southern California coast. More than twenty terraces have been identified in this series; the highest has an elevation of about 400 m (1300 ft). (John S. Shelton.)

share of the coastal zone for use as sites of nuclear power plants and oil refineries.

Wise decisions on coastal-zone management often depend partly on accurate knowledge of the operation of coastal processes, such as wave erosion and littoral drift of sediment. Engineered shoreline changes often set off unwanted retrogradation or progradation elsewhere along the coast, or result in filling of channels and harbors. These changes also profoundly affect the shallow-water ecosystems.

Transportation and deposition of sand by wind is an important process in shaping certain coastal landforms. We have made references in this chapter to coastal sand dunes derived from beach sand. In the remainder of this chapter we will investigate the transport of sand by wind and the shaping of dune forms. In so doing we complete the linkage between wind action and wave action in controlling coastal environments.

Wind Action

Wind blowing over the solid surface of the lands is one of the active agents of landform development. Ordinarily, wind is not strong enough to dislodge mineral matter from the surfaces of tightly knit rock or of moist, clay-rich soils, or of soils bound by a dense plant cover. Instead, the action of wind in eroding and transporting sediment is limited to land surfaces where small mineral and organic particles are in

the loose state. Such areas are typically deserts and semiarid lands (steppes). An exception is the coastal environment, where beaches provide abundant supplies of loose sand, even where the climate is humid and the land surface inland from the coast is well protected by a plant cover.

Landforms shaped and sustained by wind erosion and deposition represent distinctive life environments, often highly specialized with respect to the communities of animals and plants they support. In climates with barely sufficient soil water, there is a contest between wind action and the growth of plants that tend to stabilize landforms and protect them from wind action. We will find precarious balances in the life systems of certain marginal climatic zones. These balances are not only altered by natural changes in climate, but are easily upset by human activities, and often with serious consequences. To understand these environmental changes, we need to acquire a working knowledge of the physical processes of wind action on the land surfaces.

Erosion by Wind

Wind performs two kinds of erosional work. Loose particles lying on the ground surface may be lifted into the air or rolled along the ground. This process is *deflation*. Where the wind drives sand and dust particles against an exposed rock or soil surface, causing it to be worn away by the impact of the particles, the process is *wind abrasion*.

FIGURE 17.36 A blowout hollow on the plains of Nebraska, about 1890. The ground is well trampled by the hooves of cattle. The remnant column of the original soil provides a natural yardstick to measure the depth of material removed by deflation. (G. K. Gilbert, U.S. Geological Survey.)

Abrasion requires cutting tools carried by the wind; deflation is accomplished by air currents alone.

Deflation acts wherever the ground surface is thoroughly dried out and is littered with small, loose particles of soil or regolith. Dry river courses, beaches, and areas of recently formed glacial deposits are highly susceptible to deflation. In dry climates, almost the entire ground surface is subject to deflation because the soil or rock is largely bare. Wind is selective in its deflational action. The finest particles, those of clay and silt sizes, are lifted more easily and raised high into the air. Sand grains are moved only by moderately strong winds and travel close to the ground. Gravel fragments and rounded pebbles can be rolled over flat ground by strong winds, but they do not travel far. They become easily lodged in hollows or between other large grains. Consequently, where a mixture of size of particles is present on the ground, the finer sizes are removed and the coarser particles remain behind.

A landform produced by deflation is a shallow depression called a *blowout*. This depression may be from a few meters to a kilometer or more in diameter, but it is usually only a few meters deep. Blowouts form in plains regions in dry climates. Any small depression in the surface of the plain, particularly where the grass cover is broken through, may develop into a blowout. Rains fill the depression and create a shallow pond or lake. As the water evaporates, the mud bottom dries out and cracks, forming small scales or pellets of dried mud that are lifted out by the wind. In grazing lands, animals trample the margins of the depression into a mass of mud, breaking down the protective grass-root structure and facilitating removal when dry. In this way the depression is enlarged (Figure 17.36). Blowouts are also found on rock surfaces where the rock is being disintegrated by weathering.

In the great deserts of the southwestern United States, the floors of tectonic basins are vulnerable to deflation. The flat floors of shallow playas have in some places been reduced by deflation as much as several meters over areas of many square kilometers.

Where rainbeat, overland flow, and deflation have been active for a long period

FIGURE 17.37 A desert pavement in western Arizona. Fragments of many sizes litter the flat ground surface. The rock surfaces are darkened by a coating known as "desert varnish." Below, a closeup of the same surface shows that the pebbles are fitted closely together, protecting the sand layer below. (Arthur N. Strahler.)

on the gently sloping surface of a desert alluvial fan or alluvial terrace, rock fragments ranging in size from pebbles to small boulders become concentrated into a surface layer known as a *desert pavement* (Figure 17.37). As overland flow from torrential downpours carries away the fine particles, the remaining large fragments become closely fitted together, concealing the smaller particles—grains of sand, silt, and clay—that lie beneath. The pavement acts as an armor that effectively protects the finer particles from further removal by overland flow and deflation. The pavement is easily disturbed by the wheels of trucks and motorcycles, exposing the finer particles and allowing severe water erosion and deflation to occur.

The sandblast action of wind against exposed rock surfaces is limited to the basal meter or two of a rock mass rising above a flat plain, because sand grains do not rise high into the air. Wind abrasion produces pits, grooves, and hollows in the rock. Telephone poles on windswept sandy plains are quickly cut through at the base unless a protective metal sheathing or heap of large stones is placed around the base.

Dust Storms

Strong, turbulent winds blowing over desert surfaces lift great quantities of fine dust into the air, forming a dense, high cloud called a *dust storm*. In semiarid grasslands a dust storm is generated where ground surfaces have been stripped of protective vegetative cover by cultivation or grazing. A dust storm approaches as a dark cloud extending from the ground surface to heights of several thousand meters (Figure 17.38). Within the dust cloud there is deep gloom or even total darkness. Visibility is cut to a few meters, and a fine choking dust penetrates everywhere.

It has been estimated that as much as 875 metric tons of dust may be suspended in a cubic kilometer of air (4000 tons per mi³). On this basis, a large dust storm can carry more than 100 million tons of dust— enough to make a hill 30 km (100 ft) high and 3 km (2 mi) across the base. Dust travels long distances in the air. Dust from a single desert storm is often traceable as far as 4000 km (2500 mi).

Sand Dunes

A *sand dune* is any hill of loose sand shaped by the wind. Dunes may be active, when bare of vegetation and constantly changing form under wind currents, or they may be inactive, when covered by vegetation that has taken root and prevents further shifting of the sand.

Dune sand is most commonly composed of the mineral quartz, which is extremely hard and largely immune to chemical decay. The grains are beautifully rounded (see Figure 11.11) and behave as tiny elastic spheres. Under the force of strong winds the grains move in long, low leaps, rebounding after impact with other grains. Rebounding grains rarely rise more than a few centimeters above the dune surface. Grains struck by leaping grains are pushed forward and, in this way, the surface sand layer creeps downwind.

One common type of sand dune is an isolated heap of free sand called a *crescent dune*. As the name suggests, this dune has a crescentic outline; the points of the crescent are directed downwind (Figure 17.39). On

FIGURE 17.38 An approaching dust storm. The wall of dust represents a rapidly moving cold front. Coconino Plateau, Arizona. (D. L. Babenroth.)

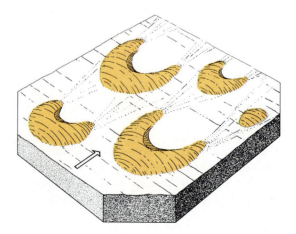

FIGURE 17.39 Crescent dunes. The arrow indicates wind direction.

FIGURE 17.40 A single, isolated crescent dune. Prevailing wind direction is from right to left. Obviously, the dune has migrated across the dry stream channels of the desert surface since the last water flood occurred. Salton Sea region, California. (John S. Shelton.)

horizontal (Figure 17.40). Sand grains slide down the steep face after being blown free of the sharp crest. When a strong wind is blowing, the flying sand makes a perceptible cloud at the crest.

Crescent dunes rest on a flat, pebble-covered ground surface. The sand heap may originate as a drift in the lee of some obstacle, such as a small hill, rock, or clump of brush. Once a sufficient mass of sand has formed, it begins to move downwind, taking the form of a crescent dune. For this reason the dunes are usually arranged in chains extending downwind from the sand source.

Where sand is so abundant that it completely covers the solid ground, dunes take the form of wavelike ridges separated by troughlike furrows. The dunes are called *transverse dunes* because, like ocean waves, their crests trend at right angles to the direction of wind (Figure 17.41). The entire area may be called a *sand sea*, because it resembles a storm-tossed sea suddenly frozen to immobility. The sand ridges have sharp crests and are asymmetrical, the gentle slope being on the windward and the steep slip face on the lee side. Deep depressions lie between the dune ridges.

the windward side of the crest the sand slope is gentle and smoothly rounded. On the lee side of the dune, within the crescent, is a steep dune slope, the *slip face*. This face maintains an angle of about 35° from the

FIGURE 17.41 Transverse dunes of a sand sea, near Yuma, Arizona. Isolated crescent dunes can be seen at the right. The view is eastward; prevailing winds are northerly. (John S. Shelton.)

FIGURE 17.42 Star dunes of reddish-yellow sand cover much of the area in this Gemini VII photo of the Sahara Desert in western Algeria. The upper bluish zone is a shallow lake. Water and sediment are brought to the lake basin by a major stream, Wadi Saura, entering from the lower right. Bedrock is exposed in narrow hogback ridges representing a deeply eroded fold structure in sedimentary rocks. The area shown is about 50 km (30 mi) across. (NASA No. S65–6380.)

Sand seas require enormous quantities of sand, often derived from weathering of a sandstone formation underlying the ground surface, or from adjacent alluvial plains. Transverse dune belts also form adjacent to beaches that supply abundant sand and have strong onshore winds.

Wind is a major agent of landscape development in the Sahara Desert. Enormous quantities of reddish dune sand have been derived from weathering of older sandstone formations. The sand is formed into a great expanse of free sand dunes, called an *erg*. Elsewhere there are vast flat-surfaced sheets of sand that are armored against further deflation by a layer of pebbles. A surface of this kind is called a *reg*.

Some of the Saharan dunes are complex forms not abundant in the western hemisphere, for example, the star dune (heaped dune), a large hill of sand whose base resembles a many-pointed star in plan (Figure 17.42). Radial ridges of sand rise toward the dune center and culminate in sharp peaks as high as 100 m (300 ft) or more above the base. Star dunes remain

fixed in position and have served for centuries as reliable landmarks for desert travelers.

Another group of dunes belongs to a family in which the curve of the dune crest is bowed convexly downwind, the opposite of the curvature of crests in the crescent dune. These dunes are parabolic in form. A common representative of this class is the *coastal blowout dune,* formed adjacent to beaches. Here large supplies of sand are available and are blown landward by prevailing winds (Figure 17.43A. A saucer-shaped depression is formed by deflation, and the sand is heaped in a curving ridge resembling a horseshoe in plan. On the landward side is a steep slip face that advances over the lower ground and buries forests, killing the trees (Figure 17.44). Coastal blowout dunes are well displayed along the southern and eastern shore of Lake Michigan. Dunes of the southern shore have been protected for public use as the Indiana Dunes State Park.

On semiarid plains, where vegetation is sparse and winds strong, groups of *parabolic blowout dunes* develop to the lee of shallow

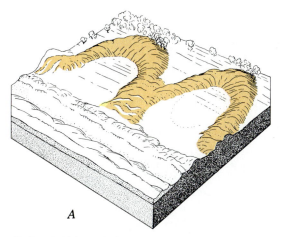

A. Coastal blowout dunes.

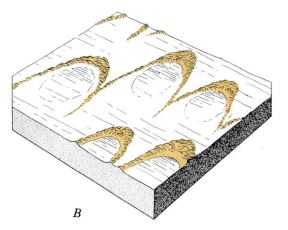

B. Parabolic dunes on a semiarid plain.

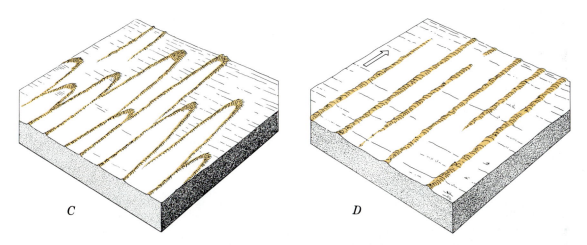

C. Parabolic dunes drawn out into hairpin forms.

D. Longitudinal dune ridges on a desert plain.

FIGURE 17.43 Four types of dunes. Types A, B, and C are of the parabolic class. The prevailing wind direction, shown by an arrow in Block D, is the same for all four examples. (Drawn by A. N. Strahler.)

FIGURE 17.44 This coastal blowout dune is advancing over a forest, with the slip face gradually burying the tree trunks. Lake Michigan coast, near Michigan City, Indiana. (Barry Voight.)

FIGURE 17.45 Long narrow hairpin dunes, stabilized by vegetation, extend landward from a beach with abundant sand supply. Where the plant cover has been breached, the loose sand has moved freely and resumed its landward advance. San Luis Obispo Bay, California. (John S. Shelton.)

FIGURE 17.46 Longitudinal sand dunes in the southern Arabian Peninsula trend northeast—southwest, parallel with the prevailing northeast winds. In the lower right-hand part of the frame is a dissected plateau of the flat-lying sedimentary strata. (NASA Landsat 1186—06381, January 1973.)

deflation hollows (Figure 17.43B). Sand is caught by low bushes and accumulates on a broad, low ridge. These dunes have no steep slip faces and may remain relatively immobile. In some cases the dune ridge migrates downwind, drawing the parabola into a long, narrow form with parallel sides resembling a hairpin in outline (Figure 17.43C). Hairpin dunes stabilized by vegetation are seen in Figure 17.45.

Another class of dunes, described as *longitudinal dunes,* consists of long, narrow ridges oriented parallel with the direction of the prevailing wind (Figure 17.43D). These dune ridges may be many kilometers long and cover vast areas of tropical and subtropical deserts in Africa and Australia (Figure 17.46).

Coastal Dunes

Landward of sand beaches we usually find a narrow belt of dunes in the form of irregularly shaped hills and depressions; these are the *foredunes.* They normally bear a cover of beachgrass and a few other species of plants capable of survival in the severe environment (Figure 17.47).

On coastal foredunes, the cover of beachgrass and other small plants, sparse as it seems to be, acts as a baffle to trap sand moving landward from the adjacent beach. As a result, the foredune ridge is built up as a barrier rising several meters above high tide level. For example, dune summits of the Landes coast of France reach heights of 90 m (300 ft) and span a belt of 10 km (6 mi) wide.

The swash of storm waves cuts away the upper part of the beach, and the dune barrier is attacked. Between storms the beach is rebuilt and, in due time, the dune ridge is also restored if plants are maintained. In this way the foredunes form a protective barrier for tidal lands lying on the landward side. If the plant cover of the dune ridge is depleted by vehicular and foot traffic or by bulldozing of sites for approach roads and buildings, a blowout will rapidly develop. The new cavity may extend as a trench across the dune ridge. With the onset of a storm with high water levels, swash is funneled through the gap and spreads out on the tidal marsh or tidal lagoon behind the ridge. Sand swept through the gap is spread over the tidal deposits (see Figure 17.23).

For many coastal communities of the eastern United States seaboard, the breaching of a dune ridge with its accompanying overwash brings a certain measure of environmental damage to the tidal marsh or estuary, and there may be extensive property damage as well.

Another form of environmental damage related to dunes is the rapid downwind movement of sand when the dune status is changed from one of fixed, plant-controlled forms to that of active dunes of free sand. When plant cover is depleted, wind rapidly reshapes the dunes to produce crests and slip faces (see Figure 17.45). Blowout dunes are developed, and the free sand slopes

FIGURE 17.47 Beachgrass thriving on coastal foredunes has trapped drifting sand to produce a ridge (left) that rises sharply above the barren surface of a blowout depression (right). Oregon Dunes National Recreation Area. (Arthur N. Strahler.)

advance on forests, roads, buildings, and agricultural lands. In the Landes region of coastal dunes on the southwestern coast of France, landward dune advance has overwhelmed houses and churches and even caused entire towns to be abandoned.

Loess

In several large midlatitude areas of the world the ground is underlain by deposits of wind-transported silt, which settled out from dust storms over many thousands of years. This form of transported regolith material is known as *loess*. It generally has a uniform yellowish to buff color and lacks any visible layering (Figure 17.48). Loess has a tendency to break away along vertical cliffs wherever it is exposed by the cutting of a stream or grading of a roadway.

The thickest deposits of loess are in northern China, where a layer over 30 m (100 ft) thick is common and a maximum of 100 m (300 ft) has been measured. It covers many hundreds of square kilometers and appears to have been derived as dust from the interior of Asia. Loess deposits are also of major importance in the United States, central Europe, central Asia, and Argentina.

In the United States, thick loess deposits lie in the Missouri–Mississippi valley (Figure 17.49). Much of the prairie plains region of Indiana, Illinois, Iowa, Missouri, Nebraska, and Kansas is underlain by a loess layer ranging in thickness from 1 to 30 m (3 to

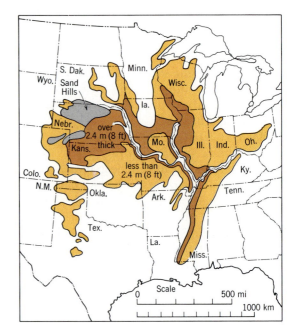

FIGURE 17.49 Map of loess distribution in the central United States. (Data from Map of Pleistocene Eolian Deposits of the United States, Geological Society of America, 1952.)

100 ft). There are also extensive deposits along lands bordering the lower Mississippi River floodplain on its east side, throughout Tennessee and Mississippi. Still other loess deposits are in the Palouse region of northeast Washington and western Idaho.

The American and European loess deposits are directly related to the continental glaciers of the Pleistocene

FIGURE 17.48 (left) This thick layer of loess near Vicksburg, Mississippi, shows no horizontal layering. The material has excellent cohesion and can stand unsupported in vertical walls for decades on end. (right) A closeup of the same loess cliff shows that the material is a soft but cohesive silt. (Orlo E. Childs.)

FIGURE 17.50 A succession of contour terraces, cut into thick loess and covered by tree plantings, is designed to prevent deep gullying and loss of soil. Arched entrances to cave dwellings can be seen at lower left and lower right. Grain fields on the flat valley floor occupy surfaces of thick sediment trapped behind dams. Shanxi Province, near Xi'an (Sian), People's Republic of China. (Alan H. Strahler.)

Epoch. At the time when the ice covered much of North America and Europe, a generally dry winter climate prevailed in the land bordering the ice sheets. Strong winds blew southward and eastward over the bare ground, picking up silt from the floodplains of braided streams that discharged the meltwater from the ice. This dust settled on the ground between streams, gradually building up to produce a smooth, level ground surface. The loess is particularly thick along the eastern sides of the valleys because of prevailing westerly winds, and it is well exposed along the bluffs of most streams flowing through the region today.

Loess is of major importance in world agricultural resources. Loess forms the parent matter of rich black soils (Mollisols) especially suited to cultivation of grains. The highly productive plains of southern Russia, the Argentine pampa, and the rich grain region of north China are underlain by loess. In the United States, corn is extensively cultivated on the loess plains in Kansas, Iowa, and Illinois, where rainfall is sufficient. Wheat is grown farther west on loess plains of Kansas and Nebraska and in the Palouse region of eastern Washington.

Loess forms vertical walls along valley sides and is able to resist sliding of flowage. Because it is easily excavated, loess has been widely used for cave dwellings both in China and in central Europe.

The thick loess deposit covering a large area of northeastern China in the province of Shanxi (Shansi) and adjacent provinces poses a difficult problem of severe soil erosion. Although the loess is capable of standing in vertical walls, it also succumbs to deep gullying during the period of torrential summer rains. From the steep walls of these great scars, fine sediment is swept into streams and carried into tributaries of the Huanghe (Hwang Ho, Yellow River). An intensive program of slope stabilization has been implemented by the Chinese government by using artificial contour terraces, seen in Figure 17.50, in combination with tree planting. Valley bottoms have been dammed so as to trap the silt to form flat patches of land suitable for cultivation.

Induced Deflation

Cultivation of vast areas of short grass prairie under a climate of substantial seasonal water shortage is a practice that invites deflation of soil surfaces. Much of the Great Plains region of the United States is such a marginal region. In past centuries these plains have experienced many dust storms generated by turbulent winds. Strong cold fronts frequently sweep over this area and lift dust high into the troposphere at times when soil moisture is low.

Human activities in the very dry, hot deserts have contributed measurably to raising of dust clouds. In the desert of northwest India and Pakistan (the Thar Desert bordering the Indus River), the continued trampling of fine-textured soils by hooves of grazing animals and by human feet produces a blanket of dusty hot air that hangs over the region for long periods and extends to a height of 9 km (30,000 ft).

In other deserts, such as those of North Africa and the southwestern United States, ground surfaces in the natural state contribute comparatively little dust because of the presence of desert pavements and sheets of coarse sand from which fines have already been winnowed. This protective layer is easily destroyed by wheeled vehicles, exposing finer-textured materials and allowing deflation to raise dust clouds. The disturbance of large expanses of North African desert by tank battles during World War II caused great dust clouds; dust from this source was identified as far away as the Caribbean region.

CHAPTER 18

GLACIAL LANDFORMS AND THE ICE AGE

GLACIERS   Glaciers are large natural bodies of slowly moving ice; they represent an enormous store of the world's fresh water.

ALPINE GLACIERS   Glaciers range greatly in size and form—from long, narrow alpine glaciers of high mountains to vast platelike continental ice sheets.

GLACIAL LANDFORMS   The distinctive scenery of alpine mountains consists of landforms eroded by glaciers moving to lower levels in converging streams.

ICE SHEETS   Ice sheets of the present, on Greenland and Antarctica, bury large continental masses under thousands of meters of ice.

THE ICE AGE   Ice sheets of the Pleistocene Epoch covered large parts of North America and Eurasia, greatly altering the landscape and leaving a host of new landforms.

GLACIATIONS   The cause of repeated advances and disappearances of ice sheets in late Cenozoic time remains a subject of vigorous scientific debate, with widely divergent hypotheses under consideration.

POSTGLACIAL TIME   In the Holocene Epoch, a brief time span following disappearance of the last Pleistocene ice sheet, many environmental changes occurred.

GLACIAL ice has played a dominant role in shaping landforms of large areas in midlatitude and subarctic zones. Glacial ice also exists today in two great accumulations of continental dimensions and in many smaller masses in high mountains. In this sense, glacial ice is an environmental agent of the present, as well as of the past, and is itself a landform.

Glacial ice of Greenland and Antarctica strongly influences the radiation and heat balances of the globe. Moreover, these enormous ice accumulations represent water in storage in the solid state; they constitute a major component of the global water balance. Changes in ice storage can have profound effects on the position of sea level with respect to the continents.

Coastal environments of today have evolved with a rising sea level following melting of ice sheets of the last ice advance in the Pleistocene Epoch, or Ice Age. When we examine evidence of the former extent of those great ice sheets, we need to keep in mind that the evolution of humans as an animal species occurred during a series of climatic changes that placed many forms of environmental stress on all terrestrial plants

FIGURE 18.1 This large Alaskan glacier displays two medial moraines, which can be traced upvalley to glacier junctions. The ice is heavily crevassed where it passes over an icefall. Can you find evidence that this glacier was formerly much thicker than today? (Charles Moore/Black Star.)

and animals in the midlatitude zone. There were also important climatic effects extending into the tropical and equatorial zones.

Glaciers

Most of us know ice only as a brittle, crystalline solid because we are accustomed to seeing it only in small quantities. Where a great thickness of ice exists—100 m (300 ft) or more—the ice at the bottom behaves as a plastic material. The ice will then slowly flow in such a way as to spread out the mass over a larger area, or to cause it to move downhill, as the case may be. This behavior characterizes a *glacier,* defined as any large natural accumulation of land ice affected by present or past motion.

Conditions necessary for the accumulation of glacial ice are simply that snowfall of the winter will, on the average, exceed the amount of ablation of snow that occurs in summer. The term *ablation* includes both evaporation and melting of snow and ice. Thus each year a layer of snow is added to what has already accumulated. As the snow compacts by surface melting and refreezing, it turns into a granular ice, then is compressed by overlying layers into hard crystalline ice. When the ice becomes so thick that the lower layers become plastic, outward or downhill flow commences, and an active glacier has come into being.

At sufficiently high altitudes, even in the tropical and equatorial zones, glaciers form because air temperature is low and mountains receive heavy orographic precipitation. Glaciers that form in high mountains are characteristically long and narrow because they occupy former stream valleys. These mountain glaciers bring the plastic ice from small collecting grounds high on the range down to lower altitudes, and consequently to warmer temperatures. Here the ice disappears by ablation. These *alpine glaciers* are a distinctive type (Figure 18.1).

In arctic and polar regions, prevailing temperatures are low enough that ice can accumulate over broad areas, wherever uplands exist to intercept heavy snowfall. As a result, the uplands become buried under enormous plates of ice whose thickness may

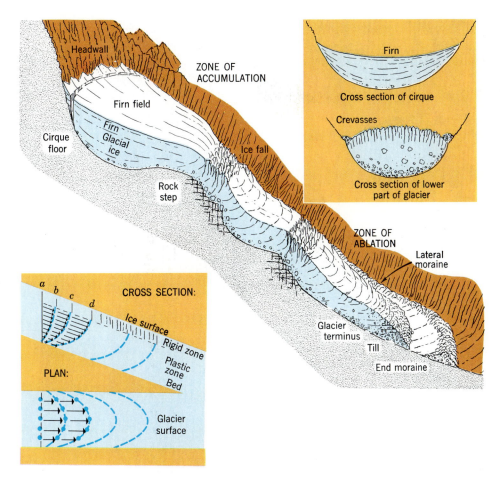

FIGURE 18.2 Structure and flowage of a simple alpine glacier. (Drawn by A. N. Strahler.)

reach several thousand meters. During glacial periods, the ice spreads over surrounding lowlands, enveloping all landforms it encounters. This extensive type of ice mass is called an *ice sheet.*

Figure 18.2 illustrates a number of features of alpine glaciers. The center illustration is of a simple glacier occupying a sloping valley between steep rock walls. Snow is collected at the upper end in a bowl-shaped depression, the *cirque.* The upper end lies in the zone of accumulation. Layers of snow in the process of compaction and recrystallization constitute the *firn.*

The smooth firn field is slightly concave up in profile (upper right). Flowage in the glacial ice beneath carries the excess ice out of the cirque, downvalley. An abrupt steepening of the grade forms a rock step, over which the rate of ice flow is accelerated, and produces deep crevasses

(gaping fractures), which form an ice fall. The lower part of the glacier lies in the zone of ablation. Here the rate of ice wastage is rapid, and old ice is exposed at the glacier surface, which is extremely rough and deeply crevassed. The glacier terminus is heavily charged with rock debris.

The uppermost layer of a glacier is brittle and fractures readily into crevasses, whereas the ice beneath behaves as a plastic substance and moves by slow flowage (lower left). If one were to place a line of stakes across the glacier surface, the glacier flow would gradually deform that line into a parabolic curve, indicating that rate of movement is fastest in the center and diminishes toward the sides.

A simple glacier readily establishes a dynamic equilibrium in which the rate of accumulation at the upper end balances the

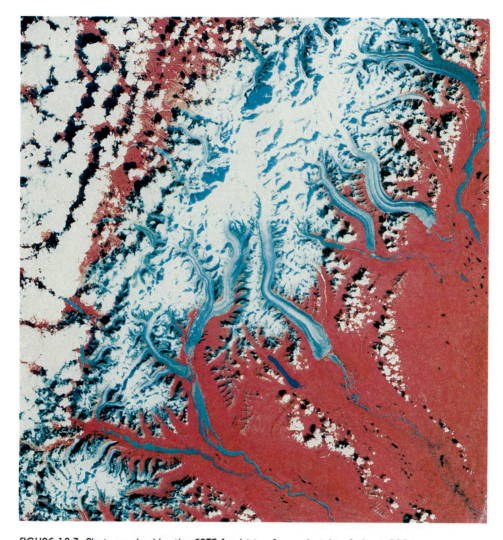

FIGURE 18.3 Photographed by the ERTS-1 orbiting from a height of about 920 km (570 mi), glaciers of the Alaska Range in south-central Alaska appear as blue curving bands, emerging from high collecting grounds on a snow-covered mountain axis trending from northeast to southwest across the center of the photo. The puffy white patches at the left and right are clouds. Darker lines running down the length of a glacier are medial moraines, formed of rock debris on the ice surface. Where these moraines have been distorted into a sinuous pattern, the glacier has experienced a rapid downvalley surge at rates up to 1.2 m (4 ft) per hour. Those glaciers with smooth moraines, paralleling the banks, are experiencing very slow, uniform flow throughout their entire length. The group of high mountain peaks in the upper central area includes Mount McKinley, highest point in North America. This is a false-color image created by combining data of three narrow spectral bands, each assigned a primary color for printing. The area shown is about 105 km (65 mi) across; north is toward the top. (NASA EROS Data Center, No. 81033210205A2.)

rate of ablation at the lower end. Equilibrium is easily upset by changes in the average annual rates of nourishment or ablation. Rate of flowage of glaciers is very slow, amounting to a few centimeters per day for large ice sheets and for the more sluggish alpine glaciers, and up to several meters per day for an active alpine glacier.

Besides showing the continual slow internal flowage we have described, some alpine glaciers experience episodes of very rapid movement, described as *surges*. When surging occurs, the glacier may develop a sinuous pattern of movement, not unlike the open bends of an alluvial river (Figure 18.3). A surging glacier (sometimes called a "galloping glacier" in the news media) may travel downvalley several kilometers in a few months.

A Glacier as an Open Flow System

Glacier equilibrium can be interpreted through an open material flow system, diagrammed in Figure 18.4. As shown in cross section, matter in the form of solid precipitation enters the system through the surface of the zone of accumulation. Downvalley flow carries the glacier ice to the zone of ablation, where it leaves the system by direct evaporation (sublimation) or as meltwater of runoff.

Glacial Erosion

Glacial ice is usually heavily charged with rock fragments ranging from pulverized rock flour to huge angular boulders of fresh rock. Some of this material is derived from the rock floor on which the ice moves. In alpine glaciers rock debris is also derived from material that slides or falls from valley walls. Glaciers are capable of great erosive work. *Glacial abrasion* is erosion caused by ice-held rock fragments that scrape and grind against the bedrock. By *plucking*, the moving ice lifts out blocks of bedrock that

have been loosened by freezing of water in joint fractures.

Rock debris obtained by erosion must eventually be left stranded at the lower end of a glacier when the ice is dissipated. Both erosion and deposition result in distinctive landforms.

Landforms Made by Alpine Glaciers

Landforms made by alpine glaciers are shown in a series of diagrams in Figure 18.5. Previously unglaciated mountains are attacked and modified by glaciers, after which the glaciers disappear and the remaining landforms are exposed to view.

Diagram *A* shows a region sculptured entirely by weathering, mass wasting, and streams. The mountains have a smooth, full-bodied appearance, with rather rounded divides. Soil and regolith are thick. Imagine now that a climatic change results in the accumulation of snow in the heads of most of the valleys high on the mountain sides.

An early stage of glaciation is shown at the right side of Diagram *B*, where snow is collecting and cirques are being carved by the outward motion of the ice and by intensive frost shattering of the rock near the masses of compacted snow.

In Diagram *B*, glaciers have filled the valleys and are integrated into a system of tributaries that feed a trunk glacier. Tributary glaciers join the main glacier with smooth, accordant junctions. The cirques grow steadily larger. Their rough, steep walls soon replace the smooth, rounded slopes of the original mountain mass. Where two cirque walls intersect from opposite sides, a jagged, knifelike ridge is formed. Where three or more cirques grow together, a sharp-pointed peak is formed. Such peaks in the Swiss Alps are called "horns" (Figure 18.6). One of the best known is the striking Matterhorn.

Glacier flow constantly deepens and widens its channel, so that after the ice has finally disappeared, a deep, steep-walled *glacial trough* remains (Figure 18.5C). The U-shape of its cross-profile is characteristic of a glacial trough (Figure 18.7). Tributary glaciers also carve U-shaped troughs, but they are smaller in cross section, with floors

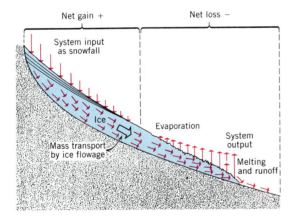

FIGURE 18.4 Schematic diagram of a glacier as a material flow system. (From A. N. Strahler, *Physical Geology*, Harper & Row, New York. Copyright 1981 by Arthur N. Strahler.)

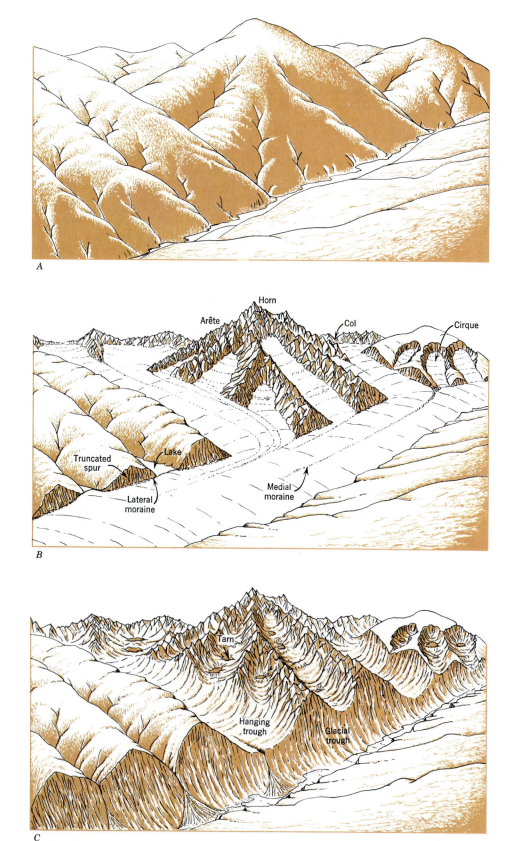

FIGURE 18.5 Landforms produced by alpine glaciers. (A) Before glaciation sets in, the region has smoothly rounded divides and narrow, V-shaped stream valleys. (B) After glaciation has been in progress for thousands of years, new erosional forms are developed. (C) With the disappearance of the ice a system of glacial troughs is exposed. (Drawn by A. N. Strahler.)

FIGURE 18.6 The Mer de Glace, a glacier in the French Alps, seen from an aerial tramway. The glacier is greatly shrunken in both depth and width as compared with its full dimensions two centuries ago, during the Little Ice Age. A massive lateral moraine runs along the left side of the glacier. To the right and left are alpine horns. (Paul W. Tappan.)

lying high above the floor level of the main trough; they are called hanging troughs. Streams, which later occupy the abandoned trough systems, form scenic waterfalls and cascades where they pass down from the lip of a hanging trough to the floor of the main trough. High up in the smaller troughs the bedrock is unevenly excavated, so that the floors of troughs and cirques contain rock basins and rock steps. The

FIGURE 18.7 A U-shaped glacial trough in the Beartooth Plateau, near Red Lodge, Montana. Talus cones have been built out from the steep trough walls. (Alan H. Strahler.)

FIGURE 18.8 A terminal moraine, shaped like the bow of a great canoe, lies at the mouth of a deep glacial trough on the east face of the Sierra Nevada. Cirques can be seen in the distance. Lee Vining, California. (John S. Shelton.)

rock basins are occupied by small lakes, called tarns. Major troughs sometimes hold large, elongated lakes.

Debris is carried by an alpine glacier within the ice and also dragged along between the ice and the valley wall, where it forms a debris ridge called a *lateral moraine* (Figures 18.2 and 18.6). Where two ice streams join, this marginal debris is dragged along to form a narrow band riding on the

FIGURE 18.9 Development of a glacial trough. (A) During maximum glaciation, the U-shaped trough is filled by ice to the level of the small tributaries. (B) After glaciation, the trough floor may be occupied by a stream and lakes. (C) If the main stream is heavily loaded, it may fill the trough floor with alluvium. (D) Should the glacial trough have been deepened below sea level, it will be occupied by an arm of the sea, or fiord. (Drawn by E. Raisz.)

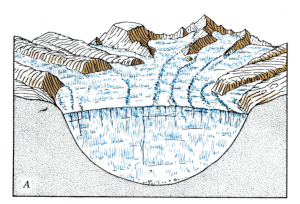

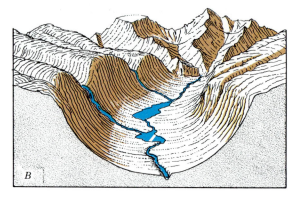

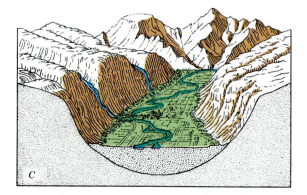

ice in midstream (Figure 18.3). At the terminus of a glacier, debris accumulates in a heap called a *terminal moraine*. This heap takes the form of an embankment curving across the valley floor and bending upvalley along each wall of the trough (Figure 18.8).

Glacial Troughs and Fiords

Many large glacial troughs now are filled with alluvium and have flat floors. Aggrading streams that issued from the receding ice front were heavily laden with rock fragments so that the deposit of alluvium extended far downvalley. Figure 18.9 shows a comparison between a trough with little or no fill and another with an alluvial-filled bottom.

When the floor of a trough open to the sea lies below sea level, the seawater enters as the ice front recedes. The result is a deep, narrow estuary known as a *fiord* (Figure 18.10). Fiords are observed to be

opening up today along the Alaskan coast, where some glaciers are melting back rapidly and ocean waters are being extended inland along the troughs. Fiords are found largely along mountainous coasts in lat. 50° to 70° N and S. On these coasts, glaciers were nourished by heavy snowfall of the orographic type, associated with the marine west-coast climate.

Environmental Aspects of Alpine Glaciation

In past centuries the high, rugged terrain produced by alpine glaciation was sparsely populated and formed impassable barriers between settlements or between nations. This very difficulty of access protected the alpine terrain from the impact of humans and has left us a legacy of wilderness areas of striking scenic beauty. Mountainsides steepened by glaciation make ideal ski slopes. Ski resorts have mushroomed by the dozens in glaciated mountains, and this form of recreation is now a major industry. In summer, these same areas are invaded by tens of thousands of campers and hikers. Rock climbing draws hundreds more to scale the near-vertical rock walls of glacial troughs, cirques, and horns.

The floors of glacial troughs have served for centuries as major lines of access deep into the heart of glaciated alpine ranges. For example, in the Italian Alps, several major flat-floored glacial troughs extend from the northern plain of Italy deep into the Alps. The principal Alpine passes between Italy on the south and Switzerland and Austria on the north lie at the heads of these troughs. An example is the Brenner Pass, located at the head of the Adige trough.

Ice Sheets of the Present

Two enormous accumulations of glacial ice are the Greenland and Antarctic ice sheets. These are huge plates of ice, a few thousand meters thick in the central areas, resting on landmasses of subcontinental size. The Greenland Ice Sheet has an area of 1,740,000 km² (670,000 mi²) and occupies about seven-eighths of the entire

FIGURE 18.10 Geirangerfjord, Norway, is a deeply carved glacial trough occupied by an arm of the sea. (M. Desjardins/Photo Researchers.)

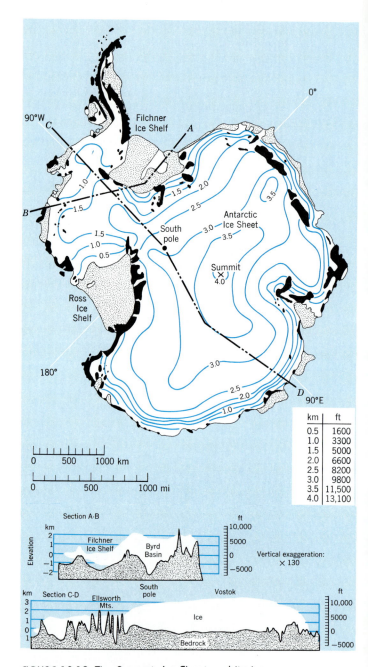

FIGURE 18.11 The Greenland Ice Sheet. (After R. F. Flint, 1957, *Glacial and Pleistocene Geology*, John Wiley & Sons, New York.)

FIGURE 18.12 The Antarctic Ice Sheet and its ice shelves. (Based on data of American Geophysical Union.)

island of Greenland (Figure 18.11). Only a narrow, mountainous coastal strip of land is exposed.

The Antarctic Ice Sheet covers 13 million km² (5 million mi²) (Figure 18.12). A significant point of difference between the two ice sheets is their position with reference to the poles. The antarctic ice rests almost squarely on the south pole, whereas the Greenland Ice Sheet is considerably offset from the north pole, with its center at about lat. 75° N. This position illustrates a fundamental principle: that a large area of high land is essential to

the accumulation of a great ice sheet. No land exists near the north pole; ice there exists only as sea ice.

The surface of the Greenland Ice Sheet has the form of a very broad, smooth dome. From a high point at an altitude of about 3000 m (10,000 ft), there is a gradual slope outward in all directions. The rock floor of the ice sheet lies near sea level under the central region, but is higher near the edges.

Accumulating snows add layers of ice to the surface, and at great depth the plastic ice slowly flows outward toward the edges. At the outer edge of the sheet the ice thins down to a few hundred meters. Continual loss through ablation keeps the position of the ice margin relatively steady where it is bordered by a coastal belt of land. Elsewhere the ice extends in long tongues, called outlet glaciers, to reach the sea at the heads of fiords. From the floating glacier edge huge masses of ice break off and drift out to open sea with tidal currents to become icebergs.

Ice thickness in Antarctica is even greater than that of Greenland (Figure 18.12). For example, on Marie Byrd Land, a thickness of 4000 m (13,000 ft) was measured. Here the rock floor lies 2000 m (6500 ft) below sea level.

An important glacial feature of Antarctica is the presence of great plates of floating glacial ice, called *ice shelves* (Figure 18.12). The largest of these is the Ross Ice

Shelf with an area of about 520,000 km^2 (200,000 mi^2) and a surface height averaging about 70 m (225 ft) above the sea. Ice shelves are fed by the ice sheet, but they also accumulate new ice through the compaction of snow.

Sea Ice and Icebergs

We can distinguish *sea ice*, formed by direct freezing of ocean water, from *icebergs*, which are bodies of land ice broken free from tide-level glaciers. Aside from differences in origin, a major difference between sea ice and floating masses of land ice is thickness. Sea ice is limited in thickness to about 5 m (15 ft).

Pack ice is sea ice that completely covers the sea surface (Figure 18.13). Under the forces of wind and currents, pack ice breaks up into individual patches, called ice floes. The narrow strips of open water between such floes are known as leads. Where ice

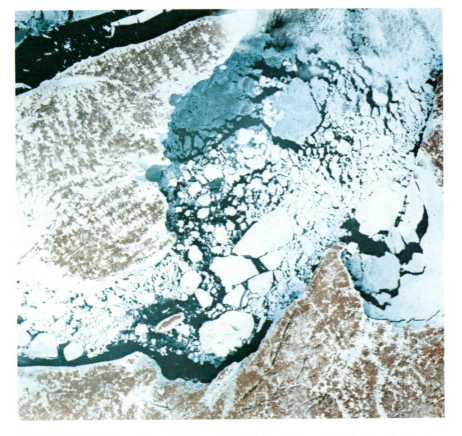

FIGURE 18.13 Ice floes along the shoreline of Prince Edward Island, Nova Scotia, in mid-January. The snow-covered ice plates appear pure white; open leads are solid blue. (NASA Landsat image 1180—14314.)

FIGURE 18.14 An iceberg of the North Atlantic Ocean. (James Holland/Black Star.)

floes are forcibly brought together by winds, the ice margins buckle and turn upward into pressure ridges resembling walls or irregular hummocks. Travel on foot across the polar sea ice is extremely difficult because of such obstacles. The surface zone of sea ice is composed of fresh water.

Icebergs are formed by the breaking off, or calving, of blocks from a valley glacier or tongue of an ice sheet. An iceberg may be as thick as several hundred meters. Being only slightly less dense than seawater, the iceberg floats very low in the water, about five-sixths of its bulk lying below water level (Figure 18.14). The ice is fresh, of course, since it is formed of compacted and recrystallized snow.

In the northern hemisphere, icebergs are derived largely from glacier tongues of the Greenland Ice Sheet. They drift slowly south with the Labrador and Greenland currents and may find their way into the North Atlantic in the vicinity of the Grand Banks of Newfoundland. Icebergs of the antarctic are distinctly different. Whereas those of the North Atlantic are irregular in shape and present rather peaked outlines above water, the antarctic icebergs are commonly tabular in form, with flat tops and steep clifflike sides. This is because tabular bergs are parts of ice shelves. A large tabular berg of the antarctic may be tens of kilometers broad and over 600 m (2000 ft) thick, with an ice wall rising 100 m (300 ft) above sea level.

The Ice Age

The period of growth and outward spreading of great ice sheets is known as a *glaciation*. We can safely assume that a glaciation is associated with a general cooling of average air temperatures over the regions where the ice sheets originated. At the same time, ample snowfall must have persisted over the growth areas to allow the ice masses to grow in volume. The opposite kind of change—a shrinkage of ice sheets in depth and volume—would result in the receding of the ice margins toward the central highland areas and eventual disappearance of the ice sheets. This period is called a *deglaciation*. Following a deglaciation, but preceding the next glaciation, is a period in which a mild climate prevails; it is called an *interglaciation*.

A succession of alternating glaciations and interglaciations, spanning a total period on the order of 1 to 10 million years or more, constitutes an *ice age*. Throughout the past 2 to 3 million years, the earth has been experiencing an ice age that has long been known simply as *The Ice Age*—a title adopted by the naturalist Louis Agassiz in the 1830s. Our present position in time is within an interglaciation, following a deglaciation that set in quite rapidly about 15,000 years ago. In the preceding glaciation, called the *Wisconsinan Glaciation*, ice sheets covered much of North America and Europe and parts of northern Asia and

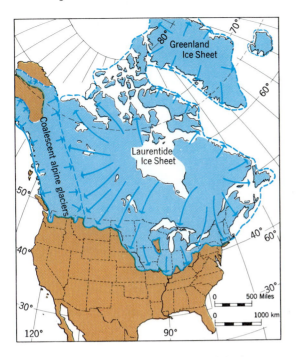

FIGURE 18.15 Pleistocene ice sheets of North America at their maximum extent reached as far south as the present Ohio and Missouri Rivers. (After R. F. Flint, 1957, *Glacial and Pleistocene Geology*, John Wiley & Sons, New York.)

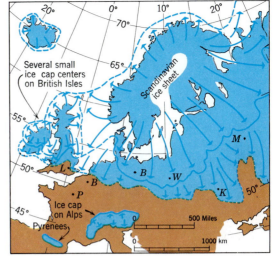

FIGURE 18.16 The Scandinavian Ice Sheet dominated northern Europe during the Pleistocene glaciations. Solid line shows limits of ice in the last glacial stage; dotted line on land shows maximum extent at any time. (After R. F. Flint, 1957, *Glacial and Pleistocene Geology*, John Wiley & Sons, New York.)

southern South America. The maximum ice advance of the Wisconsinan Glaciation was reached about 18,000 years ago.

Figures 18.15 and 18.16 show the extent to which North America and Europe were covered at the maximum known spread of the last advance of the ice. Canada was engulfed by the vast Laurentide Ice Sheet. It spread south into the United States, covering most of the land lying north of the Missouri and Ohio rivers, as well as northern Pennsylvania and all of New York and New England. Alpine glaciers of the Cordilleran ranges coalesced into a single ice sheet that spread to the Pacific shores and met the Laurentide sheet on the east.

In Europe, the Scandinavian Ice Sheet centered on the Baltic Sea, covering the Scandinavian countries. It spread south into central Germany and far eastward to cover much of Russia. In north-central Siberia, large icecaps formed over the northern Ural Mountains and highland areas farther east. At one time ice from these centers grew into a large sheet covering much of

central Siberia. The British Isles were almost covered by a small ice sheet that had several centers on highland areas and spread outward to coalesce with the Scandinavian Ice Sheet. The Alps at the same time were heavily inundated by enlarged alpine glaciers.

South America, too, had an ice sheet. This grew from icecaps on the southern Andes Range south of about latitude 40° S, and spread westward to the Pacific shore as well as eastward to cover a broad belt of Patagonia. It covered all of Tierra del Fuego, the southern tip of the continent. The South Island of New Zealand, which today has a high spine of alpine mountains with small relict glaciers, developed a massive icecap in late Pleistocene time. All high mountain areas of the world underwent greatly intensified alpine glaciation at the time of maximum ice-sheet advance. Today only small alpine glaciers remain and, in less favorable locations, the glaciers are entirely gone.

Maximum southern extent of ice in each of the last four glaciations in the north-central United States is shown in Figure 18.17. The ice fronts advanced southward

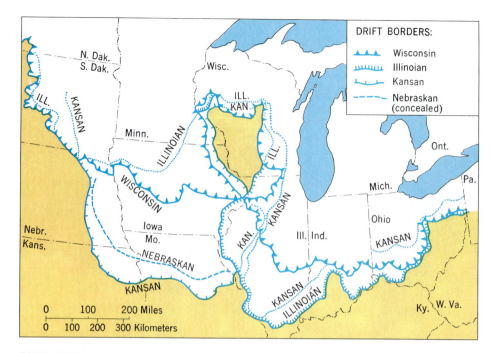

FIGURE 18.17 In each glacial stage, the ice sheet reached a different line of maximum advance. (After R. F. Flint, 1957, *Glacial and Pleistocene Geology*, John Wiley & Sons, New York.)

in great lobes. Notice that an area in southwestern Wisconsin escaped inundation by Pleistocene ice sheets. Known as the Driftless Area, it was apparently bypassed by glacial lobes moving on either side. Landforms made by the last ice advance and recession are very fresh in appearance and show little modification by erosion processes. It is to these deposits and landforms that we now turn our attention.

Erosion by Ice Sheets

Like alpine glaciers, ice sheets are effective eroding agents. The slowly moving ice scraped and ground away much solid bedrock, leaving behind smoothly rounded rock masses bearing countless grooves and scratches trending in the general direction of ice movement (Figure 18.18). The evidence of ice abrasion is common throughout glaciated regions of North America and may be seen on almost any exposed hard rock surface.

Conspicuous knobs of solid bedrock shaped by the moving ice are also common features (Figure 18.19). One side, that from

which the ice was approaching, is characteristically smoothly rounded. The lee side, where the ice plucked out angular joint blocks, is irregular and blocky.

Vastly more important than the minor abrasion forms are enormous rock excavations made by the ice sheets. These

FIGURE 18.18 Glacial striations and shallow grooves on a bedrock surface heavily abraded by Pleistocene alpine glaciers in the Boundary Ranges of British Columbia, near the Alaska border. (J. M. Ryder, University of British Columbia.)

FIGURE 18.19 A glacially abraded rock knob. (From A. N. Strahler, *The Earth Sciences*, 2nd ed., Harper & Row, New York. Copyright © 1971 by Arthur N. Strahler.)

excavations were localized where the bedrock was weak and the ice current was accentuated by the presence of a valley paralleling the direction of ice flow. Under such conditions the ice sheet behaved much as a valley glacier in scooping out a deep, U-shaped trough (Figure 18.20). The Finger Lakes of western New York State are fine examples. Here a set of former stream valleys lay parallel to the southward spread of the ice, which scooped out a series of deep troughs. Blocked at the north ends by glacial debris, the basins now hold long, deep lakes.

Many hundreds of lake basins were created by glacial erosion and deposition over the glaciated portion of North America.

Deposits Left by Ice Sheets

The term *glacial drift* includes all varieties of rock debris deposited in close association with glaciers. Drift is of two major types: (1) *Stratified drift* consists of layers of sorted and stratified clays, silts, sands, or gravels deposited by meltwater streams or in bodies of water adjacent to the ice; (2) *Till* is a heterogeneous mixture of rock fragments ranging in size from clay to boulders that is deposited directly from the ice without water transport.

Over those parts of the United States formerly covered by Pleistocene ice sheets, glacial drift thickness averages from 6 m (20 ft) over mountainous terrain, such as New England, to 15 m (50 ft) and more

FIGURE 18.20 A Pleistocene continental ice sheet carved this finger-lake trough in granite on Mount Desert Island, Maine. Beyond is the smoothly rounded ice-abraded profile of Mount Sargent. (Arthur N. Strahler.)

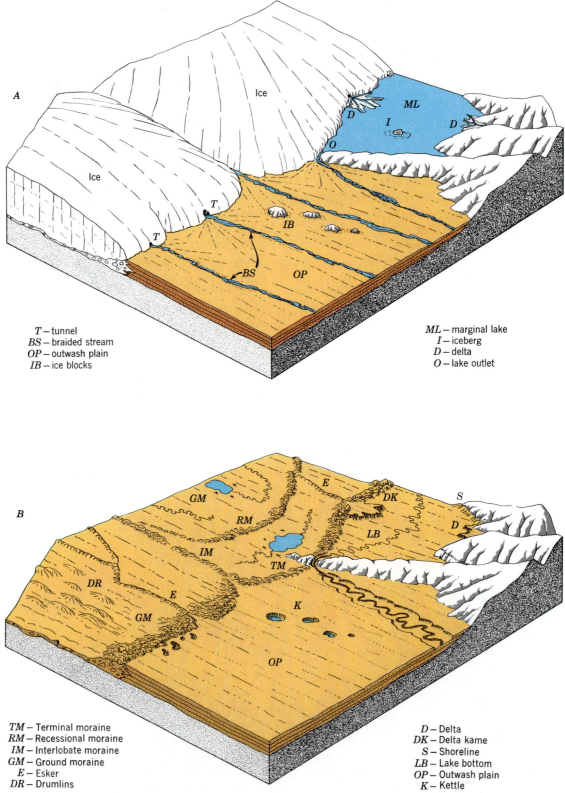

A

T — tunnel
BS — braided stream
OP — outwash plain
IB — ice blocks

ML — marginal lake
I — iceberg
D — delta
O — lake outlet

B

TM — Terminal moraine
RM — Recessional moraine
IM — Interlobate moraine
GM — Ground moraine
E — Esker
DR — Drumlins

D — Delta
DK — Delta kame
S — Shoreline
LB — Lake bottom
OP — Outwash plain
K — Kettle

FIGURE 18.21 Marginal landforms of continental glaciers. (*A*) With the ice front
stabilized and the ice in a wasting, stagnant condition, various depositional features
are built by meltwater. (*B*) The ice has wasted completely away, exposing a variety
of new landforms made under the ice. (Drawn by A. N. Strahler.)

over the lowlands of the north-central United States. Over Iowa, drift thickness is from 40 to 60 m (150 to 200 ft); over Illinois, it averages more than 30 m (100 ft). Locally, where deep stream valleys existed prior to glacial advance, as in Ohio, drift is much thicker.

To understand the form and composition of deposits left by ice sheets, you need to consider the conditions prevailing at the time of existence of the ice. Figure 18.21, Block *A*, shows a region partly covered by an ice sheet with a stationary front edge. This condition occurs when the rate of ice ablation balances the amount of ice brought forward by spreading of the ice sheet. Although the Pleistocene ice fronts advanced and receded in many minor and major fluctuations, there were long periods when the front was essentially stable and thick deposits of drift accumulated.

The transportational work of an ice sheet is like that of a huge conveyor belt. Anything carried on the belt is dumped off at the end and, if not constantly removed, will pile up in increasing quantity. Rock fragments brought within the ice are deposited at the edge as the ice evaporates or melts. There is no possibility of return transportation.

Glacial till that accumulates at the immediate ice edge forms an irregular, rubbly heap—the terminal moraine. After the ice has disappeared (Figure 18.21, Block *B*), the moraine appears as a belt of knobby hills interspersed with basinlike hollows, some of which hold small lakes. The name *knob and kettle* is often applied to morainal belts. Terminal moraines form great curving patterns; the convex curvature is southward and indicates that the ice advanced as a series of great *ice lobes*, each with a curved front (Figure 18.22). Where two lobes come together, the moraines curve back and fuse together into a single moraine pointed northward. (This deposit is an interlobate moraine.) In its general recession accompanying disappearance, the ice front paused for some time along a number of lines, causing morainal belts similar to the terminal moraine belt to be formed. (These belts are known as recessional moraines.) They run roughly parallel with the terminal moraine but are often thin and discontinuous.

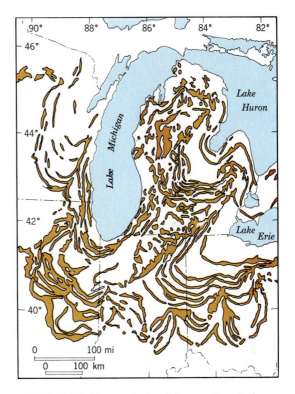

FIGURE 18.22 Moraine belts of the north-central United States have a festooned pattern left by ice lobes. (After R. F. Flint and others, *Glacial Map of North America*, Geological Society of America.)

Figure 18.21 shows a smooth, sloping plain lying in front of the ice margin. This is the *outwash plain*, formed of stratified drift left by braided streams issuing from the ice. The plain is built of layer upon layer of sands and gravels.

Large streams issue from tunnels in the ice, particularly when the ice for many kilometers back from the front has become stagnant, without forward movement. Tunnels then develop throughout the ice mass and serve to carry off the meltwater. After the ice has gone, the position of a former ice tunnel is marked by a long, sinuous ridge known as an *esker*. The esker is the deposit of sand and gravel formerly laid on the floor of the ice tunnel. After the ice has melted away, only the stream-bed deposit remains, forming a ridge (Figure 18.23). Eskers are often many kilometers long.

Another common glacial form is the *drumlin*, a smoothly rounded, oval hill

FIGURE 18.23 An esker, such as this one in Maine, is a valuable source of sand and gravel. Cobbles and boulders must first be screened out. (A. N. Strahler.)

FIGURE 18.24 This small drumlin, located south of Sodus, New York, shows a tapered form from upper right to lower left, indicating that the ice moved in that direction (north to south). (Ward's Natural Science Establishment, Inc., Rochester, N.Y.)

resembling the bowl of an inverted teaspoon. It consists of glacial till (Figure 18.24). Drumlins invariably lie in a zone behind the terminal moraine. They commonly occur in groups or swarms, which may number in the hundreds. The long axis of each drumlin parallels the direction of ice movement and the drumlins thus point toward the terminal moraines and serve as indicators of direction of ice movement. Drumlins were formed under moving ice by a plastering action in which layer upon layer of bouldery clay was spread on the drumlin.

Between moraines, the surface overridden by the ice is overspread by a cover of glacial till (ground moraine). This cover is often inconspicuous because it forms no prominent landscape feature. The moraine layer may be thick and may obscure or entirely bury the hills and valleys that existed before glaciation. Where thick and smoothly spread, the moraine layer forms a level *till plain*. Plains of this origin are widespread throughout the Middle West.

Between the ice front and rising ground, valleys that may have opened out northward were blocked by ice. Under such conditions, marginal glacial lakes formed along the ice front (see Block *A* of Figure 18.21). These lakes overflowed along the lowest available channel between the ice and the rising ground slope, or over some low pass along a divide. Streams of meltwater from the ice built *glacial deltas* into these marginal lakes. When the ice disappeared, the lakes drained away, leaving a flat floor exposed. Here layers of fine clay and silt had accumulated. Glacial lake plains often contain extensive areas of marshland.

Deltas, built with a flat top at what was formerly the lake level, are now curiously isolated, flat-topped landforms known as *delta kames* (Figure 18.21, Block *B*). Kames are built of well-washed and well-sorted sands and gravels.

Environmental and Resource Aspects of Glacial Deposits

Because much of Europe and North America was glaciated by the Pleistocene ice sheets, landforms associated with the ice are of major environmental importance, and the deposits constitute a natural resource as well. Agricultural influences of glaciation are both favorable and unfavorable, depending on preglacial topography and whether the ice eroded or deposited heavily.

In hilly or mountainous regions, such as New England, the glacial till is thinly distributed and extremely stony. Soils developed on glacial deposits of the northern United States and Canada are acid and low in fertility. Extensive bogs are floored by bog soils unsuited to agriculture unless transformed by water drainage systems. Early settlers found till cultivation difficult because of countless boulders and cobbles in the soil. Till accumulations on steep mountain slopes are subject to mass movements in the form of earthflows. Clays in the till become weakened after absorbing water from melting snows and spring rains. Where slopes have been oversteepened by excavation for highways, movement of till is a common phenomenon.

Along moraine belts, the steep slopes, irregularity of knob-and-kettle topography, and abundance of boulders conspired to prevent crop cultivation but invited use as pasture. These same features, however, make morainal belts extremely desirable as suburban residential areas. Pleasing landscapes of hills, depressions, and small lakes made ideal locations for large estates.

Flat till plains, outwash plains, and lake plains, on the other hand, comprise some of the most productive agricultural land in the world. Fertile soils have formed on these till plains and on lake beds bordering the Great Lakes. We must not lose sight of the fact that in these areas wind-deposited silt (loess) forms a blanket over clay-rich till and sandy outwash. Exposed glacial drift would be a poor parent base for soil.

Stratified drift deposits are of great economic value. The sands and gravels of outwash plains, delta kames, and eskers provide the aggregate necessary for concrete and the base courses beneath highway pavements. The purest sands may be used for molds, which are needed for metal castings.

Stratified drift, where thick, forms an excellent aquifer and is a major source of ground water supplies. Deep accumulations of stratified sands in preglacial valleys are capable of yielding ground water in adequate quantities for municipal and industrial uses.

The Late-Cenozoic Ice Age

The Ice Age occurred during the last 3 million years (m.y.) of the Cenozoic Era. As shown in the table of geologic time, Table 12.1, the Cenozoic Era has seven epochs. The Ice Age falls within the last three epochs: Pliocene, Pleistocene, and Holocene. These three epochs comprise only a small fraction—about one-twelfth—of the total duration of the Cenozoic Era.

TABLE 18.1 North American glaciations, based on continental evidence (ca. 1950)

Glaciations	Interglaciations
Wisconsinan	
	Sangamonian
Illinoian	
	Yarmouthian
Kansan	
	Aftonian
Nebraskan	

During the first half of this century, most geologists associated the Ice Age with the Pleistocene Epoch, which was then thought to have begun about one million years ago (-1 m.y.). In other words, they identified the boundary between the Pliocene and Pleistocene epochs as the starting point of the Ice Age. That supposition has now been totally invalidated by new evidence from deep-sea sediments showing that many glaciations and interglaciations occurred in late Pliocene time, and that the Ice Age probably commenced as far back in time as -3 m.y. For this reason, we can rename the Ice Age the *Late-Cenozoic Ice Age,* leaving the date of its onset to be established when more evidence becomes available.

From the middle 1800s until about 1950, the record of glaciations, deglaciations, and interglaciations was almost entirely interpreted from continental deposits. During this early period of research there emerged a history of four distinct North American glaciations in the Pleistocene Epoch. A similar and possibly equivalent four-glaciation history was also established for Europe on the basis of studies in the Alps. Names of the four glaciations and interglaciations of North America are given in Table 18.1. At present, opinion is divided as to whether this classic system of names should be retained or abandoned.

Investigating the Ice Age

Unraveling the history of glaciations and interglaciations was first attempted solely on the basis of evidence from glacial and related deposits and landforms exposed on the continental surfaces. Because ice-sheet advance and recession were evidently not synchronous over all parts of even one continent, the correlation of events proved extremely difficult. One form of evidence has come from interpretation of layered deposits (science of stratigraphy). The till of one glaciation is typically followed by a loess layer associated with deglaciation, then by formation of an ancient soil (paleosol) and by deposition of organic matter, such as bog peat, indicative of an interglaciation with its mild climate. Older tills show varying degrees of chemical alteration by weathering. Landforms of earlier glaciations, where not buried under new glacial deposits, show increasing degrees of modification by mass wasting and fluvial erosion with increasing age.

Starting in the early 1950s, the radio-carbon dating method of establishing the absolute age of sample materials became an important tool of late-Pleistocene and Holocene research. The method was used to make age determinations on such carbon-bearing substances as charcoal, invertebrate shells, wood, and peat derived from archaeological sites and glacial deposits. An example of the use of the radiocarbon method was to determine the ages of tree trunks felled by the rapid advance of the ice sheet at a locality in Wisconsin. A forest had grown up in a mild period (a minor interglaciation) in late Wisconsinan time. The advancing ice broke off the tree trunks and incorporated them into dense red clay of a moraine. Figure 18.25 shows the succession of layers at this locality. Radio-carbon analysis of the logs gave an age close to −12,000 years. This was a record of the final advance of ice in the Wisconsinan Glaciation; it was almost immediately followed by the final ice recession from that region, marking the end of the Pleistocene Epoch and the beginning of the Holocene Epoch.

A great scientific breakthrough in the study of late-Cenozoic glacial history came in the 1960s. First, it became possible to measure the absolute age of lava rock and certain types of water-laid sediments by means of paleomagnetism. The earth's magnetic field has undergone many sudden

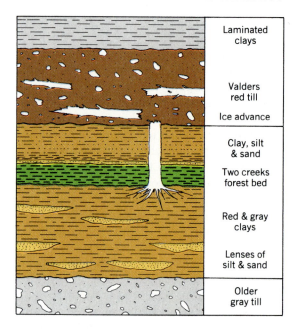

FIGURE 18.25 The Two Creeks forest bed, exposed near Manitowoc, Wisconsin, was developed in a substage of mild climate, but was overridden by ice advance in the Valderan Substage that followed. (After J. L. Hough, 1958, *Geology of the Great Lakes*, p. 102, figure 31, Univ. of Illinois Press, Urbana.)

reversals of polarity in Cenozoic time and the absolute ages of these reversals have been firmly established. Second, it became possible to take long sample cores of undisturbed fine-textured sediments of the deep ocean floor and to determine the age of sediment layers at various control points within each core by identifying magnetic polarity reversals. The cores contain materials that have settled out of the overlying ocean. One kind of material is fine clay of inorganic mineral particles. A second kind of material consists of the tests (hard parts) of microorganisms that lived (as plankton) in the near-surface zone of the overlying ocean. It was soon recognized that the relative abundances of certain species of microorganisms are indicators of the former temperature of the surface water layer—whether cold or warm. By studying the percentage compositions of species, investigators have been able to identify cycles of alternating colder and warmer global climate. They have placed the cycles in an absolute framework of age in years before present.

The most important advance in the interpretation of deep-sea cores has been the development of a method of chemical analysis of the varieties of oxygen atoms (isotopes) present in carbonate minerals of deep-sea cores. Variations in the ratios of these varieties of atoms are directly related to changes in atmospheric temperature and reveal a long history of alternating glaciations and interglaciations going back at least as far as -2 m.y. and possibly to -3 m.y. It is now possible to say with some degree of assurance that in late Cenozoic time there occurred more than 30 glaciations spaced at time intervals of about 90,000 years. How much longer this sequence will continue into the future is not known, but perhaps for one or two million years, or even longer.

What Caused the Late Cenozoic Ice Age?

What caused the Late-Cenozoic Ice Age with its numerous cycles of glaciation and interglaciation? First, we must search for an underlying or basic cause for the Ice Age. Perhaps the answer lies in plate tectonics, through the history of the breakup of Pangaea and the motions of lithospheric plates. Refer back to Figure 12.24 and study the way in which the fragments of Pangaea moved apart and the paths they followed. In Permian time, only the northern tip of the Eurasian continent projected into the polar zone. As the Atlantic basin opened up, North America moved westward and poleward to a position opposite Eurasia, while Greenland took up a position between North America and Europe. The effect of these plate motions was to bring an enormous landmass area to a high latitude and to surround a polar ocean with land. This arrangement was favorable to growth of ice sheets because the flow of warm ocean currents into the polar ocean was greatly reduced, or at times totally cut off. The polar ocean was ice covered much of the time and average temperatures in high latitudes were periodically lowered enough

to allow ice sheets to grow on the encircling continents. You should also notice that during the breakup of Pangaea, Antarctica moved southward and took up a position over the south pole, where it was ideally situated to develop a large ice sheet.

Another change in geological conditions was a worldwide withdrawal of the oceans from passive continental margins. In the Cretaceous Period (-136 to -65 m.y.) the ocean level was high relative to the continents and shallow seas spread far inland over what is now continental surface. Global climate was then mild, even at high latitudes. But starting in early Cenozoic time, sea level entered a falling trend and continental margins became increasingly exposed, and the total surface area of the oceans was reduced. It has been suggested that this change favored a gradual cooling of global air temperatures, thus favoring the occurrence of glaciations.

Another geological mechanism suggested as a basic cause of the Ice Age is increased volcanic activity on a global scope in late Cenozoic time. Eruptions of composite volcanoes produce dust veils that linger in the stratosphere and reduce the intensity of solar radiation reaching the ground (Chapter 5). Temporary cooling of near-surface air temperatures follows such eruptions. Although the geologic record shows periods of high levels of volcanic activity in the Miocene and Pliocene epochs, their role in initiating the Ice Age has not been convincingly demonstrated.

Quite different as a possible cause of the Ice Age is a long-term change in the sun's energy output. As yet, data are insufficient to evaluate this mechanism as a possible basic cause.

The Astronomical Hypothesis of Ice Ages

We turn briefly to consider what timing and triggering mechanisms may have been responsible for cycles of glaciation and interglaciation known to have been repeated many times during the Ice Age. The possible triggering of glaciations by bursts of volcanic activity has already been mentioned, but proof seems difficult to establish. We will limit this discussion to one major contender among several triggering hypotheses; it is called the *astronomical hypothesis* and has been under consideration on and off for about 40 years. It is based on facts about the motions of the earth in its orbit around the sun. These facts are well established by astronomical observations and are not a subject of debate. Two factors are involved here: (1) the changing distance between earth and sun and (2) the changing angle of tilt of the earth's axis of rotation. As to the first factor, the distance that separates the earth and sun at summer solstice, June 21, undergoes a cyclic variation. A single cycle lasts 21,000 years, during which the earth–sun distance may vary from 1 to 5 percent greater than the average distance to 1 to 5 percent less than the average distance. The axial tilt (now 23½°) undergoes a 40,000-year cycle of change in which the tilt angle may increase to as much as about 24° and decrease to as little as 22°. If we pick a point on the earth at, say, lat. 65° N, we can calculate the total cycle of change in the intensity of incoming solar radiation, or insolation, on a day in summer (June 21). This cycle of incoming solar radiation combines the cycle of changing earth–sun distance and the cycle of axial tilt.

The changing insolation, plotted against time on a graph, yields a series of peaks and valleys. The major peaks are repeated at intervals of about 80,000 to 90,000 years. It is postulated that each peak triggered a deglaciation, ending a glaciation and bringing on an interglaciation. During the longer periods of reduced insolation, ice sheets would form and grow in volume. Actually, the glacial events might lag behind the insolation peaks and valleys.

The actual mechanisms by means of which insolation changes cause ice sheets to grow or to disappear remain unknown or at least unproven. The entire subject is so complex that it is difficult for even a research scientist to grasp fully. Interactions between the atmosphere, the oceans, and the continental surfaces (including the ice sheets) are numerous and closely interrelated. Many threads of cause and effect cross and recross these earth realms. Changes that occur in one realm are fed back to the other realms in a most complex manner.

Holocene Environments

The elapsed time span of about 10,000 years since the Wisconsinan glaciation ended is called the *Holocene Epoch;* it began with rapid warming of ocean surface temperatures. Continental climate zones shifted rapidly poleward and soil-forming processes began to act on new parent matter of glacial deposits in midlatitudes. Plants became reestablished in glaciated areas in a succession of climate stages. The first of these is known as the Boreal stage. "Boreal" refers to the present subarctic region where needleleaf forests dominate the vegetation. The history of climate and vegetation throughout the Holocene time has been interpreted through a study of spores and pollens found in layered order from bottom to top in postglacial bogs. (This study is called palynology.) Plants can be identified and ages of samples can be determined. A dominant tree of the Boreal stage was spruce. Interpretation of pollens indicates that the Boreal stage in midlatitudes had a vegetation similar to that now found in the region of boreal forest climate.

There followed a general warming of climate until the Atlantic climatic stage was reached about 8000 years ago (−8000 years). The Atlantic stage lasted for about 3000 years and had average air temperatures somewhat higher than those of today—perhaps on the order of 2.5 C° (4.5 F°) higher. We call such a period a climatic optimum with reference to the midlatitude zone of North America and Europe.

Next came a period of temperatures that was below average, the Subboreal climatic stage, in which alpine glaciers showed a period of readvance. In this stage, which spanned the age range −5000 to −2000 years, sea level, drawn far down from present levels during glaciation, had returned to a position close to that of the present, and a general coastal submergence of the stable continent margins had occurred.

The past 2000 years show climatic cycles on a finer scale than those we have described as Holocene climatic stages. This refinement in detail of climatic fluctuations is a consequence of the availability of historical records and of more detailed evidence generally. A secondary climatic optimum occurred in the period A.D. 1000 to 1200 (−1000 to −800 years). This warm episode was followed by the Little Ice Age (A.D. 1450–1850; −550 to −150 years). During this time valley glaciers made new advances to lower levels. In the process, the ice overrode nearby forests and so left a mark of its maximum extent.

Human Activity and Glaciations

In Chapter 3 we examined conflicting views of the role of human activity in causing global climate change. Cycles of glaciation and interglaciation demonstrate the power of natural forces to make drastic swings from cold to warm climates. Lesser climatic cycles of the Holocene Epoch also occurred through natural causes. Only following the Industrial Revolution do we recognize possible linkages between global air temperature change and combustion of hydrocarbon fuels on a massive scale. There is general agreement that increased carbon dioxide tends to cause a rise in average temperatures. But we do not yet know to what extent observed changes in global temperatures are parts of a natural cycle and to what extent they are influenced by the impacts of an industrial society.

Landforms and Human Activity

The group of chapters we have now completed has reviewed geomorphic processes that operate on the surface of the continents. The great variety and complexity of landforms we have described are not difficult to comprehend when each agent of denudation is examined separately. Landforms produced by glaciers, waves, and wind are localized in distinctive environments—alpine and arctic regions, coastlines, and dry regions, respectively. The landforms of fluvial denudation are the most widespread features of the continental landscape. Fluvial denudation integrates weathering, mass wasting, and fluvial processes into highly complex systems of denudation responding to climate controls.

Human influence on landforms is most strongly felt on surfaces of fluvial denudation because of the severity of surface changes caused by agriculture and urbanization. Landforms shaped by wind and by waves and currents are also highly sensitive to changes induced by human activity. Only glaciers maintain their integrity and are thus far undisturbed by human activity. Perhaps even this last realm of Nature's superiority will eventually fall prey to human interference through climate changes induced by industrial activity.

THE SOIL LAYER

SOIL AND CLIMATE Soil derives its character from the prevailing climate, through infusions of heat energy and water, and from interactions with organisms that live within it.

THE DYNAMIC SOIL Many complex physical, chemical, and organic processes operate simultaneously within the soil, which may also undergo important changes with the passage of time.

PARENT MATERIAL The inorganic substance of a soil is derived from a body of parent mineral matter. Parent minerals are altered into clay minerals, which form the vital fraction of the soil.

SOIL FERTILITY An abundance of certain ions, called bases—they are plant nutrients— imparts natural fertility to the soil. Soils of high acidity lack the nutrient bases required for grasses, grains, and other crops.

SOIL ORDERS Soils of the world fall into ten great classes, called the soil orders. Within these orders are about four dozen suborders.

OLD SOILS Seven soil orders consist of soils with well-developed horizons or with fully weathered minerals; they are old soils with a long history of development, well adjusted to prevailing soil-temperature and soil-water conditions.

THE soil is the very heart of the life layer on the lands, for it is the place in which plant nutrients are produced and held. As we emphasized in Chapter 7, the soil layer also holds water in storage for plants to use. The role of climate is to vary the input of water and heat into the soil. This same heat energy and water are responsible for the breakup and chemical change of rock to produce the parent mineral body of the soil (Chapter 11). It is from this mineral matter that many of the plant nutrients are derived.

But climate acting on rock cannot make a soil layer capable of sustaining a rich plant cover. Plants themselves, together with many forms of animal life, play a major role in determining the qualities of the soil layer. Those qualities have evolved through centuries of time by the interaction of organic processes with physical and chemical soil processes. The organic processes include the synthesis of organic compounds, and these are eventually added to the body of the soil. Plants use the mineral nutrients to build complex organic molecules. Upon death of plant tissues, these nutrients are released and reenter the soil, where they are reused by living plants. So the concept of nutrient cycling by plants is one of the keys to understanding the development of the soil layer.

The Dynamic Soil

The soil is a dynamic layer in the sense that many complex physical and chemical activities are going on simultaneously within it. Because climate and plant cover vary greatly from place to place over the globe, the combined effects of soil-forming activities are expressed differently from place to place. Anyone can observe that the pale gray soil beneath a spruce forest in Maine is quite different in color and structure from the dark brown soil beneath the prairie farmlands of Iowa. In each of these localities, the soil has reached a physical and chemical state related to the climate controls and soil-forming processes prevailing there.

The geographer is keenly interested in the differences in soils from place to place over the globe. The capability of a given soil to furnish food crops largely determines which areas of the globe support the bulk of the human population. Despite changes in population distribution made possible by technology and industrialization, most of the world's inhabitants still live where the soil furnishes them food. Many of those same humans die prematurely because the soil does not furnish enough food for all.

The substance of the soil exists in all three states—solid, liquid, and gas. The solid portion consists of both inorganic (mineral) and organic substances. The liquid present in the soil is a complex solution capable of engaging in a multitude of important chemical reactions. Gases present in the open pores of the soil consist not only of the atmospheric gases, but also gases liberated by biological activity and chemical reactions within the soil.

Soil science, often called *pedology*, is obviously a highly complex body of knowledge.

The Nature of Soil

Thus far, we have used the word "soil" without giving it a precise definition. *Soil*, as the term is used in soil science, is a natural surface layer containing living matter and supporting or capable of supporting plants. Substance of the soil includes both inorganic (mineral) matter and organic matter, the latter both living and dead. Living matter in the soil consists not only of plant roots, but of many kinds of organisms, including microorganisms. The upper limit of the soil is air or shallow water. Horizontal limits of the soil may be deep water or barren areas of rock or ice.

In this strict sense, the surface layer of large expanses of the continents cannot be called soil. For example, dunes of moving sand, bare rock surfaces of deserts and high mountains, and surfaces of fresh lava near active volcanoes do not have a soil layer. In time these areas may develop true soils but, in that process, the surface layers will be greatly altered.

For most classes of soils, the bulk of the solid fraction of the soil consists of finely divided mineral matter. The term *parent matter* is applied to all forms of mineral matter suitable for transformation into soil. Parent mineral matter may be derived from the underlying bedrock, which is solid, unaltered rock (Figure 19.1). Bedrock is softened, disintegrated, and chemically changed in the process of rock weathering. Weathering processes (explained in Chapters 11 and 14) change bedrock to regolith, a residual layer of mineral matter that is one of the common forms of parent matter. Other important kinds of regolith consist of mineral particles transported to a place of rest by action of streams, glaciers, waves and water currents, or winds. For example, dunes formed of sand transported by wind are a type of regolith on which soil may be formed. The many and varied forms of transported regolith were described in Chapters 14 through 18.

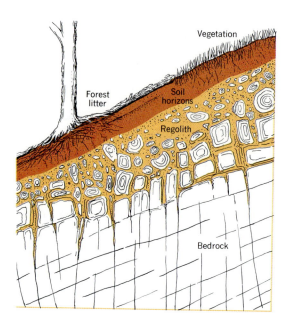

FIGURE 19.1 The true soil is characterized by distinct horizons and the capability to support plants. Below is infertile regolith, derived from the underlying bedrock.

Concept of the Pedon

Modern soil science makes use of the concept of the *polypedon*, which is the smallest distinctive division of the soil of a given area. A unique single set of properties applies to the polypedon, and this set differs from that applying to adjacent polypedons. The polypedon is visualized in terms of space geometry as being composed of pedons. A *pedon* is a soil column extending down from the surface to reach a lower limit in regolith or bedrock. As Figure 19.2 shows, soil scientists visualize a pedon as a six-sided (hexagonal) column. The surface area of a single pedon ranges from 1 to 10 m² (1 to 10 yd²).

The *soil profile* is the display of horizons on one face of the pedon. Obviously, the same soil profile is displayed on all six faces of the pedon. In practice, a soil scientist digs a deep pit, exposing a soil profile in the side of the pit. The enclosing walls of a pedon are purely imaginary. The polypedon is not, as the diagram might imply, broken into hexagonal blocks.

Figure 19.2 shows a number of *soil horizons*, which are distinctive, horizontal layers identified in terms of physical and chemical composition, organic content or

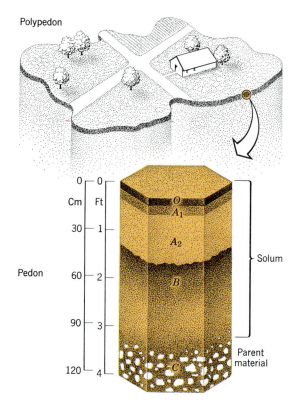

FIGURE 19.2 Concept of the pedon and polypedon.

structure, or a combination of such properties. Most horizons are visibly set apart on the basis of color. Mineral soil horizons are designated by a set of capital letters and numeral subscripts, starting with A at the top. In Figure 19.2 we see A, B, and C horizons. An organic horizon, designated by the letter O, lies on the A horizon.

The *soil solum* consists of the A and B horizons of the soil profile; these are the dynamic and distinctive layers of the profile. The C horizon, by contrast, is the parent matter. The soil solum occupies the zone in which living plant roots exert control on the soil horizons; the C horizon lies below that level of root activity. Of course, the C horizon is subject to inorganic processes of change, which may be of physical or chemical nature.

Soil Color

Color is the most obvious property of a soil seen at a distance. The dark brown to black color of the soil is conspicuous to the traveler in farmlands of Iowa and

Nebraska; the red color of soil in the Piedmont upland of Georgia quickly catches the eye. Certain color relationships are quite simple. Black color usually indicates the presence of abundant organic matter; red color usually indicates the presence of iron compounds.

The soil color may in some areas be inherited from the parent mineral matter, but more generally, it is a property generated by the soil-forming processes. For example, a white horizon in soils of dry climates often indicates the presence of mineral salts that have entered the soil. A pale, ash-gray horizon in soils of the boreal forest climate results from the leaching out of organic matter and various colored minerals originally present in the parent matter. As we explain soil-forming processes and describe the various classes of soils, soil color will take on more meaning.

Soil Texture

The mineral fraction of the soil usually spans a very wide range of particle sizes. The term *soil texture* refers to the proportion of particles falling into each of several size grades. Metric units of length are used in describing particle grades. In measuring soil texture, gravel and larger particles are eliminated, since these play no important role in soil processes. The remaining grades are classified as *sand, silt,* and *clay.* The diameter range of each of these grades is shown in Figure 19.3. Millimeters are the standard units down to 0.001 mm, when the scale shifts to microns. Each unit on the scale represents a power of ten, so that clay particles of 0.001 micron diameter are one ten-millionth as large as sand grains 1 mm in diameter.

Soil texture is described by a series of names emphasizing the dominant constituent, whether sand, silt, or clay. Figure 19.4 gives examples of five soil textures with typical percentage compositions. A *loam* is a mixture containing a substantial proportion of each of the three grades. Loams are classified as sandy, silty, or clay-rich when one of the grades is dominant.

Texture is important because it largely determines the ability of the soil to retain

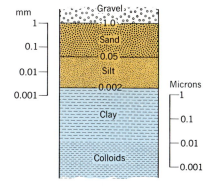

FIGURE 19.3 Size grades used in the description of soil textures. (U.S. Department of Agriculture system.)

water or to transmit water to the intermediate belt below. Recall that storage capacity of a soil is its capacity to hold water against the pull of gravity. Figure 19.5 shows how storage capacity varies with soil texture. Pure sand holds the least water, whereas pure clay holds the most. Loams hold intermediate amounts.

Sand transmits the water downward most rapidly, clay most slowly. When planning the quantity of irrigation water to be applied, these factors must be taken into account. Sand reaches its capacity very rapidly, and added water is wasted. Clay-rich loams take up water very slowly and, if irrigation is too rapid, water will be lost by surface runoff. Sandy soils require more frequent watering than clay-rich soils. The intermediate loam textures are generally best as agricultural soils because they drain well, but also have favorable water retention

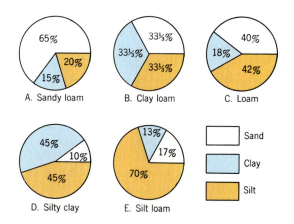

FIGURE 19.4 Typical compositions of five soil texture classes.

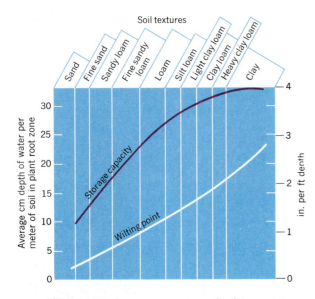

FIGURE 19.5 Storage capacity and wilting point vary according to soil texture.

properties. Finely divided organic matter in the soil can greatly increase the water-holding capacity, and must be taken into account.

Soil texture is largely an inherited feature of a given soil and depends on the composition of the parent matter. Some types of parent matter are endowed with a large spread in particle sizes, and others consist of mostly sand or mostly clay.

Agricultural soil scientists also use a measure of soil-water storage termed the *wilting point*. Soil water in an amount less than the value at the wilting point cannot be absorbed by plants rapidly enough to meet their needs. At this point, the foliage of plants not adapted to drought will wilt. As Figure 19.5 shows, the wilting point depends on texture.

Soil Colloids

Colloids play a vital role in the soil. *Soil colloids* consist of particles smaller than about 0.1 micron. These particles include not only the mineral colloids, which are inorganic, but also organic colloids. *Humus* is finely divided, partially decomposed organic matter found resting on the soil surface and mixed through the upper horizons. Humus particles of colloidal

dimensions are gradually carried down by percolating soil water to reach lower layers. These particles give the soil its brown and black coloration.

If you examine mineral colloids under the electron microscope, you will find that they consist of thin, platelike bodies (Figure 19.6). When well mixed in water, particles of this small size remain suspended indefinitely, giving the water a murky appearance.

Soil water contains a wide variety of ions. An *ion* is an electrically charged atom or group of atoms. Compounds that dissolve in water break up into ions. A familiar example is ordinary table salt, sodium chloride, with the formula NaCl. One atom of sodium (Na) is united with one atom of chlorine (Cl) in a single molecule of salt. When the salt is dissolved, the two atoms are separated and become ions. The sodium ion bears a positive charge and is referred to as a *cation*. The chlorine ion bears a negative charge.

Getting back to the soil colloids, we find that each particle holds a surface electrical charge, as shown in Figure 19.7. Colloids of

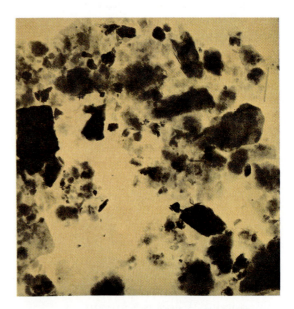

FIGURE 19.6 Seen here enlarged about 20,000 times are tiny flakes of the clay minerals of colloidal dimensions. The minerals are illite (sharp outlines) and montmorillonite (fuzzy outlines). These particles have settled from suspension in San Francisco Bay. (San Francisco District Corps of Engineers, U.S. Army.)

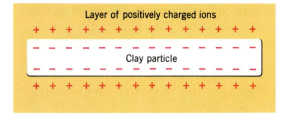

FIGURE 19.7 A colloidal particle with negative surface charges and a layer of positively charged ions.

mineral origin have negatively charged outer surfaces. These negative charges attract and hold such cations as may be present in the soil-water solution. The numerous plus signs in the diagram designate cations held by the colloidal particle. Organic colloids—humus particles, that is—are also capable of holding ions.

Among the many ions in the soil solution, one important group consists of plant nutrients. Known as *base cations,* or simply *bases,* these are principally ions of four metallic elements: calcium, magnesium, potassium, and sodium. Colloids hold these ions, but also give them up to plants; they enter the plant tissue through root membranes. Without this ion-holding ability of soil colloids, most of the vital nutrients would be carried out of the soil by percolating water and would be taken out of the region in streams, eventually reaching the sea. This leaching process goes on continually in climates having a substantial water surplus, but loss is greatly

retarded by the ion-holding capacity of soil colloids.

Soil Acidity and Alkalinity

The soil solution also contains hydrogen and aluminum ions, and they are also positively charged ions, or cations. The presence of these ions in the soil solution tends to make the solution acid in chemical balance. On the other hand, an abundance of the base cations tends to make the soil solution alkaline in balance.

An important principle of soil chemistry is that the acid ions have the power to replace the base cations clinging to the surfaces of the soil colloids. As acid ions accumulate, the base cations are carried away by leaching and are lost. When this happens, the soil acidity is increased. At the same time, loss of nutrient base cations reduces the fertility of the soil.

The degree of acidity or alkalinity of a solution is designated by the pH number. The lower the pH number, the greater the degree of acidity. A pH number of 7 represents a neutral state; higher values are in the alkaline range. Table 19.1 shows the natural range of acidity and alkalinity found in soils. High soil acidity is typical of cold, humid climates; soil alkalinity is typical of dry climates. Acidity can be corrected by the application of lime, an oxide of calcium. This is an important agricultural practice because it reduces the hydrogen ion content of the soil. Then, when fertilizers are applied, the base cations can reoccupy positions on the soil colloids. Recall from

TABLE 19.1 Soil acidity and alkalinity

pH	4.0	4.5	5.0	5.5	6.0	6.5	6.7	7.0	8.0	9.0	10.0	11.0
Acidity	Very strongly acid		Strongly acid	Moderately acid	Slightly acid		Neutral		Weakly alkaline	Alkaline	Strongly alkaline	Excessively alkaline
Lime requirements	Lime needed except for crops requiring acid soil		Lime needed for all but acid-tolerant crops		Lime generally not required			No lime needed				
Occurrence	Rare	Frequent	Very common in cultivated soils of humid climates						Common in sub-humid and arid climates		Limited areas in deserts	

Based on data of C. E. Millar, L. M. Turk, and H. D. Foth (1958), *Fundamentals of Soil Science,* 3rd ed., John Wiley & Sons, New York. See Chart 4.

our Introduction that Edmund Ruffin used lime as early as 1818.

Parent Minerals of the Soil

To understand how soils are formed and how they differ from place to place requires some knowledge of the common minerals that form the parent matter and colloids of the soil. Soil scientists recognize two classes of minerals abundant in soils: primary minerals and secondary minerals. The *primary minerals* are compounds present in unaltered rock of igneous origin. Mostly, these are silicate minerals—compounds of silicon, oxygen, and varying proportions of aluminum, calcium, sodium, iron, and magnesium. (The igneous rocks and silicate minerals are described in Chapter 11.) A common primary silicate mineral is quartz. The feldspars and micas comprise other classes of primary silicate minerals. Primary minerals form a large fraction of the solid matter of many kinds of soils, but they play no important role in processes that make a soil capable of sustaining plant life. Chemical changes during rock weathering near the earth's surface alter the primary minerals into *secondary minerals,* which are essential to soil development and to soil fertility.

Clay Minerals in the Soil

In terms of the properties of soils, perhaps the most important secondary minerals are the clay minerals, described in Chapter 11. At this point you should review the compositions and properties of kaolinite, illite, and montmorillonite.

For the soil scientist interested in soil fertility, the ability of a given clay mineral to hold base cations is a most important property. The three clay minerals we have mentioned differ greatly among themselves in this respect. Montmorillonite can hold a very large supply of base cations, making them available to plants. Thus, the presence of montmorillonite contributes greatly to high soil fertility and the soil is said to have a *high base status.* Kaolinite, in contrast, holds few base cations. A soil having only kaolinite as the principal clay mineral is low

in natural fertility and is described as a soil of *low base status.* Illite ranks intermediate between the two extremes. We should add that humus colloids have a high capacity to hold base cations, so that the presence of humus is usually associated with potentially high soil fertility.

Mineral oxides are important alteration products in many kinds of soils, particularly those exposed to atmospheric conditions over very long periods of time (hundreds of thousands of years) in areas of warm, moist climates. Under these conditions, the clay minerals are ultimately broken down chemically into simple oxides, from which the silicon of the original primary minerals has been completely removed. Oxides of aluminum and iron are the most important oxides in soils in terms of bulk. The two atoms of aluminum are combined with three atoms of oxygen to form *sesquioxide of aluminum* (chemical formula Al_2O_3). In combination with water molecules, this form of aluminum oxide forms the mineral bauxite, which occurs as hard, rocklike lumps and layers below the soil surface. Sesquioxide of iron (Fe_2O_3) in combination with water molecules is limonite, a yellowish to reddish mineral that gives its color to certain varieties of soils. Limonite and bauxite occur in close association in soils of warm, moist climates in low latitudes. Because these minerals have a very small capacity to hold base cations, their presence is associated with soils of low base status that are deficient in plant nutrients.

Soil Structure

Soil structure refers to the way in which soil grains are grouped together into larger masses, called *peds.* Peds range in size from small grains to large blocks. Small peds of roughly spherical shape give the soil a granular structure or crumb structure, illustrated in Figure 19.8. Angular blocky structure is also illustrated in Figure 19.8. One factor contributing to formation of soil aggregates is the binding effect of soil colloids. Colloid-rich clays shrink greatly in volume as they dry out, and shrinkage results in formation of soil cracks, defining the surfaces of the aggregates. Other aggregates commonly found in soils take

FIGURE 19.8 Granular structure (left) and blocky structure (right). The length of the bar is 2.5 cm (1 in.). (Division of Soil Survey, U.S. Department of Agriculture.)

the form of platelike bodies and elongate prisms.

Soils with well-developed granular or blocky structure are easy to till; this is an important agricultural factor in lands where primitive plows, drawn by animals, are still widely used. Soils in which the clay colloids are in a dispersed state lack aggregates. These soils are sticky and heavy when wet and are difficult to cultivate. When dry, they become too hard to be worked.

Soil-Forming Processes

Soil horizons are developed by the interactions through time of climate, living organisms, and the configuration of the land surface. Horizons are usually explained by either selective removal or accumulation of certain ions, colloids, and chemical compounds.

Soils formed in regions of large water surplus are subject to selective removal of matter as the excess water percolates through the soil to the ground water zone and is eventually disposed of as stream runoff. In many cases, however, matter carried downward simply accumulates at a lower level in the soil, where it forms a distinctive horizon (Figure 19.9). The downward transport process is called *eluviation;* it produces a distinct soil horizon

from which matter has been removed. This is the *A horizon.* Accumulation of matter in the underlying zone is called *illuviation;* it forms the *B horizon.*

Subtances carried downward by eluviation may consist of colloidal particles of humus, clay minerals, or oxides of iron and aluminum. Typically, quartz grains remain behind to comprise the bulk of the A horizon, which is pale in color and loose in texture. Substances accumulated by illuviation in the B horizon may be colloidal clay minerals, humus, or sesquioxides of iron and aluminum. These give the B horizon a dense, tough character and, in some cases, produce a hard rocklike layer.

An important secondary mineral, not yet mentioned in our discussion of soils, is calcium carbonate (Chapter 11). In pure

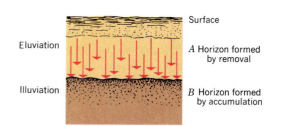

FIGURE 19.9 Downward migration of soil constituents is characteristic of cool, moist climates having a large water surplus.

432

form, it bears the mineral name calcite (chemical composition CaCO₃). In many areas, the parent material of the soil contains a substantial proportion of calcium carbonate derived from the disintegration of limestone, a common variety of bedrock. Carbonic acid, found in all rainwater and soil water, readily reacts with calcium carbonate. The products of this reaction remain in solution as ions.

In moist climates, where a large amount of surplus soil water moves downward to the intermediate zone and ground water zone, calcium carbonate is removed from the entire soil in a process called *decalcification*. Soils that have lost most of their calcium are also usually acid in chemical balance. Calcium is an important plant nutrient, so that soils that have been affected by decalcification are usually low in bases. Addition of lime or pulverized limestone not only corrects the acid condition, but restores the needed calcium as well.

In dry climates, calcium carbonate is dissolved in the upper, or A, horizon during periods of rain or snowmelt when soil-water recharge is taking place. The dissolved carbonate matter is carried down to the B horizon, where water penetration reaches its limits. Here, the carbonate matter is precipitated (deposited in crystalline form) in the B horizon, a process called *calcification*. Calcium carbonate deposition takes the form of white or pale-colored grains, nodules, or plates in the B horizon. Under somewhat different climatic conditions, calcification occurs instead in the underlying C horizon, which is beneath the solum. Deposits that form in the C horizon often form a hard, massive rocklike layer to which soil scientists have given the name *calcrete*.

Finally, we take note of a process important in the desert climate. In low, flat, poorly drained areas, continued evaporation of soil water is accompanied by the rise of water in capillary films drawn upward through the soil much as a cotton wick draws kerosene upward in an oil lamp. Usually in such cases, a saturated layer of ground water lies close to the surface and serves as the source of supply. As the rising capillary water evaporates in the upper soil layer, dissolved salts are precipitated and

accumulate as a distinctive horizon. This process is called *salinization*. A horizon rich in salt is called a *salic horizon;* it is often white and is easily recognized (see Figure 19.28). Typical of salts that accumulate in a salic horizon are compounds of sodium, of which ordinary table salt (halite) is a familiar example. Sodium in large amounts is associated with highly alkaline conditions and is toxic to many kinds of plants. When salinization occurs in irrigated lands in a desert climate, the soil can be ruined for further agricultural use. We covered this topic in Chapter 10.

From what we have presented on soil-forming processes, it must be obvious that the soil-water balance is a key to soil development. Soil temperature, along with soil water, is an important factor in determining the chemical development of soils and the formation of horizons. Temperature acts as a control over biologic activity and also influences the intensity of chemical processes affecting soil minerals. Below the freezing point (0°C, 32°F), there is no biologic activity; chemical processes affecting minerals are inactive. Root growth of most plants and germination of their seeds require soil temperatures above 5°C (41°F). For plants of the warm, wet low-latitude climates, germination of seeds requires a soil temperature of at least 24°C (75°F).

Temperature of the uppermost soil layer and the soil surface strongly affects the rate at which organic matter is decomposed by microorganisms. Thus, in cold climates, organic matter in the form of fallen leaves and stems tends to accumulate to form a thick organic horizon, the O horizon (Figure 19.2). This finely divided, black material is humus. Humus particles are carried downward by percolating soil water and enrich the solum with organic matter. Plant roots, decaying within the solum, are also an important source of organic matter.

In warm, moist climates of low latitudes, the rate of decomposition of plant material is very high, so that nearly all the fallen leaves and stems are disposed of by bacterial activity. Under these conditions the O horizon may be missing and the entire soil profile lacking in organic matter.

Soil scientists recognize a set of soil-temperature regimes, based on the mean annual soil temperature and the range of temperature throughout the year.

Landform and Soil

Configuration of the ground surface is an important factor in soil formation and can be summarized by the single word *landform*. Landform includes the varying steepness of the ground surface, or slope, as well as the compass orientation—the aspect—of a small plot of ground. Another landform property is the relief, or average elevation difference between adjacent high and low points (i.e., between hilltops and valley bottoms). Generally speaking, soil horizons are thick on gentle slopes but thin on steep slopes, because the soil is more rapidly removed by erosion processes on the steeper slopes.

Aspect influences the soil-water regime. Slopes facing north (in the northern hemisphere) are sheltered from direct insolation and tend to have moister soils than slopes facing south and exposed to direct solar rays.

Biological Processes in Soil Formation

The total role of biological processes in soil formation includes the presence and activities of living plants and animals as well as their nonliving organic products. Living plants contribute to soil formation in two basic ways. First is the production of organic matter both above the soil as stems and leaves and within the soil as roots. As already noted, these plant tissues provide the raw material for formation of humus. Second, plants take up a variety of nutrient elements from the soil and store them in plant tissues. When the plant dies and plant tissues decay, these nutrients, which include the base cations, are released to the soil where they can be taken up by new generations of plants. This process of nutrient recycling is discussed in Chapter 20. Nutrient recycling is a mechanism by means of which soil nutrients are prevented from escaping through the leaching action of surplus water moving downward.

Animals living in the soil, or entering and leaving the soil by means of excavated passageways, span a wide range in species and in the sizes of individual animals. The total role of animals is extremely important in soil formation in soils that have sufficient heat and moisture to support large animal populations. For example, earthworms continually rework the soil not only by burrowing, but also by passing the soil through their intestinal tracts. They ingest large amounts of decaying leaf matter, carrying it down from the surface and incorporating it into the mineral soil horizons. Many forms of insect larvae perform a similar function. Tubelike openings are made by larger animals—moles, gophers, rabbits, badgers, prairie dogs, and many other species. The growth of roots followed by their decay leaves tubular openings in the soil.

In moist climates, the evolution of a soil from parent mineral matter is accompanied by increasing plant growth and changes in plant species. We will examine this evolutionary process—called plant succession—in Chapter 20. Various close relationships between ecosystems and soil characteristics will become evident in descriptions of the widely varied classes of soils.

Human activity also influences the physical and chemical nature of the soil. Large areas of agricultural soils have been tilled and fertilized for centuries. Both structure and composition of these agricultural soils have undergone profound changes and can now be recognized as distinct soil classes of importance equal to natural soils.

The Global Scope of Soils

In physical geography the most important aspect of soil science is the classification of soils into major types and subtypes recognized in terms of their distribution over the earth's land surfaces. Geographers are particularly interested in ways in which the factors of climate, parent material, time, biologic process, and landform are linked with the distribution of types of soils. Geographers are also interested in the kinds of natural vegetation associated with each of the major soil classes. The geography of soils is thus an essential ingredient in determining the quality of environments of the globe—important because soil fertility, along with availability of fresh water, is a basic measure of the capability of an environmental region to produce food for the human race.

Soils of the world are classified in a system developed by scientists of the U.S. Soil Conservation Service, in cooperation with soil scientists of many other nations. The classification system consists of six categories or levels. The top level consists of ten *soil orders*. Every polypedon falls into one and only one soil order. The second level of classification consists of *suborders*, of which there are 47. The third level consists of *great groups*, numbering 185. Three lower levels of classification are used, but these do not concern us here.

Soil orders and suborders are, in many cases, distinguished on the basis of the presence of a *diagnostic horizon*. Each diagnostic horizon has some unique combination of physical properties (color, structure, texture) or chemical properties (minerals present or absent). The two basic kinds of diagnostic horizons are (1) a horizon formed at the surface and called an *epipedon* (from Greek *epi*, over or upon) and (2) a subsurface horizon formed by removal or accumulation of matter. We will refer to a number of diagnostic horizons in our descriptions of soil orders.

The Soil Orders

The ten soil orders making up the highest level of the classification system can be grouped for easier understanding into three simple classes based on overall properties and history of development:

I   Soils with well-developed horizons or with fully weathered minerals, resulting from long-continued adjustment to the prevailing soil-temperature and soil-water regimes.
  Oxisols   Very old, highly weathered soils of low latitudes, with a subsurface horizon of accumulation of mineral oxides and very low base status.
  Ultisols   Soils of equatorial, tropical, and subtropical latitude zones, with a subsurface horizon of clay accumulation and low base status.

  Vertisols   Soils of subtropical and tropical zones with high clay content and high base status, developing deep, wide cracks when dry and showing evidence of movement between soil blocks.
  Alfisols   Soils of humid and subhumid climates with a subsurface horizon of clay accumulation and high base status. Alfisols range from equatorial to subarctic latitude zones.
  Spodosols   Soils of cold moist climates, with a well-developed B horizon of illuviation, and low base status.
  Mollisols   Soils of semiarid and subhumid midlatitude grasslands, with a dark, humus-rich epipedon and very high base status.
  Aridisols   Soils of dry climates, low in organic matter and often having subsurface horizons of accumulation of carbonate minerals or soluble salts.

II   Soils with a large proportion of organic matter.
  Histosols   Soils with a thick upper layer very rich in organic matter.

III   Soils with poorly developed horizons or no horizons, and capable of further mineral alteration.
  Entisols   Soils lacking horizons, usually because of recency of accumulation of parent matter.
  Inceptisols   Soils with weakly developed horizons, having minerals capable of further alteration by weathering processes.

Our brief survey of global soils follows the same sequence as for climate, beginning with the equatorial zone and moving toward the higher latitudes. A global soils diagram, Figure 19.10, shows schematically the broad patterns of soil orders on an imaginary supercontinent. The world soils map, Figure 19.11, shows the major areas of occurrence of the soil orders. The Alfisols, in particular, have been subdivided into four important suborders that correspond well to four basic climate zones. The map should be considered as a broad generalization, indicating those areas where

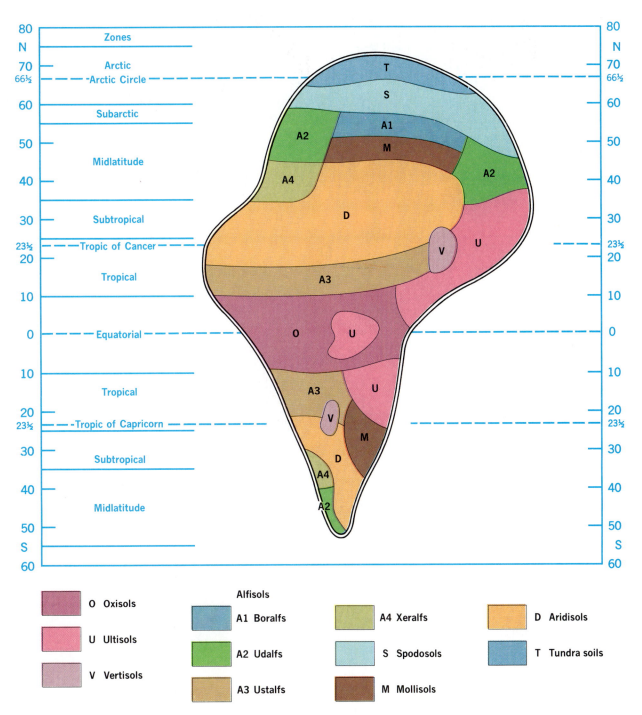

FIGURE 19.10 Schematic diagram of the soil orders and major suborders on an imaginary supercontinent. Compare with Figure 19.11.

a given soil order is likely to be found. The map does not show numerous important areas of Entisols, Inceptisols, and Histosols, since these orders are geologically controlled by local occurrences of features, such as floodplains, recent deposits left by glaciers, areas of sand dunes, marshlands, or bogs.

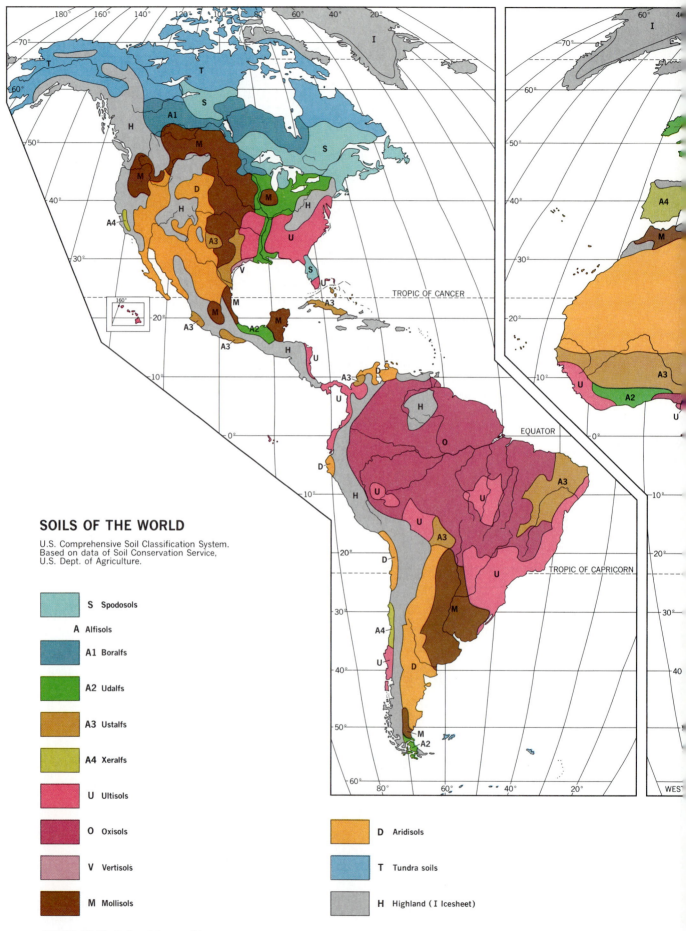

SOILS OF THE WORLD

U.S. Comprehensive Soil Classification System.
Based on data of Soil Conservation Service,
U.S. Dept. of Agriculture.

S	Spodosols	
A	Alfisols	
A1	Boralfs	
A2	Udalfs	
A3	Ustalfs	
A4	Xeralfs	
U	Ultisols	
O	Oxisols	
V	Vertisols	
M	Mollisols	
D	Aridisols	
T	Tundra soils	
H	Highland (I Icesheet)	

FIGURE 19.11 Soils of the world.

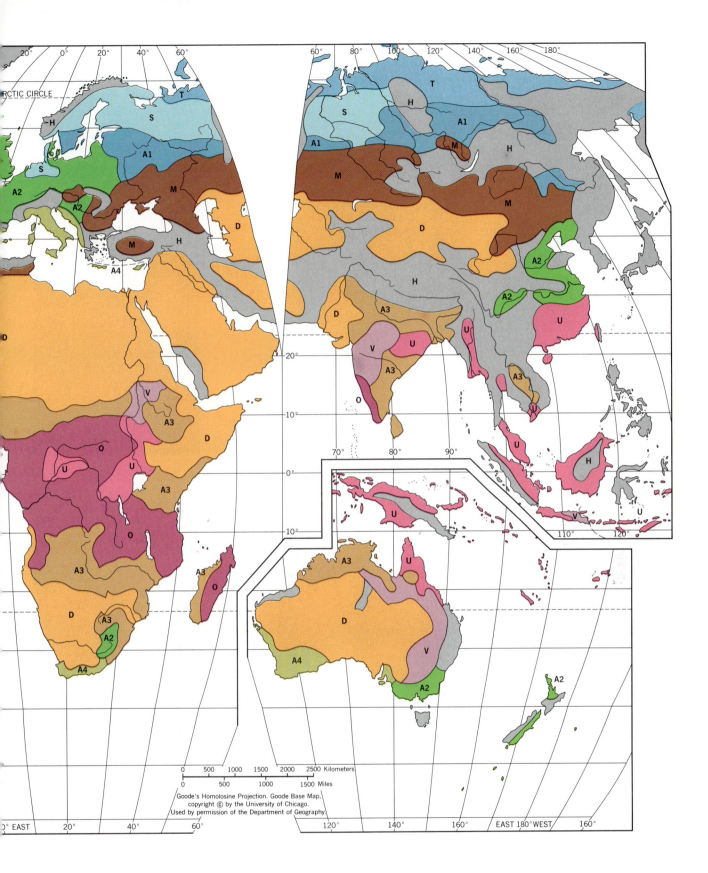

Goode's Homolosine Projection. Goode Base Map,
copyright © by the University of Chicago.
Used by permission of the Department of Geography.

OXISOLS

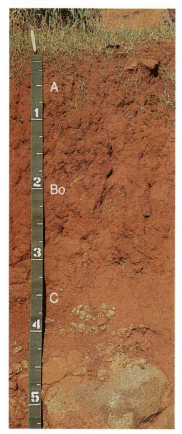

A Torrox, Hawaii

ULTISOLS

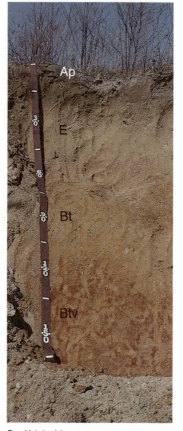

B Udult, Virginia

VERTISOLS

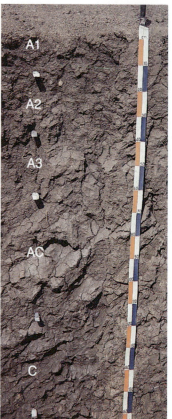

C Usert, India

ALFISOLS

D Udalf, Michigan

ALFISOLS

E Ustalf, Texas

SPODOSOLS

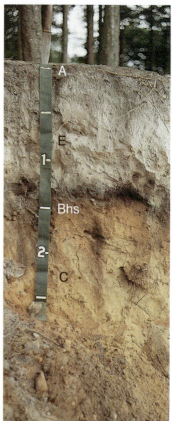

F Orthod, France

FIGURE 19.12 Soil profiles of several soil orders. (Panels A through K by Henry D. Foth. Used by permission. Panel L from U.S. Department of Agriculture, Soil Conservation Service.)

MOLLISOLS

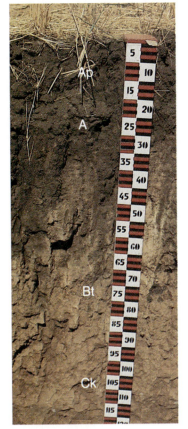

G Boroll, USSR

MOLLISOLS

H Udoll, Argentina

MOLLISOLS

I Ustoll, Colorado

MOLLISOLS

J Rendoll, Argentina

ARDISOLS

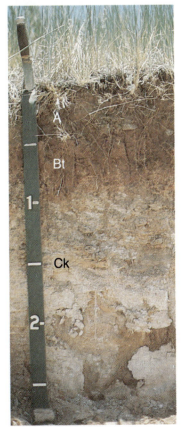

K Argid, Colorado

HISTOSOLS

L Fibrist, Minesota

440

TABLE 19.2 Formative elements in names of soil orders

Name of Order	Formative Element	Derivation of Formative Element	Pronunciation of Formative Element
Entisol	ent	Meaningless syllable	recent
Inceptisol	ept	L. *inceptum*, beginning	inept
Histosol	ist	Gr. *histos*, tissue	histology
Oxisol	ox	F. *oxide*, oxide	ox
Ultisol	ult	L. *ultimus*, last	ultimate
Vertisol	ert	L. *verto*, turn	invert
Alfisol	alf	Meaningless syllable	alfalfa
Spodosol	od	Gr. *spodos*, wood ash	odd
Mollisol	oll	L. *mollis*, soft	mollify
Aridisol	id	L. *aridus*, dry	arid

Table 19.2 explains the names of the soil orders. The formative element is a syllable used in names of suborders and great groups. Although there are several suborders within each order, we will refer to only a few.

Soils of Low Latitudes— Oxisols, Ultisols, Vertisols

Three soil orders dominate the vast land areas of low latitudes: Oxisols, Ultisols, and Vertisols. Soils of these orders have had long time spans to develop in an environment of warm soil temperatures and a soil-water surplus throughout the year or in a wet season. *Oxisols* have developed in equatorial, tropical, and subtropical zones on land surfaces that have been stable over long periods of time. During soil development, the climate has been moist, with a large water surplus. The wet equatorial climate (1) has been particularly conducive to development of Oxisols over vast areas of South America and Africa. Here the native vegetation is rainforest. The wet–dry tropical climate (3) with its large seasonal water surplus is also associated with Oxisols in South America and Africa.

Oxisols usually lack distinct horizons, except for darkened surface layers, even though they are formed in strongly weathered parent matter. Soil minerals are weathered to an extreme degree and are dominated by stable sesquioxides of aluminum and iron, and by kaolinite. Red,

yellow, and yellowish-brown colors are normal (Figures 19.12*A* and 19.13). The base status of the Oxisols is very low, since nearly all the base cations required by plants have been removed from the soil profile. Only very close to the soil surface are limited supplies of nutrient bases temporarily stored. Even though the soils consist of weathered minerals, the soil is quite easily broken apart and allows easy penetration by rainwater and plant roots.

In many areas, the Oxisol profile contains a subsurface horizon of sesquioxides capable of hardening to a rocklike material if it becomes exposed at

FIGURE 19.13 An Oxisol (suborder Torrox) in Hawaii. This soil has developed by deep chemical weathering of basalt, residual boulders of which can be seen in place near the base of the exposure. Sugarcane is being cultivated here. (Henry D. Foth. Used by permission.)

FIGURE 19.14 Ultisol profile in North Carolina. The thin, pale layer at the top is an A2 horizon, showing the effects of removal of materials by eluviation. Near the base of the thick, reddish B horizon is a mottled zone of plinthite. (Soil Conservation Service.)

the surface and is subjected to repeated wetting and drying. This material is referred to as *plinthite,* from the Greek word *plinthos,* meaning brick.

Ultisols are quite closely related to the Oxisols in outward appearance and environment of origin. Ultisols are reddish to yellowish in color (Figures 19.12*B* and 19.14). An important distinction is that the Ultisols have a subsurface horizon of clay accumulation, called an *argillic horizon,* not found in the Oxisols. In the soil profile, the argillic horizon is designated as the B horizon; it has developed by the process of illuviation. Clay minerals present in this horizon are of aluminum oxide composition, including kaolinite as a typical example. A plinthite horizon may also be present. Although forest is the characteristic native vegetation, the base status of the Ultisols is low; most of the base cations present are found in a shallow surface layer where they are taken up by trees and shrubs and recycled when decay of plant matter releases the ions to the surface layer.

Ultisols are widespread throughout Southeast Asia and the East Indies. Other important areas are in eastern Australia, Central America, South America, and the southeastern United States. Ultisols extend into the lower midlatitude zone in the United States, where they correspond quite closely in extent with the area of moist subtropical climate (6). In lower latitudes, Ultisols are identified with the wet–dry tropical climate (3) and the monsoon and trade-wind littoral climate (2). Note that these are climates with at least a short dry season.

Both Oxisols and Ultisols of low latitudes were used for centuries under shifting agriculture prior to the advent of modern agricultural technology. This primitive agricultural method, known as slash-and-burn, is still widely practiced (Chapter 20). Without fertilizers these soils can sustain crops on freshly cleared areas for only two or three years, at most, before the nutrient bases are exhausted and the garden plot must be abandoned. Substantial use of lime, fertilizers, and other industrial inputs is necessary for high, sustained crop yields. The exposed soil surface is, however, vulnerable to devastating soil erosion, particularly on steep hillslopes (Chapter 15).

Vertisols have a unique set of properties standing in sharp contrast to those of the Oxisols and Ultisols. Vertisols are black in color and have a high clay content (Figures 19.12*C* and 19.15). Much of the clay consists of the mineral montmorillonite, which shrinks and swells greatly with seasonal changes in soil-water content. Wide, deep vertical cracks develop in the soil during a dry season (Figure 19.16). As the dry soil blocks are wetted and softened by rain, some fragments of surface soil drop into the cracks before they close, so that the soil "swallows itself." As the soil blocks absorb water they expand, so that one block moves against another. These movements are capable of forming small soil mounds and will cause objects, such as fenceposts, to be tilted out of line.

Vertisols typically form under grass and savanna vegetation (Chapter 20) in subtropical and tropical climates with a

FIGURE 19.15 A Vertisol (suborder Ustert) in India. Slickenslided surfaces have been exposed in this trench, giving evidence of soil movement. (Henry D. Foth. Used by permission.)

pronounced seasonal soil-water shortage. These climates include the semiarid subtype of the dry tropical steppe climate (4s) and the wet–dry tropical climate (3). Vertisols require a parent matter capable of yielding clay minerals that shrink and swell with soil-water changes. Thus, the major areas of occurrence are scattered and show no distinctive pattern on the world map. An important region of Vertisols is the Deccan Plateau of western India, where basalt, a dark variety of igneous rock, supplies the

FIGURE 19.16 Soil cracks in Vertisol in Texas. Locally this soil is known as the Houston black clay. The crop growing here is cotton. (Soil Conservation Service.)

silicate minerals that are altered into the necessary clay minerals.

Vertisols are high in base status and are particularly rich in such nutrient bases as calcium and magnesium. The soil solution is nearly neutral in pH and there is a moderate content of organic matter distributed through the soil. The soil retains large amounts of water because of its fine texture, but much of this water is held by the montmorillonite clay particles and is not available to plants. Where soil cultivation depends on human or animal power, as it does in most of the developing nations where the soil occurs, agricultural yields are low. One problem is that the moist soil becomes highly plastic and is difficult to till with primitive tools. For this reason, many areas of Vertisols have been left in grass or shrub cover, which may provide grazing for cattle. Soil scientists think that the use of modern technology, including heavy farm machinery, could result in substantial production of food and fiber from Vertisols not now in production.

Among the soils of low latitudes, certain areas of Inceptisols are of great importance. These areas occur within some map regions shown as Ultisols and Oxisols. Especially important are the Inceptisols of river floodplains and delta plains in Southeast Asia that support dense populations of farmers subsisting on rice. The rice fields must be flooded at the time the rice is planted. Examples are the combined floodplain and deltaic plain of the Ganges–Brahmaputra river system in India and Bangladesh, the lower Irrawaddy and Sittang river plains in Burma, the lowland of the Chao Phraya in Thailand, and the Lower Mekong floodplain and delta in Cambodia and Vietnam. Here, annual river floods inundate the low-lying plains and deposit layers of fine silt rich in primary minerals from which base cations are produced by chemical alteration. This constant enrichment of the soil explains the high soil fertility in a region that otherwise might develop only Ultisols of low fertility. Inceptisols of these floodplain and delta lands are of a suborder called *Aquepts*—Inceptisols of wet places.

Much closer to home is another prime example of Aquepts within the domain of the Ultisols; the lower Mississippi River

FIGURE 19.17 Laterite, formed by hardening of plinthite, is being quarried for building stone in this scene from India. (Henry D. Foth. Used by permission.)

floodplain and deltaic plain. Important examples of Aquepts found within areas of Oxisols are the floodplains of the lower Amazon River in Brazil and the Congo River in Zaire. Here, in the wet equatorial climate, the dense rainforest and presence of marshes have not permitted agricultural development such as that found in Southeast Asia. The occurrences of Aquepts we have described are good examples of the control exerted locally by parent material on the quality and fertility of the soil.

Ancient Crusts of Plinthite

In the tropical wet–dry and monsoon climate zones of Southeast Asia, Africa, and elsewhere, a common landscape feature is a stepping or benching of the land surface caused by layers of rocklike material derived from the plinthite horizon of

Oxisols. As previously stated, plinthite has a property of hardening when exposed at the surface, where it is subjected to repeated wetting and drying. In the hardened state, plinthite is referred to as *laterite* (from the Latin *later*, brick). An alternate term is *ferricrete*, stressing the presence of iron oxide as the cementing material.

Laterite (ferricrete) represents a plinthite horizon that has become exposed to the surface environment through erosional removal of overlying soil horizons of an Oxisol. In Southeast Asia, plinthite is quarried and cut into building blocks that harden into enduring laterite blocks (Figure 19.17). Laterite crusts occur widely over the uplands of savanna regions. Figure 19.18 is a schematic cross section of a laterite crust of the African savanna. The laterite bench surrounds an isolated dome-shaped rock knob, locally called a "bornhardt" (see Figure 21.4). (Laterite crusts are not necessarily related to bornhardts or other similar landforms.)

The Alfisols

The *Alfisols* are soils of moist climates characterized by an argillic horizon; this is a B horizon enriched by accumulated silicate clay minerals with a moderate capacity to hold base cations such as calcium and magnesium. Illite is one such clay mineral. The base status of the Alfisols is therefore generally quite high. Above the B horizon of clay accumulation is a horizon of pale color, the A2 horizon, that has lost some of the original bases, clay minerals, and sesquioxides by the process of eluviation. These materials have become concentrated by illuviation in the B horizon. Alfisols also have a gray, brownish, or reddish surface horizon, called an *ochric epipedon*.

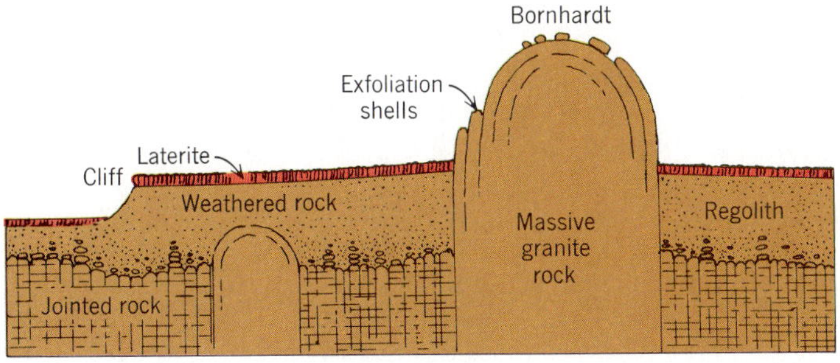

FIGURE 19.18 Schematic diagram of a rock knob (bornhardt) with its surrounding plain bearing a laterite crust.

Bornhardt

Exfoliation shells

Laterite

Cliff

Weathered rock

Massive granite rock

Regolith

Jointed rock

New World Environments

Physical factors that have favored or disfavored human occupation in past centuries are often subtle and complex. Take the case of the eastern United States, which is divided into separate environmental regions: in the northern half is a moist continental forest environment and a grasslands environment; in the southern half is the moist subtropical forest environment. Both halves have ample heat and water for the natural growth of forests or tall-grass prairie; both were occupied by Europeans who cleared the forests, drained marshlands, and planted their crops on tilled fields of bare soil.

Consider, first, that the early colonists and those emigrants who followed came from a common source, the marine west-coast environmental region of the British Isles and western Europe. Initially, they brought the same agricultural system to both North and South. Yet, in due time, two quite different agricultural systems evolved. In the North, diversified farming and dairying were carried on in small farms worked by the landowners and their families. In the South, plantation agriculture spread widely and slaves were imported to work the soil and harvest the crops on large landholdings. Eventually,

political fission occurred—Union versus Confederacy—and was resolved in a bloody civil war. Did the differences in natural environment of North and South play any part in this sequence of historical events? Let us pursue this question further.

Environmental differences between North and South are most obvious in terms of soils. Compare the Midwest (Ohio to Iowa) with the Atlantic and Gulf states (Virginia to Louisiana). (We are excluding the northern region of Spodosols, which proved generally unfavorable to cereal agriculture.) The soils map shows Alfisols and Mollisols in the northern region, whereas Ultisols cover most of the southern region. The difference in soil quality is obvious: high base status of soils in the North; low base status of soils in the South. There are exceptions, of course, to this statement. For example, the Black Belt lowland of Alabama and Mississippi has a small patch of Vertisols of high base status, and the lower Mississippi alluvial plain has Inceptisols of high base status.

History shows that the widespread southern Ultisols, which belong to the moist-climate subtype of Udults, suffered rapid depletion of nutrients. Recall that the nutrient bases in the Ultisols are mostly in a shallow upper layer. As we pointed out in a case study in the Introduction, by the early

Map of the southern Piedmont upland. (From F. Grave Morris, *Geographical Journal*, vol. 90, p. 370.)

The red soil exposed in this highway cut in North Carolina is a Udult, a suborder of the Ultisols. Beneath the red layer, and partially obscured by red soil washed down from above, is thick residual regolith of a special kind called *saprolite*. It consists of chemically decomposed bedrock and is rich in clay minerals. (Orlo E. Childs.)

Severe gullying near Ducktown, Tennessee. Sulfur oxides from a nearby smelter caused the forest that was once here to die off. Sheet erosion then removed the red soil layer, exposing the weak regolith. (A. N. Strahler.)

1800s on the Atlantic seaboard fertility of cultivated soils had declined greatly. Lime was badly needed to neutralize soil acidity. Once neutralized, the soil needed massive doses of fertilizers, rich in phosphorus, to supply the missing nutrients. In contrast, the northern Alfisols and Mollisols, rich in nutrient bases, were able to sustain their fertility over many generations of crop farming, needing only the application of animal manure.

A second point to consider in comparing the agricultural history of the two regions is that of accelerated soil erosion, which set in with great severity in the Piedmont upland region of the southeast as a result of deforestation and land tillage. Because plant nutrients are concentrated in a shallow surface layer in the Ultisols, sheet erosion is particularly devastating to soil fertility. In thick Mollisols of the Midwest tall-grass prairies, nutrients are abundant at depth in the soil. Here, the effects of sheet erosion

are not as devastating in early stages, because the remaining soil holds a large supply of nutrients. To a lesser degree, the same principle applies to the Alfisols, which, throughout the Midwest, are mostly Udalfs. Although these soils require applications of lime and fertilizers, the needs are not so demanding as for the Udults, nor are the Alfisols as vulnerable to a decline in fertility.

A cultural factor in sustained fertility of these midwestern soils should also be noted here. As one geographer, Carl O. Sauer, has stated, settlers in the corn belt were good husbandmen: "They took care of their land, and it did well by them."* As small farmers working the land they owned, they took care to protect their fields from severe soil erosion; they rotated crops to keep the soil fertility high.

* *Data Source:* Carl O. Sauer (1956), The agency of man on the earth, pp. 49–69 in *Man's Role in Changing the Face of the Earth*, W. L. Thomas, ed., University of Chicago Press, Chicago. (See p. 65.)

FIGURE 19.19 Ustalf profile in Texas. The pale upper layer is an ochric epipedon. The argillic horizon of darker color sets in at about 25 cm (10 in.). (Soil Conservation Service.)

FIGURE 19.20 Xeralf profile in California. The tape measure rests on the top of the dense argillic horizon (B horizon). (Soil Conservation Service.)

The world distribution of Alfisols is extremely wide in latitude (see Figure 19.11). Alfisols range from latitudes as high as 60° N in North America and Eurasia to the equatorial zone in South America and Africa. Obviously, the Alfisols span an enormous range in climate types and soil-water regimes. For this reason, we need to recognize four of the important suborders of Alfisols, each with its own climate affiliation. A brief description of each follows.

Boralfs are Alfisols of cold (boreal) forest lands of North America and Eurasia. They have a gray surface horizon and a brownish subsoil.

Udalfs are brownish Alfisols of the midlatitude zone and are closely associated with the moist continental climate (10) in North America, Europe, and eastern Asia

(Figure 19.12*D*). A forest of deciduous trees was the typical natural vegetation of this climate type, but large areas have been intensively cultivated for many generations. The Udalfs are highly productive when moderate amounts of lime and fertilizers are applied.

Ustalfs are brownish to reddish Alfisols of the warmer climates (Figures 19.12*E* and 19.19). They range from the subtropical zone to the equator and are found associated with the wet–dry tropical climate (3) in Southeast Asia, Africa, Australia, and South America. In Africa, Ustalfs may owe their high base status to the constant rain of fine dust carried by prevailing winds from the adjacent tropical desert. Ustalfs of north India and Pakistan are highly productive under irrigation and are major producers of wheat.

Xeralfs are Alfisols of the Mediterranean climate (7), with its cool moist winter and dry summer. The Xeralfs are typically brownish or reddish in color (Figure 19.20). Good examples are found in coastal and inland valleys of central and southern

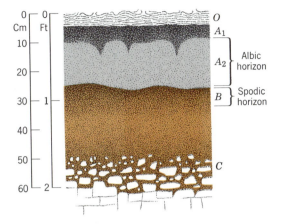

FIGURE 19.21 Diagram of a Spodosol profile.

FIGURE 19.22 A Spodosol (suborder Orthod), developed in sandy parent material, in Quebec, Canada. The strongly leached albic horizon (A₂, or E) appears almost pure white in contrast to the reddish-brown spodic horizon beneath it. (Henry D. Foth. Used by permission.)

California. These soils have a high natural fertility and support cattle-grazing grasslands and the cultivation of grape, citrus, and avocado.

Spodosols

Poleward of the Alfisols in North America and Eurasia lies a great belt of soils of the order *Spodosols,* formed in the cold boreal forest climate (11) beneath a needleleaf forest. Spodosols have a unique property: a B horizon of accumulation of reddish mineral matter with a low capacity to hold base cations (Figure 19.12*F*). This horizon is called the *spodic horizon* (Figure 19.21). The materials of the spodic horizon are amorphous. (Amorphous minerals are in a noncrystalline state, resembling glass.) The horizon is made up of a dense mixture of organic matter and compounds of aluminum and iron, all brought downward by eluviation from an overlying A2 horizon. Because of the intensive removal of matter from the A2 horizon, it has a bleached, pale gray to white appearance (Figure 19.22). This conspicuous feature led to the naming of the soil as *podzol* (ash-soil) by Russian peasants. In modern terminology this pale layer is an *albic horizon,* and is designated the E horizon. A thin, very dark layer of organic matter, the O horizon, overlies the A horizon.

Spodosols are strongly acid, are low in plant nutrients such as the base cations of calcium and magnesium, and are low in humus. Although the base status of the

Spodosols is low, forests of pine and spruce are supported through the process of recycling of the base cations.

Spodosols are closely associated with regions recently covered by the great ice sheets of the Pleistocene Epoch (Ice Age); the soils are therefore very young. Typically, the parent material is coarse sand consisting largely of the mineral quartz (silicon dioxide, SiO_2) and therefore incapable of yielding clay minerals. Spodosols are naturally poor soils in terms of agricultural productivity. Because they are acid, the application of lime is essential. Heavy applications of fertilizers are also required. With proper management and the input of the required industrial products, the Spodosols can be highly productive, where soil texture is favorable. An unusual example is the high yield of potatoes from Spodosols in Maine and New Brunswick. Another factor unfavorable to agriculture is shortness of the growing season in the more northerly parts of the Spodosol belt.

Throughout the northern regions of Spodosols are innumerable patches of *Histosols,* a soil order unique in having a very high content of organic matter in a thick, dark upper layer (Figures 19.12*L* and 19.23). Most Histosols go by such common names as peats or mucks. They have formed in shallow lakes and ponds by accumulation of partially decayed plant

448

FIGURE 19.23 Profile of a Histosol (suborder Saprist) seen in the wall of a pit cut into a bog near Belle Glade, Florida. Water is being pumped from the floor of the pit. (Henry D. Foth. Used by permission.)

matter. In time, the water is replaced by a layer of organic matter, or *peat*, and becomes a bog (Figure 19.24). Histosols are also found in low latitudes, where conditions of poor drainage have favored thick accumulations of plant matter.

Histosols that are mucks (fine black materials of sticky consistency) are agriculturally valuable in midlatitudes, where they occur as beds of former lakes in the glaciated region. After appropriate drainage and application of lime and fertilizers, these mucks are remarkably productive for truck garden vegetables (Figure 19.25). Peat bogs are extensively used for cultivation of cranberries (cranberry bogs). *Sphagnum* peat from bogs is dried and baled for sale as a mulch for use on suburban lawns and shrubbery beds. For centuries, in Europe, dried peat from bogs of glacial origin has been used as a low-grade fuel. (See also Figure 9.31).

Mollisols

Mollisols are soils of grasslands that occupy vast areas of semiarid and subhumid climates in midlatitudes. Mollisols are unique in having a very thick, dark brown to black surface horizon called the *mollic epipedon* (Figures 19.12 *G* to *J* and 19.26). This layer includes both A and B horizons and is always more than 25 cm (10 in.)

FIGURE 19.24 This peat bog in Connemara, Ireland, has been trenched to reveal the Histosol profile. Piles of peat blocks, seen in the foreground, will be dried for use as fuel. (Copyright © 1988 by Mark A. Melton.)

FIGURE 19.25 Garden crops cultivated on a Histosol of a former glacial lake bed, northern New Jersey. Special drainage is required to remove excess soil water. (Arthur N. Strahler.)

thick. The soil has a loose structure or soft consistency when dry. The typical granular structure is shown in Figure 19.8, left. Other important qualities of the Mollisols are the dominance of calcium among the base cations of the A and B horizons and the very high base status of the soil.

Most areas of Mollisols are closely associated with the semiarid subtype of the dry midlatitude climate (9s) and with the adjacent subhumid zone of the moist continental climate (10sh). In North America, Mollisols dominate the Great Plains region, the Columbia Plateau, and the northern Great Basin. In South America, a large area of Mollisols covers the Pampa region of Argentina and Uruguay. In Eurasia a great belt of Mollisols stretches from Rumania eastward across the steppes of Russia, Siberia, and Mongolia. The Russians refer to the Mollisols as *chernozems*, a term that gained widespread use throughout the western world as well.

Because of their loose texture and very high base status, Mollisols are among the naturally most fertile soils in the world. They now produce most of the grain that moves in commercial trade channels. Most of these soils have not been widely used for crop production except during the last century. Prior to that time, they were used mainly for grazing by nomadic herds. The Mollisols have favorable properties for growing cereals in large-scale mechanized farming and are relatively easy to manage. Production of grain varies considerably from one year to the next because seasonal rainfall is highly variable and the soil-water storage is the factor limiting production unless irrigation is available.

Brief mention of four suborders of the Mollisols will help you to understand important regional differences related to climate. *Borolls*, the cold-climate suborder of the Mollisols, are found in a large area extending on both sides of the United States–Canada border east of the Rocky Mountains (Figure 19.12*G*). This is an important region of spring wheat production. Borolls also occupy a wide belt across Eurasia.

Udolls are Mollisols of a relatively moist climate, as compared with the other suborders. Udolls are found in the prairie plains of the Midwest and extend from Illinois, westward across Iowa and Missouri, into eastern Nebraska and Kansas. This region lies within the moist continental climate (10sh) in a subhumid region where there is only a small soil-water shortage and a small annual water surplus. Formerly, the Udolls supported tall-grass prairie, but today they are closely identified with the corn belt, an extremely productive agricultural region where a highly fertile soil occupies a region of adequate

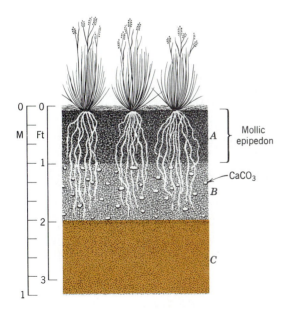

FIGURE 19.26 Schematic diagram of a Mollisol profile.

FIGURE 19.27 Exposed in this riverbank is a Ustoll developed on thick loess, which shows vertical columnar parting. (A. N. Strahler.)

FIGURE 19.28 A salic horizon, appearing as a white layer, lies close to the surface in this Aridisol profile in the Nevada desert. The scale is marked in feet; 1 ft = 30 cm. (Soil Conservation Service.)

precipitation. The thick, black Udolls have long been referred to as prairie soils (Figure 19.12*H*).

Ustolls are Mollisols of the semiarid subtype of the dry midlatitude climate (9s), with a substantial soil-water shortage in the summer months. The Ustolls underlie much of the Great Plains region east of the Rockies, a region of short-grass prairie (Figure 19.27). A horizon of calcium carbonate accumulation, called a *calcic horizon,* is typical of the Ustolls (Figure 19.12*I*). Although Ustolls have a high base status, they cannot sustain agriculture without irrigation except in the moister eastern margins, where wheat is cultivated. In Eurasia, a broad belt of Ustolls extends from the Black Sea to central Siberia.

Xerolls are Mollisols of the Mediterranean soil-water regime, with its tendency to cool, moist winters and rainless summers. Xerolls are found in the semiarid northern Great Basin and Columbia Plateau, lying between the Cascade Mountains and the Rocky Mountains.

Aridisols

Aridisols, soils of the desert climate, are dry for long periods of time. Because the climate supports only a very sparse vegetation, humus is lacking and the soil color ranges from pale gray to pale red (Figure 19.12*K*). Soil horizons are weakly developed, but there may be important subsurface horizons of accumulation of calcium carbonate (petrocalcic horizon) or soluble salts (salic horizon) (Figure 19.28). The salts, of which sodium is a dominant constituent, give the soil a very high degree of alkalinity. The Aridisols are closely correlated with the desert climate of

FIGURE 19.29 This gray desert soil, an Aridisol, has proved highly productive when cultivated and irrigated. The locality is near Palm Springs, California, in the Coachella Valley. (Ned L. Reglein.)

FIGURE 19.30 Profile of a tundra soil (Cryaquept) in northern Yukon Territory, Canada. Permafrost (perennially frozen soil water) appears at a depth of between 40 and 60 cm. (Henry D. Foth. Used by permission.)

Tundra Soils

Soils of the arctic tundra fall largely into the order of *Inceptisols*, soils with weakly developed horizons and usually associated with a moist climate. Inceptisols of the tundra climate belong to the suborder of *Aquepts*—Inceptisols of wet places. More specifically, the tundra soils can be assigned to the *Cryaquepts*, a great group within the Aquepts. The prefix *cry* is derived from the Greek word *kryos*, meaning "icy cold." We may refer to these soils simply as *tundra soils*. They are formed largely of mechanically broken primary minerals ranging in size from silt to clay. Layers of peat are often present between mineral layers (Figure 19.30). Beneath the tundra soil lies perennially frozen ground (permafrost), described in Chapter 9. Because the annual summer thaw affects only a shallow surface layer, soil water cannot easily drain away. Thus, the soil is water saturated over large areas. Repeated freezing and thawing of this shallow surface layer disrupts plant roots, so that only small, shallow-rooted plants can maintain a hold.

tropical, subtropical, and midlatitude zones (4d, 5d, 9d).

Most Aridisols are used, as they have been through the ages, for nomadic grazing. This use is dictated by the limited rainfall, which is inadequate for crops without irrigation. Locally, where water supplies from exotic streams or ground water permit, Aridisols can be highly productive for a wide variety of crops under irrigation (Figure 19.29). Great irrigation systems, such as those of the Imperial Valley of the United States, the Nile Valley of Egypt, and the Indus Valley of Pakistan, have rendered Aridisols highly productive, but not without attendant problems of salt buildup and waterlogging.

Global Soils in Review

We have painted the picture of global soils in broad brush strokes, leaving out much detail that would be meaningful.

Some important basic concepts have been illustrated with specific examples. The important role of climate in governing the soil-forming processes is a persistent theme

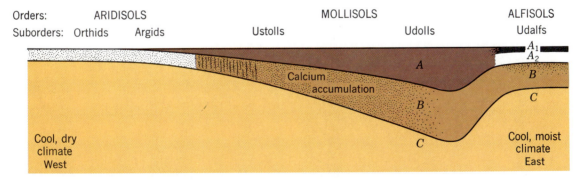

FIGURE 19.31 A schematic diagram of the changing soil profile from a cool, dry desert on the west to a cool, moist climate on the east. (Based on a diagram by C. E. Millar, L. M. Turk, and H. D. Foth, 1958, *Fundamentals of Soil Science*, 3rd ed., John Wiley & Sons, New York.)

of this chapter. It is a theme that allows us to make good use of the concepts of the soil-water budget. Soils associated with water-surplus budgets are strikingly different in structure, composition, and fertility from those associated with budgets featuring a large soil-water shortage. Surprisingly, the most fertile soils from the standpoint of plant nutrients are those of the semiarid climates—Mollisols, for example. A great abundance of water is not conducive to producing rich agricultural soils because, under such conditions, the nutrient bases are leached from the soil and can only be maintained by repeated application of fertilizers. The broad relationships between climate and soil orders are illustrated by a continuous profile crossing the United States (Figure 19.31).

CHAPTER 20

ENVIRONMENTS OF NATURAL VEGETATION

NATURAL VEGETATION   It is the native plant cover of a region, but cannot be found today over large areas densely inhabited by humans.

LIFE-FORM   Geographers classify natural vegetation according to the life-form of plants—trees, shrubs, forest, woodland, or grassland, for example.

PLANT HABITATS   These are the local environments of small units of vegetation; landform, soil, and soil water largely determine the quality of a habitat.

WATER NEED   Plant species are adapted to soil-water conditions and to the temperature of the soil and air.

PHOTOSYNTHESIS   Using solar energy, water and carbon dioxide are combined to produce carbohydrate in plants. Respiration reverses the process.

PRODUCTIVITY   Rate of production of organic matter varies greatly from one environment to another.

ECOLOGICAL SUCCESSION   This term refers to the natural progression of changing plant species and life-forms following a disturbance of the plant cover.

THE BIOMES   Ecosystems of the lands are divided into five major biomes: forest, savanna, grassland, desert, and tundra.

IN this chapter we focus attention on vegetation of the lands. Thinking of plants as stationary objects on the landscape places them in much the same frame as other physical elements of that landscape, such as landforms, soils, streams, and lakes. Plants are also consumable and renewable sources of food, medicinals, fuel, clothing, shelter, and a host of other life essentials. The ways in which humans have used this plant resource to their advantage—or have been hindered by plants in their progress—have been persistent themes in the writings of geographers.

Two concepts of the geography of world vegetation stand opposed but inseparable, like the two sides of a coin. One is the concept of *natural vegetation* (or native vegetation), a plant cover that attains its development without appreciable human interference and that is subject to natural forces of modification and destruction, such as storms or fires. The other concept is that of vegetation sustained in a modified state by human activities. Extremes of both cases are found over the world. Natural vegetation can still be seen over vast areas of the wet equatorial climate, although rainforests are being cleared at a rapid rate. Much of the arctic tundra and the forest of the subarctic zones is in a natural state. In contrast, much of the continental surface in midlatitudes is almost totally under human control through intensive agriculture, grazing, or urbanization. You can drive across an entire state, such as Ohio or Iowa, without seeing a vestige of the plant cover it bore before the coming of the European settlers. Only if you know where to look can you find a few small plots of virgin prairie or virgin forest.

Some areas of natural vegetation appear to be untouched but are dominated by human activity in a subtle manner. Certain of our national parks and national forests have been protected from fire for many decades, setting up an unnatural condition in terms of what might be expected of the natural ecosystem. When lightning starts a forest fire, the firefighters parachute down and put out the flames as fast as possible. But we realize that periodic burning of forests and grasslands must be a natural phenomenon, and it must perform some vital function in the ecosystem. One vital function is to release nutrients from storage in the biomass so that the soil can be revitalized. Already those who manage our parks and forests are experimenting with a "hands-off" policy in fire control.

Humans have influenced vegetation in yet another way—by moving plant species from their original habitats to foreign lands and foreign environments. The eucalyptus tree is a striking example. From Australia the various species of eucalypts have been transplanted to such far-off lands as California, North Africa, and India. Sometimes these exported plants thrive like weeds, forcing out natural species and becoming a major nuisance. It is said that scarcely one of the grasses seen clothing the coast ranges of California is a native species, yet the casual observer might think that these represent the native vegetation. Even so, all plants have limits of tolerance to the environmental conditions of soil water, heat and cold, and soil nutrients. Consequently, the structure and outward appearance of the plant cover conforms to the basic environmental controls, and each vegetation type stays within a characteristic geographical region, whether this be forest, grassland, or desert.

Structure and Life-form of Plants

The plant geographer classifies plants in terms of the *life-form*, which is the physical structure, size, and shape of the plant. Botanical associations are not necessarily related to life-form. Although the life-forms go by common names and are well understood by almost everyone, we will review them to establish a uniform set of meanings.

Both trees and shrubs are erect, woody plants (Figure 20.1). They are perennial,

FIGURE 20.1 Forest of the midlatitude zone is represented by this stand of beech and hemlock trees in Great Smoky Mountains National Park, Tennessee. This is a mixed deciduous—needleleaf forest. Lightning has shattered the tree at the left. The lower layer consists of scattered shrubs and seedling trees. (Donaldson Koons.)

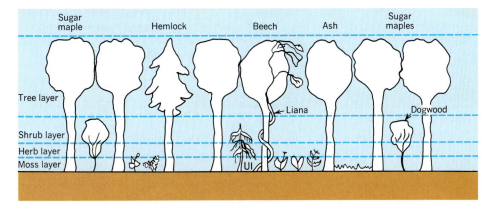

FIGURE 20.2 This schematic diagram shows the layers of a beech–maple–hemlock forest, similar to that pictured in Figure 20.1. The vertical dimensions of the lower layers are greatly exaggerated. (Modified from Pierre Dansereau, 1951, *Ecology*, vol. 32.)

meaning that the woody tissues endure from year to year, and most have life spans of many years. *Trees* are large, woody perennial plants having a single upright main trunk, often with few branches in the lower part, but branching in the upper part to form a crown. *Shrubs* are woody perennial plants having several stems branching from a base near the soil surface, so as to place the mass of foliage close to ground level.

Lianas are also woody plants, but they take the form of vines supported on trees and shrubs. Lianas include not only the tall heavy vines of the wet equatorial and tropical rainforests (see Figure 20.13), but also some woody vines of midlatitude forests.

Herbs comprise a major class of plant life-forms. *Herbs* are tender plants, lacking woody stems, and usually small. They occur in a wide range of shapes and leaf types. Some are annuals and others are perennials. Some are broad-leaved, and others are grasses. Herbs as a class share few characteristics in common except that they form a low layer as compared with shrubs and trees.

Forest is a vegetation structure in which trees grow close together with crowns in contact, so that the foliage largely shades the ground. In most forests of the humid climates the life-forms are arranged in distinct layers, and the vegetation is said to be stratified (Figure 20.2). Tree crowns form the uppermost layer, shrubs an

intermediate layer, and herbs a lower layer. The lowermost layer consists of mosses and related small plants. In *woodland*, crowns of trees are mostly separated by open areas, usually having a low herb or shrub layer. Lichens represent yet another life-form seen in a layer close to the ground. Lichens are plant forms in which algae and fungi live together to form a single plant structure. In some alpine and arctic environments lichens grow in profusion and dominate the vegetation (see Figure 9.38).

Plant Habitats

As we travel through a hilly wooded area it becomes obvious that the vegetation is strongly influenced by landform and soil. As explained in Chapter 19, landform refers to the configuration of the land surface, including features such as hills, valleys, ridges, or cliffs. Vegetation on an upland—relatively high ground with thick soil and good drainage—is quite different from that on an adjacent valley floor, where water lies near the surface much of the time. Vegetation is also strikingly different in form on rocky ridges and on steep cliffs, where water drains away rapidly and soil is thin or largely absent.

The total vegetation cover is actually a mosaic of small units reflecting the inequalities in conditions of slope, drainage, and soil type. Such subdivisions of the plant environment are described as *habitats*. In the

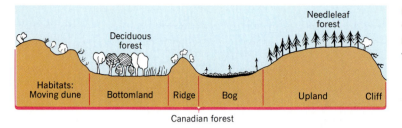

FIGURE 20.3 Habitats within the Canadian forest. (Modified from Pierre Dansereau, 1951, *Ecology*, vol. 32.)

example shown in Figure 20.3 the Canadian forest actually comprises at least six habitats: upland, bog, bottomland, ridge, cliff, and active dune. Just where each habitat is located and how large an area it occupies depends largely on the geologic history of the region and on processes of erosion and deposition that have acted in the past to shape the landscape.

Closely tied in with the effect of landform on habitat is the distribution of water in the soil and rock. Finally, each habitat has its own set of soil properties, determined not only by the conditions of slope and water but, in part, by the plants themselves.

In establishing the larger units of vegetation for generalized maps, the plant geographer usually bases the classes upon the life-form assemblages of the upland habitat, because it is here that a middle range of environmental conditions prevails.

Awareness that varied habitats present a wide spectrum of controls over plants leads us to look further into the role of soil-water availability in favoring one form of plant over another.

by which to control the excessive loss of water by transpiration at times when stored soil water is depleted and a shortage becomes severe.

The adaptation of plant structures to water budgets with large water shortages is of particular interest to the plant geographer. Transpiration occurs largely from specialized leaf pores, which provide openings in the outer cell layer. These pores allow water vapor and other gases to pass into and out of the leaf. Surrounding these openings are cells that can open and close the openings and thus, to some extent, regulate the flow of water vapor and other gases. Water vapor may also pass through the cuticle, or outermost protective layer of the leaf. This form of water loss is reduced in some plants by thickening of the outer layers of cells or by the deposition of wax or waxlike material on the leaf surface. Many desert plants have thickened cuticle or wax-coated leaves, stems, or branches.

A plant may also adapt to a desert environment by greatly reducing the leaf area or by bearing no leaves at all. Needlelike leaves and spines representing

Plants and Water Need

Because the soil-water balance has been a dominant theme of earlier chapters, we can start with an investigation of the response of plants to the degree of availability of soil water. The green plants, or leaf-bearing plants, give off large quantities of water to the atmosphere through transpiration. This water loss takes the form of evaporation from water films on exposed surfaces of certain leaf cells. (Only a very small proportion of the water consumed by plants is used in photosynthesis.) Most plants derive their water through roots in the soil zone. These plants must have mechanisms

FIGURE 20.4 Prickly pear cactus (*Opuntia*) in the Sonoran Desert, Arizona. (Robert J. Ashworth/Photo Researchers.)

leaves greatly reduce loss from transpiration. In cactus plants the foliage leaf is not present, and transpiration is limited to fleshy stems (Figure 20.4).

In addition to developing leaf structures that reduce water loss by transpiration, plants in a water-scarce environment improve their means of obtaining water and of storing it. Roots become greatly extended to reach soil moisture at increased depth. In cases where the roots reach to the ground water zone, a steady supply of water is assured. Plants drawing from such a source may be found along dry stream channels and valley floors in desert regions. Other desert plants produce a widespread but shallow root system enabling them to absorb the maximum quantity of water from sporadic desert downpours that saturate only the uppermost soil layer. Stems of desert plants are commonly greatly thickened by a spongy tissue in which much water can be stored.

A quite different adaptation to extreme aridity is seen in many small desert plants that complete a very short cycle of germination, leafing, flowering, fruiting, and seed dispersal immediately following a desert downpour.

Plants may be classified according to their water requirements. Terms associated with the water factor are built on three simple prefixes of Greek roots: *zero-*, dry; *hygro-*, wet; and *meso-*, intermediate. Plants that grow in dry habitats are *xerophytes;* those that grow in water or in wet habitats are *hygrophytes;* those of habitats of an intermediate degree of wetness and relatively uniform soil-water availability are *mesophytes.*

The xerophytes are highly tolerant of drought and can survive in habitats that dry quickly following rapid drainage of precipitation (e.g., on sand dunes, beaches, and bare rock surfaces). The plants typical of dry climates are also xerophytes; cactus is an example. The hygrophytes are tolerant of excessive water and may be found in shallow streams, lakes, marshes, swamps, and bogs (Figure 20.5). The mesophytes are found in upland habitats in regions of ample rainfall. Here the drainage of soil water is good, and moisture penetrates deeply where it can later be used by the plants.

Certain climates, such as the wet–dry tropical climate and the moist continental climate, have a yearly cycle with one season in which water is unavailable to plants because of lack of precipitation or because the soil water is frozen. This season alternates with one in which there is abundant water. Plants adapted to such regimes are called *tropophytes,* from the Greek word *trophos,* meaning change, or turn. Tropophytes meet the impact of the season of unavailable water by dropping their leaves and becoming dormant. When water is again available, they leaf out and grow at a rapid rate. Trees and shrubs that seasonally shed their leaves are *deciduous plants;* in distinction, *evergreen plants* retain most of their leaves in a green state through the year.

FIGURE 20.5 Hygrophytes are encroaching upon the borders of this small lake on the Gaspé Peninsula, Quebec. The leading edge of the bog vegetation, consisting of sedges, floats on the water. (Pierre Dansereau.)

The Mediterranean climate also has a strong seasonal wet–dry alternation, the summers being dry and the winters wet. Plants in this climate adopt the habit of xerophytic plants and characteristically have hard, thick leathery leaves. An example is the live oak, which holds most of its leaves through the dry season. Such hard-leaved evergreen trees and woody shrubs are called *sclerophylls*. The prefix *scler* is from the Greek word for "hard." Plants that hold their leaves through a dry or cold season have the advantage of being able to resume photosynthesis immediately when growing conditions become favorable, whereas the deciduous plants must grow a new set of leaves.

Plants and Temperature

Temperature, another of the important climatic factors in plant ecology, acts directly on plants through its influence on the rates at which the physiological processes take place. In general, we can say that each plant species has an optimum temperature associated with each of its functions, such as photosynthesis, flowering, fruiting, or seed germination. Some overall optimum yearly temperature conditions exist for its growth in terms of size and numbers of individuals. There are also limiting lower and upper temperatures for the individual functions of the plant and for its total survival. Temperature acts as an indirect factor in many other ways. Higher air temperatures increase the water-vapor capacity of the air, thus inducing greater transpiration as well as faster rates of direct evaporation of soil water.

In general, the colder the climate, the fewer are the species that are capable of surviving. A large number of tropical plant species cannot survive below-freezing temperatures. In the severely cold arctic and alpine environments of high latitudes and high altitudes, only a few species can survive. This principle explains why a forest in the equatorial zone has many species of trees, whereas a forest of the subarctic zone may be dominantly of one, two, or three tree species. Tolerance to cold is closely tied up with the ability of the plant to withstand the physical disruption that accompanies the

freezing of water. If the plant has no means of disposing of the excess water in its tissues, the freezing of that water will damage the cell tissue.

Plant geographers recognize that there is a critical level of climatic stress beyond which a plant species cannot survive; here, a geographical boundary will exist, that marks the limits of its distribution. Such a boundary is sometimes referred to as a frontier. Although the frontier is determined by a complex of climatic elements, it is sometimes possible to single out one climate element related to soil water or temperature that coincides with the plant frontier.

Ecology and Ecosystems

Organisms, whether belonging to the plant kingdom or to the animal kingdom, also interact wth each other. Study of these interactions—in the form of exchanges of matter, energy, and stimuli of various sorts—between life forms and the environment is the science of *ecology*, very broadly defined. The total assemblage of components entering into the interactions of a group of organisms is known as an ecological system, or more simply, an *ecosystem*. The root "eco" comes from a Greek word connoting a house in the sense of household, in which a family lives together and interacts within a functional physical structure.

Geographers view ecosystems as natural resource systems. Food, fiber, fuel, and structural material are products of ecosystems—they represent organic compounds placed in storage by organisms through the expenditure of energy basically derived from the sun. Geographers are interested in the influence of climate on ecosystem productivity. Our basic understanding of global climates and the global range of soil-water budgets will prove most useful in explaining the global pattern of ecosystems of the lands.

Ecosystems have inputs of matter and energy used to build biological structures, to reproduce, and to maintain necessary internal energy levels. Matter and energy are also exported from an ecosystem. An ecosystem tends to achieve a balance of the various processes and activities within it. For the most part these balances are quite sensitive and can be easily upset or destoyed. Physical geography meshes closely with ecology. We have already stressed the importance of organisms and their life processes in shaping the characteristics of the soil layer.

Plants as Producers of Carbohydrates

We turn next to a brief description of the way in which green plants cycle materials of the atmosphere and energy from sunlight to synthesize compounds out of which the structure of the plant is built. This process is called *photosynthesis,* defined in the simplest possible terms as the production of carbohydrate. *Carbohydrate* is a general term for a class of organic compounds consisting of the elements carbon (C), hydrogen (H), and oxygen (O).

Photosynthesis of carbohydrate requires a series of complex biochemical reactions using water (H_2O) and carbon dioxide (CO_2) as well as light energy. A simplified chemical reaction for photosynthesis can be written

$$H_2O + CO_2 + \text{light energy} \rightarrow \text{carbohydrate} + O_2$$

Oxygen in the form of gas molecules (O_2) is a by-product of photosynthesis.

Respiration is the process opposite to photosynthesis, in which carbohydrate is broken down and combined with oxygen to yield carbon dioxide and water. The overall reaction is

$$\text{Carbohydrate} + O_2 \rightarrow CO_2 + H_2O + \text{chemical energy}$$

As in the case of photosynthesis, the actual reactions are far from simple. The chemical energy released is stored in many types of energy-carrying molecules and used later to synthesize all the biological molecules necessary to sustain life.

At this point, it is helpful to link photosynthesis and respiration in a continuous cycle involving both a producer and a decomposer. Figure 20.6 shows one closed loop for hydrogen (H), one for carbon (C), and two loops for oxygen (O). We are not taking into account that there are two atoms of hydrogen in each molecule of water and carbohydrate, or that there are two atoms of oxygen in each molecule of carbon dioxide and oxygen gas. Only the flow pattern counts in this representation.

A good place to start is the soil, from

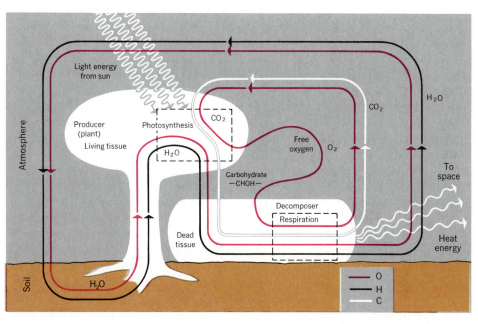

FIGURE 20.6 A simplified flow diagram of the essential components of photosynthesis and respiration.

which water is drawn up into the body of a living plant. In the green leaves of the plant, photosynthesis takes place while light energy is absorbed by the leaf cells. Carbon dioxide is being brought in from the atmosphere at this point. Here oxygen is liberated and begins its atmospheric cycle. The plant tissue then dies and falls to the ground, where it is acted on by organisms that are decomposers. Through respiration, oxygen is taken out of the atmosphere or soil air and combined with the carbohydrates. Energy is now liberated. Here both carbon dioxide and water enter the atmosphere as gases.

An important concept emerges from this flow diagram. Energy passes through the system. It comes from the sun and returns eventually to outer space. On the other hand, the material components, hydrogen, oxygen, and carbon, are recycled within the total system. Of course, many other material components are recycled in the same way. These are plant nutrients, essential in the growth of plants. Nutrients are constantly recycled. Because the earth as a planet does not lose or gain matter, the material components never leave the total system, but they can be stored in other ways and forms where they are unavailable for use by plants for prolonged periods of geologic time. We will develop this concept more fully later in this chapter.

Net Photosynthesis

Because both photosynthesis and respiration go on simultaneously in a plant, the amount of new carbohydrate placed in storage is less than the total carbohydrate being synthesized. We must thus distinguish between gross photosynthesis and net photosynthesis. *Gross photosynthesis* is the total amount of carbohydrate produced by photosynthesis; *net photosynthesis* is the amount of carbohydrate remaining after respiration has broken down sufficient carbohydrate to power the plant. Stated as an equation,

Net photosynthesis =
        gross photosynthesis − respiration

Since both photosynthesis and respiration

occur in the same cell, gross photosynthesis cannot be readily measured. Instead, we will deal with net photosynthesis. In most cases, respiration will be held constant, so that use of the net instead of the gross will show the same trends.

Light intensity sufficient to allow maximum net photosynthesis is only 10 to 30 percent of full summer sunlight for most green plants. Additional light energy is simply ineffective. The fact or of duration of daylight then becomes the important factor in the rate at which products of photosynthesis accumulate as plant tissues. On this subject, you can draw on your knowledge of the seasons and the changing angle of the sun's rays with latitude. At low latitudes, days are not far from the average 12-hour length throughout the year, whereas at high altitudes days are short in winter but long in summer. The seasonal contrast in day length increases with latitude. In subarctic latitudes, photosynthesis can obviously go on in summer during most of the 24 hours, and this factor can compensate partly for the shortness of the season.

Net Primary Production

Plant ecologists measure the accumulated net production by photosynthesis in terms of the *biomass*, which is the dry weight of the organic matter. This quantity could, of course, be stated for a single plant or species, but a more useful statement is made in terms of the biomass per unit of surface area within the ecosystem—the hectare, square meter, or square foot. Of all ecosystems, forests have the greatest biomass; that of grasslands and croplands is very small in comparison. For freshwater bodies and the oceans the biomass is even smaller—on the order of one-hundredth that of the grasslands and croplands.

The annual rate of production of organic matter is the vital information we need, not the biomass itself. Granted that forests have a very large biomass and grasslands very little, what we want to know is: Which ecosystem is the more productive? In other words, which ecosystem produces the greatest annual yield in terms of useful resources? For the answer we turn to

TABLE 20.1 Net primary production for various ecosystems

	Grams per Square Meter per Year	
	Average	Typical Range
Lands		
Rainforest of the equatorial zone	2000	1000–5000
Freshwater swamps and marshes	2000	800–4000
Midlatitude forest	1300	600–2500
Midlatitude grassland	500	150–1500
Agricultural land	650	100–4000
Lakes and streams	500	100–1500
Extreme desert	3	0–10
Oceans		
Estuaries (tidal)	2000	500–4000
Continental shelf	350	200–600
Open ocean	125	1–400

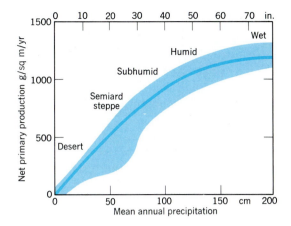

FIGURE 20.7 Net primary production increases rapidly with increasing precipitation, but levels off in the higher values. Observed values fall mostly within the shaded zone. (Data of Whittaker, 1970.)

figures on the *net primary production* of various ecosystems. Table 20.1 gives this information in units of grams of dry organic matter produced annually from 1 m² of surface. The figures are only rough estimates, but they are nevertheless highly meaningful. Note that the highest values are in two quite unlike environments: forests and estuaries. Agricultural land compares favorably with grassland, but the range is very large in agricultural land, reflecting many factors, such as availability of soil water, soil fertility, and use of fertilizers and machinery.

Net Production and Climate

The geographer is interested in the climatic factors controlling organic productivity. Besides light intensity and temperature, an obvious factor will be availability of water. Soil-water shortage or surplus might be the best climatic factor to examine, but data are not available. Ecologists have related net annual primary production to mean annual precipitation, as shown in Figure 20.7. The production values are for plant structures above the ground surface. Although the productivity increases rapidly with precipitation in the lower range from desert through semiarid to subhumid climates, it seems to level off in the humid range. Apparently, a large soil-water surplus carries with it some compensating influence, such as removal of plant nutrients by leaching.

Combining the effects of light duration and intensity, temperature, and precipitation, you would guess that the maximum net productivity year-round would be in wet equatorial climates where soil water is adequate at all times. Productivity would be low in all desert climates and would decrease generally with increasing latitude. Productivity would be reduced to near zero in the arctic zone, where the combination of a short growing season and low temperatures would act together to slow the growth process.

The Green Revolution
in Southeast Asia

Prospects for exploiting the savanna environment for increased food production rest today on concerted application of techniques embodied in the term *green revolution*. A major ingredient in that revolution has been the development of new genetic strains of rice and wheat. These strains are capable of greatly increased production per unit of area of land now under cultivation. To achieve the potential of these new plant strains requires substantial applications of fertilizers. Changes in agricultural practices and increased use of machines are also a part of the total picture. Although striking increases of rice and wheat yields were achieved in Southeast Asia by the green revolution, ensuring sharp increases in prices of fertilizer and fuel soon dealt a staggering blow to the new agriculture.

Because of the green revolution, wheat production in the Punjab of India was doubled in the five years between 1966 and 1971. It was hoped that this phenomenal increase would sustain the growing needs of India's population, which increases by some 17 million persons per year. However, wheat production leveled off and, in 1974, was forced into sharp decline by a number of negative factors. These included partial failure of the rains and the enormous increase in cost of fertilizers and fuels. Tube wells, drilled into the sand and gravel to reach the ground water, require electricity for pumps, and this energy source was also severely cut. Despite such setbacks, agricultural productivity in Asia generally has increased to the point of self-sufficiency in food production.

Although application of green revolution technology to Africa has lagged behind that in Southeast Asia and Latin America, some important advances had been made there by 1985. For example, a drought-tolerant variety of sorghum was introduced into the Sudan with prospects of more than doubling the yields of traditional varieties. Improved varieties of wheat introduced into Ethiopia are capable of doubling yields on small highland farms if given adequate fertilizer.

One hazard of the green revolution lies in the requirement that the genetic strains bred for high yields must be used to the exclusion of a variety of native strains. Should the high-yield strain prove vulnerable to an epidemic plant disease, the entire crop of a whole nation could be wiped out in one season. Current research in crop breeding is now stressing the development of more individualized varieties that respond better to local conditions. These varieties are less subject to epidemic diseases and can still produce improved yields with less fertilizer input. A second hazard lies in the need to increase the size of fields by merging many small plots into large ones to allow mechanized

Threshing wheat in northern India. The bullocks follow a circular path while treading on the wheat stalks. In the background, the chaff is being winnowed by the wind as the grain is poured from baskets. (Marc & Evelyne Bernheim/Woodfin Camp.)

agriculture to work most efficiently. In so doing, a variety of food crops is no longer grown, and dependence for survival comes to rest on the single crop. Under traditional practices, the Asiatic farmer planted several food crops to ensure that if some failed, others would yield enough food to prevent starvation. The small farmer is therefore reluctant to move to more efficient, single-crop agriculture.

It seems obvious now that future increases in yields will be achieved only by modifying the techniques of the green revolution to become less dependent on industrial midlatitude technology. New crop strains and methods of cultivation and fertilization will have to be more compatible with ancient agricultural systems and local culture patterns.

Soil scientists have pointed out that large areas of Vertisols remain to be placed under cultivation. The Vertisols are rich in nutrient bases, but will require machine cultivation to overcome difficult tillage. Primitive plows and hoes cannot work these soils. Most of the arable land of the savanna environment in Southeast Asia has already been intensively developed by existing standards. To hope for major expansion of agriculture into poor Oxisols and to semidesert zones marginal to the tropical deserts is, at best, unrealistic.

464

These deductions are illustrated by a schematic diagram of net primary productivity on an imagined supercontinent (Figure 20.8). Comparing this diagram with the world climate diagram (Figure 7.8), we can assign rough values of productivity to each of the climates (units are grams of carbon per square meter per year):

Highest (over 800)	Wet equatorial (1)
Very high (600–800)	Monsoon and trade-wind littoral (2)
High (400–600)	Wet-dry tropical (3) (S.E. Asia) Moist subtropical (6) Marine west coast (8)
Moderate (200–400)	Mediterranean (7) Moist continental (10)
Low (100–200)	Dry tropical, semiarid (4s) Dry midlatitude, semiarid (9s) Boreal forest (10)
Very low (0–100)	Dry tropical, desert (4d) Dry midlatitude, desert (9d) Boreal forest (10) Tundra (12)

The wet–dry tropical climate (3) is in a narrow transition zone between very high productivity and low productivity so that some areas are in the "very high" rating and others are in the "high" rating. Transition zones such as this are highly sensitive to alternating cycles of drought and of excessive rainfall. They pose the most difficult famine problems of all global regions because they carry dense populations, which can be adequately fed only under the best of conditions.

Cycling of Materials in Ecosystems

From an investigation of energy flow in ecosystems, we now turn to the flow of materials used by plants. The elements

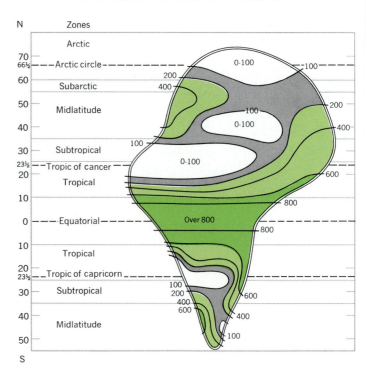

FIGURE 20.8 Schematic diagram of net primary production of organic matter on an imagined supercontinent. The units are grams of carbon per square meter per year. The quantities shown are estimates only. Compare this map with the world climate diagram, Figure 7.8. (Based on data of D. E. Reichle, 1970.)

hydrogen, carbon, and oxygen are basic materials whose pathways we have already followed in the photosynthesis–respiration circuits (Figure 20.6). At this point, a broader view of the cycling of other essential elements will give you an increased understanding of the role of geological processes in storing and releasing these elements and how humans can act to upset the natural balance of these cycles.

Take the 15 most abundant elements in living matter as a whole and designate their total mass as 100 percent. Percentages of each of the elements are given in Table 20.2. The three principal components of carbohydrate—hydrogen, carbon, and oxygen—account for almost all living matter and are called *macronutrients*. The remaining one-half percent is divided among 12 elements. Six of these are also classed as macronutrients: nitrogen, calcium, potassium, magnesium, sulfur, and phosphorus. The macronutrients are all

TABLE 20.2 Elements comprising global living matter, taking 100 percent as the total of the 15 most abundant elements

Basic Carbohydrate		
Hydrogen (M)		49.74
Carbon (M)		24.90
Oxygen (M)		24.83
	Subtotal	99.47
Other Nutrients		
Nitrogen (M)		0.272
Calcium (M)		0.072
Potassium (M)		0.044
Silicon		0.033
Magnesium (M)		0.031
Sulfur (M)		0.017
Aluminum		0.016
Phosphorus (M)		0.013
Chlorine		0.011
Sodium		0.006
Iron		0.005
Manganese		0.003

M indicates macronutrient.
Data of E. S. Deevey, Jr. (1970), *Scientific American*, vol. 223.

required in substantial quantities for organic life to thrive.

Nitrogen is much more abundant than the other five lesser macronutrients, since it is contained within each amino acid building block of protein. Almost all the nitrogen available to living organisms is obtained from the atmosphere through the efforts of nitrogen-fixing bacteria. These bacteria have a biochemical system that can convert nitrogen from its inert atmospheric form, molecular nitrogen, to other forms, such as ammonia and nitrate, that can be used directly by plants. The plants assimilate these forms of nitrogen and convert them to organic forms, which are consumed by animals grazing on the plants. In this way nitrogen is passed up the food chain.

Decomposers convert the nitrogen from organic forms in decaying plant and animal matter to simple, inorganic forms that the plants can absorb from soil or water. Certain bacteria can also return nitrogen to its inert atmospheric form. This function completes the nitrogen cycle, by which nitrogen moves from the atmosphere to organisms in the life layer and back again to the atmosphere.

Of the remaining macronutrients, you will recognize three—calcium, potassium, and magnesium—as the base cations so important in soils. They are all derived from rocks through mineral weathering, along with silicon, sodium, sulfur, phosphorus, iron, and manganese. Some of the sulfur and sodium reaches the soil as a component of rainwater, having originally been swept from breaking waves on the sea surface and carried aloft as minute suspended salt particles. Sulfur and chlorine introduced into the atmosphere by volcanic eruptions are also delivered to the soil surface by the washout process. Sulfur and nitrogen are also supplied from industrial combustion sources in the form of acid rain (Chapter 5).

Quite a number of additional elements, not among the nine macronutrients, are also vital to life processes. Their presence is needed in mere traces and, for this reason, they are called *micronutrients*. The micronutrient list includes iron, copper, zinc, boron, molybdenum, manganese, and chlorine.

Figure 20.9 is a flow diagram showing pathways taken by the more important mineral elements from the inorganic realms of the lithosphere, hydrosphere, and atmosphere, through the realm of living tissues, or biosphere, and returning to the inorganic realm. Within the large box representing the lithosphere are smaller compartments representing the parent matter of the soil and the soil itself. In the soil, nutrients are held as ions on the surface of soil colloids and are readily available to plants. These same elements are also held in enormous storage pools where they are unavailable to organisms. The storage pools include seawater (unavailable to land organisms), sediments on the sea floor, and enormous accumulations of sedimentary rock beneath both lands and oceans.

Eventually, elements held in the geologic storage pools are released into the soil by weathering. Soil particles are lifted into the atmosphere by winds and fall back to earth or are washed down by precipitation.

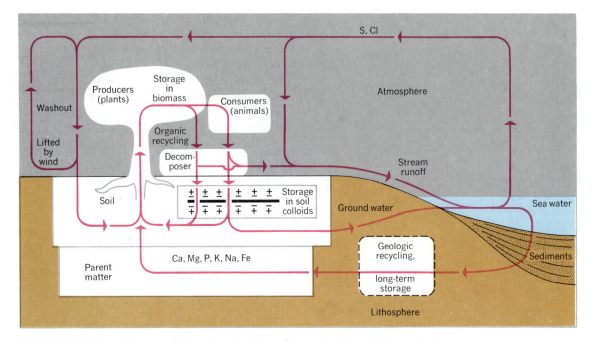

FIGURE 20.9 A flow diagram of the mineral portion of the materials cycle in and out of the biosphere and within the inorganic realm of the lithosphere, hydrosphere, and atmosphere. The four macronutrients of atmospheric origin—hydrogen, carbon, oxygen, and nitrogen—are omitted.

Chlorine and sulfur are shown as passing from the ocean into the atmosphere and entering the soil by the same mechanisms of fallout and washout. The organic realm, or biosphere, is shown in three compartments: producers, consumers, and decomposers. Considerable element recycling occurs between these organisms and the soil. However, there is a continual process of escape of these elements to the sea as ions dissolved in stream runoff and ground water flow.

Plant Succession and Human Impact on Vegetation

The phenomenon of change in ecosystems through time is familiar to us all. A drive in the country reveals patches of vegetation in many stages of development—from open, cultivated fields to grassy shrublands to forests. Clear lakes gradually fill in with sediment from the rivers that drain into them and become marshes. These kinds of changes, in which plant and animal communities succeed one another on the

way to a stable endpoint, make up an *ecological succession*.

In general, succession leads to formation of the most complex community of organisms possible in an area, given its physical controlling factors of climate, soil, and water. The stable community, which is the endpoint of succession, is the *climax*. Succession may begin on a newly constructed mineral deposit, such as a sand dune or river bar. It may also occur on a previously vegetated area that has been recently disturbed by agents, such as fire, flood, and windstorms, and by human activity.

An important case of succession is that on farmlands intensively cultivated for decades then abandoned to the natural sequence of events. An example of this "old-field" succession comes from the southeastern United States, where the moist subtropical climate provides ample soil water much of the year and the growing season is long. Figure 20.10 shows the succession, starting at the left with the bare field. The first plants to take hold in the hostile environment are called pioneers.

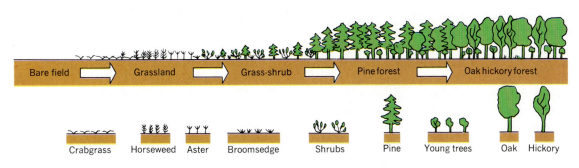

FIGURE 20.10 Old-field succession in the Piedmont region of the southeastern United States, following abandonment of cornfields and cottonfields. This is a pictorial graph of continuously changing plant composition spanning about 150 years. (After E. P. Odum, 1973, *Fundamentals of Ecology*, 3rd ed., Saunders, Philadelphia.)

They are annual herbs most persons would call weeds. Crabgrass, horseweed, and aster thrive during the first two years. These plants supply organic matter to the soil; they also begin to shade the soil and reduce the extremes of soil temperature. Next, grasses and shrubs move in, and these occupy the old field during the first two decades or so. Pine seedlings now enter the habitat and, as these grow, they develop a shade cover, greatly changing the climate near the ground.

In the next half century, as the pine forest matures, broad-leaved deciduous trees begin to come in and displace the pines. By the middle of the second century, a mature forest of oak and hickory has dominated the habitat. This climax forest has a lower shrub layer, a basal herb layer, and a substantial accumulation of organic matter in various stages of decomposition. This is an environment in which soil water tends to be conserved and the thermal environment is protected from extremes of heat or cold. Although generally stabilized in its composition and structure, serious natural upsets of equilibrium may occur through fires and severe storms. Insect hordes and disease epidemics may also radically alter the climax forest. Fire will be followed by an appropriate succession, with stages leading to the return of the oak–hickory forest.

Cutting and clearing of forest trees brings a demise to the forest climax. Introduction of a plant disease from a foreign continent can cause the extinction of a particular plant species. An example is the chestnut blight, which eliminated the American chestnut from the forests of the northeastern United States. Imported insects may also wipe out most of the mature individuals of a plant species when there is no native predator available to combat the invasion. These are just a few of the ways in which humans interfere with natural vegetation.

Biomass Energy

Net primary production represents a source of renewable energy derived from the sun that can be exploited to fill human energy needs. The use of *biomass energy* as a resource involves releasing solar energy that has been fixed in plant tissues through photosynthesis. This process can take place in a number of ways—the simplest is direct burning of plant matter as fuel, as in a campfire or a wood-burning stove. Other approaches involve the generation of intermediate fuels from plant matter— methane gas, charcoal, and alcohol, for example. Biomass energy conversion is not highly energy efficient; typical values for net annual primary production of plant communities range from one to three percent of available solar energy. However, the abundance of terrestrial biomass is so great that biomass utilization could provide the energy equivalent to three million barrels of oil per day for the United States by the year 2000.

An important use of biomass energy is the burning of firewood for cooking (and some space heating) in developing nations. In 1970, the annual growth of wood in the forests of developing countries totaled nearly half the world's energy production—plenty of firewood was thus available. However, fuelwood used now exceeds production in many areas, creating local shortages and severe strains on some forest ecosystems. The forest–desert transition areas of thorntree, savanna, and desert scrub in central Africa south of the Sahara Desert are examples.

Even in closed stoves, wood burning is not very efficient, ranging from 10 to 15 percent for cooking. However, the conversion of wood to charcoal and/or gas can boost efficiencies to values as high as 70 to 80 percent with appropriate technology. In this process, termed pyrolysis, controlled partial burning in an oxygen-deficient environment reduces carbohydrate to free carbon (charcoal), and yields flammable gases, such as carbon monoxide and hydrogen. Charcoal is more energy efficient than wood, burns more cleanly, and is easier to transport. As an added advantage, charcoal can be made from waste fibers and agricultural residues that would normally be discarded. Thus, charcoal is an efficient fuel that can help extend the firewood supply in areas where wood is in high demand.

Another use of biomass of increasing importance is the conversion of agricultural wastes to alcohol. In this process, yeast microorganisms are used to convert the carbohydrate to alcohol through fermentation. With the rising price of motor fuels, alcohol attracted attention as an extender for gasoline. Gasohol, a mixture of up to 10 percent alcohol in gasoline, can be burned in conventional engines without adjustment. Distillation of alcohol, however, requires heating, thus greatly reducing the net energy yield. A number of approaches are currently being explored to reduce the energy cost of distillation. Examples are the use of new membranes that pass alcohol faster than water, and the use of special heat-saving technology. Different strains of microorganisms, which are more efficient than yeast and work at higher temperatures, may also be helpful. Alcohol, charcoal, and firewood are all alternatives to fossil fuels that will become increasingly important as petroleum becomes scarcer and more costly in the coming decades.

Terrestrial Ecosystems—The Biomes

Ecosystems fall into two major groups: aquatic and terrestrial. The *aquatic ecosystems* include life-forms of the marine environments and the freshwater environments of the lands. Marine ecosystems include the open ocean, coastal estuaries, and coral reefs. Freshwater ecosystems include lakes, ponds, streams, marshes, and bogs. Our survey of physical geography will not include the ecology of these aquatic environments. Instead, we will focus on the *terrestrial ecosystems;* they comprise the assemblages of land plants spread widely over the upland surfaces of the continents. The terrestrial ecosystems are directly impacted by climate and interact with the soil and, in this way, are closely woven into the fabric of physical geography.

Within terrestrial ecosystems the largest recognizable subdivision is the *biome.* Although the biome includes the total assemblage of plant and animal life interacting within the life layer, the green plants dominate the biome physically because of their enormous biomass, as compared with that of other organisms. The plant geographer concentrates on the characteristic life-form of the green plants within the biome. As we have already seen, these life-forms are principally trees, shrubs, lianas, and herbs, but other life-forms are important in certain biomes. Biomes are recognized and mapped on the concept of the climax vegetation. Areas intensively modified by agriculture and urbanization are assigned to a biome according to the climax vegetation assumed to have once been present.

The following are the principal biomes, listed in order of availability of soil water and heat.

Forest	Ample soil water and heat.
Savanna	Transitional between forest and grassland.
Grassland	Moderate shortage of soil water; adequate heat.
Desert	Extreme shortage of soil water; adequate heat.
Tundra	Insufficient heat.

For the plant geographer the biomes are broken down into smaller vegetation units,

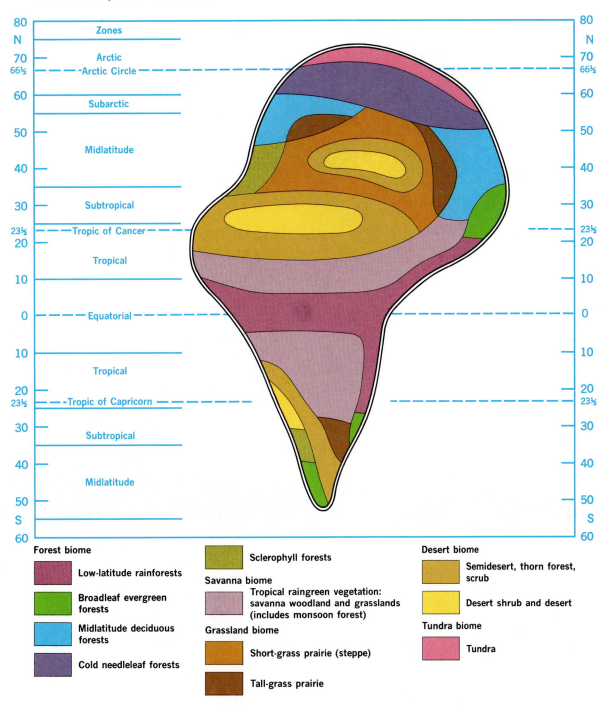

Forest biome

- Low-latitude rainforests
- Broadleaf evergreen forests
- Midlatitude deciduous forests
- Cold needleleaf forests
- Sclerophyll forests

Savanna biome

- Tropical raingreen vegetation: savanna woodland and grasslands (includes monsoon forest)

Grassland biome

- Short-grass prairie (steppe)
- Tall-grass prairie

Desert biome

- Semidesert, thorn forest, scrub
- Desert shrub and desert

Tundra biome

- Tundra

FIGURE 20.11 A schematic diagram of the vegetation formation classes on an idealized continent. Compare with the world vegetation map, figure 20.12.

called *formation classes,* concentrating on the life-form of the plants. For example, at least four and perhaps as many as six kinds of forests are easily recognizable within the forest biome. At least three kinds of grasslands are easily recognizable. Deserts, too, span a wide range in terms of the abundance and life-form of plants.

The formation classes we will introduce are major widespread types clearly

associated with twelve climate types and their soil-water budgets. Association with soil orders and suborders will also be important. In this way we survey the global scope of plant geography as a synthesis of climate with organic and physical processes of the life layer.

Figure 20.11 is a schematic diagram to show how the vegetation formation classes would be arranged on an idealized

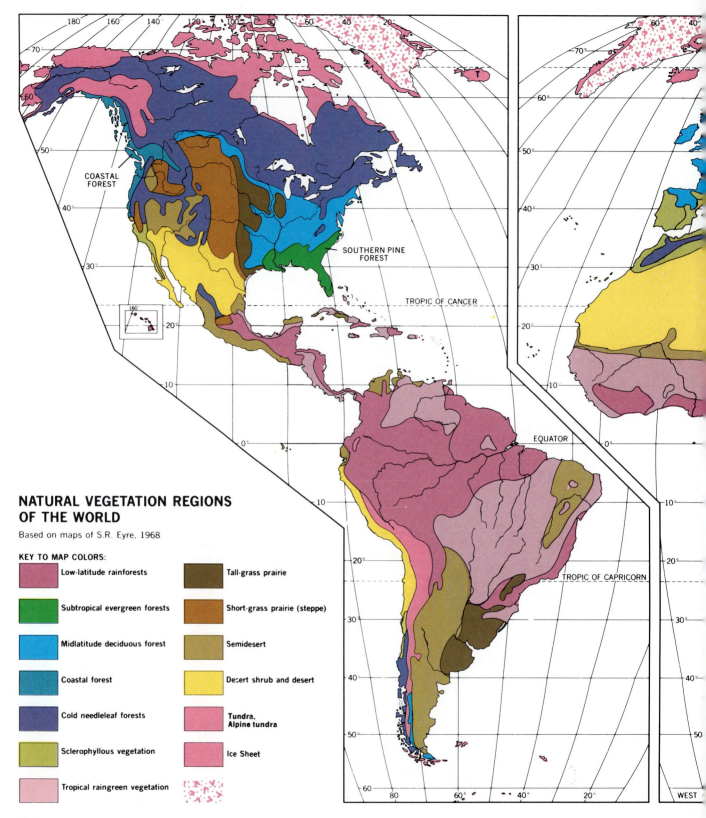

NATURAL VEGETATION REGIONS OF THE WORLD

Based on maps of S.R. Eyre, 1968

KEY TO MAP COLORS:

- Low-latitude rainforests
- Subtropical evergreen forests
- Midlatitude deciduous forest
- Coastal forest
- Cold needleleaf forests
- Sclerophyllous vegetation
- Tropical raingreen vegetation
- Tall-grass prairie
- Short-grass prairie (steppe)
- Semidesert
- Desert shrub and desert
- Tundra, Alpine tundra
- Ice Sheet

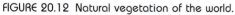

FIGURE 20.12 Natural vegetation of the world.

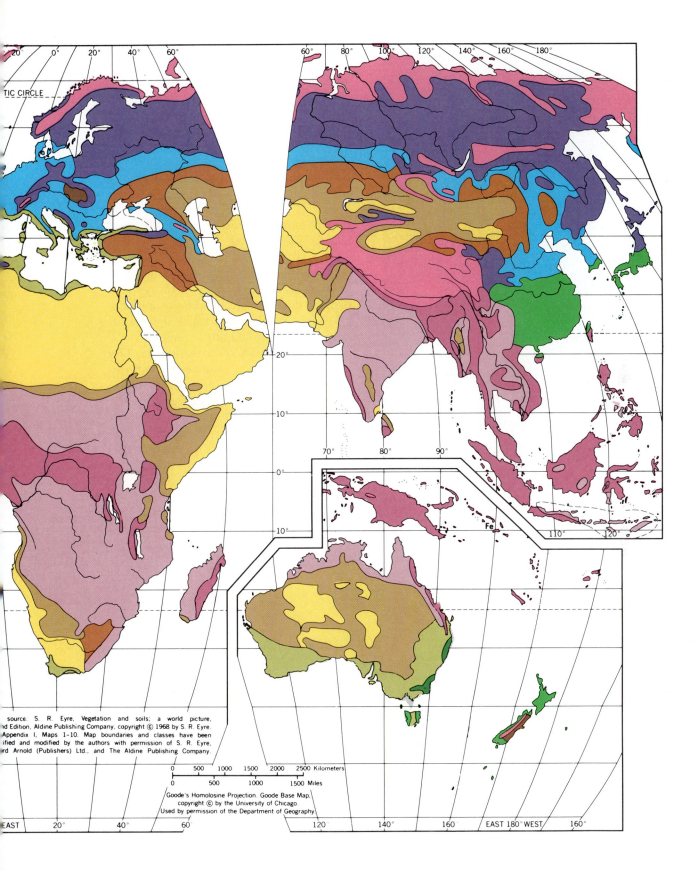

source. S. R. Eyre, Vegetation and soils; a world picture, 2nd Edition, Aldine Publishing Company, copyright © 1968 by S. R. Eyre. Appendix I, Maps 1–10. Map boundaries and classes have been modified and modified by the authors with permission of S. R. Eyre, Edward Arnold (Publishers) Ltd., and The Aldine Publishing Company.

0   500   1000   1500   2000   2500 Kilometers

0   500   1000   1500 Miles

Goode's Homolosine Projection. Goode Base Map,
copyright © by the University of Chicago.
Used by permission of the Department of Geography

FIGURE 20.13 Interior of lowland rainforest, Baiyer River, New Guinea. The trees are smooth-trunked and support thick lianas (left). (Tom McHugh/Photo Researchers.)

continent, combining Eurasia and Africa into a single supercontinent. Figure 20.12 is a generalized world map of the formation classes. It simplifies the very complex patterns of natural vegetation to create large uniform regions in which a given formation class might be expected to occur.

Forest Biome

Low-Latitude Rainforest

Low-latitude rainforest, found in the equatorial and tropical latitude zones, consists of tall, closely set trees; their crowns form a continuous canopy of foliage and provide dense shade for the ground and lower layers (Figure 20.13). The trees are characteristically smooth barked and unbranched in the lower two-thirds. Tree leaves are large and evergreen; from this characteristic the equatorial rainforest is often described as "broadleaf evergreen forest." Crowns of the trees tend to form into two or three layers, or strata, of which the highest layer consists of scattered emergent crowns rising to 40 m (130 ft) and protruding conspicuously above a second layer, 15 to 30 m (50 to 100 ft) high, which is continuous (Figure 20.14). A third, lower layer consists of small, slender trees 5 to 15 m (15 to 50 ft) high with narrow crowns.

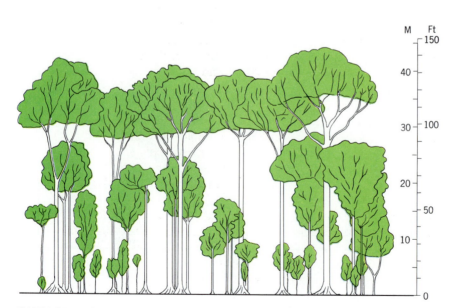

FIGURE 20.14 This diagram shows the typical structure of equatorial rainforest. (After J. S. Beard, 1946, *The Natural Vegetation of Trinidad,* Clarendon Press, Oxford.)

FIGURE 20.15 This air-plant, or epiphyte, is a bromeliad, supported on a tree branch in subtropical evergreen rainforest. Everglades National Park, Florida. (Arthur N. Strahler.)

Typical of the low-latitude rainforest are thick woody lianas supported by the trunks and branches of trees. Some are slender, like ropes; others reach thicknesses of 20 cm (8 in.). They rise to heights of the upper tree levels where light is available and may have profusely branched crowns. *Epiphytes* are numerous in the equatorial rainforest. These are plants attached to the trunk, branches, and foliage of tree and lianas, using the "host" solely as a means of physical support. Epiphytes are of many plant classes and include ferns, orchids, mosses, and lichens (Figure 20.15).

A particularly important botanical characteristic of the equatorial rainforest is the large number of species of trees that coexist. As many as 3000 species may be found in a few square kilometers. Individuals of a given species are often widely separated. Consequently, if a particular tree species is to be extracted from the forest for commercial uses, considerable labor is involved in seeking out the trees and transporting them from their isolated positions.

The floor of the low-latitude rainforest is usually so densely shaded that plant foliage is sparse close to the ground and gives the forest an open aspect, making it easy to traverse. The ground surface is covered only by a thin litter of leaves. Rapid consumption of dead plant matter by bacterial action results in the absence of humus on the soil surface and within the soil profile.

Low-latitude rainforest is a response to a climate that is continuously warm, frost-free, and has abundant precipitation in all months of the year (or, at most, only one or two months are dry). Low-latitude rainforest is thus closely correlated in extent with the wet equatorial climate (1) and the monsoon and trade-wind littoral climate (2). A large water surplus characterizes the annual water budget so that soil water is adequate at all times. In the absence of a cold or dry season, plant growth goes on continuously throughout the year. Individual species have their own seasons of leaf shedding, possibly caused by slight changes in the light period.

Soils of the equatorial rainforest are typically reddish Oxisols and Ultisols on well-drained upland surfaces. Recall from Chapter 19 that soils of these orders have low base status. The meager supply of nutrient bases is held in a thin surface layer, where it is effectively recycled by rainforest plants.

World distribution of the low-latitude rainforest is shown on the vegetation map, Figure 20.12. A large area of this rainforest lies astride the equator in the following areas: the Amazon lowland of South America; the Congo lowland of Africa and a coastal zone extending westward from Nigeria to Guinea; and the East Indian region, from Sumatra on the west to the islands of the western Pacific on the east.

The low-latitude rainforest extends poleward through the tropical zone (lat. 10° to 25° N and S) along coasts windward to the trades. The trade-wind coastal climate in which the tropical-zone rainforest thrives has a short dry season, but not intense enough to deplete the soil water.

In the southern hemisphere, belts of tropical-zone rainforest extend down the eastern Brazilian coast, the Madagascar coast, and the coast of northeastern Australia. Tropical-zone rainforest is important along east coasts in Central America and the West Indies. These highlands receive abundant orographic rainfall in the belt of trades. A good example seen by many Americans is the rainforest of the eastern mountains of Puerto Rico.

In Southeast Asia tropical-zone rainforest is extensive in coastal zones and highlands, which have heavy monsoon rainfall and a very short dry season. The western coasts of India and Burma have rainforest supported by orographic rains of the southwest monsoon.

Within the regions of low-latitude rainforest are many islandlike highlands of cooler climate in which precipitation is increased by the orographic effect. Here, rainforest extends upward on the rising mountain slopes. Between 1000 and 2000 m (3000 and 6000 ft) the rainforest gradually changes in structure and becomes *montane forest*. Several areas of montane forest are included in the world climate map—for example, in Central America and along the eastern slopes of the Andes mountains of South America. Montane forest has lower tree height and a less dense canopy than rainforest of adjacent lowlands; with increasing elevation, forest height becomes even lower. Tree ferns and bamboos are numerous; epiphytes are particularly abundant. Heavy accumulations of mosses attached to tree branches become conspicuous at higher altitudes. In these mossy forests (cloud forests), enshrouded much of the time in clouds, mist and fog are persistent. Relative humidity is high and condensed moisture keeps the plant surfaces wet.

Monsoon Forest

Monsoon forest of the tropical latitude zone is unique because of its deciduous habit. Most of the trees shed their leaves seasonally, in contrast with the low-latitude rainforest, which remains green throughout the year. The shedding of leaves in the monsoon forest results from the stress of a long dry season that occurs at time of low sun and cooler temperatures. In the dry season the forest has somewhat the dormant winter aspects of deciduous forests of midlatitudes. A representative example of a monsoon forest tree is the teakwood tree. Monsoon forest is typically open and grades into woodland with open areas occupied by shrubs and grasses (Figure 20.16). Here, it is transitional to the savanna woodland of the savanna biome.

Compared with the low-latitude evergreen rainforest, monsoon forest shows less competition among trees for light but a greater development of vegetation in the lower layers. Tree heights are less than in the low-latitude rainforest. Many tree species are present and may number 30 to 40 species in a small tract. Tree trunks are massive; the bark is often thick and rough. Branching starts at a comparatively low level and produces large, round crowns.

Monsoon forest is a response to a wet–dry tropical climate in which a long rainy season with a water surplus alternates with a dry, rather cool season having a soil-water shortage. These conditions are most strongly developed in the Asiatic monsoon climate, but are not limited to that area. Perhaps the typical regions of monsoon forest are in Burma, Thailand, and Cambodia. Large areas of deciduous

FIGURE 20.16 Monsoon woodland in the Bandipur Wild Animal Sanctuary in the Nilgiri Hills of southern India. The scene is taken in the rainy season, with trees in full leaf. (John S. Shelton.)

FIGURE 20.17 This lush rainforest of Rotorua, North Island of New Zealand, is one type of evergreen forest. It is known as the Smoking Forest because of the natural steam emissions of geothermal origin. Note the tree fern in the foreground. (F. Kenneth Hare.)

tropical forest occur in south-central Africa and in Central and South America, bordering the equatorial and tropical rainforests. Soils are largely Ultisols, Oxisols, Vertisols, and Alfisols.

On the world vegetation map (Figure 20.12) monsoon forest is given the same color as closely related forms of savanna biome vegetation. These are all grouped together under a single designation and map color as *tropical raingreen vegetation.*

Subtropical Broadleaf Evergreen Forest

Subtropical broadleaf evergreen forest is typical of moist climates of the subtropical zone and lower midlatitude zone where winters are mild and there is a large water surplus. This variety of forest differs from the low-latitude rainforests (which are also broadleaf evergreen types) in having relatively few species of trees. Thus, there are large populations of individuals of a species. Trees are not as tall as in the low-latitude rainforests; the leaves tend to be smaller and more leathery, and the leaf canopy less dense.

In the northern hemisphere, subtropical evergreen forest consists of broad-leaved trees such as evergreen oaks (see Figure 9.6) and trees of the laurel and magnolia families. The name "laurel forest" is applied to these forests, which are associated with the moist subtropical climate (6) in the southeastern United States, southern China, and southern Japan. Here the soils are

mostly Ultisols, with low base status. However, these lands are under intense crop cultivation and have been largely deforested for several centuries, so that little natural forest remains.

Another climatic association of broadleaf evergreen forest is with the marine west-coast climate (8) in the southern hemisphere, notably in New Zealand (Figure 20.17) and Chile. Here the kinds of trees are quite different from those of the northern hemisphere.

Broadleaf evergreen forest may have a well-developed lower layer of vegetation that, in different places, may include tree ferns, small palms, bamboos, shrubs, and herbaceous plants. Lianas and epiphytes are abundant.

On the world vegetation map, Figure 20.12, broadleaf evergreen forest of the southern United States is grouped with another formation class, *southern pine forest,* in a single color representing subtropical evergreen forest. The southern pine forest consists of a number of different pine species. This needleleaf forest is found in sandy soils of the coastal plain, where it appears to be a specialized vegetation type dependent on fast-draining sandy soils and frequent fires for its preservation.

Midlatitude Deciduous Forest

Midlatitude deciduous forest is familiar to inhabitants of eastern North America and western Europe as a native forest type

FIGURE 20.18 Deciduous forest of the Catskill Mountains, New York. Some needleleaf trees—pine and hemlock—are also present. (Arthur N. Strahler.)

(Figure 20.18). It is dominated by tall, broadleaf trees that provide a continuous and dense canopy in summer but shed their leaves completely in the winter. Lower layers of small trees and shrubs are weakly developed (see Figure 20.1). In the spring a luxuriant low layer of herbs quickly develops, but this is greatly reduced after the trees have reached full foliage and shaded the ground.

The deciduous forest is almost entirely limited to the midlatitude landmasses of the northern hemisphere. Common trees of the deciduous forest of eastern North America, southeastern Europe, and eastern Asia are oak, beech, birch, hickory, walnut, maple, basswood, elm, ash, tulip, sweet chestnut, and hornbeam. Hemlock, a needleleaf evergreen tree, may also be present. In western and central Europe, under a marine west-coast climate, dominant trees are mostly oak and ash, with beech in cooler and moister areas. Where the deciduous

forests have been cleared by lumbering, pines readily develop as second-growth forest.

The midlatitude deciduous forest represents a response to the moist continental climate (10), which receives adequate precipitation in all months. There is a strong annual temperature cycle with a cold winter season and a warm summer. Precipitation is markedly greater in the summer months (especially in eastern Asia) and increases at the time of year when water need is greatest. Only a small water shortage is incurred in the summer, and a large surplus normally develops in spring.

Soils most closely identified with midlatitude deciduous forests are Udalfs and Boralfs; both are suborders of the Alfisols. Because the Alfisols are soils of high base status, they proved highly productive for farming after the forests were cleared.

Needleleaf Forest

Needleleaf forest, as a general term, is a forest composed largely of straight-trunked, conical trees with relatively short branches and small, narrow, needlelike leaves. These trees are conifers. Where evergreen, the needleleaf forest provides continuous and deep shade to the ground so that lower layers of vegetation are sparse or absent, except for a thick carpet of mosses in many places. Species are few and large tracts of forest consist almost entirely of but one or two species. Our world vegetation map, Figure 20.12, shows color patterns for two distinct classes of needleleaf forest: (1) cold needleleaf forest of high latitudes and high altitudes and (2) coastal forest.

Boreal forest is a cold-climate needleleaf forest that predominates in two great continental belts, one in North America and one in Eurasia. These belts span the landmasses from west to east in latitudes 45° to 75°. The boreal forest of North America, Europe, and western Sibera is composed of evergreen conifers, such as spruce and fir (Figure 20.19). Needleleaf forest of north-central and eastern Siberia is boreal forest dominated by larch. The larch tree sheds its needles in winter and is thus a deciduous forest. Deciduous trees, such as aspen and balsam poplar, willow, and birch, tend to take over rapidly in areas of needleleaf

FIGURE 20.19 Needleleaf forest of red and black spruce in southern Ontario. (Alan H. Strahler.)

forest that have been burned over, or can be found bordering streams and in open places. Between the boreal forest and the midlatitude deciduous forest lies a broad transition zone of mixed boreal and deciduous forest.

Needleleaf evergreen forest extends into lower latitudes wherever mountain ranges and high plateaus exist. Thus, in western North America this formation class extends southward into the United States on the Sierra Nevada and Rocky Mountain ranges and over parts of the higher plateaus of the southwestern states (Figure 20.20, left). In Europe, needleleaf evergreen forests flourish on all the higher mountain ranges.

In its northernmost range, the boreal needleleaf forest grades into *cold woodland*. This form of vegetation is limited to the very cold subarctic climate. Trees are low in height and well spaced apart; a shrub layer may be well developed. The ground cover of lichens and mosses is distinctive (see Figure 9.33). Cold woodland is often referred to as *taiga*. It is transitional into the treeless tundra at its northern fringe.

Boreal forest is closely identified with the boreal forest climate (11) throughout North America and Eurasia. Both northern and southern boundaries of the formation class coincide rather well with the climatic boundaries, defined in terms of annual water need. Much of the boreal forest region is underlain by Spodosols, acid soils of low base status. In interior areas of western Canada, Russia, and Siberia the soils are Boralfs, the cold-regime suborder of the Alfisols. Also important are large areas of Histosols and Cryaquepts. The Cryaquepts are a cold-climate great soil group within the Aquepts, which is a suborder of the Inceptisols characterized by poor drainage and saturated soil.

Coastal forest is a distinctive needleleaf evergreen forest of the Pacific Northwest coastal belt, ranging in latitude from northern California to southern Alaska. Here, under a regime of heavy orographic precipitation and prevailing high humidities, are perhaps the densest of all coniferous forests, with the world's largest

FIGURE 20.20 Two kinds of needleleaf forest of the western United States. (left) Open forest of western yellow pine (ponderosa pine), in the Kaibab National Forest, Arizona. (Arthur N. Strahler.) (right) A grove of great redwood trees (*Sequoia*) in Humboldt State Park, California. (Alan H. Strahler.)

FIGURE 20.21 Chaparral, a vegetation type consisting of woody shrubs and small trees. San Gabriel Mountains, southern California. (Arthur N. Strahler.)

trees. Forests of coastal redwood (*Sequoia*), big tree (*Sequoiadendron*), and Douglas fir are particularly remarkable (Figure 20.20, right). Individuals of redwood and big tree attain heights of over 100 m (325 ft) and girths of over 20 m (65 ft).

Sclerophyll Forest

Sclerophyll forest consists of trees with small, hard, leathery leaves. Typically the trees are low-branched and gnarled, with thick bark. The formation class includes sclerophyll woodland, an open forest in which the canopy coverage is only 25 to 60 percent. Also included are extensive areas of *scrub*, a plant formation type consisting of shrubs having a canopy coverage of perhaps 50 percent. The trees and shrubs are evergreen, their thickened leaves being retained despite a severe annual drought.

Sclerophyll forest is closely associated with the Mediterranean climate (7) and quite narrowly limited in geographical extent—primarily to west coasts between 30° and 40° or 45° N and S latitude. In the Mediterranean lands the sclerophyll forest forms a narrow peripheral coastal belt. Here the Mediterranean forest consists of such trees as cork oak, live oak, Aleppo pine, stone pine, and olive. What may have once been luxuriant forests of such trees were greatly disturbed by human activity over the centuries and reduced to woodland or entirely destroyed. Today, large areas consist of dense scrub.

The other northern hemisphere region of sclerophyll forest is that of the California

coast ranges. Some of this forest or woodland is composed largely of the live oak and white oak. Grassland occupies the open ground between scattered oaks (see Figure 9.14). Much of the vegetation is sclerophyllous scrub or "dwarf forest," known as *chaparral* (Figures 20.21 and 20.22). It varies in composition with elevation and exposure. Chaparral may contain wild lilac, manzanita, mountain mahogany, "poison oak," and live oak.

Sclerophyll vegetation is represented in central Chile and in the Cape region of South Africa by scrub that is of a quite different flora from scrub of the northern hemisphere. Important areas of sclerophyll forest, woodland, and scrub are found in

FIGURE 20.22 A closeup view of the chaparral shown in figure 20.21. The rain gauge, mounted in a tilted position, rests on a patch of bare ground. San Dimas Experimental Forest. (Arthur N. Strahler.)

southeast, south-central, and southwest Australia, including several species of eucalyptus and acacia.

On the world vegetation map, Figure 20.12, all sclerophyll vegetation is shown in an olive-green color; it includes forest, woodland, and scrub.

Soils of the sclerophyll forest and woodland include the Xeralfs (a suborder of the Alfisols) and a suborder of the Mollisols (Xerolls). The prefix "xer" is used to signify the soil-water regime of the Mediterranean climate. Vertisols are also present in some areas, such as California and eastern Australia. All these soils are of high base status, with a generous supply of nutrient bases. Nevertheless, the long, severe summer drought, which brings very low soil-water storage and high soil temperatures, places a severe stress on plants. Sclerophyll vegetation is well adapted to these severe conditions.

Savanna Biome

The savanna biome, usually associated with the tropical wet–dry climate (3) of Africa and South America, includes vegetation formation classes ranging from woodland to grassland.

Savanna woodland consists of trees spaced rather widely apart, permitting development of a dense lower layer, which usually consists of grasses. The woodland has an open, parklike appearance. Here, the soil-water regime shows a seasonal decline in soil-water storage to an amount too small to sustain a closed-canopy forest. Savanna woodland typically forms a belt adjacent to equatorial rainforest in which soil-water storage remains high throughout the year.

In the tropical savanna woodland of Africa the trees are of medium height, the crowns flattened or umbrella-shaped, and the trunks have thick, rough bark (see Figure 8.13). Some species of trees are xerophytic forms with small leaves and thorns; others are broad-leaved deciduous species shedding their leaves in the dry season. In this respect, savanna woodland is akin to monsoon forest, into which it grades in some places.

Fire is a frequent occurrence in the savanna woodland during the dry season. The trees of the savanna are of species particularly resistant to fire. Many geographers hold the view that periodic burning of the savanna grasses is responsible for the maintenance of the grassland against the invasion of forest. Fire does not kill the underground parts of grass plants, but limits tree growth to a few individuals of fire-resistant species. The browsing of animals, which kills many young trees, is also a factor in maintaining grassland at the expense of forest. Many rainforest tree species that might otherwise grow in the wet–dry climate regime are prevented by fires from invading.

In Africa, the savanna woodland grades into a belt of *thorntree–tall-grass savanna,* a formation class transitional into the desert biome. The trees are largely of thorny species. Trees are more widely scattered and the open grassland is more extensive than in the savanna woodland. One characteristic tree is the flat-topped acacia, seen in Figure 8.13. Elephant grass is a common species; it may grow to a height of 5 m (16 ft) to form a thicket impenetrable to humans.

The thorntree–tall–grass savanna is closely identified with the semiarid subtype of the dry tropical and subtropical climates (4s, 5s). In Africa and India, the formation class includes areas of the semidesert climate subtype (4sd, 5sd). In the semiarid climate, soil-water storage is very low in ten out of the twelve months. Only during the brief rainy season is soil water adequate for the needs of plants. Onset of the rains is quickly followed by greening of the trees and grasses. For this reason, vegetation of the savanna biome is described as raingreen, an adjective that applies also to the monsoon forest. As we noted earlier, the world vegetation map (Figure 20.12) uses a single color for all tropical raingreen vegetation.

Soils of the savanna biome include Ustalfs, Ultisols, Oxisols, and Vertisols. The distribution of these orders varies according to parent matter of the soil and past climatic history.

A different expression of the savanna biome is found in eastern Australia. Here, in a north–south belt, the trees are evergreen species of eucalyptus.

The Sahelian Drought and Desertification

The savanna environment is subject to devastating drought years as well as to years of abnormally high rainfall that can result in severe floods. *Drought* is the occurrence of lower-than-average precipitation in that season in which ample precipitation usually occurs. The term drought is usually used with respect to the growing season of plants that provide food for humans and their domesticated animals. Rainfall of the wet season of the savanna environment can be expected to vary greatly from one year to the next. Typically, climate records show that two or three successive years of abnormally low rainfall (a drought) alternate with several successive years of average or higher-than-average rainfall.

The drier savanna zone of North Africa, including the semidesert zone adjoining it on the north, provides a lesson in the devastating impact of human activity on a delicate ecological system. Countries of this perilous belt, called the Sahel, or Sahelian zone, are shown in the accompanying figure. All these countries were struck a severe blow by a drought that extended from 1968 through 1974. Both nomadic cattle herders and grain farmers share this zone. As we stated in the Introduction, during that drought, grain crops failed and cattle could find no forage. In the worst stages of the 1968–1974 Sahel drought, nomads were forced to sell the remaining cattle that were their sole means of subsistence. Some five million cattle perished and many thousands of people died of starvation and disease.

Following the severe 1968–1974 drought, the Sahel experienced a somewhat lessened rainfall deficiency in 1975 and 1980; severe drought returned in 1980 and continued through 1984. The effects were particularly severe in Ethiopia. Abundant rainfall was received in the period of July through mid-September of 1985. Another severe drought was anticipated for 1988. Rainfall data, revealing 14 consecutive years of below-normal rainfall with respect to a 30-year period, have suggested to climatologists that a long-term trend to increasing aridity may be in progress.

Periodic droughts in past decades are well documented in the Sahel, as they are in other world regions of the savanna environment. Shifts in the position of the subtropical high pressure belt and the intertropical convergence zone can be called on to explain these droughts and the intervening periods of more-than-average rainfall. Associated with the Sahelian drought of 1968–1974 is a special phenomenon, recently given the name *desertification*—the transformation of the land surface by human activities superimposed on a natural drought situation. Desertification consists of a change in appearance of the land surface toward closer resemblance to a desert, largely through the destruction of grasses, shrubs, and trees that previously existed. Also visible are the effects of accelerated soil erosion, such as rilling and gullying of slopes, and accumulations of sediment in stream channels. Deflation, too, is

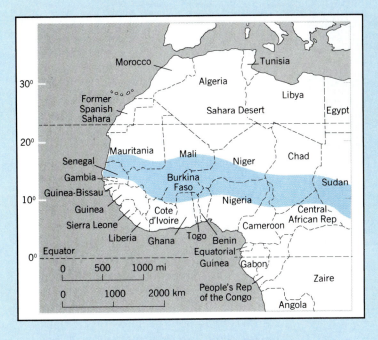

The Sahel, or Sahelian Zone, in western North Africa lies just south of the great Sahara Desert.

intensified and may be seen in scour of soil surfaces and the accumulations of soil drifts in the lee of obstacles.

Desertification in the African savanna environment can be attributed to greatly increased numbers of humans and their cattle. European managers of African colonies contributed to population increases by supplying food in time of famine, by reducing the death rate from disease, and by developing ground water supplies for crop irrigation and livestock. With each succeeding drought, a heavier impact was made on the vegetation and soil by the increased population. Ultimately, land degradation was too sustained and too severe to permit recovery of the plant cover in the rainy season or over a moister period

of ample rainfall seasons. Thus, as long as the population pressure persists, desertification is permanent, and takes on the outward appearances of a change toward a desert climate.

Today, many groups of scientists and many organizations, such as the Food and Agriculture Organization of the United Nations, are investigating the phenomenon of desertification. Regions of present and potential desertification are being mapped and various proposals for halting and reversing the process are being studied. Although the process has been plaguing the human race in one place or another for centuries, only now has it received an exciting new name and achieved the proportions of an international crusade.

FIGURE 20.23 Closeup view of virgin tall-grass prairie preserved in Kalsow Prairie, a State Botanical Monument in Iowa. Among the tall leaves of grass are forbs in flower. (Gene Ramsay, Webb Publishing Company.)

Grassland Biome

Tall-grass prairie consists largely of tall grasses. *Forbs,* which are broad-leaved herbs, are also present, but they are secondary in abundance. Trees and shrubs are almost totally absent but may occur in the same region as narrow patches of forest in valleys. The grasses are deeply rooted and form a continuous and dense sward (Figure 20.23).

Prairie grasslands are best developed in the midlatitude and subtropical zones where winter and summer seasons are well developed. The grasses flower in spring and early summer and the forbs in late summer. The tall-grass prairies are closely associated with the moist continental climate (10). They are found in the subhumid belt of that climate in which water need exceeds precipitation during the summer months. The North American tall-grass prairies are found in a broad belt extending from Illinois northwestward to southern Alberta and Saskatchewan. In the Midwest, prairie extends well into the humid subtype of the same climate. Here, areas of forest are mixed with areas of prairie in a transitional belt between forest and prairie regions. Figure 21.11 is a composite map showing world distribution of tall-grass prairie and steppe.

Another major area of tall-grass prairie is the Pampa region of South America, occupying parts of Uruguay and eastern Argentina. The Pampa region falls into the moist subtropical climate (6) with mild winters and a substantial water surplus.

Tall-grass prairie is closely identified with the Mollisols, particularly the suborder Udolls. These are soils of high base status that are rich in the nutrient bases needed by grasses.

Steppe, also called *short-grass prairie,* is a formation class consisting of short grasses occurring in sparsely distributed clumps or bunches (Figure 20.24). Scattered shrubs and low trees may also be found in the steppe. The plant cover is poor and much bare soil is exposed. Many species of grasses and other herbs occur; a typical grass of the American steppe is buffalo grass. Other typical plants are the sunflower and loco weed. Steppe grades into semidesert in dry environments and into prairie where rainfall is higher.

The world vegetation map (Figure 20.12) shows that steppe grassland is largely concentrated in vast midlatitude areas of North America and Eurasia. The only southern hemisphere occurrence shown on the map is the *veldt* of South Africa, formed on a highland surface in Orange Free State and Transvaal.

Steppe grasslands coincide quite closely with the semiarid subtype of the dry continental climate (9s). The soil-water

FIGURE 20.24 Cattle grazing on short-grass prairie in the northern Great Plains. (Charlton Photos.)

budget shows a substantial soil-water shortage and there is no water surplus. Soil-water storage is far below the soil storage capacity in all months. During a spring period of soil-water recharge, a substantial amount of water is made available to grasses and results in rapid growth into early summer. By midsummer the grasses are usually dormant, although occasional summer rainstorms cause periods of revived growth.

Soils of the steppe grasslands are largely within the order of Mollisols. Borolls are found in the colder northerly areas, and Ustolls in the warmer southerly areas.

Desert Biome

The desert biome includes several formation classes that are transitional from grassland and savanna biomes into vegetation of the more severe desert environment. We recognize two basic formation classes: semidesert and dry desert.

Semidesert is a transitional formation class found in a very wide latitude range—from the tropical zone to the midlatitude zone—and is identified with the semidesert and desert subtypes of all three dry climates.

Semidesert is a sparse xerophytic shrub vegetation. One example is the sagebrush vegetation of the middle and southern Rocky Mountain region and Colorado Plateau (Figure 20.25). Semidesert shrub

FIGURE 20.25 Sagebrush semidesert along the base of the White Cliffs in southern Utah. The reddish soil is an Aridisol. The classic automobile is an air-cooled Franklin, ca. 1934. (Donald L. Babenroth.)

vegetation seems recently to have expanded widely into areas of the western United States that were formerly steppe grasslands, as a result of overgrazing and trampling by livestock. Soils of the semidesert are largely Aridisols.

Thorntree semidesert of the tropical zone consists of xerophytic trees and shrubs adapted to a climate with a very long, hot dry season and only a very brief but intense rainy season. These conditions are found in the semidesert and desert subtypes of the dry tropical and dry subtropical climate (4sd, 4d, 5sd, 5d). The thorny trees and shrubs are locally known as thorn forest, thornbush, and thornwoods (see Figure 8.23). Many of these are deciduous plants that shed their leaves in the dry season. The shrubs may be closely intergrown to form impenetrable thickets. Cactus plants are present in some localities. A lower layer of herbs may consist of annuals, which largely disappear in the dry season, or of grasses that become dormant in the dry season. Aridisols are the dominant soil order of the thorntree semidesert, but Ustalfs are present in some areas.

Dry desert is a formation class of xerophytic plants widely dispersed and providing no important degree of ground cover. The visible vegetation of dry desert consists of small hard-leaved or spiny shrubs, succulent plants (cacti), or hard grasses. Many species of small annuals may be present, but appear only after a rare but heavy desert downpour.

Desert plants differ greatly in appearance from one part of the world to another. In the Mojave–Sonoran deserts of the southwestern United States, plants are often large and in places give a near-woodland appearance (Figure 20.26). Examples are the treelike saguaro cactus, the prickly pear cactus, the ocotillo, creosote bush, and smoke tree (Figure 20.4).

In the Sahara Desert (most of it very much drier than the American desert) a typical plant is *Stipa*, a hard grass; another, found along the dry beds of watercourses, is the tamarisk tree.

Much of the world map area assigned to desert vegetation has no plants of visible dimensions, because the surface consists of shifting dune sands or sterile salt flats. Soils of the dry desert belong largely to the order

FIGURE 20.26 A desert scene near Pheonix, Arizona. The tall, columnar plant is saguaro cactus; the delicate wandlike plant is ocotillo. Small clumps of prickly pear cactus are seen between groups of hard-leaved shrubs. (Alan H. Strahler.)

of Aridisols. There are many areas of Psamments, a suborder of the Entisols consisting of loose dune sands.

The soil-water budget of the dry desert shows a large annual total soil-water shortage. When rain falls in rare, torrential rainstorms, the soil water is used to maximum advantage by the plants and is rapidly depleted. During long periods when soil-water storage is near zero, the plants must maintain a nearly dormant state to survive.

Tundra Biome

Arctic tundra is a formation class of the tundra climate (12). Tundra of the arctic regions flourishes under a regime of long summer days during which time the ground ice melts only in a shallow surface layer. The frozen ground beneath (permafrost) remains impervious and meltwater cannot readily escape. Consequently, in summer a marshy condition prevails for at least a short time over wide areas. Humus accumulates in well-developed layers because bacterial action is very slow. Size of plants is partly limited by the mechanical rupture of roots during freeze and thaw of the surface layer of soil, producing shallow-rooted plants. In winter, drying winds and mechanical abrasion by wind-driven snow tend to reduce any portions of a plant that project above the snow.

Plants of the arctic tundra are low and mostly herbaceous, although dwarf willow occurs in places. Sedges, grasses, mosses, and lichens dominate the tundra in a low layer (Figure 9.37). Typical plant species are ridge sedge, arctic meadow grass, cotton grasses, and snow lichen. There are also

FIGURE 20.27 Alpine tundra near the summit of the Snowy Range, Wyoming. Flag-shaped spruce trees (left) represent the upper limit of tree growth. (Arthur N. Strahler.)

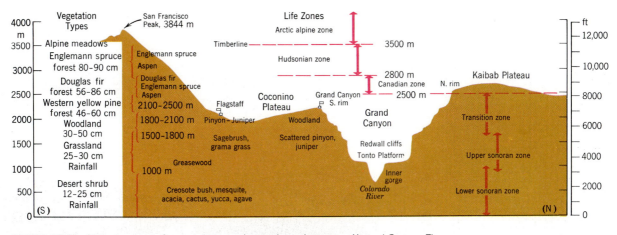

FIGURE 20.28 Altitude zone of vegetation in the arid southwestern United States. The profile shows the Grand Canyon–San Francisco Mountain district of northern Arizona. (Based on data of G. A. Pearson, C. H. Merriam, and A. N. Strahler.)

many species of forbs that flower brightly in the summer. Tundra composition varies greatly as soils range from wet to well-drained. One form of tundra consists of sturdy hummocks of plants with low, water-covered ground between. Some areas of arctic scrub vegetation composed of willows and birches are also found in tundra.

In all latitudes, where altitude is sufficiently high, *alpine tundra* is developed above the limit of tree growth and below the vegetation-free zone of barren rock and perpetual snow (Figure 20.27). Alpine tundra resembles arctic tundra in many physical respects.

Soils of the arctic tundra include representatives from three soil orders: Inceptisols, Entisols, and Histosols. Large areas of arctic soils are classified as Cryaquepts (a great soil group within the suborder of Aquepts, order Inceptisols). Cryaquepts are poorly drained mineral soils.

Altitude Zones of Vegetation

In earlier chapters we described the effects of increasing elevation on climatic factors, particularly air temperature and precipitation. Vegetation also responds to an increase in elevation, as we showed in the case of the transition of rainforest into montane forest in low latitudes. To round out this concept, we turn to the dry midlatitude climate of the southwestern United States, for here altitude zonation of vegetation is particularly striking.

Figure 20.28 shows the vegetation zones of the Colorado Plateau region of northern Arizona and adjacent states. Zone names, elevations, dominant forest trees, and annual precipitation data are given in the figure. Ecologists have set up a series of life zones, whose names suggest the similarities of these zones with latitude zones encountered in poleward travel on a meridian. The Hudsonian zone, 2900 to 3500 m (9500 to 11,500 ft), bears a needleleaf forest quite similar to needleleaf boreal forest of the subarctic zone. Here Spodosols lie beneath the forest. A zone of alpine tundra lies above tree line. The snow line is encountered at about 2750 to 3000 m (9000 to 10,000 ft) in midlatitudes, which is of course much lower than at the equator.

Global Vegetation in Review

The leading theme of this chapter has been that the structure of the natural plant cover is a response to environments dictated by climate. Each of the life-forms favored by green plants has a capability for survival and a limit to survival as well. Forests exist where the environment is most favorable to plant growth because of the abundance of heat and of soil water during a long growth season. Where soil water is in short supply but heat remains adequate, forest gives way to the savanna structure in which shrubs

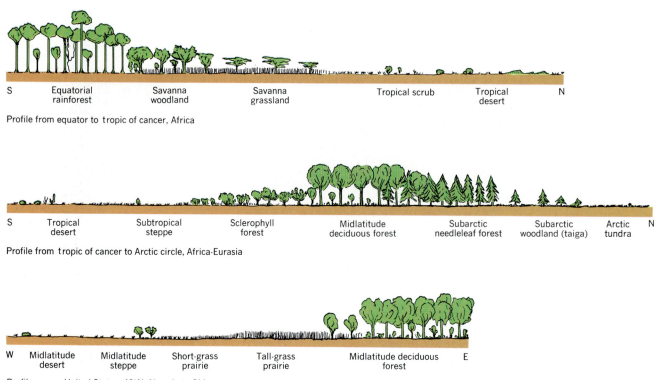

S   Equatorial   Savanna   Savanna   Tropical scrub   Tropical   N
     rainforest   woodland   grassland   desert

Profile from equator to tropic of cancer, Africa

S   Tropical   Subtropical   Sclerophyll   Midlatitude   Subarctic   Subarctic   Arctic   N
     desert   steppe   forest   deciduous forest   needleleaf forest woodland (taiga) tundra

Profile from tropic of cancer to Arctic circle, Africa-Eurasia

W   Midlatitude   Midlatitude   Short-grass   Tall-grass   Midlatitude deciduous   E
     desert   steppe   prairie   prairie   forest

Profile across United States, 40°N, Nevada to Ohio

FIGURE 20.29 Three schematic profiles showing the succession of plant formation classes across climatic gradients.

and herbs have the upper hand over trees (Figure 20.29). The savanna woodland grades into grassland and this, in turn, grades into desert, where only those plants capable of living through long drought periods can survive. Traced into colder, higher latitudes, where heat supply becomes reduced below the optimum level, the forest gives way to arctic tundra. Here plants contend with the effects of prolonged freezing of soil water and have only a short cold annual period of growth.

We have found a significant correspondence of vegetation structure with soil type because soil-forming processes are also strongly dominated by climate. But organic processes also shape soil character, so that the close associations of vegetation formation classes and soil types are partly a result of interaction.

With this chapter we conclude the systematic, structured study of the separate ingredients of physical geography: climate, soil, vegetation, and landform. In the remaining chapter we attempt a synthesis of this information through the recognition of several important global environmental regions.

CHAPTER

GLOBAL ENVIRONMENTAL REGIONS—A SYNTHESIS

ENVIRONMENTAL REGIONS   They can be recognized through distinctive combinations of climate, soils, vegetation, and landforms, but geologic factors introduce another dimension of landscape variation from place to place.

THE RAINFOREST ENVIRONMENT   Despite its rich forest of great trees and an abundance of water, the rainforest offers the disadvantage of poor soils to humans who would transform it into an agricultural landscape.

THE SAVANNA ENVIRONMENT   It supports an enormous human population in severely limited expanses of fertile alluvial soils; elsewhere, a combination of poor soils and frequent droughts poses a formidable problem to the human occupants.

THE TROPICAL DESERT ENVIRONMENT   It is sparsely inhabited, except where irrigation is available, but irrigation itself brings a new set of difficulties to plague the pursuit of agriculture.

MIDLATITUDE ENVIRONMENTS   They have been greatly transformed by agricultural and industrial activity in North America and Eurasia.

BOREAL FOREST AND ARCTIC TUNDRA   In the high-latitude environments, severely cold winters and a short growing season limit or rule out agricultural production.

PHYSICAL geography contributes useful knowledge through the recognition of a global system of *environmental regions*. Each region presents its own unique set of physical environmental qualities and supports a distinctive ecosystem. Physical and biological factors interact to produce the total set of environmental qualities that make up each region.

Distinctive environmental regions existed for eons of geologic time before the human species (*Homo sapiens*) came on the scene. Until late in the Pleistocene Epoch, the sparse populations of genus *Homo* functioned environmentally in a manner little different from other higher mammals. Throughout Holocene time, human modification of natural ecosystems increased in intensity and total impact. Through use of fire, grazing animals, and land tillage, vegetation of large areas became radically changed and geomorphic processes were significantly altered in intensity. In this respect, some ecosystems fared quite differently from others. For example, the Mediterranean lands, under a climate precariously balanced between dry and moist regimes, suffered widespread destruction of the forest and woodland cover, along with devastating soil erosion and valley sedimentation. In central and western Europe, the deciduous forest biome was largely destroyed through agricultural activities and little remains today to show us what that landscape might have looked like without human intervention. On the other hand, the equatorial rainforest biome remained, up to very recent decades, in nearly pristine condition over large areas. Until machinery, powered with petroleum energy, could be used to mount a massive attack on the great trees of the rainforest, Amazonia held out as a vast natural ecologic fortress. Deserts also resisted human occupation on a vast scale, except in floodplains close to exotic rivers.

Thus, each environmental region has shown its own set of opportunities for and restraints on expanded utilization. Especially crucial are the opportunities that humans have to modify further the natural terrestrial ecosystems into food-producing ecosystems. We have already made the point that this transformation has been remarkably successful in some parts of the world, but highly unsuccessful in others. Can the production of food be further extended into regions now clothed in equatorial rainforests or savanna grasslands? Can more deserts be made to bloom by massive applications of irrigation water? Partial answers to these questions lie in the environmental factors relating to physical geography. Other parts of the answers lie in the application of technology and the availability of supplies of energy. Ethical issues and political policies are also factors to be weighed in seeking these answers.

Environmental Regions

An environmental region is closely identified in location and extent in terms of climate types and subtypes, following our system of climate classification based on the soil-water balance (Chapter 7). An environmental region may also be closely identified with a biome, or with one or more vegetation formation classes. We have already established the role of climate as the dominant driving factor in shaping the characteristics of vegetation. Climate has an important role in governing soil characteristics and the intensity of various solar-powered geomorphic processes. We are thus attempting a synthesis of physical and biological factors and forms within a climatic framework. The product of this synthesis serves both as a review of physical geography and as a point of departure to evaluate questions of planning for expanded exploitation and management of natural resources.

Although environmental regions can be climatically defined, they are most effectively treated by combining and rearranging the 13 climate types and their subtypes to maximize correspondence of the environmental regions with the major biomes and vegetation formation classes. Table 21.1 lists 12 environmental regions, designated for convenience by capital letters A through L. We have, in some instances, modified the names of environmental regions to combine climate and vegetation names. The table lists climate types and subtypes, biomes and vegetation formation classes, and soil orders and suborders associated with each region.

We introduce each environmental region with a brief summary of climate, vegetation, and soils. You may wish to review the pertinent background information given in detail in earlier chapters. We will comment on the relative intensity of various geomorphic processes within each environmental region. Other topics of interest, described in earlier chapters, are plant resources and agriculture and the special problems associated with further development of agricultural and water resources.

TABLE 21.1 Environmental regions

	Name of Region	Climate	Biome	Formation Class	Soil Order
LOW LATITUDE	A Rainforest	(1) Wet equatorial (2) Monsoon and tradewind littoral	Forest	Equatorial and tropical rainforest Montane forest	Oxisols Ultisols
	B Savanna	(3) Wet–dry tropical (4s) Semiarid subtype of dry tropical	Savanna	Monsoon forest Savanna woodland Thorntree–tall-grass savanna	Oxisols Ultisols Ustalfs Vertisols
	C Tropical desert	(4sd, 4d, 4dw) Semidesert, desert, and western littoral subtypes of dry tropical (5sd, 5d) Semidesert and desert subtypes of dry subtropical (7sd) Semidesert subtype of Mediterranean	Desert	Thorn forest and thorn woodland Thorntree–desert grass savanna Semidesert scrub and woodland Desert	Aridisols Psamments
MIDLATITUDE	D Moist subtropical forest	(6h, 6p) Humid and perhumid subtypes of moist subtropical	Forest	Broadleaf evergreen forest Southern pine forest Midlatitude deciduous forest	Ultisols
	E Mediterranean	(7s, 7sh, 7h) Semiarid, subhumid, and subhumid subtypes of Mediterranean	Forest	Mediterranean mixed evergreen forest Sclerophyllous scrub Australian sclerophyll forest	Xeralfs Xerolls Xerults
	F Marine west-coast forest	(8h, 8p) Humid and perhumid subtypes of marine west-coast	Forest	Coastal forest Midlatitude deciduous forest Broadleaf evergreen forest	Spodosols Alfisols Entisols Inceptisols
	G Moist continental forest	(10h, 10p) Humid and perhumid subtypes of moist continental	Forest	Midlatitude deciduous forest Mixed boreal and deciduous forest Lake forest	Udalfs Spodosols
	H Prairie and steppe grassland	(10sh, 10h) Subhumid and humid subtypes of moist continental (9s, 5s) Steppe subtype of dry midlatitude and dry subtropical (6h) Humid subtype of moist subtropical (South America)	Grassland	Tall-grass prairie Short-grass prairie (steppe)	Mollisols Aridisols
	I Midlatitude desert	(9sd, 9d) Semidesert and desert subtypes of dry midlatitude	Desert	Semidesert scrub and woodland Desert	Aridisols
HIGH LATITUDE	J Boreal forest	(11) Boreal forest	Forest	Boreal forest Arctic woodland	Spodosols Histosols Boralfs
	K Tundra	(12) Tundra	Tundra	Arctic tundra	Cryaquents Cryorthents Histosols
	L Ice sheet	(13) Ice sheet			

FIGURE 21.1 World map of equatorial and tropical rainforest and montane forest. Stations for which climographs and soil-water budgets are given as examples in Chapters 8 and 9 are shown by labeled dots. (Data source same as in figure 20.12. Based on Goode Base Map.)

I THE LOW-LATITUDE ENVIRONMENTAL REGIONS

The low-latitude environmental regions are identified with climates of Group I, occupying the equatorial belt of doldrums and intertropical convergence zone (ITC), the belt of tropical easterlies (northeast and southeast trades), and large portions of the oceanic subtropical highs. Three low-latitude environmental regions are recognized: (A) rainforest, (B) savanna, and (C) tropical desert. These cover the climatic range from extremely dry to extremely moist; they also span the range from a climate of extreme seasonality of precipitation to one with heavy precipitation throughout the year. Environmental characteristics are correspondingly different to an extreme degree in terms of natural vegetation, soils, and landforms.

Rainforest Environment (A)

Climate, Soils, and Vegetation

The rainforest environment is represented by world areas of low-latitude rainforests, which are closely identified with the wet equatorial climate (1) and the monsoon and trade-wind littoral climate (2). Figure 21.1 shows these rainforest areas and associated montane forest of highlands. The soil-water regimes of these climates show very large annual total rainfall, very large annual water surplus, and high soil-water storage throughout all or most months of the year.

In the rainforest environment, the large water surplus and prevailing warm soil temperatures promote decay and decomposition of rock to great depths, so

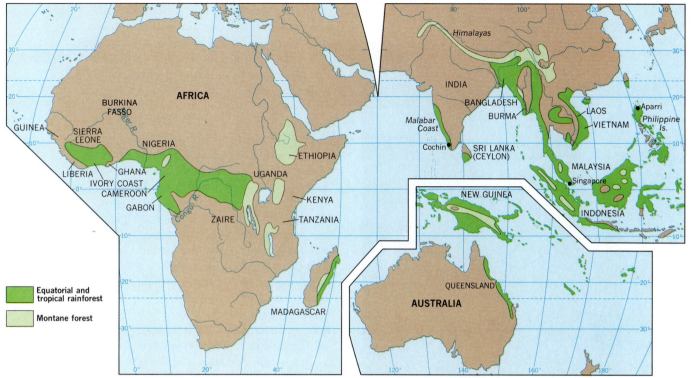

Equatorial and tropical rainforest

Montane forest

that a thick regolith is present over large areas. Mineral alteration has reached an advanced stage, in which clays capable of holding nutrient base cations have almost entirely disappeared from the soil and regolith. Silica has also been depleted. There remains a heavy concentration of sesquioxides of aluminum and iron, and the soils have low base status. Oxisols are thus the dominant soil order of the rainforest environment, along with Ultisols of somewhat less advanced state of oxide accumulation.

Because of the warm and extremely uniform temperature and high levels of soil-water storage, broadleaf rainforest is the dominant formation class of natural vegetation. Net primary production of the equatorial and tropical rainforest is extremely high, exceeding that of any other terrestrial biome and comparing favorably with the most productive of the aquatic ecosystems (Chapter 20).

Geomorphic Processes

Under the rainforest environment, chemical weathering is dominant in intensity over physical weathering. Note that this is a frost-free climate, so that mechanical weathering by frost action is totally absent. A thick cover of vegetation, soil, and regolith protects bedrock from direct insolation and from extremes of daily temperature change.

Of the total mineral load carried by streams, by far the greater portion is in the

FIGURE 21.2 Called the Needle, this shaft of basaltic lava and the cliffs around it are wasted by carbonic and organic acids under a warm, moist climate on the windward side of the island of Maui, Hawaii. A plant cover clings to even the steepest rock walls. (Arthur N. Strahler.)

form of dissolved solids. Little coarse sediment moves as bed load, because the cover of vegetation holds back all but the finest soil particles from transport by overland flow. Small streams of steep gradient have bouldery channels but the boulders experience little downstream movement. Mafic rock, particularly basaltic lava, undergoes rapid removal under attack by soil acids and produces landforms quite similar to those formed by carbonic acid on massive limestones in moist climates of higher latitudes. The effects of solution removal of basaltic lava are displayed in spectacular grooves, fins, and spires on the walls of deep alcoves in parts of the Hawaiian Islands (Figure 21.2).

Mass wasting, involving rapid sliding and flowage of moist soil and regolith, is

especially conspicuous as a denudation process where the terrain is mountainous and slopes are steep. Slides and avalanches of soil are common and can strip away the forest and soil to expose regolith and bedrock. Such conditions are characteristic of those parts of the rainforest that happen to lie in the recent alpine and volcanic chains where subduction is active on plate boundaries. Rugged, mountainous terrain is thus associated with rainforest in Central America, the West Indies, the Andes range of South America, Indonesia, and New Guinea. On stable shield areas, slopes are low or moderate over vast areas; examples are the Amazon–Orinoco lowland of South America and the Congo Basin of Africa. Gradients of major rivers crossing these lowlands are extremely low. For example, at Iquitos, Peru, the elevation of the Amazon River is only 116 m (280 ft) above sea level, yet the airline distance from Iquitos to the mouth of the Amazon is almost 3000 km (1850 mi).

Savanna Environment (B)

Climate, Vegetation, and Soils

The savanna environment is represented by world areas of tropical raingreen vegetation, which are closely identified with the wet–dry tropical climate (3) and adjacent belts of the semiarid and semidesert subtypes of the dry tropical climate (4s, 4sd) (Figure 21.3). Savanna areas of major importance are found in South America, Africa, India, Indochina, and northern Australia. Here the savanna environment typically occupies a position between the rainforest on the equatorward side and a great belt of tropical desert on the poleward side.

The outstanding feature of the soil-water regime of the savanna environment is the alternation of a very wet season, in the high-sun period, with a very dry season, in the low-sun period. The temperature regime of the savanna environment is characterized by a very hot period immediately preceding the onset of the rainy season, and by a comparatively cool season at time of low sun. Alternation of a wet-season water surplus with a dry-season soil-water deficit is the unique mark of the savanna environment in its moister equatorward portion, where monsoon forest and savanna woodland are the typical plant formation classes. The semiarid portion, across which there is a very rapid transition to semidesert, has too short a wet season to generate a water surplus. Here the vegetation consists of thorntree–tall-grass savanna.

The natural vegetation of the savanna environment is classified in general terms as raingreen vegetation. The broad-leaved savanna trees shed their leaves in the cool, dry season, and the grasses turn to dry straw. This long period of dormancy terminates almost abruptly with the first torrential rains of the wet season. Quickly the grasses become a green carpet and the trees put forth new leaves.

Soils of the savanna region belong to several soil orders. Large upland areas are underlain by very old Oxisols and Ultisols. Vertisols, soils of high base status, are important where parent materials favor their development (see western India). Ustalfs, the dry regime suborder of the Alfisols, are found over large areas of savanna in West Africa, eastern Africa, India, Indochina, northeastern Brazil, and northern Australia. In contrast to the Oxisols and Ultisols, the Ustalfs are soils of high base status, basically capable of high agricultural yields. Also important are Aquepts (suborder of Inceptisols) found over agriculturally productive river floodplains and deltaic plains, for example, plains of the Ganges and Mekong river systems in Southeast Asia.

Geomorphic Processes and Landforms

Geomorphic processes of the savanna environment are closely tied to pedogenic processes—not only processes acting under present climates but also processes that may have acted during long periods in the late Cenozoic Era in which a moister climate may have prevailed. Large areas of savanna landscape lie on stable shield areas of the former continent of Gondwana (Chapter 12). Bedrock of these shield areas is extremely complex in rock composition and structure. Exposures of plutonic igneous masses (batholiths) are numerous, and metamorphic belts of great complexity are also abundant. Some parts of the shields bear sedimentary covers; elsewhere there

occur expanses of ancient flood basalts. Thus the landforms of the savanna regions are to a large degree structurally controlled, and this makes it difficult for us to recognize distinctive landforms governed solely by processes unique to the wet–dry climate.

The nature and intensity of weathering processes acting under the savanna soil-water regime are poorly understood. Chemical weathering during the wet season is considered by some investigators to be a very important process affecting igneous and metamorphic rocks rich in aluminosilicate minerals. The intensity of erosion by running water is also considered to be extremely high, because of the rather poor protection afforded to the soil surface by savanna vegetation, particularly where it has been burned over prior to the rains and heavily grazed and trampled by large herds of wild herbivores or cattle. Sediment loads of streams—both suspended load and bed load—are very high during periods of peak flow. In contrast, stream discharges fall to low values or to zero during the long dry season, when broad braided channels of sand and silt are exposed to view.

Two landscape features of the savanna region deserve notice: (1) prominent, isolated rock knobs and (2) benchlike upland surfaces. The rock knobs are of a special type, called bornhardts, found throughout the savanna region of East Africa and South Africa (Figure 21.4). Typically, a bornhardt is a steep-sided knob of granite or similar plutonic rock, with rounded summit and often showing exfoliation shells (Figure 19.18). Surrounding the bornhardt is a more-or-less level land surface or plain, underlain by regolith. Some geomorphologists consider this type of plain to be a pediplain, formed by processes similar to pediment-producing processes acting in deserts (Chapter 15). Figure 19.18 is a schematic cross section showing one interpretation of the origin of the rock knob. Deep chemical weathering has taken place beneath thick regolith surrounding the knob, where the bedrock is closely jointed or consists of more susceptible rock. Related to bornhardts are *tors*, small hills consisting of rounded joint blocks. Heaps of rounded boulders rising above a plain are a common landform of

FIGURE 21.3 World map of tropical raingreen vegetation. (Data same as in Figure 20.12. Based on Goode Base Map.)

the savanna landscape of the Indian peninsula.

For us to imply that bornhardts and tors are unique to the savanna environment, as some authors have done, would be inaccurate. Similar landforms can be found in a wide range of climates including the

desert and the moist midlatitude climates. For example, the rock tors of the Dartmoor region of southwestern England bear a close resemblance to the tors of Africa.

Two, three, or more laterite crusts may be formed in succession at lower levels, as denudation proceeds. The benched

FIGURE 21.4 The rounded rock hills at left and right are bornhardts. This savanna woodland scene is in the Transvaal province of the Union of South Africa. (G. Douglas, courtesy of SATOUR.)

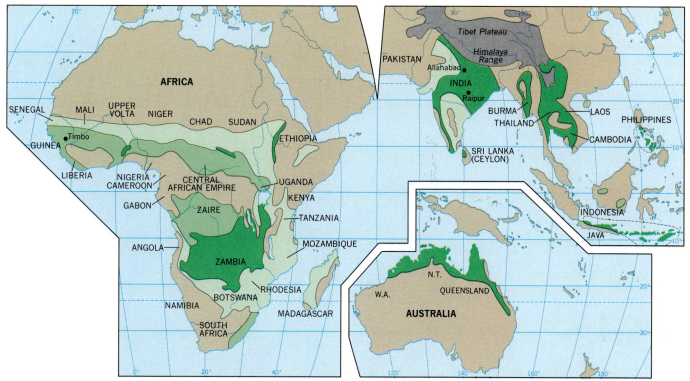

landscape resembles in some ways the landscape of plateaus, cliffs, and mesas developed in flat-lying sedimentary strata of covered shields (Chapter 16). Elsewhere in tropical environments, crusts consist of weathered materials of soil or regolith cemented by silica; this material is called *silcrete* to distinguish it from ferricrete.

A general hypothesis for the development of crusts of ferricrete and silcrete calls for their origin as soil horizons. These horizons formed during long periods of moister climate representing a former geographical expansion of the wet monsoon climate into what is today the region of wet–dry tropical climate. Exposure of the soil horizon and its conversion into a laterite crust accompanied a change to the existing climate. This change occurred as the belt of wet equatorial climate contracted and withdrew into lower latitudes.

Climate, Vegetation, and Soils

The tropical deserts comprise a global environmental region sustained by subsiding air masses of the continental high-pressure cells. These cells dominate much of the earth's land area in a latitude belt between 15° and 30° N and S, roughly centered on the tropics of cancer and capricorn. The vast desert belt extending across North Africa (Sahara Desert), Arabia, and Iran to Pakistan (Thar Desert) best exemplifies the tropical desert (Figure 21.5). Other important continental deserts of this belt are the Sonoran Desert of the southwestern United States and northern Mexico, the interior Kalahari Desert of South Africa, and the Australian Desert.

Also included in this environmental region are the western littoral deserts of

FIGURE 21.5 World map of the desert biome, including desert and semidesert formation classes. (Data source same as in Figure 20.12. Based on Goode Base Map.)

tropical and equatorial latitudes. The Atacama Desert of Chile and the Namib Desert of coastal southwest Africa are perhaps the most celebrated of these cool, foggy deserts, but they also exist in Lower (Baja) California, the Atlantic coast of North Africa, and the west coast of Australia.

The global extent of the tropical desert environment is represented by world areas of desert and semidesert vegetation, closely associated with the semidesert and desert subtypes of the dry tropical and subtropical climates (4sd, 4d, 5sd, 5d), the western

littoral subtype of the dry tropical climate (4wd), and the semidesert subtype of the Mediterranean climate (7sd). The soil-water balance of these dry climates is characterized by a very large annual soil-water deficiency, absence of any water surplus, and very low amounts of soil-water storage throughout the entire year. Annual potential evapotranspiration (water need) is very large, particularly in the dry tropical climate. High-sun temperatures are extremely high (Figure 8.17). There is a strong annual temperature cycle in the continental interiors so that the subtropical

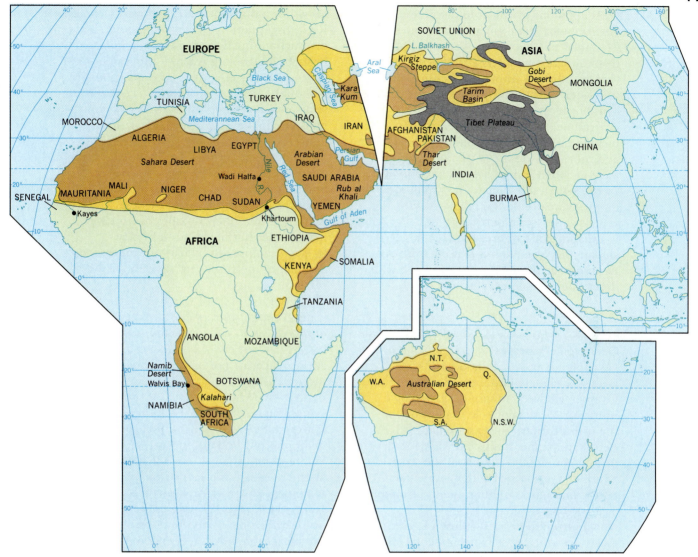

desert experiences a pronounced cool season at time of low sun. The western littoral deserts are distinctive because of the comparative coolness of the air temperatures throughout the year and the very small annual temperature range. Figure 21.6 shows the monthly mean temperatures for a desert station on the Atlantic coast of Africa compared with a station in the heart of the Sahara Desert. Both are at the same latitude. Both have about the same minimum monthly temperature, but the summer maximum is much higher in the interior desert.

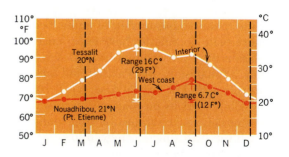

FIGURE 21.6 These two temperature graphs for stations in the Republic of Mauritania, North Africa, show the low annual range of the west-coast desert (Port Etienne) in contrast with the large annual range within the interior desert (Tessalit).

Adaptations of plants to the desert environment (xeric environment) have been explained in Chapter 20. Vegetation of the tropical desert is closely correlated with climate subtypes through differences in the soil-water storage. In the semidesert climate belt, bordering the tropical savanna environment, plant formation classes include thorn forest and thorn woodland and thorntree–desert savanna. These we have included in the savanna environment. In the semidesert climate belt transitional to moister climates on the poleward side, the vegetation is semidesert scrub and woodland. For vast areas of true desert climate the vegetation is classed as desert. Aridisols are the soil order associated with desert and semidesert climates.

Geomorphic Processes

Geomorphic processes associated with aridity of climate have been described in Chapter 15. Keep in mind that geologic history plays a major role in determining the overall appearance and relief of the desert landscape. Basin and Range structure of the mountainous desert of the American Southwest, associated with continental rifting and block faulting, is almost unique as a geologic type. The largest of the world's tropical deserts—those of Africa and Australia—occupy stable continental shields that are the fragments of the ancient continent of Gondwana. Many kinds of igneous and metamorphic rocks and structures are exposed in these shields, whereas some large areas have covers of relatively undisturbed sedimentary strata (see Figure 12.6). Other deserts are found in the mountainous terrain of alpine belts, where plate boundaries have experienced collision in recent geologic time. Deserts of Iran, Afghanistan, and Pakistan fall into this category. The coastal desert of Peru and Chile owes its position and narrow width to an active subduction zone along which the Nazca plate is descending beneath the American plate.

A common denominator of desert landscape evolution exists through a unique set of weathering processes, fluvial processes, and the work of wind. Mechanical weathering by salt-crystal growth is a dominant and pervasive process. Rock disruption by thermal expansion and contraction during the diurnal cycle of heating and cooling is also a universal process, but its effectiveness is questionable. Although chemical weathering is considered relatively less important in deserts than in moister climates, evidence is clear that hydrolysis and oxidation affect the silicate minerals wherever they are exposed throughout the desert environment. Surfaces of pebbles, boulders, and rock outcrops become darkened by desert varnish, a thin film of oxides of iron and manganese. In the absence of below-freezing temperatures, frost action can be considered ineffective in the warmer desert regions.

Evaporation of soil water and ground water, brought toward the surface of soil and bedrock by capillary film movement, results in a variety of rocklike crusts. The surface layer of porous sandstone becomes indurated into a hard crust by carbonate deposition. Where the crust is broken away, crumbling of the softer rock inside through salt-crystal growth leads to formation of deep pits and hollows and these may be enlarged into shallow caves (Chapter 14). Within Aridisols developed on alluvial materials, a petrocalcic horizon is a common feature (Chapter 19). Where the overlying soil horizons have been removed by erosion, the petrocalcic horizon is exposed as a hard rocklike crust. Geographers refer to this layer as calcrete (also *caliche*) to distinguish it from ferricrete and silcrete crusts found in the savanna environment. As calcrete crusts are eroded, they give rise to stepped landforms resembling the cliffs, mesas, and buttes of eroded sedimentary strata.

Poorly drained, shallow basins (playas) accumulate highly soluble salts. These form sterile, white salt flats. Saline lakes and salt flats (playas) are described in Chapter 10.

Fluvial processes are, as we have seen in Chapter 15, of major importance in shaping desert landforms, even though production of runoff in large volumes is sporadic. Large volumes of alluvium accumulate wherever topographic basins exist.

Landforms of the Sahara Desert illustrate some of the features to be expected in tropical deserts occupying stable shield areas. Structural features are brought out into sharp relief. These may be mesas and buttes, hogbacks, volcanic necks, or granite domes, depending on the composition of the rock and its structure. Pediments are developed locally at the bases of these residual relief features. Broad, shallow stream channels conduct alluvium and

dissolved salts to shallow basins of accumulation. Here crusts of gypsum and other salts are formed through evaporation.

Dune forms of the Sahara Desert are described in Chapter 17, desert oases in Chapter 8, and desert irrigation in Chapter 10.

II THE MIDLATITUDE ENVIRONMENTAL REGIONS

The midlatitude environmental regions are identified with climates of Group II, occupying the polar front zone in which both tropical and polar air masses play an important climatic role. Strong climatic seasons are a characteristic of the midlatitude climates—seasons in which temperatures as well as precipitation show strong annual cycles.

Those who have not studied physical geography refer to the midlatitude region of the globe as the "temperate zone." Nothing could be more misleading. "Temperate" means a regime marked by moderation—not extreme or excessive. But intemperance is the rule of most of the midlatitude zone. Not only do seasons swing from one extreme to the other in heat and cold or in dryness and wetness, but the day-to-day swings in weather as cyclones pass by provide a pattern of intemperance often more dramatic than that of the seasons. Intemperance in the midlatitude zone is partly due to the occurrence of vast continents in that zone in the northern hemisphere. Continentality is synonymous with intemperance of climate. Only along the west coasts do we find narrow strips of land where a truly temperate climate prevails.

Moist Subtropical Forest Environment (D)

Climate, Vegetation, and Soils
The moist subtropical forest environment is closely identified with areas of humid and perhumid subtypes of the moist subtropical climate (6h, 6p) in which the forest biome prevails (Figure 21.7). The major areas are in the northern hemisphere: southeastern

United States, southern China, northern Taiwan, and southern portions of Japan (Kyushu and southern Honshu). A narrow coastal strip of southeastern Australia qualifies for inclusion, but in some respects is not typical. Representative natural vegetation is broadleaf evergreen forest (laurel forest). The typical soils are Ultisols (particularly the Udults, which are a moist-climate suborder of the Ultisols).

The moist subtropical climate, by its very name, borders on the tropical zone and is transitional between the low-latitude and midlatitude environments. The climate has a mild winter season, but toward its low-latitude borders the winter season weakens greatly in intensity. Here it is not easy to decide where the change to the trade-wind littoral climate sets in. This transitional character of climate is also reflected in soils and vegetation. In North America, Ultisols of the moist subtropical climate are transitional between Spodosols of the cold climates and Oxisols of the warm, wet climates. In North America and Southeast Asia, broadleaf evergreen forest gives way in the colder northern part to midlatitude deciduous forest. While being aware of the transitional nature of the environment, we can try to make some broad generalizations about the more typical parts of the moist subtropical region.

The moist subtropical climate is found on the eastern sides of landmasses at lat. 25° to 30° N and S. The adjacent oceanic high-pressure cells pump moist, unstable air into the eastern sides of these continents. This effect is strongest in the summer, when the oceanic highs are strong and persistent. The maritime tropical air mass involved in this motion is unstable and produces ample summer rainfall. In Southeast Asia, under

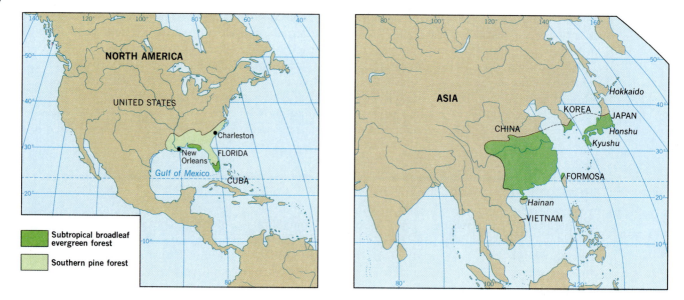

FIGURE 21.7 World map of the northern hemisphere subtropical evergreen forests. (Same data source as in Figure 20.12. Based on Goode Base Map.)

monsoon influence, rainfall peaks sharply in the summer. Tropical cyclones arrive on these shores after their long westward paths through the trades. As these storms turn inland, they bring very heavy rainfall and flooding conditions.

The soil-water regime of the moist subtropical environment shows a large annual water surplus, whereas there is only a very small soil-water shortage, or none at all. The high levels of soil-water storage through the long warm summer favor intensive plant growth and support the forest biome.

Geomorphic Processes and Landforms

Because the moist subtropical forest environment is transitional from the wet low-latitude environment, many of the statements made earlier in this chapter about geomorphic processes in the latter region apply here as well. Intensity of chemical weathering is high and produces deep residual regolith where the bedrock is rich in silicate minerals. Frost action as a form of mechanical weathering should be included as a process in the colder, northern limits of the region. Mass wasting in the form of flowage and sliding of moist soil and regolith is conspicuous. Dominance of suspended load over coarse bed load is typical of the fluvial process.

Landform evolution in the moist subtropical environment conforms closely to the idealized model presented in Chapter 15 for the uplift, rejuvenation, and dissection of a peneplain in a humid climate. The larger graded rivers occupy forested, marshy floodplains and show sinuous meanders, with development of natural levees and bar-and-swale landforms. Base flow is an important contributor to the total stream flow, so that most larger streams are perennial.

Geologic controls over landform development are highly varied from place to place in this environmental region. In the southeastern United States, a stable shield region with a continental shelf margin, a low, flat coastal plain contrasts with subdued hill and mountain terrain developed on Appalachian mountain roots. In contrast, the mountainous islands of Japan show very steep slopes, typical of a mountain arc adjacent to deep trenches over an active subduction zone.

Mediterranean Environment (E)

Climate, Vegetation, and Soils

The Mediterranean environment is identified with those global areas of sclerophyll vegetation associated with the

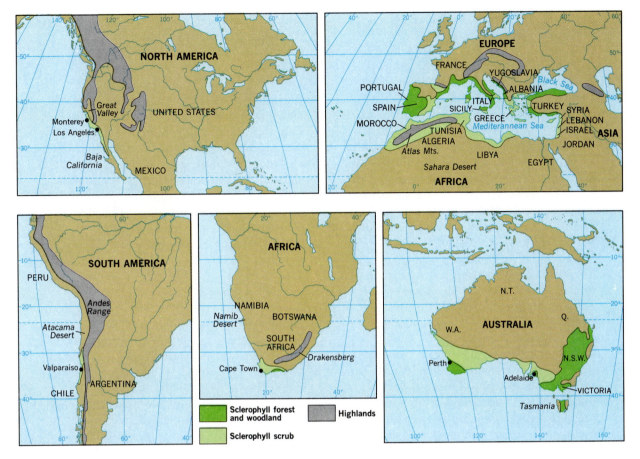

FIGURE 21.8 World map of sclerophyll vegetation associated with the Mediterranean environment. (Same data source as in Figure 20.12. Based on Goode Base Map.)

semiarid, subhumid, and humid subtypes of the Mediterranean climate (7s, 7sh, 7h) (Figure 21.8). This dry-summer wet-winter climate is named for its occurrence in lands bordering the Mediterranean Sea. A second important occurrence is in central and southern California, extending from the Pacific shores inland into the Great Valley and other valleys within the coast ranges. In the southern hemisphere, the Mediterranean climate occurs in coastal Chile, the Cape Town area of South Africa, and coastal zones of western and southern Australia.

The Mediterranean climate spans a wide range from moist to dry. Whereas the soil-water shortage developed in summer is always substantial, the water surplus may be zero in the semiarid subtype or as great as 30 cm in the humid subtype. This spread in water surplus results in a corresponding variety in sclerophyll plant formation

classes, ranging from sclerophyll forest to woodland and scrub. In Australia, the sclerophyll forest consists largely of eucalyptus trees. The thermal regime is also quite different in the narrow western littorals—where summers are cool and the annual temperature range small—than in interior valleys—where summers are hot and the annual range is substantial.

Soils of the Mediterranean climate are assigned to the suborders of the Mediterranean soil-water regime, with the prefix "xeric." Most are Xeralfs, Xerolls, and Xerults. Of these the Xeralfs (Alfisols) and Xerolls (Mollisols) are of high base status and are fertile soils when soil water is adequate.

Geomorphic Processes

In the Mediterranean environment, mechanical weathering by salt-crystal growth is probably important through the dry summer. Chemical weathering is particularly active in the rainy winter, but because this is the time of cool temperatures, mineral alteration is probably less rapid than in moist-summer climates. Chemical weathering has caused widespread occurrences of thick residual regolith over bodies of igneous and metamorphic rock rich in silicate minerals. Mass wasting in the form of winter and spring earthflows and landslides occurs in such materials and in clay-rich bedrock where slopes are steep. These forms of mass wasting are a serious hazard in California, where urbanization has spread over steep slopes.

Because of the sparseness of plant cover in woodland and scrub plant formations, fluvial processes of the Mediterranean environment are closely akin to those of semiarid and semidesert environments. When torrential rains occur in winter, large quantities of debris are swept downslope by overland flow and carried long distances by streams in flood. Mudflows and debris floods are produced at such times. Alluvial fans are a prominent landform of the Mediterranean environment in California. Fan surfaces and canyon floors are high-hazard areas, where destruction of houses and roadways by stream floods, debris floods, and mudflows are an ever-present threat. The hazards of flooding are greatly increased when fire has consumed the plant cover of a mountain watershed. Chaparral of southern California hillsides is extremely flammable during the long dry summer.

As explained in Chapter 15, accelerated soil erosion, induced by human activity over the past 2000 years or longer, has been a dominant geomorphic process throughout the Mediterranean lands. Many hillsides have been denuded of residual soils and present a rocky aspect today. Correspondingly, thick sediment deposits have built up in valley floors and, in places, aggradation has reached depths of several meters.

Marine West-Coast Forest Environment (F)

Climate, Vegetation, and Soils

The marine west-coast forest environment is represented by forested areas of the humid and perhumid subtypes of the marine west-coast climate (8h, 8p) (Figure 21.9). This climate is found along the western coast of North America from Oregon to British Columbia, in the British Isles, western Europe, Victoria, Tasmania, New Zealand, and southern Chile. The climate is quite mild in winter temperatures for the comparatively high latitude in which it lies (35° to 60°). Ample precipitation occurs in all months. The soil-water balance usually shows a moderate to large annual water surplus and there is only a small soil-water shortage in summer, or no shortage.

Precipitation is strongly affected by topography. Whereas lowlands of Europe receive as little as 75 cm (30 in.) annually, coastal mountains, such as those of the northern Pacific coast, receive totals over 200 cm (80 in.). These perhumid mountainous areas support dense needleleaf forests (coastal forest); the typical soils are Entisols and Inceptisols related to the Spodosols. Under the lower precipitation regime of the British Isles and western Europe, midlatitude deciduous forest is the prevailing formation class, and Alfisols (suborder Udalfs) are the dominant soil class. In the southern hemisphere broadleaf evergreen forest is a dominant formation class in Tasmania and New Zealand, whereas mixed boreal and deciduous forest is found in the mountainous coastal belt of southern Chile.

Geomorphic Processes

Weathering processes of the marine west-coast forest environment include limited frost action as a mechanical process acting in winter in the colder and higher areas. Although lower in total intensity than in warmer climates, chemical weathering is a dominant process. Because of the recency of Pleistocene glaciation, however, bedrock subjected to ice abrasion shows little decomposition, while the parent materials of the soil have had insufficient time to undergo extensive alteration. Many landforms of glacial erosion and deposition are as yet little changed from their original configurations. Glacial troughs and fiords are striking landforms of the rugged coasts of British Columbia, Scotland, and Norway. Extensive lowlands of northwestern Europe consist of till plains, moraines, and outwash

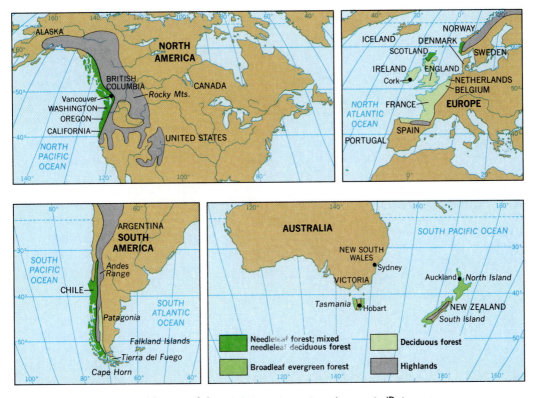

FIGURE 21.9 World map of forests of the marine west-coast environment. (Data source same as in Figure 20.12. Based on Goode Base Map.)

plains left by Pleistocene ice sheets. Where forest cover is largely undisturbed, sediment yield is low and denudation rates of mountain watersheds are probably very low. Locally, however, where clear-cutting of forest is practiced, severe soil erosion and high sediment yields occur.

Moist Continental Forest Environment (G)

Climate, Vegetation, and Soils

The moist continental forest environment is defined by forested areas of the humid and perhumid subtypes of the moist continental climate (10h, 10p) (Figure 21.10). Also included are some areas of the subhumid subtype (10sh), where the forest biome prevails. Limited to the northern hemisphere landmasses, this environmental region occupies a large part of northeastern United States and southeastern Canada in the latitude range 35° to 48° N. In Europe, this environmental region extends from western Europe across central Europe and Russia at least as far as the Urals. In eastern Asia a large area covers much of northern China, Manchuria, Korea, and northern Japan.

The climate shows a high degree of continentality, with cold winters, warm summers, and a large annual temperature range. Precipitation is substantial in all months but typically shows a stronger summer maximum. Soil water is frozen throughout one to three winter months, at which time water need is effectively zero. Annual water surplus is moderate to large and soil-water shortage in summer is small, or may be zero. Forest is of two formation classes. Midlatitude deciduous forest is typical of the southerly portions with less severe winters and longer summers; mixed boreal and deciduous forest is typical of the northerly portions with colder winters and cooler summers. Spodosols are the dominant soil order in areas of mixed boreal and deciduous forest in North America, whereas Alfisols are associated

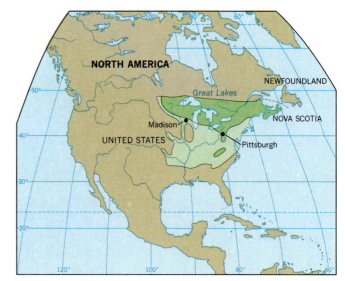

FIGURE 21.10 World map of midlatitude deciduous forests of the moist continental forest environment. (Data source same as in Figure 20.12. Based on Goode Base Map.)

with most other areas in North America and throughout Europe and east Asia. Suborders of the Alfisols are Boralfs and Udalfs, the former being typical of the colder northerly portions.

Geomorphic Processes

Weathering, mass wasting, and fluvial processes of the moist continental forest environment are generally similar to those of the moist subtropical environment to the south, except that frost action is more intense and chemical weathering less intense. Under natural conditions of undisturbed forest cover, fluvial erosion would be of low intensity, but little or no native forest remains over large parts of this environmental region, and induced soil erosion is locally intense. Pleistocene continental glaciation has dominated the landforms over much of the region in North America and Europe; these include till plains and moraines, lake basins and lake plains, outwash plains, and sheets of loess over uplands. Many stream valleys in unglaciated areas lying south of the ice limits in North America and Europe underwent aggradation during glacial stages, followed by trenching and alluvial terrace formation in postglacial time.

Prairie and Steppe Environment (H)

Climate, Vegetation, and Soils

The prairie and steppe environmental region consists of the grassland biome in middle latitudes (Figure 21.11). These grasslands are correlated with areas of the subhumid subtype of the moist continental climate (10sh) and the steppe subtype of the dry midlatitude and dry subtropical climates (9s, 5s). In North America a small portion of the humid subtype of the moist continental climate (10h) is included as an area of tall-grass prairie. In South America grassland of the Pampa occupies an area classified as humid subtype of the moist subtropical climate (6h).

In North America, areas of tall-grass prairie and steppe grasslands extend continuously from the Mississippi River valley to the Rocky Mountains, and a more westerly region occupies the Columbia River basin (Figure 21.12). Tall-grass prairie forms the eastern portion of this vast area; steppe, or short-grass prairie, the western portion. In the Ukraine of Russia there is a large expanse of steppe grassland that projects eastward in a long narrow belt reaching into Siberia as far east as long. 90° E. Other areas of steppe occur in

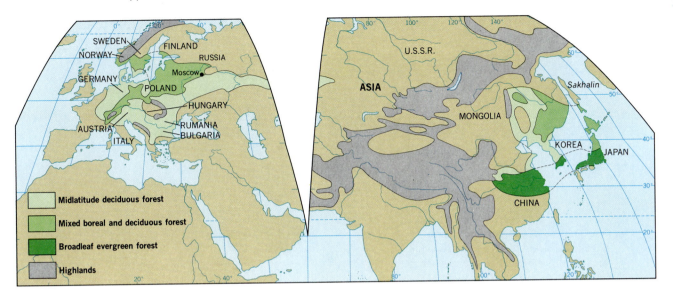

Midlatitude deciduous forest

Mixed boreal and deciduous forest

Broadleaf evergreen forest

Highlands

Mongolia and northern China. In South America the Pampa of eastern Argentina, Uruguay, and southern Brazil is a large expanse of tall-grass prairie.

In the northern hemisphere, climates associated with grasslands show extreme continentality, with cold winters and warm to hot summers and a very large annual temperature range. A substantial soil-water shortage (greater than 15 cm) is always present in the semiarid climate subtype and there is no water surplus. In the subhumid subtype of the moist continental climate (10sh) the shortage ranges from 0 to 15 cm and there is little or no surplus. Tall-grass prairie also extends eastward into the humid subtype of the moist continental climate (10h), where a substantial water surplus occurs.

The Pampa of South America, lying in the humid subtype of the moist subtropical climate (6h), shows a water surplus of from 20 to 30 cm, and it is clear that this grassland area occupies a much moister environment than any of the northern hemisphere grasslands except the tall-grass prairie lying east of the Mississippi River. The subtropical climate of the Pampa region has exceptionally mild winters and

lacks continentality because no large landmass lies on the poleward side.

The dominant soils of the grassland biome are suborders of the Mollisols. Borolls occupy the colder northerly portions in North America and Eurasia; Udolls and Ustolls occupy the warmer portions in North America. Udolls underlie much of the tall-grass Pampa of South America. Aridisols are associated with fringe areas of steppe bordering on semidesert.

Geomorphic Processes

Because the grasslands extend from moist to dry climates no single statement about geomorphic processes and landforms can apply to the entire environmental region. We shall limit our comments to the semiarid climate subtype, where short-grass prairie occurs. Because of the large annual soil-water shortage and lack of water surplus, two processes are particularly important: disruption of rock by salt crystallization and accumulation of deposits of calcium carbonate. Carbonate matter accumulates in the soil as a calcic horizon and this may harden into a petrocalcic horizon, called caliche in the American Southwest (see Figure 9.22). Where the upper soil horizons

FIGURE 21.11 World map of the grassland biome in the subtropical and midlatitude zones. (Data source same as in Figure 20.12. Based on Goode Base Map.)

FIGURE 21.12 Farmlands, marked off in a square township grid, extend for tens of kilometers across a former glacial lake bed near Aberdeen, South Dakota. The agricultural soils here are Borolls, a cold-climate suborder of the Mollisols. A river floodplain with meanders occupies the foreground. (John S. Shelton.)

Tall-grass prairie

Short-grass prairie (steppe)

Major highlands

are stripped away, crusts of calcrete resist erosion and tend to produce platforms, cliffs, and mesas on a small scale.

Locally, the grass cover and soil are removed entirely by trenching of streams, so that areas of badlands are formed and maintained. Sediment yield can rise to extremely high values from these rapidly eroding areas. Stream channels are of the broad, shallow form with braided appearance. Stream flow is highly seasonal and many channels are dry much of the year. Wind action is locally very important. Deflation raises dust storms, even on surfaces bearing a grass cover. Locally, blowout depressions are formed and maintained. Enlargement of these depressions is carried out by a combination of water erosion on the exposed side slopes of the depression and deflational removal of dried particles of mud and sand. Trampling and wallowing by grazing animals contributes to enlargement of the depressions.

Fixed sand dunes occur over large areas of the American short-grass prairie. These dunes were active in late Pleistocene time during climatic episodes in which the plant cover was depleted. An outstanding example is the Sand Hills region of Nebraska, where fixed sand dunes occupy an area of about one quarter of the state. This area is now given over to cattle ranching. Water-table ponds in the deeper depressions supply fresh water. The grass cover is easily broken, with rapid development of small blowouts.

Midlatitude Desert Environment (I)

Brief mention, only, is appropriate for the midlatitude desert environment because most of the basic properties of climate, vegetation, soils, and geomorphic processes of the tropical desert environment apply here with little modification. The midlatitude desert occupies large areas only in central Asia: Turkestan, the Tarim Basin, and the Gobi Desert of Mongolia (see Figure 21.5). In North America a narrow belt of semidesert runs west–east across the

Great Basin in southeastern California, southern Nevada, and western Utah.

The important distinguishing feature of the climate of the Asiatic midlatitude desert and semidesert is the extreme cold of the winters. Because of the lower temperatures, water need is much less than in the hot low-latitude deserts. Further agricultural development of the midlatitude desert environment by irrigation faces much the same set of problems as in the low-latitude desert environment.

III THE HIGH-LATITUDE ENVIRONMENTS

The high-latitude environments are identified with the climates of Group III, controlled by polar and arctic air masses. As a group, these climates have low air temperatures, a severe winter season, and, over large inland areas, only small amounts of precipitation. A condition of frozen soil water for several consecutive months of winter is also typical, so that there is usually a long annual period with near-zero water use and near-zero water need.

Boreal Forest Environment (J)

The boreal forest environment is identified with global areas of boreal forest and the boreal forest climate (11); these are found only in the northern hemisphere (Figure 21.13). The boreal forest climate occupies two great zones, one extending across North America from Alaska to Labrador, the other across Eurasia from the Scandinavian Peninsula to the Bering Sea. The salient feature of this climate is the extreme continentality of the thermal regime. Long, bitterly cold winters dominate the climate and summers are brief. Soil water is frozen for five to seven consecutive months of the year, when water need is effectively zero. Precipitation is low in continental interiors, but accumulates as snow throughout the winter and is rapidly released as surface water in the spring thaw. Boreal forest, the dominant plant formation class, consists of needleleaf evergreen trees, except in Siberia, where deciduous forests of larch occupy semiarid areas of the boreal climate. Dominant soil orders are Spodosols and Histosols.

Most of the boreal forest region is dominated by landforms shaped beneath great Pleistocene ice sheets, which had their centers over the Hudson Bay–Labrador region, the northern Cordilleran Ranges, the Baltic region, and highland centers in Siberia. Much of the boreal forest region lies on the continental shield areas of varied rock type and complex geologic structure. Severe ice erosion exposed hard bedrock over vast areas and created numerous irregularly shaped rock basins. Thin,

FIGURE 21.13 World map of the cold needleleaf forests of the boreal forest environment. (Data source same as in Figure 20.12. Based on Goode Base Map.)

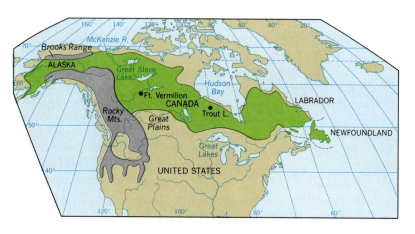

bouldery till mantles the bedrock in places. Here and there are isolated glacio-fluvial deposits—eskers and kames, and sometimes drumlins. Many of the shallower lake basins have been filled by bog materials and converted into muskeg. Peat bogs have provided low-grade fuel in northern Europe. These materials are classed as Histosols. Discontinuous permafrost lies beneath the surface of most of the colder northern part of the boreal forest region, whereas sporadic permafrost is found throughout the less cold southern part.

Tundra Environment (K)

The tundra environmental region is identified with global areas of tundra climate (12) and the tundra biome. The tundra climate occupies northern continental fringes of North America and Eurasia from the arctic circle to about the 75th parallel (Figure 9.36). The distinctive feature of the tundra climate is coldness through all months of the year. Although there is a brief milder season corresponding to summer of lower latitudes, mean monthly temperatures are usually below 7°C (45°F). Winters are very long and extremely cold. Soil water is solidly frozen for eight or more consecutive months. Precipitation is low in all months, but because of poor soil drainage, saturated conditions follow thawing of the surface layer.

Perennially frozen ground, or permafrost, prevails over the tundra region and a wide bordering area of boreal forest climate. The active layer of seasonal thaw is from 0.6 to 4 m (2 to 14 ft) thick, depending on latitude and the nature of the ground. Continuous permafrost, which extends without gaps or interruptions under all surface features, coincides largely with the tundra climate, but also includes a large part of the boreal forest climate in Siberia.

Soils of the tundra are poorly developed with respect to horizons and show little or no mineral alteration of parent materials. Much of the tundra region is underlain by Cryaquepts and Cryorthents; these are great groups of the suborders of Aquents and Orthents, respectively. Bog areas are classified as Histosols. Many areas of broken rock have no identifiable soil.

Vegetation of the treeless tundra consists of grasses, sedges, and lichens, along with shrubs of willow. Traced southward the vegetation changes into birch–lichen woodland, then into the needleleaf forest. In some places a distinct tree line separates the forest and tundra. Coinciding approximately with the 10°C (50°F) isotherm of the warmest month, this line has been used by geographers as a boundary between boreal forest and tundra.

Ice-Sheet Environment (L)

The ice-sheet environment is represented by areas of the ice-sheet climate (13) over the Greenland and Antarctic ice sheets and sea ice of the Arctic Ocean. Physical features of ice sheets and floating sea ice

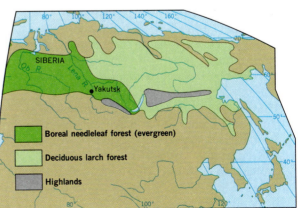

Boreal needleleaf forest (evergreen)

Deciduous larch forest

Highlands

were explained in Chapter 18. Because of low monthly mean temperatures throughout the year over the ice sheets, this environment is devoid of vegetation and soils. The few species of animals found on the ice margins are associated with a marine habitat.

Global Environments In Review

Our survey of physical geography has brought together four major ingredients that define the physical environment: climate, soils, vegetation, and landforms. We used climate as the basis for recognizing global environmental regions. We made this choice because climate provides the basic input of heat and water into the life layer where soil is formed and plants develop their distinctive formation classes.

Landforms add another dimension of variation within the broad climatic regions. Geologic processes play an important, independent role in landform evolution through tectonic and volcanic activities that have shaped the continents and the major landform units within the continents. Climate is also a powerful factor in shaping landforms. In this way, each of the global environmental regions has some distinctive quality of landform instilled by climate through the same inputs of heat and water that vary the quality of soil and plant cover.

In this final chapter we have surveyed the principal environmental regions and their food resources. Each type of climate, together with its characteristic soils, natural vegetation, and landforms, comprises a unique natural environmental region. Perhaps the impression most deeply implanted by this survey is the enormous range in environmental quality, judged in terms of plant resources and the potential for production of plant food for human consumption. It may be a bit disconcerting to consider that vast areas of the continents are almost total deserts, either too dry or too cold to support human populations. In the desert wastes of the Sahara or the snow-covered plateau of Antarctica, the average American family might stay alive for at most a few days without a supply of water or heat, or both, brought in from more favorable environments.

Another distinct impression our survey might leave is that the human family has already done a remarkably thorough job of exploiting the favorable environments of the globe. For certain environmental regions we have made an assessment of the prospects for increased food production by using massive inputs of irrigation water, fertilizers, fuels, and pesticides in concert with the newest in genetic strains of staple crops and the best in management techniques.

Interaction of humans with the natural environment has been a persistent theme of our study of physical geography. Here we are in the heartland of geography as a science—a study in human ecology and the spatial distribution of the varied ingredients. Many forms of interaction appear harmful to the natural environment and pose threats to the very sources of water and food on which humans and lesser life-forms depend. In closing this survey of physical geography we point out that physical geography can provide an important part of the science background needed to analyze environmental problems in a sound perspective.

Energy and Environment

With the availability of enormous supplies of energy and a high level of technology, twentieth-century civilization has a greatly broadened choice of environments that it can occupy. No finer example of this principle can be found on a vast scale than in the Los Angeles basin of southern California. With no flowing rivers of fresh water close at hand, no coal to speak of, only modest local reserves of petroleum and natural gas, and no good natural harbor, a great metropolis has arisen. Water importation by aqueduct systems on a vast scale solved the water problem; ships and pipelines solved the fuel problem; rail and truck transport solved the food problem; construction machinery and dredges solved the harbor problem.

A more extreme example of this type is the maintenance of human habitation on the ice shelves and ice sheet of Antarctica. If a nation chooses to pay the high cost of sending by ship and plane all food, fuel, and structural materials needed to keep a group of humans alive and well, even the most severe of global environments is not barred from occupation. As the cost of energy rises steadily, the allocation of national resources to such projects is given increasing attention in terms of benefits to be derived. To overcome severe environmental shortcomings by sheer force of energy and technology we must pay an increasingly higher price. Would it not be wise, instead, to optimize the benefits and resources of those naturally favorable environments in which the sun provides an unending source of energy and both heat and water are abundantly received by the soil under a favorable climate?

Environmental Preservation

We must take note of a factor entering with increasing force into decisions regarding what courses of human activity are possible and most probable. Throughout the Industrial Revolution and its extension into European colonial expansion and growth of modern industrialized nations, human activity was dominated by self-serving motives—a better way of life and the acquisition of wealth and power. Destruction of the natural environment and its ecosystems proceeded unchecked well into the twentieth century with only a few voices raised in warning or protest.

Today, however, great concern is loudly expressed by many citizens for preservation of the remaining fragments of undamaged ecosystems. Many courses of action that were once viable possibilities have been ruled out or have come under severe criticism. The setting up of new national parks and wilderness areas is an example of a series of free choices guided by the philosophy of environmental preservation. Another, more recent example is the criticism leveled by ecologists at Brazil's plans to destroy the rainforest of Amazonia and replace it with farmland, cultured forest, and towns. This criticism led to a decision by the Brazilian government to save a part of the nation's rainforest. Perhaps, in time (and if not too late), forces of environmental preservation will prevail over those who would allow removal of the rainforest biome from the face of the earth.

In deciding such environmental issues as the destruction of a biome or the switch to coal and oil shale as our dominant energy sources, the crucial arguments may no longer be limited to ethical and economic considerations. Today, a new specter has come to haunt our decision makers. If the world's forests and woodlands are soon destroyed and the combustion of fossil fuels increases by leaps and bounds, there may follow a global climate change highly unfavorable to the human race through the disruption of its food-producing capability. As if the human race did not already have enough disasters to avoid, it must face yet another threat—possible self-destruction as the consequence of growth.

APPENDIX

FLOW SYSTEMS OF MATTER AND ENERGY

Matter and Energy

The earth realms with which physical geography is concerned consist of two components: matter and energy. Physics and chemistry are basic sciences that deal with the nature of matter and energy and the formulation of laws that govern their behavior. To define the terms *matter* and *energy* is not an easy task because they represent concepts that include everything in the real world. To start, we can substitute the word "substance" for the word "matter," but this only postpones the problem.

Substance, in turn, has the attribute of occupying space. Matter is a tangible substance because it can be seen, felt, tasted, measured, weighed, or stored. Matter has the mysterious property of gravitation, the mutual attraction acting between any two particles or aggregates (lumps, groups, or pieces) of matter.

Energy is often defined in terms of its effects. Perhaps the most common definition is that "energy is the ability to do work." Energy is somehow always involved in the motion of matter, but energy can be stored in matter that does not appear outwardly to have motion. A brick poised on a window ledge holds a store of energy even though the brick is at rest. That stored energy will rapidly be released in the form of motion if the brick is nudged slightly and falls to the sidewalk below. Whatever energy is, it can move or flow from place to place and it can also be held in storage in various ways. We frequently refer to energy as something expendable. For example, someone may say "I used up a lot of energy playing those two sets of tennis." Actually, energy cannot be destroyed by being "used"; it can only change from one form to another and move from one place to another. The same statement applies to matter, which cannot be destroyed, but only changed or moved from place to place. By "destroyed" we mean "removed from existence," or "removed from the universe."

The Nature of Matter—A Review

Several kinds of classes of matter exist in the universe. A system devised by Professor John R. Holum, a chemist, is illustrated in Figure A.1. All matter is collectively described by the word "substances." Next, *pure substances* are distinguished from *mixtures,* which consist of two or more pure substances mixed together in indefinite proportions. The pure substances prove to be either *elements* or *compounds.* Examples of elements are metals in the pure state, such as iron, gold, copper, or silver. Compounds are also familiar to everyone, for example, pure water, carbon dioxide gas, or crystals of quartz. Elements and compounds exist "in the realm of things where samples can be seen and weighed directly," to use Professor Holum's words. The lower box in Figure A.1 leads into "the realm where individual particles cannot be directly seen and weighed." The smallest representative samples of pure substances are of two classes: *atoms* and *compound particles.* The latter consist of *molecules,* which are groups of atoms. A single molecule may contain as few as two atoms or (in the case of some complex organic molecules) as many as several thousand atoms.

The atom consists of subatomic particles. Each atom has a single nucleus, which is surrounded by one or more electrons. The nucleus itself consists of several kinds of particles.

States of Matter—A Review

The physical condition, or state, in which we find matter at a given place and time is a subject of great importance in physical

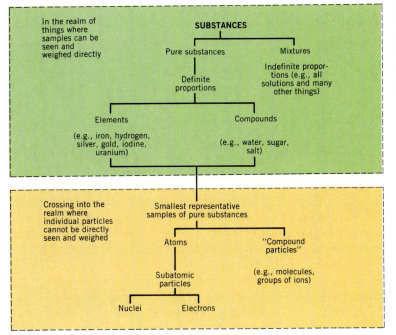

FIGURE A.1 A classification of the kinds of substances. (From John R. Holum 1972, *Elements of General and Biological Chemistry,* 3rd ed., Figure 2.2, John Wiley & Sons, New York. Reproduced by permission.)

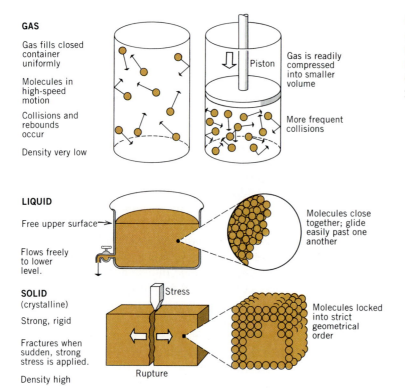

GAS

Gas fills closed container uniformly

Molecules in high-speed motion

Collisions and rebounds occur

Density very low

Piston

Gas is readily compressed into smaller volume

More frequent collisions

LIQUID

Free upper surface →

Flows freely to lower level.

Molecules close together; glide easily past one another

SOLID
(crystalline)

Stress

Strong, rigid

Fractures when sudden, strong stress is applied.

Density high

Rupture

Molecules locked into strict geometrical order

FIGURE A.2 Some properties of gases, liquids, and solids. (From A. N. Strahler, *Physical Geology*, Harper & Row, New York. Copyright © 1981 by Arthur N. Strahler.)

geography. Certain *states of matter* are well known to everyone. Every day we drink water in its *liquid state*. Ice cubes represent water in the *solid state*. The *gaseous state* of water, called water vapor, can't be seen, but is easily sensed by the human skin in summer when the humidity of the air is relatively high.

The three common states of matter—solid, liquid, and gaseous—apply to both pure substances (elements and compounds) and to mixtures. Using only the simplest concepts of atoms and compounds we can describe the three states of matter in terms of observable behavior. For this purpose, the atoms or molecules that comprise matter can be visualized as uniform spheres, all physically alike.

A *gas* is a substance that expands easily and rapidly to fill any small empty container. Atoms or molecules of the gas, as the case may be, are in high-speed motion (Figure A.2). Empty space between the atoms or molecules is vast in comparison with the dimensions of those particles. Particle motions take random directions; collisions between particles are frequent. The particles rebound like perfect spheres

at each impact, changing direction abruptly. The particles also strike and rebound from the walls of the container. A gas is usually very much less dense than a liquid or solid consisting of the same chemical substance. For example, the gaseous water vapor in warm, moist air has a density only about one one-hundred-thousandth that of liquid water.

A *liquid* is a substance that flows freely in response to unbalanced forces but maintains a free upper surface so long as it does not fill the container or cavity in which it is held. The molecules of a liquid compound—water, for example—move more or less freely past one another as individuals or small groups. Under rather strong confining pressures (such as would exist at the bottom of the deep ocean) liquids are compressed only slightly into smaller volume. For many practical purposes, liquids can be considered to be incompressible (not capable of being compressed).

Both gases and liquids are classed as *fluids* because both substances freely undergo flowage. Put simply, both of these substances flow toward lower levels under the force of gravity wherever possible. As a result, fluids of different density tend to come to rest in layers with the fluid of greatest density at the bottom and that of least density at the top. This principle has several very important applications in sciences of the atmosphere and oceans.

A *solid* is a substance that resists changes of shape and volume. Solids are typically capable of withstanding large unbalanced forces (i.e., strong stresses) without yielding permanently, although they undergo a small amount of elastic bending. When yielding does occur, it is usually by sudden breakage (rupture). These principles have important applications in the study of glacier ice and rock, both of which are in the solid state.

Changes of state are accompanied by either an input or an output of energy into the substances undergoing change. We are all familiar with this principle in the preparation of food. To boil water and produce steam (change to vapor state) a great deal of heat must be applied. To freeze water a great deal of heat must be removed from the water. Our next step will be to examine the nature and forms of energy that can be easily observed to cause changes in pure substances and mixtures behaving as gases, liquids, and solids.

Forms of Energy—A Review

We stated earlier that energy is commonly defined as the ability to do work. In strict terms of physics, energy is the product of force acting through distance. Thus, energy is the ability to move an object (exert a force) for a certain distance. Energy is stored and transported in a variety of ways. Some of the recognized forms of energy are: mechanical energy, heat energy, energy transmitted by radiation through space (electromagnetic energy), chemical energy, electrical energy, and nuclear energy.

Mechanical energy is energy associated with the motion of matter. There are two forms of mechanical energy: kinetic energy and potential energy. *Kinetic energy* is the ability of a mass in motion to do work. Thus, an automobile traveling down a highway possesses kinetic energy because it is a mass in motion. Should this mass strike a telephone pole, its ability to do work upon its own body and upon the telephone pole will become quite obvious. The energy it will release in collision will increase with the weight (mass) of the car, and it will also increase with the square of the auto's speed. Kinetic energy, then, is proportional to the quantity of mass in motion multiplied by the square of its velocity. Kinetic energy is obvious in many kinds of natural processes acting at the earth's surface—a rolling boulder, a flowing stream, or a pounding surf.

Potential energy, or energy of position, is equal to the kinetic energy an object would attain if it were allowed to fall under the influence of gravity. Suppose a brick is balanced on the edge of a tabletop, then falls. The kinetic energy the brick possesses at the moment it hits the floor, as we have seen, is proportional to the mass of the brick multiplied by the square of its velocity. If the brick is again lifted to the tabletop, the work done in lifting gives the brick a quantity of potential energy. This energy will be released when the brick is again allowed to fall. It should be obvious at this point that the floor is merely a convenient stopping place for the brick; if a hole were opened in the floor, allowing the brick to fall farther, it would possess even more kinetic energy at its impact on the floor below. Therefore, potential energy must always be valued with respect to a given reference level, or base level. Looking around outdoors, we can spot many examples of the existence of potential energy in a landscape. A boulder poised at the top of a steep mountain face is a simple example. In fact, the entire mountain represents a large reservoir of potential energy judged in reference to the level of the floor of an adjacent valley.

Mechanical energy can be transmitted from one place to another in the form of *wave motion,* in which kinetic energy is carried through matter by an impulse passed along from one particle to the next. A sound wave is one example—a push on air molecules at one point will be transmitted outward in all directions. Another example is the phenomenon of earthquake waves, which carry large amounts of energy for great distances, not only over the ground surface, but also in paths deep within the solid earth. The familiar Richter scale of earthquake intensity measures the quantity of energy released by an earthquake. In all mechanical forms of wave motion matter is displaced (up-and-down, sideways, or forword-and-backward) in a rhythmic manner. Frictional resistance within the moving substance withdraws energy and the waves gradually die out as they travel farther from the source.

Sensible heat is another form of energy of paramount importance. Kinetic energy can readily be converted into sensible heat energy through the mechanism of friction. A familiar example is the braking action of a moving automobile. As the automobile slows to a stop (losing kinetic energy), the brake drums become intensely hot.

Sensible heat represents kinetic energy, but it is of an internal form, rather than the external form seen in moving masses. Thus, a cupful of water resting completely motionless on the table has internal energy because of constant motion of the water molecules on a scale too small to be visible. This internal motion is the sensible heat of the substance and its level of intensity is measured by the thermometer. For gases, the internal motion is in the form of high-speed travel of free molecules in space but with frequent collisions with other molecules. The energy level in the gaseous state of water (water vapor) is thus higher than within the liquid water. Ice, on the other hand, represents a lower energy level than liquid water, for here the molecules are locked into place in a fixed geometrical arrangement (see Figure A.2). For these molecules in the solid state the motion is one of vibration without relative motion.

Sensible heat moves through gases, liquids, or solids by the process of *conduction*. The direction of heat flow by conduction is always in the direction from higher temperature to lower temperature. In the process of conduction the faster-moving molecules of the warmer matter pass along part of their kinetic energy to the adjacent cooler matter, causing an increase in the speed of molecular motion along the path of energy travel. In this way heat moves through the matter. By conduction, heat can pass from a gas into a liquid or solid, from a liquid into a solid or gas, and from a solid into a liquid or gas.

Sensible heat can also be transported within a layer of gas or liquid by *convection*, a process in which currents redistribute heat by mixing warmer and cooler parts of the fluid.

When ice melts, work must be done to overcome the crystalline bonds between molecules. This work requires an input of energy, but it does not raise the temperature of the substance. Instead, the energy seems mysteriously to disappear. Since energy cannot be lost, it is actually placed in storage in a form known as *latent heat*. Should the water freeze and again become ice, the latent heat will be released as sensible heat. A similar transformation from sensible to latent heat takes place when a liquid evaporates into a gas, because work must be done to overcome the bonds between molecules of the liquid. When water vapor returns to the liquid state, a process of condensation, latent heat is released as sensible heat. Both sensible heat and latent heat represent forms of stored energy (as does potential energy).

Sensible heat stored in matter may be lost directly to the surroundings through conduction, but even in a vacuum objects can lose heat. A fundamental law of physics states that all matter at temperatures above absolute zero gives off *electromagnetic energy*. The process of giving off energy is called *radiation*. We can think of this radiation as taking the form of waves traveling in straight lines through space. The waves come in a very wide range of lengths, but all travel at the same speed—300,000 km (186,000 mi) per second—regardless of their length. Together, the total assemblage of waves of all lengths constitutes the *electromagnetic spectrum*. It includes visible light with all its rainbow colors and also invisible shorter waves, such as ultraviolet rays, X rays, and gamma rays. Besides these, the spectrum includes invisible long waves known as infrared rays (sometimes called heat rays), and still longer microwaves and radio waves.

Electromagnetic energy received from the sun powers a group of important natural processes constantly at work at the earth's surface. Electromagnetic energy, upon arriving at the earth's surface, is continually converted into mechanical energy and sensible heat, which in turn are transformed in such activities as winds in the atmosphere or the breakup of exposed rock, and in the transport of the resulting rock particles to new places of rest.

Yet another form of energy is *chemical energy*, which is absorbed or released by matter when chemical reactions take place. These reactions involve the coming together of atoms to form molecules, the recombining of molecules into new compounds, and the reverse changes back to simpler forms of matter. Green plants use the sun's electromagnetic energy to produce chemical energy, which can be stored within the leaves and stems of a plant in the form of complex organic molecules.

Finally, we take note of *electrical energy* and *nuclear energy*. The lightning bolt is a spectacular display of a release of electrical energy. Nuclear energy is produced by the spontaneous disruption of atoms of certain elements said to be radioactive. Nuclear energy is an important natural process deep within the solid earth, where it continually generates heat. This accumulated heat is believed to be responsible for volcanic activity, dislocations of the earth's outer crust, and the slow motions of vast lithospheric plates.

Flow Systems of Matter and Energy

Understanding the varied processes that affect the atmosphere, hydrosphere, lithosphere, and biosphere requires a physical geographer to think in terms of *flow systems* of both matter and energy. A flow system is simply a series of connected pathways through which energy and/or matter move more or less continuously. We will deal first with the energy flow system. An *energy flow system* traces the flow path of energy from a point of entry to a point of emergence. As energy flows through such a system, it may change form many times and may be detained temporarily in storage from place to place. In this process the energy flow makes use of matter as the medium of motion and of storage.

Energy flow systems set in motion and sustain *material flow systems* (flow systems of matter). The matter involved in such systems is not only transported from place to place along certain pathways, but it can also undergo changes of state and chemical changes. The matter traveling through the system can also be held temporarily in storage at certain points.

Flow systems of energy and matter can be visualized by means of simple schematic diagrams making use of a set of standard symbols. Three such diagrams, using familiar household examples, illustrate the three basic kinds of flow systems (Figure A.3). First is the energy flow system, here illustrated in diagram *A* by a container of food being heated over an electric heating coil. The outer surface of the container represents the *system boundary*, shown by a rectangular outer box in the diagram. Input of energy into the system is shown as coming from an *energy source*, for which a circle is our symbol. An arrow shows the flow path of energy. Let us assume that the energy supplied by the electric heating coil is in the radiant form and consists of infrared rays, which are part of the electromagnetic energy spectrum. Thus the energy is transferred by radiation from the heater to the container, which absorbs the energy. Here an *energy transformation* occurs, in which the radiant energy is transformed into sensible heat, raising the temperature within the container. A rectangle is our symbol for an energy transformation. The sensible heat enters temporary *energy storage* in the container and its contents. A figure shaped like a storage bin indicates storage of either energy or matter within a flow system. Energy continues its flow by again changing its form back to infrared radiation emitted by the surface of the container. This process represents the output of energy through the system boundary. (Other forms of energy flowing both into and out of the system may be operating simultaneously, but these have been omitted for the sake of simplicity.)

With the rate of energy input held constant, a point in time is reached at which the amount of energy in storage becomes constant, as indicated by a constant temperature within the container. The rate of output of energy must then equal the rate of input. We say that the system is in a *steady state* of operation when the rates of input and output are equal and constant and the storage is also a constant quantity.

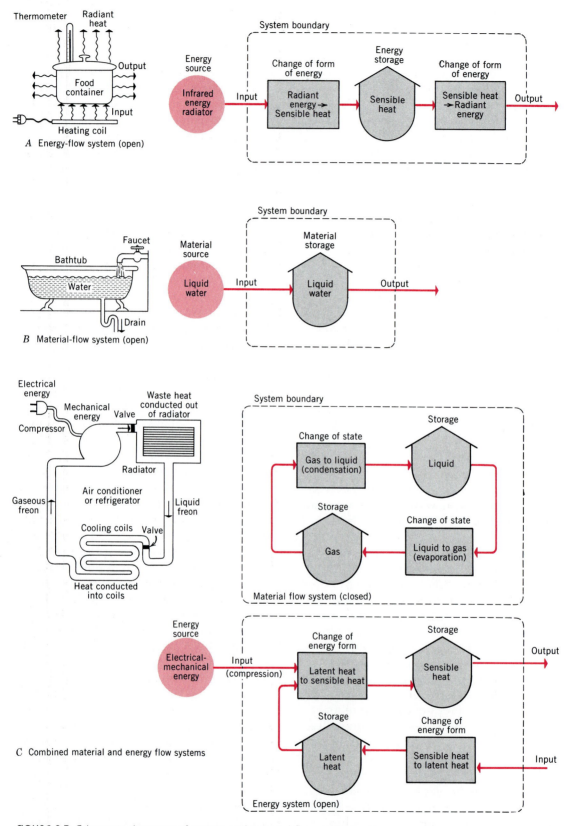

FIGURE A.3 Schematic diagrams of energy and material flow systems using common household mechanical devices as examples.

If we were to increase the rate of energy input, the storage would increase (giving a higher temperature) and so would the rate of energy output. After a short time steady state would be reestablished at a new level of activity. When the energy input is turned off, the flow system simply collapses and ceases to exist. We have illustrated an *open system*, defined as a system that requires both an input and an output of energy (or matter) through the system boundary. An important principle is that all energy flow systems are open systems. No real energy system can operate as a closed system, entirely within the system boundary, because no perfect insulator exists to prevent the outflow of energy as heat through the enclosing boundary.

Diagram *B* of Figure A.3 illustrates a simple material flow system, represented by a bathtub with its faucet and open drain. The system boundary is the outer surface of contact of the water with the walls of the tub and the free liquid surface. Input of water by means of the faucet can be regulated in such a way that the water level rises rapidly in the tub. As this happens, the rate at which water exits through the drain is steadily increased (because of the increasing hydraulic pressure of the deepening water). A point in time is reached at which the water level is constant and the quantity of matter in storage is thus a constant value. Rates of input and output are also constant and equal and the system is in steady state. This material flow system is an open system. When the input of water is shut off, the water drains completely out of the tub and the system simply disappears.

Diagram *C* of Figure A.3 illustrates systems of matter flow and energy flow working simultaneously. The total system consists of the essential working parts of an air conditioner or a refrigerator using a circulating coolant fluid, such as Freon, to carry out the cooling process. A compressor, driven by an electric motor, receives coolant gas at low pressure and pumps it through a valve, placing the gas under high pressure. The compression process heats the gas to a temperature well above the air surrounding a radiator. In the radiator the gas cools and condenses to a liquid. As the liquid enters the cooling coils, it passes through another

valve, which sprays it into the low-pressure environment of the cooling coils. At low pressure, the coolant droplets quickly evaporate, becoming a very cold gas in the cooling coils and chilling the air near the coils. The gaseous coolant returns to the compressor to complete the cycle. The net effect has been to extract heat from the environment near the cooling coils, store the heat in the gaseous Freon, then release the heat to the external surroundings during condensation in the radiator. In this way, heat is "pumped" from the cooling coils to the radiator.

The upper diagram shows the material flow system consisting of flow of coolant and its changes of state. The changes of state are shown by the same rectangular boxes used as symbols of energy transformation in the energy system. Similarly, storages in the liquid and gaseous states are shown by the bin symbol. Beginning at the upper left corner, compression forces condensation of the gaseous coolant, a change from the gaseous to the liquid state. Liquid coolant then expands and evaporates, returning to the gaseous state. We have shown here a *closed system* because no matter enters or leaves the system.

The lower diagram illustrates the energy flow system that acts simultaneously with the material flow system. For each change in material state there is a change in form of energy. If the diagrams were printed on transparent acetate film, you could place one diagram over the other and see that the basic circuits of material flow and energy flow are perfectly superimposed. However, the energy system must have inputs and outputs of energy in addition to an internal closed flow circuit. Electrical energy is converted into mechanical energy of the compressor, but this transformation takes place outside the system. Mechanical energy is used to compress the gas, raising its internal temperature and adding to the store of sensible heat. At high pressure the gas condenses, releasing latent heat to add to the store of sensible heat. Part of the stored sensible heat is disposed of to the outside environment by conduction through the radiator and represents the energy output of the system. As the coolant evaporates, heat is drawn into the system from the outside and much of the sensible heat it holds is transformed into latent heat, which goes into temporary storage in the gaseous coolant. Condensation reverses the process and the latent heat is transformed back into sensible heat. Like the first example of an energy system, this one is an open system and must have both an energy input and an energy output.

The refrigerating machine is thus a combination of a closed material flow system and an open energy flow system. Most of the natural systems encountered in physical geography are also combinations of material and energy flow systems. No material system, whether open or closed, can operate without the expenditure of energy. The reverse of this statement is not true, however. It is possible to have an energy flow system without mass motion of matter being involved, because energy can flow through a body of matter by conduction or radiation, or a combination of the two, without any place-to-place motion of the matter itself being required.

These concepts of flow systems of energy and matter can be put to good use at various points in a study of physical geography where the system concept is important.

The Global Radiation Balance as an Open Energy System

A simplified and generalized model of the global radiation balance as an open energy system is shown in Figure A.4. The arbitrary boundary of this system lies at the outer limit of the earth's atmosphere.

Input of solar energy is in the form of shortwave radiation, part of which is reflected back directly into space without being transformed or stored. That which is absorbed by atmosphere, oceans, and lands is transformed into sensible heat. Stored energy is then emitted in the form of longwave radiation, involving another energy transformation. System output is by longwave radiation into outer space.

Over long periods of time, this system is approximately in a steady state of operation, in which incoming and outgoing forms of radiation are equal and the quantity of stored energy remains constant.

An in-depth model of the global radiation balance as consisting of two subsystems is shown in Figure A.5. An atmospheric subsystem and a ground subsystem are interconnected by energy flow paths. Figures give the same energy flow units and quantities used in Chapter 2. (See Figure 2.17.)

A Convective Storm as a Flow System of Energy and Matter

A convective storm in which precipitation is occurring can be visualized as an open energy system, shown in Figure A.6. The storm may consist of a single thunderstorm cell or a group of cells. The lateral boundary of the system is arbitrarily taken as an imaginary vertical surface surrounding the storm on the sides. The

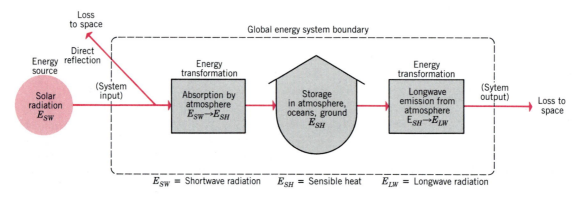

FIGURE A.4 A simplified diagram of the global radiation balance as an open energy system.

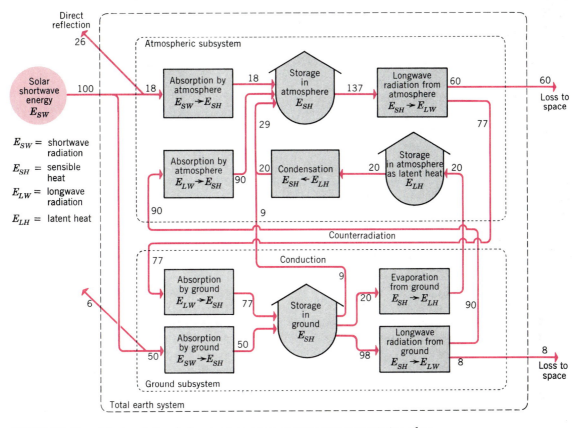

FIGURE A.5 The global radiation balance as an open energy system consisting of an atmospheric subsystem and a ground subsystem.

ground surface and the tropopause may be taken as lower and upper boundaries.

Two forms of energy input can be recognized. First is direct incoming solar radiation, which warms the air, increasing the sensible heat in storage. Second is the importation of latent heat by evaporation from a moist ground surface or an ocean surface below. To simplify our analysis we have ignored the possible importation of both sensible and latent heat by horizontal air motion through the vertical side boundary. It can be assumed that an equal quantity of energy leaves the system in the same manner. Condensation within the rising air releases sensible heat, increasing the total air-mass store of sensible heat. From storage, energy leaves the system either by longwave radiation or by mass transport of sensible heat in falling raindrops or snowflakes. The small quantity of heat conducted directly to the ground from the overlying air is not shown.

An open material flow system, shown in the lower diagram of Figure A.6, can be coordinated with the energy flow system. (Ice is included with the liquid state.) The major activity in the storm is the change of state of water from vapor to liquid, producing precipitation. Evaporation of the falling precipitation may take place, thus returning some of the liquid water to the vapor state, as shown by the return circuit in the diagram.

Enlarging our perspective, we can think of this system as representing the entire global precipitation process, averaged over long spans of time. The global energy system would have an input limited to solar shortwave energy and an output limited to longwave radiation to outer space. The global material flow system would show input by evaporation from liquid water storages on the continents and oceans and a return of liquid water to those storages. It would closely resemble the diagram of the

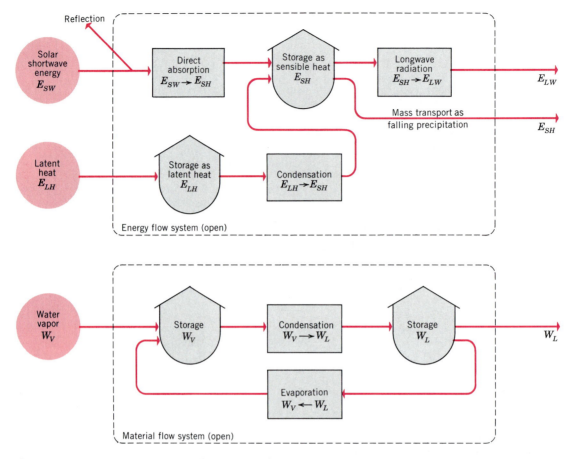

FIGURE A.6 A convective storm as a flow system of energy and matter.

hydrologic cycle (see Figure A.7) except that transfer of runoff from continents to oceans would be excluded from the system.

The Hydrologic Cycle as a Closed Material Flow System

The global water balance can be treated as a material flow system. Figure A.7 shows a closed system representing the total global hydrosphere. Three subsystems are present: atmosphere, continents, and oceans. (For the sake of simplicity, glacier ice is included with liquid water as a single form of water.) In the atmosphere subsystem the input is in the form of water vapor derived by evaporation from the ocean surface and by evaporation and evapotranspiration from the continental surface. Stored water vapor

FIGURE A.7 The hydrologic cycle as a closed material system.

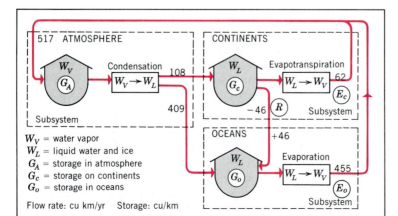

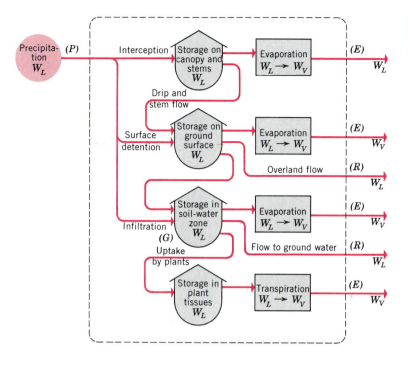

FIGURE A.8 The soil-water balance as an open material flow system.

undergoes condensation to become precipitation in the liquid form as rain (or solid state as snow), leaving the atmosphere subsystem and entering either the continent subsystem or the ocean subsystem. By runoff, water in storage on the continents is transferred to the oceans without change of state.

The Soil–Water Balance as an Open Material Flow System

The soil-water balance can be treated as an open material flow system, as shown in Figure A.8. System input is by precipitation. Rain or snow falling on plant foliage and stems is intercepted and goes into temporary storage; it may be lost by evaporation (change of state), or it may drip or flow down to the ground surface. Some water is temporarily stored as *surface detention* in depressions on the soil surface, from which it may be lost by evaporation, or may leave the system as overland flow. By infiltration, water enters storage in the soil-water zone. Some of this stored water may evaporate from the soil surface, leaving the system as water vapor. Uptake of water through the roots of plants is another way in which soil-water storage is depleted. Transpiration disposes of this water in the vapor state. Surplus soil water may infiltrate

downward through the intermediate zone, entering the ground water zone and leaving the system as ground water flow.

The Runoff Energy System

Runoff of fresh water on and beneath the land surface is powered by gravity; it is an open energy system in which work is done as the water moves to lower levels under a hydraulic gradient and encounters frictional resistance. Figure A.9 shows the principal energy storages and transformations that make up the system. The total open system of runoff has as its energy input the quantity of available potential energy present in precipitation that has arrived at the land surface. This water mass occupies an elevated position above a base level of flow (in this case, the ocean surface).

Two subsystems are shown. The surface water subsystem consists of overland flow and stream flow; the ground water subsystem consists of subsurface water moving under a hydraulic gradient, whether the paths of motion be directed downward or upward.

In the surface-water subsystem potential energy is initally in storage at the ground surface, where it is transformed into kinetic energy as overland flow and stream flow move to lower levels. Kinetic energy is

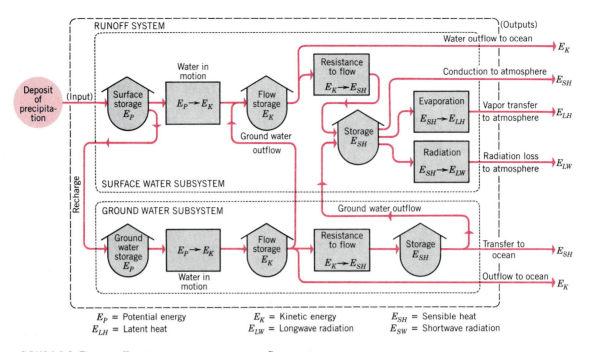

E_P = Potential energy E_K = Kinetic energy E_{SH} = Sensible heat
E_{LH} = Latent heat E_{LW} = Longwave radiation E_{SW} = Shortwave radiation

FIGURE A.9 The runoff system as an open energy flow system.

temporarily stored in the water flow, just as a moving automobile holds a store of kinetic energy. Flowage is met by internal friction of turbulent water motions and by friction along the boundary of the flowing water body with the ground beneath and the air above. Thus kinetic energy is transformed into sensible heat, which is temporarily held in storage and is then lost to the atmosphere by conduction, or is transformed to longwave radiation. Surface-water evaporation provides another energy output to the atmosphere as latent heat. Some kinetic energy of the flowing water passes directly out of the system where the trunk stream enters the ocean.

The ground water subsystem receives potential energy during recharge, and this also undergoes transformation into kinetic energy. As some of the ground water flow emerges at the land surface in springs, it adds to the kinetic energy of surface flow. Also shown is transfer of sensible heat from the ground water subsystem to the surface-water subsystem during the seepage and spring-flow process. Otherwise, ground water flows to the interface of fresh water and salt water beneath the coastline and emerges through the sea floor, transporting both sensible heat and kinetic energy of motion out of the system.

We realize, of course, that the initial

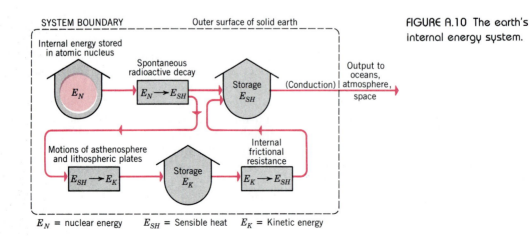

E_N = nuclear energy E_{SH} = Sensible heat E_K = Kinetic energy

FIGURE A.10 The earth's internal energy system.

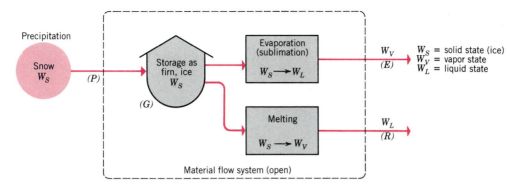

FIGURE A.11 A glacier as a material flow system.

supply of potential energy for the runoff system is furnished by solar energy, which evaporates water from the ocean surface, carries it high into the atmosphere as water vapor, and allows it to fall as precipitation. All this is part of the total hydrologic energy system.

The Earth's Internal Energy System

Tectonic and volcanic activity represent expenditures of internal energy stored within the nuclei of atoms of radioactive elements, such as uranium and thorium. As Figure A.10 shows, this energy source lies within the system boundary. It is an inheritance from the time of accretion of the planet some 4.6 billion years ago. Spontaneous decay of these atoms transforms the stored atomic energy into sensible heat stored in the surrounding rock, which may be in the solid state as crystalline rock or in the liquid state as magma. Most of this sensible heat is slowly conducted to the earth's surface, where it is lost to the oceans and atmosphere, and ultimately to outer space.

Some of the sensible heat in the mantle is, however, used to drive slow currents within the plastic asthenosphere and move the brittle lithospheric plates. Thus some sensible heat is transformed into kinetic energy of matter in motion. Kinetic energy is converted back into sensible heat through internal friction as the asthenosphere is dragged against the underlying strong mantle rock and the overlying rigid plates. This sensible heat also enters storage and follows the conduction pathway to the surface.

Because there is no external energy input in this system, the total system energy decreases with time; it is called an *exponential decay system*. We have not taken into account the possibility that some energy enters the system through tidal flexing of the earth. There is also the possibility that new energy inputs have been made from time to time by infall and impact of asteroids and large meteoroids.

A Glacier as an Open Material Flow System

Glacier equilibrium can be interpreted through an open material flow system, diagrammed in Figure A.11. Matter in the form of solid precipitation enters the system through the surface of the zone of accumulation. Downvalley flow carries the glacier ice to the zone of ablation, where it leaves the system by direct evaporation (sublimation) or as meltwater of runoff.

Matter enters directly into storage in the solid state, W_s undergoes either of two changes of state, and leaves the system in both vapor and liquid states. Glacier equilibrium requires the flow system to be in a steady state, in which rate of precipitation input (P) balances the sum of the output rates of evaporation (E) and runoff (R), while the quantity of matter in storage (G) remains constant.

The Energy Flow System of a Green Plant

The energy flow system of a green plant is shown in Diagram A of Figure A.12. Solar shortwave energy falling on a green leaf is

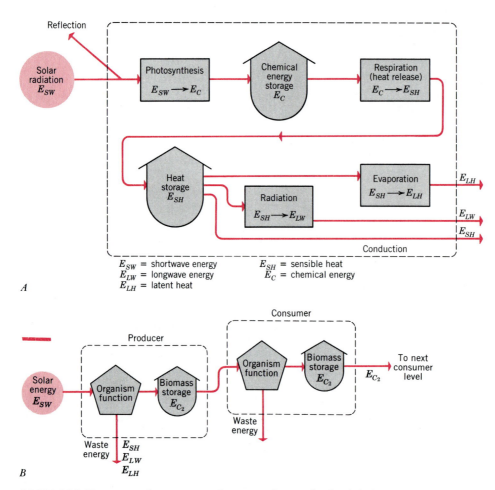

E_{SW} = shortwave energy
E_{LW} = longwave energy
E_{LH} = latent heat

E_{SH} = sensible heat
E_C = chemical energy

FIGURE A.12 The energy flow systems of a green plant and a food chain.

partly reflected and partly absorbed. Disregarding conversion to sensible heat within the leaf, we concentrate on the process of photosynthesis in which shortwave energy is converted into chemical energy and stored within molecules in plant cells. After further complex chemical changes have occurred, the chemical energy becomes *biomass energy* stored in the tissues of the plant. Respiration and spontaneous oxidation of complex molecules convert chemical energy into sensible heat, which is held in temporary storage. Stored sensible heat is disposed of through evaporation, thus leaving the plant as latent heat, and by longwave radiation or direct conduction to the atmosphere.

Diagram *B* shows energy flow up the food chain. We have used a pentagon as a shorthand symbol for the energy changes that are shown in Diagram *A* as leading to biomass storage. Stored biomass energy within the primary producer provides an input to the first level of consumers. These diagrams are repeated for higher levels in the food chain.

GLOSSARY

This Glossary contains definitions of terms italicized in the text.
Italicized terms within the definitions will be found as individual entries elsewhere in the Glossary.

ablation   Wastage of glacial ice by both melting and *evaporation*.

abrasion   Erosion of *bedrock* of a *stream channel* by impact of particles carried in a stream and by rolling of larger rock fragments over the stream bed. Abrasion is also an activity of glacial ice, waves, and wind.

abrasion platform   Sloping, nearly flat *bedrock* surface extending out from foot of *marine cliff* under shallow water of breaker zone.

absorption of radiation   Transfer of *electromagnetic energy* into heat energy within a *gas* or *liquid* through which the radiation is passing.

abyssal plain   Large expanse of very smooth, flat ocean floor found at depths of 4600 to 5500 m (15,000 to 18,000 ft).

accelerated erosion   *Soil erosion* occurring at a rate much faster than *soil horizons* can be formed form the parent *regolith*.

accretion of lithosphere   Production of new oceanic lithosphere at an active spreading plate boundary by the rise and solidification of magma of basaltic composition.

accretionary prism   Mass of deformed trench sediments and ocean floor sediments accumulated in wedgelike slices on the underside of the overlying plate above a plate undergoing subduction.

acid deposition   (See *acid rain*.)

acid mine drainage   Sulfuric acid effluent from *coal* mines, mine tailings, or spoil ridges made by *strip mining*.

acid rain   Rainwater having an abnormally high content of the sulfate ion and showing a pH between 2 and 5 as a result of air pollution by combustion products of fuels having high sulfur content.

active continental margins   Continental margins that coincide with tectonically active plate boundaries. (See also *continental margins*, *passive continental margins*.)

active systems   Remote sensor systems that emit a beam of wave energy at a source and measure the intensity of that energy reflected back to the source.

actual evapotranspiration (water use)   Actual rate of *evapotranspiration* at a given time and place.

adiabatic lapse rate   (See *dry adiabatic lapse rate, wet adiabatic lapse rate*.)

adiabatic process   Change of temperature within a *gas* because of compression or expansion, without gain or loss of heat from the outside.

advection fog   *Fog* produced by *condensation* within a moist basal air layer moving over a cold land or water surface.

aggradation   Raising of *stream channel* altitude by continued deposition of *bed load*.

A horizon   *Mineral* horizon of the *soil*, overlying the *B horizon* and often characterized in the lower part by loss of *clay minerals* and oxides of iron and aluminum.

air mass   Extensive body of air within which upward gradients of temperature and moisture are fairly uniform over a large area.

albedo   Percentage of *electromagnetic energy* reflected from a surface.

Alfisols   A *soil order* consisting of *soils* of humid and subhumid climates, with high *base status* and an *argillic horizon*.

alluvial fan   Gently sloping, conical accumulation of coarse *alluvium* deposited by a *braided stream* undergoing *aggradation* below the point of emergence of the channel from a narrow gorge or canyon.

alluvial meanders   Sinuous bends of a *graded stream* flowing in the alluvial deposit of a *floodplain*.

alluvial river   *Stream* of low gradient flowing upon thick deposits of *alluvium* and experiencing approximately annual overbank flooding of the adjacent *floodplain*.

alluvial terrace   Benchlike landform carved in *alluvium* by a *stream* during *degradation*.

alluvium   Any stream-laid *sediment* deposit found in a *stream channel* and in low parts of a stream valley subject to flooding.

alpine chains   High mountain ranges that are narrow belts of *tectonic activity* severely deformed by folding and thrusting in comparatively recent geologic time.

alpine glacier   Long narrow mountain *glacier* on a steep downgrade, occupying the floor of a troughlike valley.

Alpine system   Global system of alpine chains, mostly of Cenozoic tectonic activity, but including some adjacent inactive belts affected by Mesozoic tectonic activity.

alpine tundra   A plant *formation class* within the *tundra biome*, found at high altitudes above the limit of *tree* growth.

amphibole group   Complex *silicate minerals* rich in calcium, magnesium, and iron, dark in color, high in *density*, and classed as *mafic minerals*.

andesite   Extrusive igneous rock of diorite composition, dominated by plagioclase feldspar; the extrusive equivalent of diorite.

anemometer   *Weather* instrument used to indicate *wind* speed.

annual temperature range   Difference between *mean monthly temperature* of warmest and coldest months of the year.

annular drainage pattern   A stream network dominated by concentric (ringlike) major *subsequent streams*.

antarctic circle   *Parallel of latitude* at 66½°S.

antarctic zone   Latitude zone in the *latitude* range 60° to 75°S (more or less), centered about on the *antarctic circle*, and lying between the *subantarctic zone* and the *polar zone*.

anticline   Upfold of *strata* or other layered rock in an archlike structure; a class of *folds*. (See also *syncline*.)

anticyclone   Center of high *atmospheric pressure*.

aquatic ecosystems   *Ecosystems* consisting of life-forms of the marine environments and the freshwater environments of the lands.

Aquepts   Suborder of the soil order *Inceptisols;* includes Inceptisols of wet places, seasonally saturated with water.

aquiclude   Rock mass or layer that impedes or prevents the movement of *ground water.*

aquifer   Rock mass or layer that readily transmits and holds *ground water.*

arc–continent collision   Collision of a volcanic arc with continental lithosphere along a subduction boundary.

arctic circle   *Parallel of latitude* at 66½°N.

arctic front zone   Frontal zone of interaction between arctic *air masses* and polar air masses.

arctic tundra   Plant *formation class* within the *tundra biome,* consisting of low, mostly herbaceous plants, but with some very small stunted *trees,* associated with the *tundra climate.*

arctic zone   Latitude zone in the latitude range 60° to 75° N (more or less), centered about on the *arctic circle,* and lying between the *subarctic zone* and the *polar zone.*

argillic horizon   *Soil horizon,* usually the *B horizon,* in which *clay minerals* have accumulated by *illuviation.*

Aridisols   A *soil order* consisting of soils of dry climates, with or without *argillic horizons,* and with accumulations of *carbonates* or soluble salts.

artesian well   Drilled well in which water rises under hydraulic pressure above the level of the surrounding *water table* and may reach the surface.

artificial levee   Earth embankment built parallel with an *alluvial river* channel, usually on the crest of the *natural levee,* to contain the river in *flood stage.*

asthenosphere   Soft layer of the upper *mantle,* beneath the rigid *lithosphere.*

astronomical hypothesis   Explanation for glaciations and interglaciations making use of cyclic variations in the form of the earth's orbit and the angle of inclination of the earth's axis as controls of cyclic variations in the intensity of solar energy received at the earth's surface.

astronomical seasons   Spring, summer, autumn, and winter; subdivisions of the *tropical year* defined by the successive occurrences of *vernal*

equinox, summer solstice, autumnal equinox, and *winter solstice.*

atmosphere   Envelope of gases surrounding the earth, held by *gravity.*

atmospheric pressure   Pressure exerted by the atmosphere because of the force of gravity acting upon the overlying column of air.

atoll   Circular or closed-loop *coral reef* enclosing an open *lagoon* with no island inside.

autumnal equinox   *Equinox* occurring on September 22 or 23.

axial rift   Narrow, trenchlike depression situated along the center line of the *mid-oceanic ridge* and identified with active seafloor spreading.

backarc basin   Comparatively small ocean basin, underlain by oceanic crust and lithosphere, lying between an island arc and the continental mainland, or between two island arcs. Examples: Sea of Japan, Bering Sea.

backswamp   Area of low, swampy ground on the *floodplain* of an *alluvial river* between the *natural levee* and the *bluffs.*

backwash   Return flow of *swash* water under influence of gravity.

badlands   Rugged land surface of steep *slopes,* resembling miniature mountains, developed on weak *clay* formations or clay-rich *regolith* by fluvial erosion too rapid to permit plant growth and soil formation.

bar   Low ridge of sand built above water level across the mouth of a bay or in shallow water paralleling the shoreline. May also refer to embankment of sand or gravel on floor of a *stream channel.*

barometer   Instrument for measurement of *atmospheric pressure.*

barrier island   Long narrow island, built largely of beach sand and dune sand, parallel with the mainland and separated from it by a *lagoon.*

barrier-island coast   *Coastline* with broad zone of shallow water offshore (a *lagoon*) shut off from the ocean by a *barrier island.*

barrier reef   *Coral reef* separated from mainland *shoreline* by a *lagoon.*

basalt   *Extrusive igneous rock* of *gabbro* composition; occurs as *lava.*

base cations (bases)   Certain *cations* in the soil that are also plant nutrients; the most important are cations of

calcium, magnesium, potassium, and sodium.

base flow   That portion of the *discharge* of a stream contributed by *ground water* seepage.

base level   Lower limiting surface or level that can ultimately be attained by a stream under conditions of stability of the earth's crust and sea level; an imaginary surface equivalent to sea level projected inland.

base status of soils   Quality of a *soil* as measured by the presence or absence of clay minerals capable of holding large numbers of *base cations.* Soils of *high base status* are rich in cation-holding clay minerals, such as *montmorillonite;* soils of *low base status* are deficient in such minerals.

batholith   Large, deep-seated body of *intrusive igneous rock,* usually with an area of surface exposure greater than 100 km² (40 mi²).

bauxite   Mixture of several *clay minerals,* consisting largely of aluminum oxide and water with impurities; a principal ore of aluminum.

beach   Thick, wedge-shaped accumulation of sand, gravel, or cobbles in the zone of breaking waves.

beach drift   Transport of sand on a *beach* parallel with a *shoreline* by a succession of landward and seaward water movements at times when *swash* approaches obliquely.

bed load   That portion of the *stream load* moving close to the stream bed by rolling, sliding, and low leaps.

bedrock   Solid rock in place with respect to the surrounding and underlying rock and relatively unchanged by *weathering* processes.

B horizon   Mineral *soil horizon* located beneath the *A horizon,* and usually characterized by a gain of mineral matter (such as *clay minerals* and oxides of aluminum and iron) and organic matter (*humus*).

biomass   Dry weight of living organic matter in an *ecosystem* within a designated surface area; units are grams of organic matter per square meter.

biomass energy   Useful energy derived from natural biomass sources, such as combustible plant tissues or the fermentation of plant matter to produce alcohol or methane gas.

biome   Largest recognizable subdivision of the *terrestrial ecosystems,* including the total assemblage of plant and animal life interacting within the *life layer.*

biosphere　All living organisms of the earth and the environments with which they interact.

bitumen　Combustible mixture of hydrocarbons that is highly viscous and will flow only when heated; considered a form of petroleum. Enclosed in sand, the mixture is known as bituminous sand or tar sand.

block mountains　Class of mountains produced by block faulting and usually bounded by normal faults.

blowout　Shallow depression produced by continued *deflation.*

bluffs　Steeply rising ground slopes marking the outer limits of a *floodplain.*

Boralfs　Suborder of the soil order *Alfisols;* includes Alfisols of *boreal forests* or high mountains.

boreal forest　Variety of *needleleaf forest* found in the *boreal forest climate* regions of North America and Eurasia.

boreal forest climate　Cold climate of the *subarctic zone* in the northern hemisphere with long, extremely severe winters and several consecutive months of zero *potential evapotranspiration (water need).*

bornhardt　Prominent knob of massive *granite* or similar *intrusive igneous rock* with rounded summit and often showing exfoliation shells.

Borolls　Suborder of the soil order *Mollisols;* includes Mollisols of cold-winter semiarid plants *(steppes)* or high mountains.

braided stream　Stream with shallow channel in coarse *alluvium* carrying multiple threads of fast flow, subdividing and rejoining repeatedly and continually shifting in position.

breaker　Sudden collapse of a steepened water wave as it approaches the shoreline.

broadleaf evergreen forest　*Formation class* in the *forest biome* consisting of broadleaf evergreen trees and found in the *moist subtropical climate* and in parts of the *marine west-coast climate.*

bush-fallow farming　Agricultural system practiced in the African savanna woodland in which trees are cut and burned to provide cultivation plots.

butte　Prominent, steep-sided hill or peak, often representing the final remnant of a resistant layer in a region of flat-lying *strata.*

calcic horizon　*Soil horizon* of accumulation of *calcium carbonate* or magnesium carbonate.

calcification　Accumulation of *calcium carbonate* in a *soil,* usually occurring in the *B horizon* or in the C horizon below the *soil solum.*

calcite　Mineral having the composition *calcium carbonate.*

calcium carbonate　Compound consisting of calcium (Ca) and carbonate (CO₃) *ions,* formula CaCO₃, occurring naturally as the mineral *calcite.*

calcrete　Rocklike layer rich in *calcium carbonate,* formed below the *soil solum* in the C horizon. (See also *caliche.*)

caldera　Large, steep-sided circular depression resulting from the explosion and subsidence of a *composite volcano.*

caliche　Name applied in the southwestern United States to the *calcium carbonate* horizon of the *soil,* associated with *Aridisols.* (See also *calcrete.*)

canyon　(See *gorge.*)

carbohydrate　Class of organic compounds consisting of the elements carbon, hydrogen, and oxygen.

carbonates (carbonate minerals, carbonate rocks)　*Minerals* that are carbonate compounds of calcium or magnesium or both, i.e., *calcium carbonate* or magnesium carbonate. (See also *calcite.*)

carbonic acid action　Chemical reaction of carbonic acid in rainwater, soil water, and ground water with minerals; most strongly affects carbonate minerals and rocks, such as limestone and marble; an activity of chemical weathering.

cartography　The science and art of making maps.

cation　Positively charged *ion.*

caverns　(See *limestone caverns.*)

Celsius scale　Temperature scale in which the freezing point of water is 0°, the boiling point 100°.

Cenozoic Era　Last (youngest) of the eras of geologic time.

central eye　Cloud-free central vortex of a *tropical cyclone.*

channel　(See *stream channel.*)

chaparral　Sclerophyll scrub and dwarf *forest* plant *formation class* found throughout the coastal mountain ranges and hills of central and southern California.

chemical energy　Energy stored within an organic molecule and capable of being transformed into heat during metabolism.

chemically precipitated sediment　*Sediment* consisting of mineral matter precipitated from a water solution in which the matter has been transported in the dissolved state as *ions.*

chemical pollutants　Gases introduced into the atmosphere from industrial activities, fuel combustion, and other human activities; not included are the normal gaseous constituents of pure dry air.

chemical weathering　Chemical change in rock-forming *minerals* through exposure to atmospheric conditions in the presence of water; mainly involving *oxidation, hydrolysis, carbonic acid action,* or direct solution.

chert　*Sedimentary rock* composed largely of silicon dioxide and various impurities, in form of nodules and layers, often occurring with *limestone* layers.

chinook wind　A *local wind* occurring at certain times to the lee of the Rocky Mountains; a very dry wind with a high capacity to evaporate *snow.*

cinder cone　Conical hill built of coarse *tephra* ejected from a narrow volcanic vent; a type of *volcano.*

circle of illumination　Great circle that divides the globe at all times into a sunlit hemisphere and a shadowed hemisphere.

circum-Pacific belt　Chains of andesite *volcanoes* making up mountain belts and *island arcs* surrounding the Pacific Ocean basin.

cirque　Bowl-shaped depression carved in rock by glacial processes and holding the *firn* of the upper end of an *alpine glacier.*

clastic sediment　*Sediment* consisting of particles broken away physically from a parent rock source.

clay　*Sediment* particles smaller than 0.004 mm in diameter.

clay minerals　Class of *minerals* produced by alteration of *silicate minerals,* having plastic properties when moist.

cliff　Sheer, near-vertical rock wall formed from flat-lying resistant layered rocks, usually *sandstone, limestone,* or *lava* flows. Cliff may refer to any near-vertical rock wall. (See also *marine cliff.*)

climate　Generalized statement of the prevailing *weather* conditions at a given place, based on statistics of a long period of record and including mean values, departures from those means, and the probabilities associated with those departures.

climate types   Varieties of *climate* recognized under a system of climate classification.

climatology   Science of *climate.*

climax   Stable community of plants and animals reached at the end point of *ecological succession.*

climograph   A graph on which two or more climatic variables, such as monthly mean temperature and monthly mean precipitation, are plotted for each month of the year.

closed system   Idealized system enclosed by a boundary through which there is no flow or exchange of energy, or matter, or both.

cloud reflection   Reflection of incoming *shortwave radiation* from the upper surfaces of clouds to space.

clouds   Dense concentrations of suspended water or ice particles in diameter range 20 to 50 *microns.* (See *cumuliform clouds, stratiform clouds.*)

cloud seeding   Fall of ice crystals from the anvil top of a *cumulonimbus cloud,* serving as *nuclei* of *condensation* at lower levels. (Seeding may also be carried out artificially.)

coal   Rock consisting of hydrocarbon compounds, formed of compacted, lithified, and altered accumulations of plant remains *(peat).*

coast   (See *coastline.*)

coastal blowout dune   High *sand dune* of the *parabolic dunes* class formed adjacent to a *beach,* usually with a deep deflation hollow *(blowout)* enclosed within the dune ridge.

coastal plain   Coastal belt, emerged from beneath the sea as a former *continental shelf,* underlain by *strata* with gentle *dip* seaward.

coastline (coast)   Zone in which coastal processes operate or have a strong influence.

cold front   Moving weather *front* along which a cold *air mass* is forcing itself beneath a warm air mass, causing the latter to be lifted.

cold woodland   Plant *formation class* consisting of *woodland* with low, widely spaced trees and a ground cover of lichens and mosses, found along the northern fringes of the region of *boreal forest climate:* also called *taiga.*

colluvium   Deposit of sediment or rock particles accumulating from overland flow at the base of a slope and originating from higher slopes where sheet erosion is in progress. (See also *alluvium.*)

composite volcano   *Volcano* constructed of alternate layers of *lava* and *tephra* (volcanic ash).

compound (chemical)   Substance consisting of two or more elements, always occurring in the same combination with respect to kinds of atoms and their proportions.

condensation   Process of change of matter in the gaseous state *(water vapor)* to the liquid state (liquid water) or solid state (ice).

conduction of heat   Transmission of sensible heat through matter by transfer of energy from one atom or molecule to the next in the direction of decreasing temperature.

conformal projection   Map projection that preserves without shearing the true shape or outline of any small surface feature of the earth.

conic projections   A group of map projections in which the geographic grid is transformed to lie on the surface of a developed cone.

consequent stream   *Stream* that takes its course down the slope of an *initial landform,* such as a newly emerged *coastal plain* or a *volcano.*

consumers   Animals in the *food chain* that live on organic matter formed by *primary producers* or by other consumers.

consumption of plate   Destruction or disappearance of a subducting lithospheric plate in the asthenosphere, in part by melting of the upper surface, but largely by softening because of heating to the temperature of the surrounding mantle rock.

continental collision   Event in *plate tectonics* in which *subduction* brings two segments of the *continental lithosphere* into contact, leading to formation of a *suture.*

continental drift   Hypothesis, introduced by Alfred Wegener and others early in the 1900s, of the breakup of a parent continent, *Pangaea,* starting near the close of the Mesozoic Era, and resulting in the present arrangement of *continental shields* and intervening *ocean-basin floors.*

continentality   Tendency of large land areas in midlatitudes and high latitudes to impose a large *annual temperature range* on the air temperature cycle.

continental lithosphere   *Lithosphere* bearing continental crust of *felsic igneous rock.*

continental margins   (1) Topographic: one of three major divisions of the ocean basins, being the zones directly adjacent to the continent and including the continental shelf, continental slope, and continental rise. (2) Tectonic: marginal belt of continental crust and lithosphere that is in contact with oceanic crust and lithosphere, with or without an active plate boundary being present at the contact. (See also *active continental margins, passive continental margins.*)

continental rise   Gently sloping seafloor lying at the foot of the *continental slope* and leading gradually into the *abyssal plain.*

continental rupture   Crustal spreading apart affecting the *continental lithosphere,* so as to cause a *rift valley* to appear and to widen, eventually creating a new belt of *oceanic lithosphere.*

continental shelf   Shallow, gently sloping belt of seafloor adjacent to the continental shoreline and terminating at its outer edge in the *continental slope.*

continental shields   Ancient crustal rock masses of the continents, largely *igneous rock* and *metamorphic rock,* and mostly of Precambrian age.

continental slope   Steeply descending belt of seafloor between the *continental shelf* and the *continental rise.*

continental suture   Long, narrow zone of crustal deformation, including underthrusting and intense folding with *nappes,* produced by a *continental collision.* Examples: Himalayan Range, European Alps.

continent–continent collision   Collision between two large masses of continental lithosphere along a subduction plate boundary, resulting in a continental suture.

convection (atmospheric)   Air motion consisting of strong updrafts taking place within a *convection cell.*

convection cell   Individual column of strong updrafts produced by atmospheric *convection.*

converging plate boundary   Boundary along which two lithospheric plates are coming together, requiring one plate to pass beneath the other by *subduction;* same as subduction boundary.

coral reef   Rocklike accumulation of *carbonates* secreted by corals and algae in shallow water along a marine shoreline.

coral-reef coast   *Coast* built out by accumulations of *limestone* in *coral reefs.*

core of earth   Spherical central mass of the earth composed largely of iron and consisting of an outer liquid zone and an interior solid zone.

Coriolis effect   Effect of the earth's rotation tending to turn the direction of motion of any object or fluid toward the right in the northern hemisphere and to the left in the southern hemisphere.

corrosion   Erosion of *bedrock* of a *stream channel* (or other rock surface) by chemical reactions between solutions in stream water and *mineral* surfaces.

counterradiation   *Longwave radiation* of atmosphere directed downward to the earth's surface.

covered shields   Areas of *continental shields* in which the ancient rocks are covered beneath a veneer of sedimentary *strata.*

crater   Central summit depression associated with the principal vent of a *volcano.*

crescent dune   *Sand dune* of crescentic base outline with sharp crest and steep lee *slip face*, with crescent points (horns) pointing downwind.

crude oil   Liquid fraction of *petroleum.*

crust of earth   Outermost solid shell or layer of the earth, composed largely of *silicate minerals.*

Cryaquepts   Great group within the soil suborder of *Aquepts;* includes Aquepts of cold climate regions and particularly the *tundra climate.*

cuesta   *Erosional landform* developed on resistant *strata* having low to moderate *dip* and taking the form of an asymmetrical low ridge or hill belt with one side a steep scarp and the other a gentle slope; usually associated with a *coastal plain.*

cumuliform clouds   *Clouds* of globular shape, often with extended vertical development.

cumulonimbus cloud   Large, dense *cumuliform cloud* yielding *precipitation.*

cumulus   Cloud type consisting of low-lying, white cloud masses of

globular shape well separated from one another.

cutoff   Cutting-through of a narrow neck of land, so as to bypass the stream flow in an *alluvial meander* and cause it to be abandoned.

cycle of rock transformation   (See *rock transformation cycle.*)

cyclone   Center of low *atmospheric pressure*. (See *tropical cyclone, wave cyclone.*)

cyclone family   Succession of *wave cyclones* tracking eastward along the *polar front* while developing from open stage to occluded stage.

cyclonic storm   Intense weather disturbance within a moving *cyclone* generating strong winds, cloudiness, and precipitation.

cylindric projections   Group of map projections in which the geographic grid is transformed to lie on the surface of a developed cylinder.

daylight saving time   Time system under which time is advanced by one hour with respect to the *standard time* of the prevailing *standard meridian.*

debris flood   Streamlike flow of muddy water heavily charged with *sediment* of a wide range of size grades, including boulders, generated by sporadic torrential rains upon steep mountain watersheds.

decalcification   Removal of *calcium carbonate* from a *soil horizon* or *soil solum* as carbonic acid reacts with *carbonate* mineral matter.

deciduous plant   *Tree* or *shrub* that sheds its leaves seasonally, i.e., a *tropophyte.*

deflation   Lifting and transport in *turbulent suspension* by wind of loose particles of *soil* or *regolith* from dry ground surfaces.

deglaciation   Widespread recession of ice sheets during a period of warming global climate, leading to an interglaciation. (See also *glaciation, interglaciation.*)

degradation   Lowering or downcutting of a *stream channel* by *stream erosion* in *alluvium* or *bedrock.*

delta   *Sediment* deposit built by a *stream* entering a body of standing water and formed of the *stream load.*

delta coast   *Coast* bordered by a *delta.*

delta kame   Flat-topped hill of *stratified drift* representing a glacial *delta* constructed adjacent to an *ice sheet* in a marginal glacial lake.

dendritic drainage pattern   Drainage

pattern of treelike branched form, in which the smaller streams take a wide variety of directions and show no parallelism or dominant trend.

density of matter   Quantity of mass per unit of volume, stated in gm/cc.

denudation   Total action of all processes whereby the exposed rocks of the continents are worn down and the resulting *sediments* are transported to the sea by the *fluid agents;* includes also *weathering* and *mass wasting.*

deposition   (See *stream deposition.*)

depositional landforms   *Landforms* that are *sequential landforms* created by the deposition of *sediment* from a fluid medium.

desert biome   *Biome* of the dry climates consisting of thinly dispersed plants that may be *shrubs*, grasses, or perennial *herbs*, but lacking in *trees.*

desert climate subtype   Subtype of the *dry climate* in which no month has *soil-water storage* exceeding 2 cm.

desertification   Degradation of the quality of plant cover and soil as a result of overuse by humans and their domesticated animals, especially during periods of *drought.*

desert pavement   Surface layer of closely fitted pebbles or coarse sand from which finer particles have been removed by *deflation.*

dew-point temperature Temperature of air at the *saturation point.*

diagnostic horizons   Certain *soil horizons*, rigorously defined, that are used as diagnostic criteria in classifying *soils.*

diffuse reflection   Form of *scattering* in which solar rays are deflected or reflected by minute dust particles or cloud particles.

digital image   Numeric representation of a picture consisting of a collection of numeric brightness values *(pixels)* arrayed in a fine grid pattern.

dike   Thin layer of *intrusive igneous rock*, often near-vertical or with steep *dip*, occupying a widened fracture in the rock and typically cutting across older rock planes.

diorite   Intrusive igneous rock consisting dominantly of plagioclase feldspar and pyroxene; a felsic igneous rock.

dip   Acute angle between an inclined natural rock plane or surface and an imaginary horizontal plane of reference; always measured perpendicular to the *strike*. Also a verb, meaning to incline toward.

discharge   Volume of flow moving through a given cross section of a stream in a given unit of time; commonly given in cubic meters (feet) per second.

distributaries   Branching *stream channels* that cross a *delta* to discharge into open water.

doldrums   Belt of calms and variable winds occurring at times along the *equatorial trough.*

dome   (See *sedimentary dome.*)

down-scatter   Scattered *shortwave radiation* directed earthward within the atmosphere.

drainage basin   Total land surface occupied by a *drainage system,* bounded by a drainage divide or watershed.

drainage system   A branched network of *stream channels* and adjacent land *slopes,* bounded by a drainage divide and converging to a single channel at the outlet.

drainage winds   Winds, usually cold, that flow from higher to lower regions under the direct influence of gravity.

drought   Occurrence of substantially lower-than-average *precipitation* in a season that normally has ample precipitation for the support of food-producing plants.

drumlin   Hill of glacial *till,* oval or elliptical in basal outline and with smoothly rounded summit, formed by plastering of till beneath moving, debris-laden glacial ice.

dry adiabatic lapse rate   Rate at which rising air is cooled by expansion when no *condensation* is occurring; 1.0 C°/100 m (5.5 F°/1000 ft).

dry climate   Climate in which the total annual *soil-water shortage* is 15 cm (5.9 in.) or greater, and there is no *water surplus.*

dry desert   Plant *formation class* in the *desert biome* consisting of widely dispersed xerophytic plants that may be small, hardleaved or spiny *shrubs,* succulent plants (cacti), or hard grasses.

dry midlatitude climate   Dry *climate* of the *midlatitude zone* with a strong annual cycle of *potential evapotranspiration (water need)* and cold winters.

dry subtropical climate   Dry *climate* of the *subtropical zone,* transitional

between the *dry tropical climate* and the *dry midlatitude climate.*

dry tropical climate   Dry *climate* of the *tropical zone* with large total annual *potential evapotranspiration (water need).*

dune   (See *sand dune.*)

dust storm   Heavy concentration of dust in a turbulent *air mass,* often associated with a *cold front.*

earthflow   Moderately rapid downhill flowage of masses of water-saturated *soil, regolith,* or weak *shale,* typically forming a steplike terrace at the top and a bulging toe at the base.

earthquake   A trembling or shaking of the ground produced by the passage of *seismic waves.*

easterly wave   Weak, slowly moving trough of low pressure within the belt of *tropical easterlies;* causes a weather disturbance with rain showers.

ecological succession   Time-succession (sequence) of distinctive plant and animal communities occuring within a given area of newly formed land or land cleared of plant cover by burning, clear cutting, or other agents.

ecology   Science of interactions between *life-forms* and their environment; the science of *ecosystems.*

ecosystem   Group of organisms and the environment with which the organisms interact.

electromagnetic energy   Wavelike form of *energy* radiated by any substance possessing heat; travels through space at the speed of light.

electromagnetic spectrum   The total wavelength range of *electromagnetic energy.*

El Niño   Episodic cessation of the typical upwelling of cold deep water off the coast of Peru; literally, "The Christ Child," for its occurrence in the Christmas season once every few years.

eluviation   Soil-forming process consisting of the downward transport of fine particles, particularly the *soil colloids* (both mineral and organic), carrying them out of an upper *soil horizon.*

emergence   Exposure of submarine landforms by a lowering of sea level or a rise of the crust, or both.

energy   The capacity to do work, that is, to affect a change in the state or motion of matter. Forms of energy include *kinetic energy,* potential energy, *chemical energy,* radiant energy

(*electromagnetic energy*), nuclear energy, *sensible heat,* and *latent heat.*

energy deficit   Condition in which rate of outgoing radiant energy exceeds rate of incoming radiant energy at a given time and place.

energy flow system   Open system that receives an input of energy, undergoes internal energy flow, energy transformation, and energy storage, and has an energy output.

energy subsystem   Energy system completely contained within a larger *energy system.*

energy surplus   Condition in which rate of incoming radiant *energy* exceeds the rate of outgoing radiant energy at a given time and place.

energy system   (See *energy flow system.*)

Entisols   A *soil order* consisting of mineral soils lacking *soil horizons* that would persist after normal plowing.

environmental region   Region of the continental surface that has a unique set of physical environmental factors and supports a distinctive *ecosystem, biome,* or *formation class.*

environmental temperature lapse rate   Rate of temperature decrease upward through the *troposphere;* standard value is 6.4 C°/km (3½ F°/1000 ft).

epipedon   *Soil horizon* that forms at the surface.

epiphytes   Plants that live above ground level out of contact with the soil, usually growing on the limbs of *trees* or *shrubs;* also called "air plants."

equal-area projections   Class of map projections on which any given area of the earth's surface is shown to correct relative areal extent, regardless of position on the globe.

equator   That *parallel of latitude* occupying a position midway between the earth's poles of *rotation;* the largest of the parallels, designated as *latitude* 0°.

equatorial current   West-flowing *ocean current* in the belt of *trade winds.*

equatorial rainforest   Plant *formation class* within the *forest biome,* consisting of tall, closely set broadleaf trees of evergreen or semideciduous habit.

equatorial trough   Low-pressure trough centered more or less over the *equator* and situated between the two belts of *trade winds.*

equatorial zone   *Latitude* zone lying

between lat. 10° S and 10° N (more or less) and centered upon the *equator*.

equinox   Instant in time when the *subsolar point* falls on the earth's *equator* and the *circle of illumination* passes through both poles. *Vernal equinox* occurs on March 20 or 21; *autumnal equinox* on September 22 or 23.

erg   Large expanse of active *sand dunes* in the Sahara Desert of North Africa.

erosional landforms   Class of the *sequential landforms* shaped by the removal of *regolith* or *bedrock* by agents of erosion. Examples: *canyon*, *glacial cirque*, *marine cliff*.

esker   Narrow, often sinuous embankment of coarse gravel and boulders deposited in the bed of a meltwater stream enclosed in a tunnel within stagnant ice of an ice sheet.

Eurasian–Indonesian belt   Major *mountain arc* system extending from southern Europe across southern Asia and Indonesia.

eutrophication   Excessive growth of algae and other *primary producers* in a stream or lake as a result of the input of large amounts of nutrient *ions*, especially phosphate and nitrate.

evaporation   Process in which water in liquid state or solid state passes into the vapor state.

evaporites   Class of *chemically precipitated sediment* and *sedimentary rock* composed of soluble salts deposited from saltwater bodies.

evapotranspiration   Combined water loss to the atmosphere by *evaporation* from the soil and *transpiration* from plants.

evergreen plant   *Tree* or *shrub* that holds most of its green leaves throughout the year.

exfoliation dome   Smoothly rounded rock knob or hilltop bearing rock sheets or shells produced by spontaneous expansion accompanying *unloading*.

exotic river   *Stream* that flows across a region of *dry climate* and derives its *discharge* from adjacent uplands where a *water surplus* exists.

exponential decay system   Program of decrease in a variable quantity at a rate such that the quantity is halved in a constant time interval.

exposed shields   Areas of *continental shields* in which the ancient basement rock, usually of Precambrian age, is exposed to the surface.

extrusive igneous rock   Rock produced by the solidification of *lava* or ejected fragmental *igneous rock* (*tephra*).

Fahrenheit scale   Temperature scale in which the freezing point of water is 32°, the boiling point 212°. (See also *Kelvin scale*.)

fallout   *Gravity* fall of atmospheric particles of *particulate matter* reaching the ground.

fault   Sharp break in rock with displacement (slippage) of block on one side with respect to adjacent block. (See *normal fault*, *overthrust fault*, *transcurrent fault*, *transform fault*.)

fault coast   *Coast* formed when *shoreline* comes to rest against a *fault scarp*.

fault line   Surface trace of a *fault*.

fault-line scarp   Erosion scarp developed upon an inactive *fault line*.

fault plane   Surface of slippage between two earth blocks moving relative to each other during faulting.

fault scarp   Clifflike surface feature produced by faulting and exposing the *fault plane*; commonly associated with a *normal fault*.

feldspar   Group of *silicate minerals* consisting of silicate of aluminum and one or more of the metals potassium, sodium, or calcium. (See *plagioclase feldspar*, *potash feldspar*.)

felsic igneous rock   *Igneous rock* dominantly composed of *felsic minerals*.

felsic minerals, felsic mineral group   *Quartz* and *feldspars* treated as a mineral group of light color and relatively low *density*. (See also *mafic minerals*.)

ferricrete   Rocklike surface layer rich in sesquioxide of iron (hematite); essentially the same as an exposed layer of *laterite*.

fiord   Narrow, deep ocean embayment partially filling a *glacial trough*.

fiord coast   Deeply embayed, rugged *coast* formed by partial *submergence* of *glacial troughs*.

firn   Granular old snow forming a surface layer in the zone of accumulation of a *glacier*.

flocculation   The clotting together of colloidal mineral particles to form larger particles.

flood   Stream flow at a stream stage so high that it cannot be

accommodated within the *stream channel* and must spread over the banks to inundate the adjacent *floodplain*.

flood basalts   Large-scale outpourings of basalt lava to produce thick accumulations of *basalt* over large areas.

floodplain   Belt of low, flat ground, present on one or both sides of a *stream channel*, subject to inundation by a *flood* about once annually and underlain by *alluvium*.

flood stage   Designated stream-surface level for a particular point on a *stream*, higher than which overbank flooding may be expected.

flow map   Map using bands of varying width scaled proportionately to the magnitude of flow of some entity in a given path.

fluid   Substance that flows readily when subjected to unbalanced stresses; may exist as a *gas* or a *liquid*.

fluid agents   *Fluids* that erode, transport, and deposit mineral matter and organic matter; they are running water, waves and currents, glacial ice, and wind.

fluvial landforms   *Landforms* shaped by running water.

fluvial processes   Geomorphic processes in which running water is the dominant *fluid agent*, acting as *overland flow* and *stream flow*.

fog   Cloud layer in contact with land or sea surface, or very close to that surface. (See *advection fog*, *radiation fog*.)

folding   Process by which *folds* are produced; a form of *tectonic activity*.

folds   Wavelike corrugations of *strata* (or other layered rock masses) as a result of crustal compression.

forb   Broad-leaved *herb*, as distinguished from the grasses.

forearc trough   Long, narrow submarine trough lying between a tectonic arc and a volcanic arc and receiving sediments from either or both arcs; it is associated with an active subduction boundary along which an accretionary prism is being formed. Example: Savu Sea.

foredunes   Ridge of irregular *sand dunes* typically found adjacent to *beaches* on low-lying *coasts* and bearing a partial cover of plants.

foreland folds   Folds produced by continental collision in strata of a passive continental margin.

forest   Assemblage of *trees* growing close together, their crowns forming a layer of foliage that largely shades the ground.

forest biome   *Biome* that includes all regions of *forest* over the lands of the earth.

formation classes   Subdivisions within a *biome* based on the size, shape, and structure of the plants that dominate the vegetation.

fossil fuels   Naturally occurring hydrocarbon compounds that represent the altered remains of organic materials enclosed in rock; specifically, *coal*, *lignite*, *petroleum (crude oil)*, *natural gas*, *kerogen*, and *shale oil*.

fractional scale   Ratio of distance between two points on a map or a globe to the actual distance between the same two points on the earth's surface.

fracture zone (oceanic)   Linear ocean-floor scarps or ridges offsetting the mid-oceanic ridge and its axial rift. Most of these features are now recognized as active *transform faults* or *transform scars*.

freezing   Change from liquid state to solid state accompanied by release of *latent heat*, becoming *sensible heat*.

fringing reef   *Coral reef* directly attached to land with no intervening *lagoon* of open water.

front   Surface of contact between two unlike *air masses*. (See *cold front*, *occluded front*, *polar front*, *warm front*.)

frost   (See *killing frost*.)

frost action   Rock breakup by forces accompanying the *freezing* of water.

funnel cloud   Long, narrow cloud of tubular or funnel shape hanging from the base of a *cumulonimbus cloud*; it represents a *tornado* or a waterspout.

gabbro   *Intrusive igneous rock* consisting largely of pyroxene and *plagioclase feldspar*, with variable amounts of *olivine*; a *mafic igneous rock*.

gas (gaseous state)   *Fluid* of very low *density* (as compared with a *liquid* of the same chemical composition) that expands to fill uniformly any small container and is readily compressed.

geographic grid   Complete network of *parallels* and *meridians* on the surface of the globe, used to fix the locations of surface points.

geologic norm   Stable natural condition in a moist climate in which slow *soil erosion* is paced by

maintenance of *soil horizons* bearing a plant community in an equilibrium state.

geology   Science of the solid earth, including the earth's origin and history, materials comprising the earth, and the processes acting within the earth and upon its surface.

geomorphology   Science of *landforms*, including their history and processes of origin.

geostationary orbit   Satellite orbit that holds a fixed position over a selected point on the earth's equator.

geothermal energy   Heat energy of igneous origin drawn from steam, hot water, or dry hot rock beneath the earth's surface.

geothermal locality   Place where geothermal heat reaches the earth's surface, emanating from a deep-seated magma body in the crust.

geyser   Periodic jetlike emission of hot water and steam from a narrow vent at a *geothermal locality*.

glacial abrasion   *Abrasion* by a moving *glacier* of the *bedrock* floor beneath it.

glacial delta   *Delta* built by meltwater streams of a *glacier* into standing water of a marginal glacial lake.

glacial drift   General term for all varieties and forms of rock debris deposited in close association with *ice sheets* of the *Pleistocene Epoch*.

glacial plucking   Removal of masses of *bedrock* from beneath an *alpine glacier* or *ice sheets* as ice moves forward suddenly.

glacial trough   Deep, steep-sided rock trench of a U-shaped cross section formed by *alpine glacier* erosion.

glaciation   (1) General term for the total process of glacier growth and *landform* modification by *glaciers*. (2) Single episode or time period in which *ice sheets* formed, spread, and disappeared.

glacier   Large natural accumulation of land ice affected by present or past flowage. (See *alpine glacier*.)

glaze   Ice layer accumulated upon solid surfaces by the *freezing* of falling *rain* or drizzle.

global water balance   Balance among the three hydrologic components—*precipitation*, *evaporation*, and *runoff*—for the earth as a whole.

gneiss   Variety of *metamorphic rock*

showing banding and commonly rich in *quartz* and *feldspar*.

gorge (canyon)   Steep-sided *bedrock* valley with a narrow floor limited to the width of a *stream channel*.

graben   Trenchlike depression representing the surface of a crustal block dropped down between two opposed, infacing *normal faults*. (See *rift valley*.)

graded profile   Smoothly descending profile displayed by a *graded stream*.

graded stream   Stream (or *stream channel*) with *stream gradient* so adjusted as to achieve a balanced state in which average *bed load* transport is matched to average bed load input; an average condition over periods of many years' duration.

granite   *Intrusive igneous rock* consisting largely of *quartz*, *potash feldspar*, and *plagioclase feldspar*, with minor amounts of biotite and hornblende; a *felsic igneous rock*.

granitic rock   General term for rock of the upper layer of the continental crust, composed largely of felsic igneous and metamorphic rock; rock of composition similar to that of granite.

graphic scale   Map scale as shown by a line divided into equal parts.

grassland biome   *Biome* consisting largely or entirely of *herbs*, which may include grasses, grasslike plants, and *forbs*.

gravitation   Mutual attraction between any two masses.

gravity   Gravitational attraction of the earth upon any small mass near the earth's surface. (See *gravitation*.)

gravity gliding   Forward motion of a thrust sheet on a gently downsloping fault plane under the force of gravity.

gravity percolation   Downward movement of water under the force of gravity through the *soil-water belt* and *intermediate belt*, eventually arriving at the *water table*.

great circle   Circle formed by passing a plane through the exact center of a perfect sphere; the largest circle that can be drawn on the surface of a sphere.

great soil groups   Third level of *soil* classification, following *soil orders* and *soil suborders*.

greenhouse effect   Accumulation of heat in the lower *atmosphere* through the absorption of *longwave radiation* from the earth's surface.

green revolution   Major advance in securing increased agricultural production in the developing nations through improved genetic strains of wheat and rice and the application of fossil-fuel energy.

groin   Human-made wall or embankment built out into the water at right angles to the *shoreline.*

gross photosynthesis   Total amount of *carbohydrate* produced by *photosynthesis* by a given organism or group of organisms in a given unit of time.

ground radiation   *Longwave radiation* emitted by land or water surfaces and passing upward into the overlying atmosphere.

ground water   *Subsurface water* occupying the *saturated zone* and moving under the force of gravity.

gullies   Deep, V-shaped trenches carved by newly formed *streams* in rapid headward growth during advanced stages of accelerated soil erosion.

gyres   Large circular *ocean current* systems centered upon the oceanic subtropical high-pressure cells.

habitat   Subdivision of the plant environment having a certain combination of *slope,* drainage, *soil* type, and other controlling physical factors.

Hadley cell   Atmospheric circulation cell in low latitudes involving rising air over an *equatorial trough* and sinking air over *subtropical high-pressure belts.*

hail   Form of *precipitation* consisting of pellets or spheres of ice with a concentric layered structure.

halocarbons   Synthetic compounds containing carbon, fluorine, and chlorine atoms; used as aerosol propellants and refrigerants.

haze   Minor concentration of pollutants or natural forms of *particulate matter* in the atmosphere causing a reduction in visibility.

heat   (See *sensible heat, latent heat.*)

heat engine   Mechanical system in which motion is powered by heat energy.

heat island   Persistent region of higher air temperatures centered over a city.

herb   Tender plant, lacking woody stems, usually small or low; it may be annual or perennial.

high base status   (See *base status of soils.*)

high-latitude climates   Group of climates in the *subarctic zone, arctic zone,* and *polar zone,* dominated by arctic *air masses* and polar air masses.

high-pressure cell   Center of high barometric pressure; an *anticyclone.*

Histosols   *Soil order* consisting of *soils* with a thick upper layer of organic matter.

hogback   Sharp-crested, often sawtooth ridge formed of the upturned edge of a resistant rock layer of *sandstone, limestone,* or *lava.*

Holocene Epoch   Last epoch of geologic time, commencing about 10,000 years ago; it followed the *Pleistocene Epoch* and includes the present.

horse latitudes   *Subtropical high-pressure belt* of the North Atlantic Ocean, coincident with the central region of the Azores high; a belt of weak, variable winds and frequent calms.

horst   Crustal block uplifted between two *normal faults.*

hot spot   Center of intrusive igneous and volcanic activity thought to be located over a rising *mantle plume.*

hot springs   Springs discharging heated ground water at a temperature close to the boiling point; found in geothermal areas and thought to be related to a magma body at depth.

humid climate subtype   Subtype of the *moist climate* in which the annual *water surplus* is 1 mm or greater but less than 60 cm, with annual water surplus always greater than the annual *soil-water shortage.*

humidity   General term for the amount of *water vapor* present in the air. (See *relative humidity, specific humidity.*)

humus   Dark brown to black organic matter on or in the *soil,* consisting of fragmented plant tissues partly oxidized by *consumer* organisms.

hurricane   *Tropical cyclone* of the western North Atlantic and Caribbean Sea.

hydraulic action   *Stream erosion* by impact force of the flowing water upon the bed and banks of the *stream channel.*

hydrograph   Graphic presentation of the variation in stream discharge with elapsed time, based on data of stream gauging at a given station on a stream.

hydrologic cycle   Total plan of movement, exchange, and storage of the earth's free water in gaseous state, liquid state, and solid state.

hydrology   Science of the earth's water and its motions through the *hydrologic cycle.*

hydrolysis   Chemical union of water molecules with *minerals* to form different, more stable mineral *compounds.*

hydrosphere   Total water realm of the earth's surface zone, including the oceans, surface waters of the lands, *ground water,* and water held in the atmosphere.

hygrometer   Instrument that measures the *water vapor* content of the *atmosphere;* some types measure *relative humidity* directly.

hygrophytes   Plants adapted to a wet environment on the lands.

ice age   Span of geologic time, usually on the order of one to three million years, or longer, in which glaciations alternate with interglaciations repeatedly in rhythm with cyclic global climate changes. (See also *interglaciation, glaciation.*)

iceberg   Mass of glacial ice floating in the ocean, derived from a *glacier* that extends into tidal water.

ice lobes (glacial lobes)   Broad tonguelike extensions of an *ice sheet* resulting from more rapid ice motion where terrain was more favorable.

ice sheet   Large thick plate of glacial ice moving outward in all directions from a central region of accumulation.

ice sheet climate   Severely cold climate, found on the Greenland and Antarctic ice sheets, with *potential evapotranspiration (water need)* effectively zero throughout the year.

ice shelf   Thick plate of floating glacial ice attached to an *ice sheet* and fed by the ice sheet and by snow accumulation.

ice storm   Occurrence of heavy *glaze* of ice on solid surfaces.

ice wedge   Vertical, wall-like body of ground ice, often tapering downward, occupying a shrinkage crack in *silt* of *permafrost* areas.

ice-wedge polygons   Polygonal networks of *ice wedges.*

igneous rock   Rock solidified from a high-temperature molten state; rock formed by cooling of *magma*. (See *extrusive igneous rock, felsic igneous rock, intrusive igneous rock, mafic igneous rock, ultramafic igneous rock.*)

illite　*Clay mineral* derived by *chemical weathering* from such *silicate minerals* as *feldspar* and muscovite mica.

illuviation　Accumulation in a lower *soil horizon* (typically, the *B horizon*) of materials brought down from a higher horizon; a soil-forming process.

image processing　Mathematical manipulation of *digital images*, for example, to enhance contrast or edges.

Inceptisols　*Soil order* consisting of *soils* having weakly developed *soil horizons* and containing weatherable *minerals*.

infiltration　Absorption and downward movement of *precipitation* into the *soil* and *regolith*.

infrared imagery　Images formed by *infrared radiation* emanating from the ground surface as recorded by a remote sensor.

infrared radiation　*Electromagnetic energy* in the *wavelength* range of 0.7 to about 200 *microns*.

initial landforms　*Landforms* produced directly by internal earth processes of *volcanism* and *tectonic activity*. Examples: *volcano, fault scarp*.

inner lowerland　On a *coastal plain*, a shallow valley lying between the first *cuesta* and the area of older rock (oldland).

insolation　Interception of solar energy (*shortwave radiation*) by an exposed surface.

interglaciation　Within an ice age, a time interval of mild global climate in which continental ice sheets were largely absent or were limited to the Greenland and Antarctic ice sheets; the interval between two glaciations. (See also *deglaciation, glaciation*.)

intermediate belt　Zone below the *soil-water belt* too deep to supply capillary water to plants; i.e., too deep to be reached by plant roots.

International Date Line　The 180° *meridian of longitude*, together with deviations east and west of that meridian, forming the time boundary between adjacent *standard time* zones that are 12 hours fast and 12 hours slow with respect to Greenwich standard time.

interrupted projection　Projection subdivided into a number of sectors (gores), each of which is centered on a different central meridian.

intertropical convergence zone (ITC)　Zone of convergence of *air masses* of *tropical easterlies* (trade winds) along the axis of the *equatorial trough*.

intrusive igneous rock　*Igneous rock* body produced by solidification of *magma* beneath the surface, surrounded by preexisting rock.

inversion　(See *temperature inversion*.)

ion　Atom or group of atoms bearing an electrical charge as the result of a gain or loss of one or more electrons. (See also *cation*.)

isarithm　(See *isopleth*.)

island arcs　Curved lines of volcanic islands associated with active *subduction* zones along the boundaries of *lithospheric plates*.

isobar　Line on map passing through all points having the same *atmospheric pressure*.

isobaric surface　Surface of equal *atmospheric pressure*.

isohyet　Line on a map drawn through all points having the same numerical value of *precipitation*.

isopleth　Line drawn on the surface of a map or globe so as to pass through all points having the same value of a selected property or entity.

isotherm　Line drawn on a map to pass through all points having the same air temperature.

jet stream　High-speed air flow in narrow bands within the *upper-air westerlies* and along certain other global *latitude* zones at high levels.

joints　Fractures within *bedrock*, usually occurring in parallel and intersecting sets of planes.

kaolinite　*Clay mineral* typically formed by *hydrolysis* from *potash feldspar* (also from micas).

karst　Landscape or topography dominated by surface features of *limestone* solution and underlain by a *limestone cavern* system.

Kelvin scale (K)　Temperature scale on which the starting point is absolute zero, equivalent to −273°C.

kerogen　Waxy substance of hydrocarbon composition held in oil shale and capable of yielding liquid *petroleum* upon distillation.

killing frost　Occurrence of below-freezing temperature in the air layer near the ground, capable of damaging frost-sensitive plants.

kinetic energy　Form of energy represented by matter (mass) in motion.

knob and kettle　Terrain of numerous small knobs of *glacial drift* and deep depressions usually situated along the *moraine* belt of a former *ice sheet*.

lagoon　Shallow body of open water lying between a *barrier island* or a *barrier reef* and the mainland.

lag time　Interval of time between occurrence of *precipitation* and peak *discharge* of a *stream*.

lahar　Rapid downslope or downvalley movement of a tonguelike mass of water-saturated tephra (volcanic ash) originating high up on a steep-sided volcanic cone; a variety of mudflow.

land breeze　*Local wind* blowing from land to water during the night.

landform (soil science)　General term for the configuration of the ground surface as a factor in *soil* formation; it includes *slope* steepness and aspect, as well as relief.

landforms　Configurations of the land surface taking distinctive forms and produced by natural processes. Examples: hill, valley, plateau. (See *depositional landforms, erosional landforms, initial landforms, sequential landforms*.)

landmass　Large area of continental crust lying above sea level (base level) and thus available for removal by *denudation*.

landmass rejuvenation　Episode of rapid fluvial *denudation* set off by a rapid crustal rise, increasing the available *landmass*.

landslide　Rapid sliding of large masses of *bedrock* on steep mountain *slopes* or from high *cliffs*.

langley (ly)　Unit of intensity of solar radiation equal to one gram-calorie per square centimeter.

lapse rate　(See *environmental temperature lapse rate, dry adiabatic lapse rate, wet adiabatic lapse rate*.)

latent heat　Heat absorbed and held in storage in a *gas* or *liquid* during the processes of *evaporation* or melting, respectively; distinguished from *sensible heat*.

lateral cutting　Sidewise shifting of a *stream channel* caused by undercutting of steep banks on the outsides of bends.

lateral moraine　*Moraine* forming an embankment between the ice of an *alpine glacier* and the adjacent valley wall.

laterite   Rocklike layer rich in *sequioxide of aluminum* and iron, including the minerals *bauxite* and *limonite*, found in low latitudes in association with *Ultisols* and *Oxisols*.

latitude   Arc of a meridian between the *equator* and a given point on the globe.

lava   *Magma* emerging on the earth's solid surface, exposed to air or water.

liana   Woody vine supported on the trunk or branches of a *tree*.

life-form   Characteristic physical structure, size, and shape of a plant or of an assemblage of plants.

life layer   Shallow surface zone containing the biosphere; a zone of interaction between atmosphere and land surface, and between atmosphere and ocean surface.

limestone   Nonclastic *sedimentary rock* in which *calcite* is the predominant mineral, and with varying minor amounts of other minerals, and *clay*.

limestone caverns   Interconnected subterranean cavities formed in *limestone* by *carbonic acid action* occurring in slowly moving *ground water*.

limonite   Mineral or group of minerals consisting largely of iron oxide and water, produced by *chemical weathering* of other iron-bearing minerals.

liquid   *Fluid* that maintains a free upper surface and is only very slightly compressible, as compared with a *gas*.

liquid state   Fluid state of matter having the properties of a liquid.

lithosphere   (1) General term for the entire solid earth realm. (2) In *plate tectonics*, it is the strong, brittle outermost rock layer lying above the *asthenosphere*.

lithospheric plate   Segment of *lithosphere* moving as a unit, in contact with adjacent lithospheric plates along plate boundaries.

littoral drift   Transport of *sediment* parallel with the *shoreline* by the combined action of *beach drift* and longshore current transport.

loam   Soil-texture class in which no one of the three size grades (sand, silt, clay) dominates over the other two.

local winds   General term for *winds* generated as direct or immediate effects of the local terrain.

loess   Accumulation of yellowish to buff-colored, fine-grained *sediment*, largely of *silt* grade, upon upland surfaces after transport in the air in *turbulent suspension* (i.e., carried in a *dust storm*).

longitude   Arc of a parallel between the *prime meridian* and a given point on the globe.

longitudinal dunes   Class of *sand dunes* in which the dune ridges are oriented parallel with the prevailing wind.

longshore drift   *Littoral drift* caused by action of a longshore current.

longwave radiation   *Electromagnetic energy* emitted by the earth, largely in the range from 3 to 50 *microns*.

low base status   (See *base status of soils*.)

lowland   Broad, open valley between two *cuestas* of a *coastal plain*. (Lowland may refer to any low area of land surface.)

low-latitude climates   Group of climates of the *equatorial zone* and *tropical zone* dominated by the *subtropical high-pressure belt* and the *equatorial trough*.

low-latitude rainforest   Evergreen broadleaf forest of the wet equatorial and tropical climate zones.

low-level temperature inversion   Reversal of normal *environmental temperature lapse rate* in an air layer near the ground.

loxodrome   Line of constant compass bearing drawn on a map or navigational chart; also known as a *rhumb line*.

macronutrients   Nine elements required in abundance for organic growth, including *primary production* by green plants. (See also *micronutrients*.)

mafic igneous rock   *Igneous rock* dominantly composed of *mafic minerals*.

mafic minerals, mafic mineral group   *Minerals*, largely *silicates*, rich in magnesium and iron, dark in color, and of relatively great density.

magma   Mobile, high-temperature molten state of *rock*, usually of *silicate mineral* composition and with dissolved gases.

mantle   Rock layer or shell of the earth beneath the *crust* and surrounding the *core*, composed of *ultramafic igneous rock* of silicate mineral composition.

mantle plume   A columnlike rising of heated *mantle* rock, thought to be the cause of a *hot spot* in the overlying *lithospheric plate*.

map projection   Any orderly system of parallels and meridians drawn on a flat surface to represent the earth's surface.

marble   Variety of *metamorphic rock* derived from *limestone* or dolomite by recrystallization under pressure.

marine cliff   *Rock* cliff shaped and maintained by the undermining action of breaking waves.

marine scarp   Steep seaward *slope* in poorly consolidated *alluvium, glacial drift*, or other forms of *regolith*, produced along a coastline by the undermining action of waves.

marine terrace   Former *abrasion platform* elevated to become a steplike coastal *landform*.

marine west-coast climate   Cool, *moist climate* of west coasts in the *midlatitude zone*, usually with a substantial annual *water surplus* and a distinct winter *precipitation* maximum.

mass wasting   Spontaneous downward movement of *soil, regolith*, and *bedrock* under the influence of *gravity*; does not include the action of *fluid agents*.

material flow system   System of interconnected flowpaths of material (matter) that may comprise a *closed system* or an *open system*.

maximum–minimum thermometer   Pair of thermometers recording the maximum and minimum air temperatures since last reset.

mean annual temperature   Mean of daily air temperature means for a given year or succession of years.

mean daily temperature   Sum of daily maximum and minimum air temperature readings divided by two.

meanders   (See *alluvial meanders*.)

mean monthly temperature   Mean of daily air temperature means for a given calendar month.

mean solar day   Average time required for the earth to complete one *rotation* with respect to the sun; time elapsed between one solar noon and the next, averaged over the period of one *year*.

mean velocity   Mean, or average, speed of flow of water through an entire *stream* cross section.

Mediterranean climate   Climate type of the *subtropical zone*, characterized by the alternation of a very dry summer and a mild, rainy winter.

Mercator projection   Conformal map projection with horizontal parallels and vertical meridians and with map scale rapidly increasing with increase in latitude.

mercurial barometer   *Barometer* using the Torricelli principle, in which *atmospheric pressure* counterbalances a column of mercury in a tube.

meridian of longitude   North–south line on the surface of the global *oblate ellipsoid*, connecting the *north pole* and *south pole*.

meridional transport   Flow of energy (heat) or matter (water) across the *parallels of latitude*, either poleward or equatorward.

mesa   Table-topped *plateau* of comparatively small extent bounded by *cliffs* and occurring in a region of flat-lying *strata*.

mesophytes   Plants adapted to a *habitat* of intermediate degree of wetness and to uniform *soil water* availability.

mesosphere   Atmospheric layer of upwardly diminishing temperature, situated above the *stratopause* and below the *mesopause*.

metamorphic rock   *Rock* altered in physical structure and/or chemical (*mineral*) composition by action of heat, pressure, shearing stress, or infusion of elements, all taking place at substantial depth beneath the surface.

meteorology   Science of the *atmosphere*; particularly the physics of the lower or inner atmosphere.

mica group   Aluminum-silicate *mineral* group of complex chemical formula having perfect cleavage into thin sheets.

microcontinent   Fragment of continental crust and its lithosphere of subcontinental dimensions that is embedded in an expanse of oceanic lithosphere.

micron   Length unit; one micron equals 0.0001 cm.

micronutrients   Elements essential to organic growth, but only in very small amounts. (See also *macronutrients*.)

microwaves   Waves of the electromagnetic radiation spectrum in the wavelength band from about 0.03 cm to about 1 cm.

midlatitude climates   Group of climates of the *midlatitude zone* and *subtropical zone*, located in the *polar front zone* and dominated by both tropical air masses and polar air masses.

midlatitude deciduous forest   *Formation class* within the *forest biome* dominated by tall, broadleaf deciduous trees, found mostly in the *moist continental climate* and *marine west-coast climate*.

midlatitude zones   Latitude zones occupying the *latitude* range 35° to 55° N and S (more or less) and lying between the *subtropical zones* and the *subarctic (subantarctic) zones*.

mid-oceanic ridge   One of three major divisions of the ocean basins, being the central belt of submarine mountain topography with a characteristic *axial rift*.

millibar   Unit of *atmospheric pressure*; one-thousandth of a bar. Bar is a force of one million dynes per square centimeter.

mineral   Naturally occurring inorganic substance, usually having a definite chemical composition and a characteristic atomic structure. (See *felsic minerals, mafic minerals, silicate minerals*.)

mineral alteration   Chemical change of *minerals* to more stable compounds upon exposure to atmospheric conditions; same as *chemical weathering*.

mineral density   (See *density of matter*.)

mistral   Local drainage wind of cold air affecting the Rhone Valley of southern France.

Moho   Contact surface between the earth's *crust* and *mantle*; a contraction of Mohorovič, the seismologist who discovered this feature.

moist climate   *Climate* in which the annual *soil-water shortage* is less than 15 cm (5.9 in.).

moist continental climate   *Moist climate* of the *midlatitude zone* with strongly defined winter and summer seasons, adequate *precipitation* throughout the year, and a substantial annual water surplus.

moist subtropical climate   *Moist climate* of the *subtropical zone*, characterized by a moderate to large annual *water surplus* and a strongly seasonal cycle of *potential· evapotranspiration (water need)*.

mollic epipedon   Relatively thick, dark-colored surface *soil horizon*, or *epipedon*, containing substantial amounts of organic matter (*humus*) and usually rich in *base cations*.

Mollisols   *Soil order* consisting of *soils* with a *mollic epipedon* and *high base status*.

monadnock   Prominent, isolated mountain or large hill rising conspicuously above a surrounding peneplain and composed of a rock more resistant than that underlying the peneplain; a landform of denudation in moist climates. (See also *bornhardt*.)

monsoon forest   Plant *formation class* within the *forest biome* consisting in part of deciduous trees adapted to a long dry season in the *wet–dry tropical climate*.

monsoon and trade-wind littoral climate   *Moist climate* of low latitudes showing a strong rainfall peak in the season of high sun and a short period of reduced rainfall.

montane forest   Plant *formation class* of the *forest biome* found in cool upland environments of the *tropical zone* and *equatorial zone*.

montmorillonite   *Clay mineral* derived by the chemical alteration of *silicate minerals* in various *igneous rocks*.

moraine   Accumulation of rock debris carried by an *alpine glacier* or an *ice sheet* and deposited by the ice to become a *depositional landform*. (See *lateral moraine, terminal moraine*.)

mountain arc   Curved (arcuate) segment of an *alpine chain*.

mountain roots   Erosional remnants of deep portions of ancient *sutures* that were once *alpine chains*.

mountain winds   Daytime movements of air up the gradient of valleys and mountain slopes; alternating with nocturnal *valley winds*.

mud   Sediment consisting of a mixture of clay and silt with water, often with minor amounts of sand and sometimes with organic matter.

mudflow   A form of *mass wasting* consisting of the downslope flowage of a mixture of water and *mineral* fragments (*soil, regolith, disintegrated bedrock*), usually following a natural drainage line or *stream channel*.

multispectral image   Image consisting of two or more images, each of which is taken from a different portion of the spectrum (e.g., blue, green, red, infrared).

multispectral scanner   Remote-sensing instrument, flown on an aircraft or spacecraft, that simultaneously collects multiple *digital images (multispectral images)* of the ground. Typically, images are collected in four to eight spectral bands.

nappe Overturned recumbent fold of strata, usually associated with *thrust sheets* in a collision *orogen*.

natural gas Naturally occurring mixture of hydrocarbon compounds (principally methane) in the gaseous state held within certain porous rocks.

natural levee Belt of higher ground paralleling a meandering *alluvial river* on both sides of the *stream channel* and built up by deposition of fine *sediment* during periods of overbank flooding.

natural vegetation (native vegetation) Stable, mature plant cover characteristic of a given area of land surface free from the influences and impacts of human activities.

needleleaf forest Plant *formation class* within the *forest biome*, consisting largely of needleleaf evergreen trees. (See also *boreal forest*.)

net photosynthesis *Carbohydrate* remaining in an organism after *respiration* has broken down sufficient carbohydrate to power the metabolism of the organism.

net primary production Rate at which *carbohydrate* is accumulated in the tissues of plants within a given *ecosystem;* units are grams of dry organic matter per year per square meter of surface area.

net radiation Difference in intensity between all incoming energy (positive quantity) and all outgoing energy (negative quantity) carried by both *shortwave radiation* and *longwave radiation*.

noon (See *solar noon*.)

normal fault Variety of *fault* in which the *fault plane* inclines (*dips*) toward the downthrown block and a major component of the motion is vertical.

north pole Point at which the northern end of the earth's axis of *rotation* emerges from the earth's surface.

nuclei (atmospheric) Minute particles of solid matter suspended in the atmosphere and serving as cores for *condensation* of water or ice.

occluded front Weather *front* along which a moving *cold front* has overtaken a *warm front*, forcing the warm *air mass* aloft.

ocean-basin floors One of the major divisions of the ocean basins, comprising the deep portions consisting of *abyssal plains* and low hills.

ocean current Persistent, dominantly horizontal flow of ocean water.

oceanic lithosphere *Lithosphere* bearing oceanic crust.

oceanic trench Narrow, deep depression in the seafloor representing the line of *subduction* of an oceanic *lithospheric plate* beneath the margin of a continental lithospheric plate; often associated with an *island arc*.

ocean tide Periodic rise and fall of the ocean level induced by gravitational attraction between the earth and moon in combination with earth *rotation*.

ochric epipedon Surface *soil horizon* (*epipedon*) that is light in color and contains less than one percent organic matter.

oil sand (See *bitumen*.)

oil shale *Shale* or other types of *sedimentary rock* through which hydrocarbon compounds of organic origin are disseminated.

olivine *Silicate mineral* with magnesium and iron but no aluminum, usually olive-green or grayish-green; a *mafic mineral*.

open system System of interconnected flow paths of energy or matter with a boundary through which that energy or matter can enter and leave the system.

organic sediment *Sediment* consisting of the organic remains of plants or animals.

orogen The mass of tectonically deformed rocks and related igneous rocks produced during an orogeny.

orogeny Major episode of *tectonic activity* resulting in *strata* being deformed by folding and faulting.

orographic precipitation *Precipitation* induced by the forced rise of moist air over a mountain barrier.

outcrop Surface exposure of *bedrock*.

outwash plain Flat, gently sloping plain built up of *sand* and gravel by the *aggradation* of meltwater *streams* in front of the margin of an *ice sheet*.

overland flow Motion of a surface layer of water over a sloping ground surface at times when the *infiltration* rate is exceeded by the *precipitation* rate; a form of *runoff*.

overthrust fault *Fault* characterized by the overriding of one crustal block (or *thrust sheet*) over another along a gently inclined *fault plane;* associated with crustal compression.

oxbow lake, oxbow swamp Crescent-shaped lake or swamp representing the abandoned channel left by the *cutoff* of an *alluvial meander*.

oxidation Chemical union of free oxygen with metallic elements in *minerals*.

Oxisols *Soil order* consisting of very old, highly weathered *soils* of low latitudes, with an oxic horizon and *low base status*.

ozone Gas with a molecule consisting of three atoms of oxygen, O_3.

ozone layer Layer in the *stratosphere*, mostly in the altitude range 20 to 35 km (12 to 31 mi), in which a concentration of *ozone* is produced by the action of solar *ultraviolet radiation*.

pack ice Floating *sea ice* that completely covers the sea surface.

Pangaea Hypothetical parent continent, enduring until near the close of the Mesozoic Era, consisting of the *continental shields* of Laurasia and Gondwana joined into a single unit. (See *continental drift*.)

parabolic blowout dune Type of *parabolic dune* formed to the lee of a shallow deflation hollow, usually found in interior plains in a *dry climate*.

parabolic dunes Isolated low *sand dunes* of parabolic outline, with points directed into the prevailing wind.

parallel of latitude East–west circle on the earth's surface, lying in a plane parallel with the *equator* and at right angles to the axis of rotation.

parent matter of soil Inorganic, *mineral* base from which the *soil* is formed; usually consists of *regolith*.

particulate matter *Solid* and *liquid* particles capable of being suspended for long periods in the *atmosphere*.

passive continental margins Continental margins lacking active plate boundaries at the contact of continental crust with oceanic crust. A passive margin thus lies within a single *lithospheric plate*. Example: Atlantic continental margin of North America. (See also *continental margins, active continental margins*.)

passive systems Electromagnetic remote sensor systems that measure radiant energy reflected or emitted by an object or surface.

peat Partially decomposed, compacted accumulation of plant remains occurring in a bog environment.

ped   Individual natural *soil* aggregate.

pediment   Gently sloping, rock-floored land surface found at the base of a mountain mass or *cliff* in an arid region.

pediplain   Desert land surface of low relief composed in part of *pediment* surfaces and in part of *alluvial fan* surfaces.

pedology   Science of the *soil* as a natural surface layer capable of supporting living plants; synonymous with *soil science*.

pedon   *Soil* column extending down from the surface to reach a lower limit in some form of *regolith* or *bedrock*.

peneplain   Land surface of low elevation and slight relief produced in the late stages of *denudation* of a *landmass*.

perhumid climate subtype   Subtype of the *moist climate* in which the *soil-water surplus* is 60 cm or greater.

peridotite   Igneous rock consisting largely of olivine and pyroxene; an *ultramafic igneous rock* occurring as a pluton, also thought to compose much of the upper mantle.

permafrost   Condition of permanently frozen water in the *soil*, *regolith*, and *bedrock* in cold climates of subarctic and arctic regions.

petroleum (crude oil)   Natural liquid mixture of many complex hydrocarbon compounds of organic origin, found in accumulations (oil pools) within certain *sedimentary rocks*.

pH   Measure of the concentration of hydrogen *ions* in a solution. (The number represents the logarithm to the base 10 of the reciprocal of the weight in grams of hydrogen ions per liter of water.)

photochemical reactions   Chemical reactions occurring in polluted air through the action of sunlight upon pollutant gases to synthesize new toxic compounds or gases.

photosynthesis   Production of *carbohydrate* by the union of water with carbon dioxide while absorbing light *energy*. (See *gross photosynthesis, net photosynthesis*.)

physical geography   The study and synthesis of selected subject areas from the natural sciences—especially atmospheric science, hydrology, physical oceanography, geology, geomorphology, soil science, and plant ecology—in order to gain a complete picture of the physical environment of humans and to examine the interactions of humans with that environment.

physical oceanography   Physical science of the oceans, as distinguished from biological oceanography.

physical weathering   Breakup of massive rock (*bedrock*) into small particles through the action of physical forces acting at or near the earth's surface. (See *weathering*.)

pixel   Individual brightness value within a *digital image*.

plagioclase feldspar   Aluminum-silicate *mineral* with sodium or calcium or both.

plane of the ecliptic   Imaginary plane in which the earth's orbit lies.

plateau   Upland surface, more or less flat and horizontal, upheld by resistant beds of *sedimentary rock* or *lava* flows and bounded by a steep *cliff*.

plate tectonics   Theory of *tectonic activity* dealing with *lithospheric plates* and their activity.

playa   Flat land surface underlain by fine *sediment* or evaporite minerals deposited from shallow lake waters in a *dry climate* in the floor of a closed topographic depression.

Pleistocene Epoch   Epoch of the *Cenozoic Era*, often identified as the Ice Age; it preceded the *Holocene Epoch*.

plinthite   Iron-rich concentrations present in some kinds of *soils* in deeper *soil horizons* and capable of hardening into rocklike material with repeated wetting and drying.

plucking   (See *glacial plucking*.)

pluton   Any body of *intrusive igneous rock* that has solidified below the surface, enclosed in preexisting *rock*.

pocket beach   *Beach* of crescentic outline located at a bay head.

polar easterlies   System of easterly surface winds at high latitude, best developed in the southern hemisphere, over Antarctica.

polar front   *Front* lying between cold polar *air masses* and warm tropical air masses, often situated along a *jet stream* within the *upper-air westerlies*.

polar front zone   Broad zone in midlatitudes and higher latitudes, occupied by the shifting *polar front*.

polar high   Persistent low-level center of high *atmospheric pressure* located over the *polar zone* of Antarctica.

polar outbreak   Tongue of cold polar air, preceded by a *cold front*, penetrating far into the *tropical zone* and often reaching the *equatorial zone*; it brings rain squalls and unusual cold.

polar projection   Map projection centered on earth's north or south pole.

polar stereographic projection   Polar map projection that is a stereographic grid.

polar zones   *Latitude* zones lying between 75° and 90° N and S.

pollutants   In air pollution studies, foreign matter injected into the lower atmosphere as *particulate matter* or as *chemical pollutants*.

pollution dome   Broad, low dome-shaped layer of polluted air, formed over an urban area at times when winds are weak or calm prevails.

pollution plume   (1) The trace or path of pollutant substances, moving along the flow paths of *ground water*. (2) Trail of polluted air carried downwind from a pollution source by strong winds.

polypedon   Smallest distinctive geographic unit of the *soil* of a given area; it consists of *pedons*.

potash feldspar   Aluminum-silicate *mineral* with potassium the dominant metal.

potential evapotranspiration (water need)   Ideal or hypothetical rate of *evapotranspiration* estimated to occur from a complete canopy of green foliage of growing plants continuously supplied with all the *soil water* they can use; a real condition reached in those situations where *precipitation* is sufficiently great or irrigation water is supplied in sufficient amounts.

prairie   Plant *formation class* of the *grassland biome*, consisting of dominant tall grasses and subdominant *forbs*, widespread in subhumid continental climate regions of the *subtropical zone* and *midlatitude zone*. (See *short-grass prairie, tall-grass prairie*.)

Precambrian time   All of geologic time older than the beginning of the Cambrian Period, i.e., older than 600 million years.

precipitation   Particles of liquid water or ice that fall from the atmosphere and may reach the ground. (See *orographic precipitation*.)

pressure cell   (See *high-pressure cell*.)

pressure gradient   Change of *atmospheric pressure* measured along a line at right angles to the *isobars*.

pressure-gradient force   Force acting horizontally, tending to move air in the direction of lower *atmospheric pressure.*

prevailing westerly winds (westerlies) Surface winds blowing from a generally southwesterly direction in the *midlatitude zone,* but varying greatly in direction and intensity.

primary minerals In *pedology* (soil science), the original, unaltered *silicate minerals* of the *igneous rocks* and *metamorphic rocks.*

primary production (See *net primary production.*)

prime meridian Reference meridian of zero *longitude;* universally accepted as the Greenwich meridian.

producers Organisms that use light *energy* to convert carbon dioxide and water to *carbohydrates* through the process of *photosynthesis.*

progradation Shoreward building of a *beach, bar,* or *sandspit* by addition of coarse *sediment* carried by *littoral drift* or brought from deeper water offshore.

pyroxene group Complex aluminum-silicate *minerals* rich in calcium, magnesium, and iron, dark in color, high in density, classed as *mafic minerals.*

quartz Mineral of silicon dioxide composition.

quartzite *Metamorphic rock* consisting largely of the mineral *quartz.*

radar Wavelength band within the microwave region, beginning at about 0.1 cm and extending to about 100 cm (1 m).

radial drainage pattern Stream pattern consisting of *streams* radiating outward from a central peak or highland, such as a sedimentary *dome* or a *volcano.*

radiation (See *electromagnetic energy.*)

radiation balance Condition of balance between incoming energy of solar *shortwave radiation* and outgoing *longwave radiation* emitted by the earth into space.

radiation fog *Fog* produced by radiational cooling of the basal air layer.

rain Form of *precipitation* consisting of falling water drops, usually 0.5 mm or larger in diameter.

rain gauge Instrument used to measure the amount of *rain* that has fallen.

raingreen vegetation Vegetation that puts out green foliage in the wet season, but becomes largely dormant in the dry season; found in the *tropical zone,* it includes the *savanna biome* and *monsoon forest.*

rainshadow desert Belt of arid climate to lee of a mountain barrier, produced as a result of adiabatic warming of descending air.

raised shoreline Former *shoreline* lifted above the limit of wave action; also called an elevated *shoreline.*

rapids Steep-gradient reaches of a *stream channel* in which *stream* velocity is high.

reg Desert surface armored with a pebble layer, resulting from long-continued *deflation;* found in the Sahara Desert of North Africa.

regolith Layer of *mineral* particles overlying the *bedrock;* may be derived by *weathering* of underlying bedrock or transported from other locations by *fluid agents.* (See *residual regolith, transported regolith.*)

relative humidity Ratio of *water vapor* present in the air to the maximum quantity possible for saturated air at the same temperature.

remote sensing Measurement of some property of an object or surface by means other than direct contact; usually refers to the gathering of scientific information about the earth's surface from great heights and over broad areas, using instruments mounted on aircraft or orbiting space vehicles.

representative fraction (R.F.) Fractional scale stated as a simple fraction.

residual regolith *Regolith* formed in place by alteration of the *bedrock* directly beneath it.

resolution On a map, power to resolve minute objects present on the ground.

respiration Metabolic process in which organic compounds are oxidized within living cells to yield *chemical energy* and waste heat.

retrogradation Cutting back (retreat) of a *shoreline, beach, marine cliff,* or *marine scarp* by wave action.

revolution Motion of a planet in its orbit around the sun, or of a planetary satellite around a planet.

rhumb line (See *loxodrome.*)

rhyolite Extrusive igneous rock of granite composition; it occurs as lava or tephra.

ria coast Deeply embayed *coast* formed by partial *submergence* of a *landmass* previously shaped by fluvial *denudation.*

Richter scale Scale of magnitude numbers describing the quantity of energy released by an *earthquake.*

ridge-and-valley landscape Assemblage of *landforms* developed by *denudation* of a system of open *folds* of *strata* and consisting of long, narrow ridges and valleys arranged in parallel or zigzag patterns.

rift valley Trenchlike valley with steep, parallel sides; essentially a *graben* between two *normal faults;* associated with crustal spreading.

rill erosion Form of *accelerated erosion* in which numerous, closely spaced miniature channels (rills) are scored into the surface of exposed *soil* or *regolith.*

rock Natural aggregate of *minerals* in the solid state; usually hard and consisting of one, two, or more mineral varieties.

rock transformation cycle Total cycle of changes in which *rock* of any one of the three major rock classes—*igneous rock, sedimentary rock, metamorphic rock*—is transformed into rock of one of the other classes.

Rossby waves Horizontal undulations in the flow path of the *upper-air westerlies;* also known as upper-air waves.

rotation Spinning of a spherical object around an axis.

runoff Flow of water from continents to oceans by way of *stream flow* and *ground water* flow; a term in the water balance of the *hydrologic cycle.* In a more restricted sense, runoff refers to surface flow by *overland flow* and channel flow.

salic horizon *Soil horizon* enriched by soluble salts.

salinization Precipitation of soluble salts within the *soil*

salt marsh Peat-covered expanse of *sediment* built up to the level of high tide over a previously formed tidal mud flat.

sand *Sediment* particles between 0.06 and 2 mm in diameter.

sand dune Hill or ridge of loose, well-sorted *sand* shaped by wind and usually capable of downwind motion.

sand sea Field of *transverse dunes.*

sandspit Narrow, fingerlike

embankment of *sand* constructed by *littoral drift* into the open water of a bay.

sandstone   Variety of *sedimentary rock* consisting largely of mineral particles of *sand* grade size.

sanitary landfill   Disposal of solid wastes by burial beneath a cover of *soil* or *regolith*.

Santa Ana   Easterly wind, often hot and dry, that blows from the interior desert region of southern California and passes over the coastal mountain ranges to reach the Pacific Ocean.

saturated zone   Zone beneath the land surface in which all pores of the *bedrock* or *regolith* are filled with *ground water*.

saturation point   That point at which rising *relative humidity* in a mass of air reaches 100 percent and the air holds its full capacity of *water vapor* at the given temperature.

savanna biome   *Biome* that consists of a combination of *trees* and grassland in various proportions.

savanna woodland   Plant *formation class* of the *savanna biome* consisting of a *woodland* of widely spaced *trees* and a grass layer, found throughout the *wet–dry tropical climate* regions in a belt adjacent to the *monsoon forest* and *equatorial rainforest*.

scale of map   Ratio of distance between two points on a map and the same two points on the ground. (See *fractional scale, graphic scale*.)

scanning systems   Remote-sensing systems that make use of a scanning beam to generate images over the frame of surveillance.

scarification   General environmental impact term for artificial excavations and other land disturbances produced for purposes of extracting or processing mineral resources.

scattering   Turning aside by reflection of solar *shortwave radiation* by gas molecules of the atmosphere.

schist   Foliated *metamorphic rock* in which mica flakes are typically found oriented parallel with foliation surfaces.

sclerophyll forest   Plant *formation class* of the *forest biome*, consisting of low sclerophyll trees, and often including sclerophyll woodland or *scrub*, associated with regions of *Mediterranean climate*.

sclerophylls   Hardleaved evergreen *trees* and *shrubs* capable of enduring a long, dry summer.

scrub   Plant *formation class* or subclass consisting of *shrubs* and having a canopy coverage of about 50 percent.

sea breeze   Local wind blowing from sea to land during the day.

sea cliff   (See *marine cliff*.)

sea ice   Floating ice of the oceans formed by direct freezing of ocean water.

secondary minerals   In *soil science*, *minerals* that are stable in the surface environment, derived by *mineral alteration* of the *primary minerals*.

sediment   Finely divided *mineral* matter and organic matter derived directly or indirectly from preexisting *rock* and from life processes. (See *chemically precipitated sediment, organic sediment*.)

sedimentary dome   Up-arched *strata* forming a circular structure with domed summit and flanks with moderate to steep outward *dip*.

sedimentary rock   Rock formed from accumulation of *sediment*.

sediment yield   Quantity of *sediment* removed by *overland flow* from a land surface of a given unit area in a given unit of time.

seismic sea wave (tsunami)   Train of sea waves set off by an *earthquake* (or other seafloor disturbance) traveling over the ocean surface with a velocity proportional to the square root of the ocean depth.

seismic waves   Waves sent out during an *earthquake* by faulting or other crustal disturbance from an earthquake focus and propagated through the solid earth.

semiarid (steppe) climate subtype   Subtype of the *dry climate* in which *soil-water storage* equals or exceeds 6 cm in at least two months of the year.

semidesert   Plant *formation class* of the *desert biome*, consisting of xerophytic *shrub* vegetation with a poorly developed herbaceous lower layer; subtypes are semidesert scrub and woodland and semidesert scrub.

semidesert climate subtype   Subtype of the dry climate in which soil-water storage exceeds 6 cm in fewer than two months, but is greater than 2 cm in at least one month.

sensible heat   Heat measurable by a thermometer; an indication of the intensity of *kinetic energy* of molecular motion within a substance.

sequential landforms   *Landforms* produced by external earth processes

in the total activity of *denudation*. Examples: *canyon, alluvial fan, floodplain*.

sesquioxide of aluminum   Oxide of aluminum with a ratio of two atoms of aluminum to three atoms of oxygen.

shale   Fissile, *sedimentary rock* of *mud* or *clay* composition, showing lamination.

sheet erosion   Phase of *accelerated soil erosion* in which thin layers of *soil* are removed without formation of rills or *gullies*.

sheeting structure   Thick, subparallel layers of massive *bedrock* formed by spontaneous expansion accompanying *unloading*.

shield volcano   Low, often large, domelike accumulation of basalt lava flows emerging from long radial fissures on flanks.

shoreline   Shifting line of contact between water and land.

short-grass prairie   (See *steppe*.)

shortwave radiation   *Electromagnetic energy* in the range from 0.2 to 3 *microns*, including most of the energy spectrum of solar radiation.

shrub   Woody perennial plant, usually small or low, with several low-branching stems and a foliage mass close to the ground.

side-looking airborne radar (SLAR)   Remote sensing using radar sensor systems that send their impulses toward either side of an aircraft.

silcrete   Rocklike surface layer cemented largely with silicon dioxide; it is widespread in the *tropical zone*.

silicates, silicate minerals   *Minerals* containing silicon and oxygen atoms, linked in the crystal space lattice in units of four oxygen atoms to each silicon atom.

sill   *Intrusive igneous rock* in the form of a plate where *magma* was forced into a natural parting in the *bedrock*, such as a bedding surface in a sequence of *sedimentary rocks*.

silt   *Sediment* particles between 0.004 and 0.06 mm in diameter.

sinkhole   Surface depression in *limestone*, leading down into *limestone caverns*.

slash-and-burn   Agricultural system, practiced in the low-latitude rainforest, in which small areas are cleared and the trees burned, forming plots that can be cultivated for brief periods.

slate   Compact, fine-grained variety of *metamorphic rock*, derived from *shale*, showing well-developed cleavage.

sleet Form of *precipitation* consisting of ice pellets, which may be frozen raindrops.

slip face Steep face of an active *sand dune*, receiving sand by saltation over the dune crest and repeatedly sliding because of oversteepening.

slope (1) Degree of inclination from the horizontal of an element of ground surface, analogous to *dip* in the geologic sense. (2) Any portion or element of the earth's solid surface, as in *hillslope*. (3) Verb meaning "to incline."

smog Mixture of *particulate matter* and *chemical pollutants* in the lower atmosphere over urban areas.

snow Form of *precipitation* consisting of ice particles.

soft layer of mantle Layer within the *mantle* in which the temperature is close to the melting point, causing the mantle rock to be weak; synonym for *asthenosphere*.

soil Natural terrestrial surface layer containing living matter and supporting or capable of supporting plants.

soil colloids Mineral particles of extremely small size, capable of remaining suspended indefinitely in water; typically they have the form of thin plates or scales.

soil creep Extremely slow downhill movement of *soil* and *regolith* as a result of continued agitation and disturbance of the particles by such activities as *frost action*, temperature changes, or wetting and drying of the soil.

soil erosion Erosional removal of material from the *soil* surface.

soil horizon Distinctive layer of the *soil*, more or less horizontal, set apart from other soil zones or layers by differences in physical and chemical composition, organic content, structure, or a combination of those properties, produced by soil-forming processes.

soil orders Those ten *soil* classes forming the highest category in the classification of soils.

soil profile Display of *soil horizons* on the face of a *pedon*, or on any freshly cut vertical exposure through the *soil*.

soil science (See *pedology*.)

soil solum That part of the *soil* made up of the A and B *soil horizons*; the soil zone in which living plant roots exert a control on the soil horizons.

soil structure Presence, size, and form of aggregations (lumps or clusters) of *soil* particles.

soil suborders Second level of classification of *soils*.

soil texture Descriptive property of the *mineral* portion of the *soil* based on varying proportions of *sand*, *silt*, and *clay*.

soil water Water held in the *soil* and available to plants through their root systems; a form of *subsurface water* (Synonymous with *soil moisture*.)

soil-water balance Balance among the component terms of the *soil-water budget*; namely, *precipitation*, *evapotranspiration*, change in *soil-water storage*, and *water surplus*.

soil-water belt *Soil* layer from which plants draw *soil water*.

soil-water budget Accounting system evaluating the daily, monthly, or yearly amounts of *precipitation*, *evapotranspiration*, *soil-water storage*, *water deficit*, and *water surplus*.

soil-water recharge Restoring of depleted *soil water* by *infiltration* of *precipitation*.

soil-water shortage Difference between *potential evapotranspiration* (water need) and *actual evapotranspiration* (water use), representing the quantity of irrigation water that would be required to sustain maximum plant growth.

soil-water storage Actual quantity of water held in the *soil-water belt* at any given instant; usually applied to a soil layer of given depth, such as 300 cm (12 in.).

solar collectors Mechanical devices for absorbing direct solar energy and allowing the energy to be transported for use or storage.

solar constant Intensity of solar radiation falling upon a unit area of surface held at right angles to the sun's rays at a point outside the earth's atmosphere; equal to about 2 gram-calories per square centimeter per minute (2 cal/cm²/min), or 2 *langleys* per minute (2 ly/min).

solar day (See *mean solar day*.)

solar energy Energy arriving as electromagnetic radiation from the sun, including such energy stored as heat in air, soil, or water, and chemical energy stored in the biomass of plants through photosynthesis.

solar noon Instant at which the sun crosses the celestial meridian of a given point on the earth; instant at which the sun's shadow points exactly due north or due south.

solids Substances in the solid state; they resist changes in shape and volume, are usually capable of withstanding large unbalanced forces without yielding, but will ultimately yield by sudden breakage.

solifluction Tundra (arctic) variety of *earthflow* in which the saturated thawed layer over *permafrost* flows slowly downhill to produce multiple terraces and solifluction lobes.

sorting Separation of one grade size of *sediment* particles from another by the action of currents of air or water.

source region Extensive land or ocean surface over which an *air mass* derives its temperature and moisture characteristics.

south pole Point at which the southern end of the earth's axis of *rotation* emerges from the earth's surface.

specific humidity Mass of *water vapor* contained in a unit mass of air.

spit (See *sandspit*.)

splash erosion *Soil erosion* caused by direct impact of falling raindrops on a wet surface of *soil* or *regolith*.

spodic horizon *Soil horizon* containing precipitated amorphous materials composited of organic matter and *sesquioxides* of aluminum, with or without iron.

Spodosols *Soil order* consisting of *soils* with a *spodic horizon*, an *albic horizon*, with *low base status*, and lacking in *carbonate* materials.

spreading plate boundary Lithospheric plate boundary along which two plates of oceanic lithosphere are undergoing separation, while at the same time new lithosphere is being formed by accretion. (See also *converging plate boundary*, *transform plate boundary*.)

standard meridians Standard time meridians separated by 15° of *longitude* and having values that are multiples of 15°. (In some cases meridians are used that are multiples of 7½°.)

standard time Time system based on the local time of a *standard meridian* and applied to belts of *longitude* extending 7½° (more or less) on either side of that meridian.

steppe Plant *formation class* in the *grassland biome* consisting of short grasses sparsely distributed in clumps and bunches and some *shrubs*, widespread in areas of semiarid climate in continental interiors of North America and Eurasia; also called *short-grass prairie*.

steppe climate   (See *semiarid (steppe) climate subtype.*)

stone rings   Linked ringlike ridges of cobbles or boulders lying at the surface of the ground in arctic and alpine tundra regions.

storage capacity   Maximum capacity of *soil* to hold water against the pull of gravity.

storage recharge   Restoration of stored *soil water* during periods when *precipitation* exceeds *potential evapotranspiration (water need).*

storage withdrawal   Depletion of stored *soil water* during periods when *evapotranspiration* exceeds *precipitation*, calculated as the difference between *actual evapotranspiration (water use)* and *precipitation.*

storm surge   Rapid rise of coastal water level accompanying the onshore arrival of a *tropical cyclone.*

strata   Layers of *sediment* or *sedimentary rock* in which individual beds are separated from one another along bedding planes.

stratified drift   *Glacial drift* made up of sorted and layered *clay, silt, sand,* or gravel deposited from meltwater in *stream channels*, or marginal lakes close to the ice front.

stratiform clouds   Clouds of layered, blanketlike form.

stratosphere   Layer of *atmosphere* lying directly above the *troposphere.*

stratus   Cloud type of the low-height family formed into a dense, dark gray layer.

stream   Long, narrow body of flowing water occupying a *stream channel* and moving to lower levels under the force of *gravity*. (See *consequent stream, graded stream, subsequent stream.*)

stream capacity   Maximum *stream load* of solid matter that can be carried by a *stream* for a given *discharge.*

stream channel   Long, narrow, troughlike depression occupied and shaped by a *stream* moving to progressively lower levels.

stream deposition   Accumulation of transported particles on a *stream* bed, upon the adjacent *floodplain*, or in a body of standing water.

stream erosion   Progressive removal of mineral particles from the floor or sides of a *stream channel* by drag force of the moving water, or by *abrasion*, or by *corrosion.*

stream flow   Water flow in a *stream channel;* same as channel flow.

stream gradient   Rate of descent to lower elevations along the length of a *stream channel*, stated in m/km, ft/mi, degrees, or percent.

stream load   Solid matter carried by a *stream* in dissolved form (as *ions*), in turbulent suspension, and as *bed load.*

stream transportation   Downvalley movement of eroded particles in a *stream channel* in solution, in *turbulent suspension*, or as *bed load.*

strike   Compass direction of the line of intersection of an inclined rock plane and a horizontal plane of reference. (See *dip.*)

strip mining   Mining method in which overburden is first removed from a seam of *coal*, or a sedimentary ore, allowing the coal or ore to be extracted.

subantarctic low-pressure belt   Persistent belt of *low atmospheric pressure* centered about at lat. 65° S over the Southern Ocean.

subantarctic zone   *Latitude* zone lying between lat. 55° and 60° S (more or less) and occupying a region between the *midlatitude zone* and the *antarctic zone.*

subarctic zone   *Latitude* zone between lat. 55° and 60° N (more or less), occupying a region between the *midlatitude zone* and the *arctic zone.*

subduction   Descent of the downbent edge of a *lithospheric plate* into the *asthenosphere* so as to pass beneath the edge of the adjoining plate.

subhumid climate subtype   Subtype of the *mosit climate* in which annual *soil-water shortage* is greater than zero but less than 15 cm.

sublimation   Process of change of *water vapor* (gaseous state) to ice (solid state) or vice versa.

submarine canyon   Narrow V-shaped submarine valley cut into the continental slope, usually attributed to erosion by turbidity currents.

submarine cone   (See *submarine fan.*)

submarine fan (cone)   Submarine accumulation of coarse-textured sediment carried by turbidity currents to form a large fan-shaped deposit on the deep ocean floor, usually situated at the lower end of a submarine canyon system leading down the outer slopes of a major delta on the continental shelf.

submergence   Inundation or partial drowning of a former land surface by a rise of sea level or a sinking of the *crust* or both.

subsequent stream   *Stream* that develops its course by *stream erosion* along a band or belt of weaker *rock.*

subsolar point   Point at which solar rays are perpendicular to the earth's surface.

subsurface water   Water of the lands held in *soil, regolith,* or *bedrock* below the surface.

subtropical broadleaf evergreen forest   Broadleaf evergreen forest of the moist subtropical climate regions.

subtropical high-pressure belts   Belts of persistent high *atmospheric pressure* trending east–west and centered about on lat. 30° N and S.

subtropical zones   *Latitude* zones occupying the region of lat. 25° to 35° N and S (more or less) and lying between the *tropical zones* and the *midlatitude zones.*

summer monsoon   Inflow of maritime air at low levels from the Indian Ocean toward the Asiatic low pressure center in the season of high sun; associated with the rainy season of the *wet–dry tropical climate* and the Asiatic monsoon climate.

summer solstice   Solstice occurring on June 21 or 22, when the *subsolar point* is located at 23½° N.

sun-synchronous orbit   Satellite orbit in which the orbital plane remains fixed in position with respect to the sun.

supercooled water   Water existing in the *liquid state* at a temperature lower than the normal freezing point.

surface detention   Temporary holding of precipitation in minor surface depressions.

surface water   Water of the lands flowing exposed (as *streams*) or impounded (as ponds, lakes, marshes).

surges   Episodes of very rapid downvalley movement within an *alpine glacier.*

suspended load   That part of the *stream load* carried in *turbulent suspension.*

suspension   (See *turbulent suspension.*)

suture   (See *continental suture.*)

swash   Surge of water up the *beach* slope (landward) following collapse of a breaker.

syncline   Downfold of *strata* (or other layered rock) in a troughlike structure; a class of *folds.* (See also *anticline.*)

system boundary   Bounding surface, real or imagined, limiting the extent of a flow system of matter or energy.

taiga   (See *cold woodland.*)

tall-grass prairie   (See *prairie.*)

talus   Accumulation of loose rock fragments derived by fall of rock from a *cliff.*

talus slope   *Slope* formed of *talus.*

tectonic activity   Crustal process of bending (folding) and breaking (faulting), concentrated on or near active *lithospheric plate* boundaries.

tectonic arc   Long, narrow chain of islands or mountains or a narrow submarine ridge adjacent to a subduction boundary and its trench, formed by tectonic processes, such as the construction and rise of an accretionary prism.

tectonic crest   Ridgelike summit line of a tectonic arc associated with an accretionary prism.

tectonic erosion   Removal of masses of rock from the lower edge of a lithospheric plate by downdrag exerted by a subducting plate passing beneath it.

tectonics   Branch of geology relating to tectonic activity and the features it produces. (See also *plate tectonics, tectonic activity.*)

temperature inversion   Upward reversal of the normal *environmental temperature lapse rate,* so that the air temperature increases upward. (See *low-level temperature inversion.*)

tephra   Collective term for all size grades of solid *igneous rock* particles blown out under gas pressure from a volcanic vent.

terminal moraine   *Moraine* deposited as an embankment at the glacier terminus of an *alpine glacier* or at the leading edge of an *ice sheet.*

terrane   Continental crustal rock unit having a distinctive set of lithologic properties, reflecting its geologic history, that distinguish it from adjacent or surrounding continental crust.

terrestrial ecosystems   *Ecosystems* of land plants and animals found on upland surfaces of the continents.

thematic map   Map devoted to presentation of only a single data category.

thermal environment   Total influence of heat and cold upon living organisms in the *life layer.*

thermal infrared   Electromagnetic radiation in the infrared radiation wavelength band, approximately from 1 to 20 microns.

thermal pollution   Form of water pollution in which heated water is discharged into a stream or lake from the cooling system of a power plant or other industrial heat source.

thermocline   Water layer in which temperature changes rapidly in the vertical direction.

thorntree semidesert   *Formation class* within the *desert biome,* transitional from *grassland biome* and *savanna biome* and consisting of xerophytic *trees* and *shrubs.*

thorntree–tall-grass savanna   Plant *formation class,* transitional between the *savanna biome* and the *grassland biome,* consisting of widely scattered *trees* in an open grassland.

thrust sheet   Sheetlike mass of rock moving forward over a low-angle *overthrust fault.*

thunderstorm   Intense, local convectional storm associated with a *cumulonimbus cloud* and yielding heavy *precipitation,* along with lightning and thunder, and sometimes the fall of *hail.*

tidal current   Current set in motion by the *ocean tide.*

tidal inlet   Narrow opening in a *barrier island* or baymouth *bar* through which *tidal currents* flow.

tidal power   Power derived from tidal currents moving through constricted coastal passages, usually modified by dam structures.

tide   (See *ocean tide.*)

tide curve   Graphical presentation of the rhythmic rise and fall of ocean water because of *ocean tides.*

till   Heterogeneous mixture of rock fragments ranging in size from clay to boulders, deposited beneath moving glacial ice or directly from the melting in place of stagnant glacial ice.

till plain   Undulating, plainlike land surface underlain by glacial *till.*

time zones   Zones or belts of given east–west (longitudinal) extent within which *standard time* is applied according to a uniform system.

tor   Group of boulders of joint blocks forming a small but conspicuous hill.

tornado   Small, very intense wind vortex with extremely low air pressure in center, formed beneath a dense *cumulonimbus cloud* in proximity to a *cold front.*

trade winds   Surface winds in low latitudes, representing the low-level airflow within the *tropical easterlies.*

transcurrent fault   Variety of *fault* on which the motion is dominantly horizontal along a near-vertical *fault plane.*

transform fault   Special case of a *transcurrent fault* making up the boundary of two moving *lithospheric plates;* usually found along an offset of the *mid-oceanic ridge* where seafloor spreading is in progress.

transform plate boundary   Lithospheric plate boundary along which two plates are in contact on a transform fault; the relative motion is that of a transcurrent fault.

transform scar   Linear topographic feature of the ocean floor taking the form of an irregular scarp or ridge and originating at the offset axial rift of the mid-oceanic ridge; it represents a former transform fault but is no longer a plate boundary.

transpiration   Evaporative loss of water to the atmosphere from leaf pores of plants.

transportation   (See *stream transportation.*)

transported regolith   *Regolith* formed of mineral matter carried by fluid agents from a distant source and deposited upon the *bedrock* or upon older regolith. Examples: floodplain silt, lake clay, beach sand.

transverse dunes   Field of wavelike *sand dunes* with crests running at right angles to the direction of the prevailing wind.

travertine   Carbonate mineral matter, usually *calcite,* accumulating upon *limestone cavern* surfaces situated in the *unsaturated zone.*

tree   Large erect woody perennial plant typically having a single main trunk, few branches in the lower part, and a branching crown.

trellis drainage pattern   Drainage pattern characterized by a dominant parallel set of major *subsequent streams,* joined at right angles by numerous short tributaries; typical of *coastal plains* and belts of eroded folds.

trend surface On a map or graph, a surface fitted to the general trend of the data.

tropical cyclone Intense traveling *cyclone* of tropical and subtropical latitudes, accompanied by high winds and heavy rainfall.

tropical easterlies Low-latitude wind system of persistent air flow from east to west between the two *subtropical high-pressure belts.*

tropical rainforest Plant *formation class* similar in structure to the *equatorial rainforest,* but extending into the *tropical zone* along coasts windward to the *trade winds.*

tropical raingreen vegetation (See *raingreen vegetation.*)

tropical year *Year* defined as the time elapsing between one *vernal equinox* and the next.

tropical zones *Latitude* zones centered on the *tropic of cancer* and the *tropic of capricorn,* within the latitude ranges 10° to 25° N and 10° to 25° S, respectively.

tropic of cancer *Parallel of latitude* at 23½° N.

tropic of capricorn *Parallel of latitude* at 23½° S.

tropopause Boundary between *troposphere* and *stratosphere.*

tropophyte Plant that sheds its leaves and enters a dormant state during a dry or cold season when little *soil water* is available.

troposphere Lowermost layer of the *atmosphere* in which air temperature falls steadily with increasing altitude.

tsunami (See *seismic sea wave.*)

tundra biome *Biome* of the cold regions of *arctic tundra* and *alpine tundra,* consisting of grasses, grasslike plants, flowering *herbs,* dwarf *shrubs,* mosses, and lichens.

tundra climate Cold climate of the *arctic zone* with eight or more consecutive months of zero *potential evapotranspiration (water need).*

tundra soils Soils of the arctic tundra climate regions.

turbidity current Rapid streamlike flow of turbid (muddy) seawater close to the seabed, often confined within a submarine canyon on the continental slope or flowing down the inner wall of an oceanic trench.

turbulence In fluid flow, the motion of individual water particles in complex eddies, superimposed on the average downstream flow path.

turbulent suspension *Stream transportation* in which particles of *sediment* are held in the body of the *stream* by turbulent eddies. (Also applies to wind transportation.)

typhoon *Tropical cyclone* of the western North Pacific and coastal waters of Southeast Asia.

Udalfs Suborder of the *soil order Alfisols;* includes Alfisols of moist regions, usually in the *midlatitude zone,* with deciduous forest as the natural vegetation.

Udolls Suborder of the *soil order Mollisols;* includes Mollisols of the moist soil-water regime in the *midlatitude zone* and with no horizon of *calcium carbonate* accumulation.

Ultisols *Soil order* consisting of *soils* of warm soil temperatures with an *argillic horizon* and *low base status.*

ultramafic igneous rock *Igneous rock* composed almost entirely of *mafic minerals,* usually *olivine* or *pyroxene* group.

ultraviolet radiation *Electromagnetic energy* in the *wavelength* range of 0.2 to 0.4 *microns.*

unloading Process of removal of overlying rock load from *bedrock* by processes of *denudation,* accompanied by expansion and often leading to the development of *sheeting structure.*

unsaturated zone *Subsurface water* zone in which pores are not fully saturated, except at times when *infiltration* is very rapid; lies above the *saturated zone.*

unstable air mass *Air mass* with substantial content of *water vapor,* capable of breaking into spontaneous convectional activity leading to the development of heavy showers and *thunderstorms.*

upper-air westerlies System of westerly winds in the upper *atmosphere* over middle and high latitudes.

Ustalfs Suborder of the *soil order Alfisols;* includes Alfisols of semiarid and seasonally dry climates in which the soil is dry for a long period in most years.

Ustolls Suborder of the *soil order Mollisols;* includes Mollisols of the semiarid climate in the *midlatitude zone,* with a horizon of *calcium carbonate* accumulation.

valley winds Air movement at night down the gradient of valleys and the enclosing mountainsides; alternating with daytime *mountain winds.*

veldt Region of *steppe* grassland in Orange Free State and the Transvaal of South Africa.

vernal equinox *Equinox* occurring on March 20 or 21, when the *subsolar point* is at the *equator.*

Vertisols *Soil order* consisting of *soils* of the *subtropical zone* and the *tropical zone* with high *clay* content, developing deep, wide cracks when dry, and showing evidence of movement between aggregates.

visible light *Electromagnetic energy* in the *wavelength* range of 0.4 to 0.7 *microns.*

volcanic neck Isolated, narrow steep-sided peak formed by erosion of *igneous rock* previously solidified in the feeder pipe of an extinct *volcano.*

volcanism General term for *volcano* building and related forms of extrusive igneous activity.

volcano Conical, circular structure built by accumulation of *lava* flows and *tephra.* (See *composite volcano, shield volcano.*)

volcano coast *Coast* formed by *volcanoes* and *lava* flows built partly below and partly above sea level.

warm front Moving weather *front* along which a warm *air mass* is sliding up over a cold air mass, leading to production of *stratiform clouds* and *precipitation.*

washout Downsweeping of atmospheric *particulate matter* by *precipitation.*

water deficit Difference between *soil water* present in the *soil* (actual *soil-water storage*) and the *storage capacity* of the soil.

waterfall Abrupt descent of a *stream* over a *bedrock* downstep in the *stream channel.*

waterlogging Rise of a *water table* in *alluvium* to bring the zone of saturation into the root zone of plants.

water need (See *potential evapotranspiration.*)

water surplus Water disposed of by *runoff,* or percolation to the ground water zone after the *storage capacity* of the *soil* is full.

water table Upper boundary surface of the *saturated zone;* the upper limit of the *ground water* body.

water use (See *actual evapotranspiration.*)

water vapor The gaseous state of water.

wave cyclone Traveling, vortexlike *cyclone* involving interaction of cold and warm *air masses* along sharply defined *fronts*.

wavelength Distance separating one wave crest from the next in any uniform succession of traveling waves.

wave theory Ideal model of development of the *wave cyclone* as put forward by J. Bjerknes.

weak equatorial low Weak, slowly moving low-pressure center *(cyclone)* accompanied by numerous convectional showers and *thunderstorms;* it forms close to the *intertropical convergence zone* in the rainy season, or *summer monsoon*.

weather Physical state of the *atmosphere* at a given time and place.

weathering Total of all processes acting at or near the earth's surface to cause physical disruption and chemical decomposition of *rock*. (See *chemical weathering, physical weathering*.)

westerlies (See *prevailing westerly winds, upper-air westerlies*.)

west-wind drift Drift current moving eastward in zone of *prevailing westerlies*.

wet adiabatic lapse rate Reduced *adiabatic lapse rate* when *condensation* is taking place in rising air; value ranges from 3 to 6 C°/1000 m (2 to 3 F°/1000 ft).

wet–dry tropical climate Climate of the *tropical zone* characterized by a very wet season alternating with a very dry season.

wet equatorial climate *Moist climate* of the *equatorial zone* with a large annual *water surplus*, and with uniformly warm temperatures and high values of *soil-water storage* throughout the year.

wilting point Quantity of stored *soil water*, less than which the foliage of plants not adapted to drought will wilt.

wind Air motion, dominantly horizontal relative to the earth's surface.

wind abrasion Mechanical wearing action of wind-driven *mineral* particles striking exposed *rock* surfaces.

windows Certain *wavelength* bands of the *electromagnetic spectrum* within which *energy* is radiated through the atmosphere to escape into outer space.

wind vane Weather instrument used to indicate *wind* direction.

winter monsoon Outflow of continental air at low levels from the Siberian high passing over Southeast Asia as a dry, cool, northerly *wind*.

winter solstice Solstice occurring on December 21 or 22, when the *subsolar point* is at 23½° S.

Wisconsinan Stage (glaciation) Last glaciation of the *Pleistocene Epoch*.

woodland Plant *formation class*, transitional between *forest biome* and *savanna biome*, consisting of widely spaced trees with canopy coverage between 25 and 60 percent.

Xeralfs Suborder of the *soil order Alfisols:* includes Alfisols of the Mediterranean climate.

Xerolls Suborder of the *soil order Mollisols;* includes Mollisols of the Mediterranean climate.

xerophytes Plants adapted to a dry environment.

X rays High-energy form of radiation at the extreme short *wavelength* (high-frequency) end of the *electromagnetic spectrum*.

year Period of time required for one complete *revolution* of a planet in its orbit around the sun. (See *tropical year*.)

INDEX

Numbers in *italics* refer to definitions or explanations of terms.

549